CW00802080

THE LIVERPOOL PRIVATEERS

AND

THE LIVERPOOL SLAVE TRADE

HISTORY

OF THE

LIVERPOOL PRIVATEERS

AND

LETTERS OF MARQUE

WITH AN ACCOUNT OF THE

LIVERPOOL SLAVE TRADE,

1744–1812

GOMER WILLIAMS

INTRODUCTION BY DAVID ELTIS

LIVERPOOL UNIVERSITY PRESS

McGILL-QUEEN'S UNIVERSITY PRESS
MONTREAL & KINGSTON · LONDON · ITHACA

ISBN 0-7735-2745-1 (cloth)
ISBN 0-7735-2746-x (paper)

Legal deposit third quarter 2004
Bibliothèque nationale du Québec

Published in the European Union by
Liverpool University Press
ISBN 0-85323-709-3 (cased)
ISBN 0-85323-789-1 (limp)

First published in 1897 by William Heinemann (London) and
Edward Howell (Liverpool)

Printed in Canada on acid-free paper that is 100 per cent ancient forest free (100 per cent post-consumer recycled), processed chlorine free.

McGill-Queen's University Press acknowledges the support of the Canada Council for the Arts for our publishing program. We also acknowledge the financial support of the Government of Canada through the Book Publishing Industry Development Program (BPIDP) for its publishing activities.

National Library of Canada Cataloguing in Publication

Williams, Gomer
History of the Liverpool privateers and letters of marque : with an account of the Liverpool slave trade, 1744–1812 / by Gomer Williams.

Reprint, with a new introd. by David Eltis, of the ed. published:
London : W. Heinemann, 1897.
ISBN 0-7735-2745-1 (bnd)
ISBN 0-7735-2746-x (pbk)

1. Privateering – England – Liverpool – History. 2. Slave trade – England – Liverpool – History. I. Title.

DA77.W54 2004 359.4092242753 C2004-901641-5

British Library Cataloguing-in-Publication Data
A British Library CIP record is available for this book

AFFECTIONATELY INSCRIBED

TO

HALL CAINE

BY HIS FRIEND, THE AUTHOR.

CONTENTS.

PART I. — PRIVATEERING.

CHAPTER I.

CHAPTER II.

CHAPTER III.

CHAPTER IV.

CHAPTER V.

CHAPTER VI.

PART II.—THE LIVERPOOL SLAVE TRADE.

CHAPTER I.

CHAPTER II.

CHAPTER III.

CHAPTER IV.

CHAPTER V.

CHAPTER VI.

CHAPTER VII.

CHAPTER VIII.

APPENDIX TO PRIVATEERS.

APPENDIX TO SLAVE TRADE.

LIST OF ILLUSTRATIONS.

INTRODUCTION, 2004.

GOMER WILLIAMS, the man, remains something of a mystery. By the late nineteenth century, when he compiled this volume, slave trading from the port of Liverpool had been over for the better part of a century, though privateering had re-emerged as recently as the American Civil War. The slave trade to West Africa had eventually been replaced by traffic in palm oil and palm kernels, centered on Liverpool, whose eventual value greatly exceeded what West Africa's slave trade had generated. A Gomer Williams was employed on a Liverpool palm oil vessel trading at Old Calabar in the mid-nineteenth century, but some basic research on the available censuses and Liverpool city directories suggests that this is not the author of *Liverpool Privateers...* That Gomer Williams was not born until 1847 and spent his working life as an "author," "publisher," "journalist," "advertiser," and then "book-keeper" – to use the designations found chronologically in the above sources – rather than as a sailor or merchant. One of his journalistic endeavours was an interesting series of essays in the *Liverpool Weekly Courier* on "Female Masquerading," or women passing as men throughout history, but, as the above range of occupations suggests, he does not appear to have been hugely successful in career terms. He probably collected material on pirateering and the slave-trade in Liverpool during the early and mid 1890s, a time that he designated himself first as "author" and then as "publisher" and "journalist." He does not appear to have made a very good living from his writing, however, and, notwithstanding a longish review in the *Times*, the present volume was probably the only book he ever published.[1] Yet this book is the best private compilation of primary materials we have on privateering and slave trading, not only from the port of Liverpool but from any British port. There are indeed few fields in which that can be said of a volume published over a century ago, and surely no authors today expect their work to have anything like such longevity, let alone status.

We now know that by 1783 three ports in England – London, Bristol, and Liverpool – accounted for over sixty per cent of privateering commissions, over three-quarters if considered in terms of the tonnage of those commissioned vessels. London and Bristol privateers have their twentieth-century historians.[2] For published material on Liverpool's role, however, there is still only Gomer Williams. The situation is not very different for British slave trading. Bristol, London, and even Lancaster and other minor ports have had recent books and essays written about their respective slave trades. But for Liverpool, the major British slave-trading port, the second half of Gomer Williams' book is still the only extensive port-based treatment available in printed form. No one writing about any branch of the slave trade can ignore its stories, numerous assessments about the traffic, appendices, and not least, indexes.

For modern audiences, the book presents a few problems, none of which should detract from its value. First, it was written in a period of high imperialism immediately before the war in South Africa. One should not expect to find any questioning of British victories at sea, anymore than one should read William's contemporary, Leopold von Ranke, for regret at Prussia's rise to great power status, or Thomas Babington Macaulay for a critique of the Whig ascendancy. That the English should prevail over the French in their numerous maritime conflicts was, for Williams, as necessary to the victory of good over evil as was the triumph of abolitionism over the slave trade. Second, for his time Williams provides good information on sources, but he was not as careful as modern scholars would like him to have been. For this, there is no complete solution except a great deal of painstaking searching. He made heavy use of newspapers and drew on a considerable amount of private correspondence. As not all the material he used is still around more than a century later, Williams is now our only source for some of it – most noticeably the invaluable letters of slave trader Captain Ambrose Lace. We can at least take comfort from the fact that, based on the material that we do have, his reports are reliable and the unfamiliar material, by and large, checks out with available information. There are now extensive databases for eighteenth-century privateering voyages and the slave trade that operated from British ports. The ships he mentions can be verified in these, and more generally, with a few qualifications, the broad trends of the rise and fall of the activities he chronicles agree well with the patterns revealed by both these databases and by other modern research, as explained below. For a modern author, citing Williams is almost as secure as citing an official record such as a Parliamentary Paper.

A third issue for some modern readers might be the relative lack of broad context and analysis in the book. Why did privateering develop and how did it compare to other hazards and ventures that merchants might face? How important was privateering and slave trading compared to other trades or to the economy of the port as a whole? What was the connection between the two trades? Who was involved? What reaction was there in the port to the abolition movement and what adjustments occurred as a result? How important was Liverpool compared to other major international centres of privateering and the slave trade? These questions are neither posed nor answered in these pages. It is worth pointing out, however, that any attempt to address most of these questions a century ago would likely not have been very successful. The research of the last few decades makes such a task much easier and the primary material that Gomer Williams accumulated provides vital building blocks for the construction of answers, a few of which are attempted in this introduction.

It is worth remembering that first contact between any overseas communities usually involved voyages where plunder and trade were alternative possibilities. The ships merchants sent to Africa or the western Atlantic were invariably heavily armed and were prepared to trade or to steal depending on the circumstances. Trade took precedence wherever the potential trading partner was of approximately equal strength and/or repeated contact was likely to ensue. But wherever nationals of separate powers were in contact the prospect of violent expropriation always lurked in the background. Indeed, in transoceanic environments and in the absence of law-enforcement agencies, violence and theft between peoples of the same nationality were also always likely. For the latter, the aggrieved parties could at least have recourse to their own system of law when they sailed back within range of that system, or they could rely on ships of their respective states on the rare occasion that these were in the vicinity. But what if the threat came from citizens of other states? For north-western Europe in the Dark Ages and for Europeans sailing to Africa and the Caribbean from the fourteenth to seventeenth centuries, the only sanction was brute force except during short periods when a temporary and fragile agreement to trade held. The famous Hawkins voyages that raided villages near Sierra Leone for slaves in the sixteenth century and the hostilities between the Spanish and other European powers in the Caribbean are examples of "normal" relations between aliens before a rule of law was established. Under such circumstances, the terms "privateering," "piracy," and "slave trading" could be meaningful only where there were some formal, widely observed, and sometimes enforceable rules of conduct.

What follows from this is that the prevalence of both privateering and slave trading, and, indeed, even the currency of the terms themselves, should be taken not as an indication of anarchy and the absence of the rule of law but more often as a sign that societies with overseas interests had stepped away from a maritime equivalent of the survival of the fittest. The eighteenth century is the classic period of privateering because by that time the European state had developed a basic functioning bureaucracy but naval power had not yet developed to the point that commerce raiding had become almost exclusively a matter for the state's maritime forces. Yet even before 1800 naval vessels normally took more prizes than privateers. Thereafter, the ratio of prizes taken by privateers to prizes taken by the navy has fallen steadily and privateering has been inconsequential in twentieth-century wars. As a significant activity, it has not survived as long as piracy. As this broad perspective implies, one way to understand privateering is to locate it midway between piracy and naval action. Privateering had several origins but two of the most important were the ancient right of redress and the concept of a private man-of-war – a cheap method by which the state could supplement its naval forces.[3] Neither of these could develop without a functioning bureaucracy to keep track of and adjudicate actions on the high seas. The right of redress was a right granted by the state to a merchant who could prove loss or injury at the hands of citizens of another state to raid the vessels of that state until the loss was made good. The raiding purpose of a private man-of-war was much more generalized. There were no limits on the property that could be captured and the phrase "private man-of-war" might cover both a "channel raider" – a small vessel based in the Channel Islands with a crew of six – and large ships leaving London in private expeditionary forces to hunt for enemy treasure vessels in the South Seas. The most common privateer in the eighteenth century operated under a "letter of marque," which was essentially a license to capture enemy ships that was issued to a merchant ship whose primary goal was trade but whose captain was prepared to raid, or indeed cruise looking for prey, en route to a commercial destination. Because slave ships had very high crew per ton ratios (to control the slaves) and were usually well armed in any event, many applied for letters of marque and took prizes, especially on their outward voyages. By the eighteenth century all raiding was confined entirely to periods of war. Moreover, like the slave trade and a hundred other branches of international commerce, it was above all an enterprise, an opportunity to risk capital and make profits. Its financing was the responsibility of the licensed owners and the benefits, if any, accrued entirely to them and the crew, at least after 1708 in the British case, when the state gave up its claim to a share of the prizes.

The system was administered by a separate branch of civil law that drew on the long-established Admiralty courts. No benefits accrued until the property had been adjudicated by one of these courts, which were eventually established in British possessions around the world. Adjudication conferred prize status on the captured property, at which point it could be sold for the benefit of the captors. The junior tribunals were called Vice- Admiralty Courts and the appellate court, the High Court of Admiralty. Although Williams does not mention it, these courts eventually played a major role in the suppression of the slave trade in the nineteenth century as they proved a more effective vehicle for condemning slave ships than the better-known international Courts of Mixed Commission. The role of the state in sanctioning and directing raiding thus clearly separates privateering from piracy.

For slave trading, the counterpart of state action in shaping and controlling privateering was the establishment of sites at various points on the African coast where markets for slaves could operate under the rule of law with sanctions against theft or failure to meet credit obligations. Random

European raiding for slaves along the coast very quickly became counter-productive. The Portuguese created slave markets at coastal centres at Luanda, and later Benguela in Angola, but most of the eight or nine locations where over half of all slaves destined for markets in the Americas were first traded and then embarked were under the control of African authorities. The law involved was thus African rather than European. As a consequence, the violence normally preceding the enslavement of a person – the counterpart of piracy and privateering on the high seas – usually took place out of sight of Europeans and, increasingly, away from the coast itself. Slave routes linked the site of enslavement with the coastal market. Some African slave trading polities were able to enforce peace among Europeans in the adjacent sea-roads as well as on land, even in time of European war – the best known example being Whydah in the Bight of Benin prior to its conquest at the hands of the King of Dahomey. Unless a formal state of war had been declared, by the late seventeenth century it was possible for a merchant to expect attack from pirates or from port authorities where he was trading illegally but not, as a matter of course, from any other foreign vessel that he might meet on the open sea or at one of the major trading sites around the Atlantic – including those on the western coast of Africa. Rules of engagement and trade had been established and applied, and at one level Gomer Williams' book is about the operation of those rules as experienced by vessels operating out of Liverpool.

Williams' discussions and narratives provide illustrations of such trends, as well as basic raw material. They also show clearly that privateering and slave trading were alternative investment opportunities among hundreds of others and were not the foundations of the port's rise to major status. Liverpool came late to both activities relative to London and Bristol. It is now possible to estimate that until 1785 the port had sent out about 2,700 slaving ventures and 746 privateering expeditions. At the peak of the Liverpool slave trade in the early 1790s, about three percent of ship departures from the port were slavers (about ten percent by aggregate tonnage). As late as the 1780s, Liverpool still lagged behind London in the number of slave trading ventures dispatched and was well behind both London and Bristol as a centre of privateering activity. Its share of so-called "deep-water" privateering expeditions was particularly small.[4] The high profile of both privateering and slave trading may be due in part to the tendency of eighteenth-century newspapers to give over a disproportionate share of their shipping reports to revolts on slave ships and the exploits of privateers as merchants' decisions about whether or not to get involved in these enterprises were made on the basis of profit and loss alone. An international perspective reduces its significance to the port even further. By 1807, when the British traffic was abolished, Liverpool had become the leading British centre for slave trading. But we now know that even then it lagged behind the major Brazilian ports, Rio de Janeiro, Bahia, and possibly Pernambuco in the total number of slave voyages organized.[5] There is little room here for an argument that the slave trade (or indeed slavery, when we look at the size of the Brazilian plantation sector) fuelled industrialization in the Atlantic world.

Even at its peak and even when British sea-power was at its greatest, privateering never compensated for the drop-off in international trade during wartime. The High Court of Admiralty issued between twelve and thirteen thousand privateering commissions between 1702 and 1783, but the number of non-privateering ventures that were abandoned because of war would have been many times greater than this. Moreover, a license to raid did not guarantee prizes, any more than a prize guaranteed profits for a particular venture. Over the same period court records indicate just over three thousand prizes condemned to privateers, as opposed to prizes taken by the navy. Offsetting these captures, of course, were the successes of foreign privateers against British shipping. In the eighteenth-century conflict in which the British achieved their greatest naval success – the Seven Years War – the number of enemy vessels captured fell as foreign merchants simply stayed in port. Despite issuing 2,105 commissions between 1756 and 1762, the courts adjudicated only 382 privateering prizes. The value of these prizes varied, but some of them comprised only the vessels and equipment or a low value cargo that would not have offset the cost of mounting the expedition. And prior to 1777 Liverpool's share of these prizes fell well below its share of total British international trade. Privateering might make some individuals rich, but it did little to counteract the negative effect of war, least of all in Liverpool. More generally, the more successful the privateers were, the lower the revenue from Customs – the major source of income for any eighteenth-century government – and privateering also diverted resources from the navy.

The patterns of development of both privateering and the slave trade were rather different from what a casual reading of Williams might suggest. First, he devotes more space to privateering than to slave trading, though the latter was the more important of the two trades in economic terms. This is in sharp contrast to modern scholarship, which, I suppose, must generate several hundred books and articles on the slave trade for every work on privateering. Further, no one would suspect from his book that over the course of the eighteenth century London was three times more important than Liverpool as a centre of privateering. Williams does, however, convey a clear sense of the increasing importance of privateering in the War of American Independence, the time when Liverpool finally became a major privateering port. In these years, privateering came closest to offsetting losses resulting from a rapidly declining slave trade as British seapower came under threat. It is also clear from Williams that few merchants specialized in only one trade in any event, and the huge increase in commerce-raiding ventures leaving from Liverpool after 1777 as compared to earlier wars suggests at least a partially successful search for alternative investment strategies. As already suggested, prior to 1777 privateering in Liverpool never came close to offering an alternative to peace time trade, despite some high profile captures and condemnations. For the slave trade, Gomer Williams managed to severely underestimate early activity in the port, and in the process misled several generations of historians. His claim that there was only a single Liverpool-based transatlantic slave voyage before 1730 is probably a comment on the scarcity of newspapers – his most used source – in this early period rather than an accurate reflection of the slave trade. Liverpool port books (held, it should be noted in London rather than his beloved Liverpool), among other sources, provide a basis for estimating that Liverpool vessels had already carried nearly 30,000 slaves from Africa by 1730, with the first couple of expeditions undertaken as early as 1695. We now know that the Liverpool slave trade was founded about the same time as its Bristol counterpart and was initially bigger but developed more slowly for forty years after 1705. The dimensions of the port's later dominance are also clearer now. It reached its peak in the fifteen years prior to abolition and, but for abolition, would no doubt have sent out more slave voyages than any other port in the Atlantic world. While some Brazilian ports sent out more transatlantic slaving voyages, they did so over a longer period of time. Only Rio de Janeiro in the 1820s appears to have matched the intensity of Liverpool's slaving activity – with between two and three slave ships leaving these ports for Africa each week on average in the course of a year.

Overall, the bracketing of these two business enterprises in a single volume, as well as the tone of William's discussion of them, reflects an unequivocal moral position. There were strong elements of theft and violence and therefore immorality in both privateering and slave trading, though both were legal. Neither could occur without forcing people to do what they were not inclined to do voluntarily and both created conflict between peoples of very different back-grounds with all the associated human drama and tragedy. However financial rewards could be huge and privateers, at least, could feel that they were being both heroic and patriotic. Readers should not forget what is as hard to appreciate today in the case of slave trading as it was over a hundred years ago when Gomer Williams wrote his book, that both were legitimate endeavours in the eyes of domestic and emerging international law, and, more important, neither was viewed as in any way immoral: before the late eighteenth century, slave trading and privateering were seen as ethically indistinguishable from trading in Baltic timber or Canadian furs.

DAVID ELTIS, *Emory University*

NOTES

I would like to thank David Richardson and Stephen Behrendt for comments on an earlier draft of this introduction, and Susan Eltis for assistance with the research.

[1] *Times*, 4 December 1897. For the palm oil reference see "List of British Traders at Old Calabar," Calabar Museum, Calabar, Nigeria. I thank David Richardson of the University of Hull for this reference.
[2] John W. Damer Powell, *Bristol Privateers and Ships of War* (Bristol: Arrowsmith, 1930); David J. Starkey, "British Privateering, 1702-1783, with particular reference to London," (Unpublished PhD thesis, University of Exeter, 1985) and *British Privateering Enterprise in the Eighteenth Century* (Exeter: Exeter University Press, 1990), which for Liverpool draws somewhat on the author's unpublished MA thesis.

[3] For a fuller discussion of origins see Starkey, *British Privateering Enterprise*, 20–58.
[4] See the various tables in Starkey, *British Privateering Enterprise*, showing privateering commissions and ventures for each of the eighteenth century wars prior to 1784.
[5] All data on the slave trade are taken from David Eltis, Stephen Behrendt, and David Richardson, *The Atlantic Slave Trade: A New Census* (forthcoming, Cambridge University Press).

PREFACE.

In tracing the history of Liverpool privateering and slave trading—upon which the greatness of "the good old town" was suckled—the author has had access to original sources of information never before tapped, and gratefully acknowledges his indebtedness to Sir Thomas Brocklebank, Bart., the Messrs. MacIver, Mr. C. K. Lace, Mr. Thomas Hampson (Ruabon), and Mr. T. H. Dixon (Gresford), for permission to inspect and copy rare documents in their possession, which greatly enhance the value of the book, and to his friend, Mr. J. S. Arthur, for his good offices in this direction. The other and principal sources drawn upon are the files of old newspapers, magazines, and other contemporary records requiring no little patience and enthusiasm to ransack. In a few instances, where information on special points has been derived from other authors, as in Professor Laughton's admirable account of Captain Fortunatus Wright, it is acknowledged either in the text or in a foot-note. To quote the prospectus, "this work is not a mere compilation, but the fruit of laborious and exhaustive original research." The utmost pains have been taken to ensure accuracy, and the reader will find the statements of more than one authority corrected. To this cause, and the sifting of much curious original matter so kindly placed at the author's disposal, is to be attributed the delay in publication, by which the reader, and especially the original subscribers, gain considerably, the promised 600 pages being largely extended, with illustrations thrown in.

In dealing with the delicate subject of the Liverpool Slave Trade—a subject which, for reasons that may be guessed, has been lightly touched upon by most local writers—the author has endeavoured to confine himself to a plain statement of facts—facts which need no comment or exposition. He has directed his indignation against the system, or national sin, rather than against individuals, for many of the slave-merchants and slave-captains of old Liverpool claim our regard as patriots and worthies of no common order.

Though we are on the threshold of the Twentieth Century, with its tremendous possibilities, there are indications that white men still exist who would gladly revert to the iniquitous system of a bygone age, and enslave the African in his own land. If anything in this book should help to awaken the public conscience to jealously watch that under no specious pretext shall the bodies and souls of "African labourers" be again handed over to the tender mercies of greedy and unscrupulous adventurers, the author will rejoice.

LIVERPOOL,

November, 1897.

HISTORY

OF THE

LIVERPOOL PRIVATEERS

AND

LETTERS OF MARQUE

WITH AN ACCOUNT OF THE

LIVERPOOL SLAVE TRADE

BY

GOMER WILLIAMS

WITH ILLUSTRATIONS

LONDON
WILLIAM HEINEMANN
LIVERPOOL
EDWARD HOWELL CHURCH STREET
1897

HISTORY

OF THE

LIVERPOOL PRIVATEERS

AND

LETTERS OF MARQUE, &c.

CHAPTER I.

A Peep Behind the Scenes—the Ancient Mariner and the Ancient Merchant.

We assume, for the sake of illustration, that the reader wishes to become practically acquainted with the method of fitting out, arming, manning, and manœuvring privateers and letter-of-marque ships in ancient Liverpool. We cannot do better than follow the advice of Captain William Hutchinson, an experienced privateer commander, originally trained in that finest of all nurseries for seamen, the Newcastle colliers, who afterwards became dockmaster at Liverpool. "Safety as well as success, in my opinion," he says, "depends greatly on the manner these ships are fitted out. Trading ships, designed more for defence than offence, I would recommend to be made to look as big, powerful, and warlike as possible, in order to intimidate; but privateers the contrary, to look as little and defenceless and conceal their power as much as possible, till there is a real occasion for it, and then as suddenly as possible to make it known to give the greater surprise, which I can say from experience may often give great advantages. As to the size and number of great guns, the dimensions, strength, and properties of the ship should point out what she will be able to bear without being too crank for a sailing and fighting ship ; and though it must be allowed that the advantages in a sea fight are greatly in favour of the heaviest shot, yet the many storms a ship may have to contend with in a winter's passage, or a cruise in a turbulent ocean, where the great guns may be often rendered a useless and dangerous incumbrance by the waves running so high, that nothing but small arms can be used against the enemy, so a ship should not be overcrowded or overburdened with too heavy cannon." With regard to shot, "the first and principal," he observes, "both for quantity and quality, is the round iron cannon ball, because it will go and penetrate farther and with greater velocity than any other to do execution when engaging with a superior force; but when come to a close fight with a ship of inferior force, expecting to make her a prize, then the endeavours should be not to destroy the ship if it can possibly be avoided, but to distress them to make submission; therefore, some suitable shot that will answer that purpose best should be provided. And I would recommend round tin cases, to fit the bore of the guns, filled with musket ball ; and square bar iron, cut about fourteen inches long, tied in bundles with rope yarns just to fit the guns; or cast iron bars about the same length, a square one about an inch diameter in the middle and four others quartering, rounded on the outside, to fit the bore of the guns, when tied with rope yarns." These rude missiles will doubtless make the naval heroes of the present day laugh scornfully, but the persons upon whom Captain Hutchinson experimented with similar preparations appear to have been perfectly satisfied with their efficacy.

The following is a Quarter Bill for a privateer of twenty guns, 9-pounders, and four 3-pounders on the quarter-deck and forecastle:—

ON THE QUARTER-DECK.

The Captain to command the whole - -	1
The Master to assist and work the ship according to orders - - - - - -	1
A Midshipman to pass the word of command fore and aft - - - - - -	1
A Quarter-master at the gun and another at the helm - - - - - - -	2
The First Marine Officer with 24 musketeers -	25
Three men for the two 3-pounders and a boy to fetch powder - - - - - -	4

ON THE MAIN DECK.

The First Lieutenant to command the ten foremost guns - - - - - -	1
The Second Lieutenant to command the ten aftermost guns - - - - -	1
The Gunner to assist and attend all the great guns fore and aft - - - - - -	1
The two Master's Mates to attend the fore-topsail braces, and work the ship forward, according to orders - - - - -	2
The Boatswain's Mate, with two seamen, to assist in working the ship and to repair the main rigging - - - - - -	3
The Carpenter and his crew to attend the pumps and the wings about the water's edge, fore and aft, with shot plugs, &c. - - -	4
Six men to each of the ten guns on a side and its opposite, and a boy to fetch powder -	70

ON THE FORECASTLE.

The Boatswain to command, with two seamen to work the ship and repair the fore rigging	3

Three men and a boy to fetch powder for the two 3-pounders - - - - - - -	4
The Second Marine Officer with nine musketeers	10
In the barge upon the booms, the Third Marine Officer with eight musketeers - - -	9
In the maintop, five men with a Midshipman at small arms, and to observe the conduct and condition of the enemy - - -	6
In the foretop, five men at small arms, and to repair the rigging - - - - - -	5
In the mizentop, three men at small arms, and to repair the rigging - - - -	3
In the powder room, the Gunner's Mate with an assistant to fill and hand powder to the boys, carriers - - - - - - -	2
In the cockpit, the Doctor and his mate - -	2
	160

"The people," observes Captain Hutchinson, "should be quartered to fight nearest to where they are stationed to work the ship, that is, the afterguard on the quarter deck, the waisters in the waist, forecastlemen that are necessary on the forecastle, &c. The quarter bill and discipline of the crew should be kept from disorder as long as possible, and when occasional duty requires people to be let go from their quarters, it should not be done at random, but with judgment, such as will suit the occasion, from the musketeers, or a man from each great gun, &c., where they can be best spared to continue in, or be brought to action in the most regular order that is possible."

But what of the men who formed the crews of these vessels? "An Old Stager," speaking of privateers from personal observation, as a genuine "Dicky Sam" says: "Liverpool was famous for this kind of craft. The fastest sailing vessels were, of course, selected for this service, and as the men shipped on board of them were safe, in virtue of the letter of marque, from impressment, the most dashing and daring of the sailors came out of their hiding holes to take service in them. On the day when such a vessel left the dock, the captain or owner generally gave a grand dinner to his friends, and it was a great treat to be of the party. While the good things were being discussed in the cabin, toasts given, speeches made, and all the rest of it, she continued to cruise in the river, with music playing, colours flying, the centre of attraction and admiration, 'the observed of all observers' as she dashed like a flying fish through the water. And then the crew! The captain was always some brave, daring man, who had fought his way to his position. The officers were selected for the same qualities; and the men—what a reckless, dreadnought, dare-devil collection of human beings, half-disciplined, but yet ready to obey every order, the more desperate the better. Your true privateersman was a sort of half-horse, half-alligator, with a streak of lightning in his composition—something like a man-of-war's man, but much more like a pirate—generally with a super-abundance of whisker, as if he held with Sampson that his strength was in the quantity of his hair. And how they would cheer, and be cheered, as we passed any other vessel in the river; and when the eating and drinking and speaking and toasting were over, and the boat was lowered, and the guests were in it, how they would cheer again, more lustily than ever, as the rope was cast off, and, as the landsmen were got rid of, put about their own vessel, with fortune and the world before them, and French West Indiamen and Spanish galleons in hope and prospect. Those were jolly days to some people, but we trust we may never see the like of them again. The dashing man-of-war and the daring privateer dazzled the eyes of the understanding, and kindled wild and fierce enthusiasm on all sides. The Park and Tower guns and the extraordinary Gazette confirmed the madness, and kept up a constant fever of excitement. But count the cost. Lift up the veil, and peep at the hideous features of the demon of war. Look at the mouldering corruption beneath the whited sepulchre of glory! But no sermons, if you please."

Having got our ship, and her crew of dare-devils on board, let us consult Captain Hutchinson as to the best way of managing both ship and men. "As soon as the ship has got to sea," he tells us, "take the first opportunity to have all hands called to quarters; the officers in their stations to have everything made properly ready and fit for action; to have a general exercise, not only of the great guns and small arms, but the method of working and managing the ship, to take the advantage of the openings that often occur in attacking and being attacked by another single ship; and the designed manœuvres should be taught the people in their general exercise that they may know how to act without confusion. When a ship of nearly equal force brought to with a design to fight us, my intention was not to run directly alongside and lie to like a log, and depend upon mere battering with one side only, nor upon the stern chase guns. When it is found that there is no choice of running from a ship of much superior force chasing us, and when their best sailing is upon a wind, it is a common practice for them to run up and bring to under the lee in a triumphant manner, depending on their superior power, and commonly demanding immediate submission without expecting any resistance. The designed manner of resisting or attacking I always endeavoured to conceal as long as possible, and these two cases give all the advantages desired by my method.

"Begin the attack upon the weather quarter, shooting the ship upon the wind with the helm a-lee, till the after-lee gun, with which we begin, can be pointed upon the enemy's stern ; then fire the lee broadside, as it may be called. The ship begins the attack upon the enemy when the topsails are thrown aback, with the helm a-lee, boxing the ship round on her heels, so as to bring the wind so far aft that the ship may immediately be steered close under the enemy's stern, with particular orders to begin with the foremost gun, to rake them right fore and aft with the great guns, as they pass in that line of direction, all aiming and firing to break the neck or cheeks of the rudder head, the tiller ropes, blocks, &c., so as, if possible, to destroy the steerage tackle, which design, if it proves successful, takes the management of their ship from them, so that she must lie helpless for a time, in spite of their endeavours. When the aftermost gun is fired, put the helm hard-a-weather to bring the ship by the wind ; and then stand off on the other tack, to keep clear of their lee broadside and act according to their motions, and the experience of the effect your attack has had upon them. If they continue to lie-to, either renew the attack again in the same manner as soon as the ship will fetch the weather quarter again, or make sail off to escape, if it is found that the great inequality of their superior force admits of no possible chance of conquering them. And although this manœuvre may not have given this advantage (which in my opinion ought always to be attempted, and not to submit tamely, though a ship is above double the force), yet the power of their broadsides may be chiefly avoided by it.

"But when the inequality of force is not so great, but there is a possibility of conquering, and if the success of the first attack is perceived to oblige the enemy to continue lying-to, in order to repair the damage done their rudder or tiller, &c., then the blow should be followed by renewing the attack again with all possible expedition, in the same manner which gives the opening, not only to fire the whole round of great guns to advantage, but also to the marines and topmen to fire their small arms at the same time to great advantage, so as to do the most execution possible by firing and raking them fore and aft through their most open and tender part, the stern, with the least risk possible from the enemy's guns, and, therefore, gives the greatest possible chance that I know of to make an easy conquest, especially if so lucky as to destroy and prevent the recovery of their steerage. A ship of much superior force may be brought to such a distressed condition as to be obliged to make a submission for want of the helm to command her.

"But suppose the enemy, laid to as above mentioned, find themselves not much hurt by this manœuvre, and that you had not succeeded in destroying their steerage, and therefore you may expect that they will immediately tack, or ware ship, and stand after you, depending on their advantages of sailing faster and superior force, shall run up along your lee side, expecting by making a general discharge of their small arms and great guns (charged with suitable shot) on your deck, which lies open to them by the ship's heeling, to destroy your people and to make you submit? When this is likely to be their design, orders should be given to your people to keep themselves as snug under shelter as possible from their small shot, till their general discharge is over; then if the ship is found not so disabled, but that the topsails can be thrown aback, make a general discharge from the lee side, and the great guns, loaded with round shot only, pointed to the weather-side of the enemy's bottom, amidships to one point, at the water-edge, and box-haul the ship to run close under their stern, aiming at raking and destroying their steerage with the other broadside; then stand off on the other tack, as before mentioned, and act according to the circumstance and the condition you find yourselves in. Compare with the appearance of that of the enemy and their motions, who may be obliged to continue on the other tack to repair damages about their rudder, or to stop their leaks in the weather-side of their bottom, if your aim has proved successful.

"But when an enemy's ship of force makes only a running fight, if there is no necessity to cut them off from the shore or from the shelter of other ships, etc., and you have the advantage of sailing faster, the most sure and likely method to make an easy conquest with the least hurt to yourselves or their ship (your expected prize) is to run close up and shoot or sheer your ship across their stern each way, making a general discharge of all your force, first with one broadside, then the other, always aiming with the great guns at the rudder-head and steerage tackling, for the reasons given—that if the shots miss the rudder, etc., by raking the ship fore and aft through the stern, they may do the greatest execution possible to distress them so as to make a submission. On this occasion, when it blows fresh and obliges to carry a pressing sail large, or before the wind, to make the great guns as ready as possible, and prevent their being fired too low, all their breeches should be laid quite down in the carriage, and if your ship is crank, the yards should be braced so as to shiver the sails at the time each broadside is fired. In all these manœuvres, when the whole round of great guns are designed to be fired, care should be always taken to leave two or more men, as it may require, to charge each gun again when fired on one side, whilst the others move over to fire the guns on the opposite side, that neither side may be left unguarded; all which, with every other advantageous manœuvre that may be designed to be put in practice in action, should be taught the people along with the general exercise of great guns and small arms, by throwing a light, empty beef-cask overboard, making it the object of attack, for all the guns to be pointed at, when performing the above-described or other intended manœuvres about it; first, by running a little way large from it, then haul the wind, tack ship, and stand towards it, keeping it about three points on the lee bow till within a half-cable's length or musket shot of it; then put the helm a-lee and shoot the ship up in the wind with the top-sails aback, till the after gun can be pointed to the cask; then give the word of command to fire when there is a fair opening to make a general discharge, both below and aloft on that side. When you have box-hauled your ship, and run close past the cask to make a general discharge from the other side, then bear round away from it, ware, and haul the wind on the other tack till you can tack and fetch up to it again to repeat this, or perform any other manœuvres that may give an advantage to attack or defend a ship laid to or sailing upon a wind, as above mentioned.

"To perform the manœuvres of attacking an enemy to make a running fight large, or before the wind, you have only to turn far enough to windward of the cask to give room in sailing down to it to bring the ship's broadside to point to it each way. But to perform this manœuvre to the greatest advantage, with the least loss of time, and the ship's way through the water (which may be of great importance on this occasion to keep close up with the enemy), all the great guns should be run out close to the after part of the ports, that they may be pointed as far forward as the sides of the ports will admit, and elevated as the heeling of the ship, when brought to, to fire, may require, as above mentioned; and particular orders should be given for the aftermost guns on each side to be fired first, as soon as they can be brought to bear upon the enemy, because then the ship need not be brought any more to, but steered in that direction till the other guns are fired; then shift the helm to ware, to bring the other broadside to bear, etc.

"After the people have been thus disciplined, it is necessary to let them smell powder, as it is termed, and a little ammunition spent in exercise, it is allowed, may be the means to save a great deal expended to little or no purpose in action; therefore, I used to allow a small charge of powder for the round of great guns, with stone ballast for shot, and the musketeers two charges with balls each, and give them a fair chance by these manœuvres to fire both broadsides and small arms at the cask. If they sunk it, all hands to have an allowance of grog, as it is called, but if they did not sink it, to have the trouble and mortification to hoist out the boat and fetch it on board to serve another time."

In his observations on preparing for exercise or action, Captain Hutchinson says: "When all hands are called to quarters, every man should bring his hammock, well lashed up, and stow it to the greatest advantage, to give shelter from small arms, nearest to his own quarters; or to give them to some of his messmates where they are the most wanted, that they may know readily where to find them when exercise or action is over. When the hammocks are properly stowed, the officers, according to their stations and duties, are to see the ship effectually cleared of all incumbrances, and everything prepared, so that nothing may be wanting that is necessary for exercise or action. The lieutenants or mates, with the gunner on the gun deck, are to get all the hatches laid, except that where the powder is to be handed up. A match-tub, half filled with water, and four matches in the notches, placed as near midship as possible, to serve two guns and their opposites; also swabs to wet the decks, to prevent the fatal consequences that may attend the scattered and blown powder from the priming of the guns making a train fore and aft, which I have known take fire from the firing of the guns and do great damage, and which, in my opinion, has often been the cause of blowing ships up; and they should see that the captain of each gun has his men, powder-horn, rope, sponge, rammer, crows, handspikes, and train tackles all ready in their proper places. The boatswain must get the yards slung, the topsail sheets stoppered, and marline-spikes ready to repair the standing and running rigging that may be damaged. The carpenters are to get the pumps rigged, and shot plugs with all that is necessary ready in their proper places to stop leaks and repair damages. The gunner, when preparing for action, is to see that the charges in the guns are dry, and that there is a sufficient quantity of wads and shot of all sorts and cartridges ready filled. The marine officers are to see all the musketeers at their quarters, with their arms and ammunition in good order for exercise or action."

The reader, doubtless, feels that Captain Hutchinson was a thorough master of his profession, and observant of the minutest details. He was in truth a man of a very practical and scientific turn of mind, claiming our respect on many grounds, and that of sailors in particular, as the father of the Liverpool Marine Society. But we have not finished our lessons yet in the School of Privateering. Speaking of fortifying the quarter deck, Captain Hutchinson says: "Whatever may contribute to shelter and save the people must be allowed to deserve notice. Various methods and things have been tried for this purpose. I was in a ship that had bags of ox-hair that were said to resist even cannon shot; but in fighting with a French frigate I saw one of her shot go through eighteen inches of hair, through the middle of an eighteen-inch mast, and a long way over our ship afterwards; which proves no fence can be made about a ship against cannon shot; but against small and musket shots a fence may be made many ways. However, this fence or breastwork may be made to shelter the people from small shot; in common they are no more than breast-high, so that the musketeers can fire fairly over them upon the enemy. But from experience in fighting I have observed among new fighting men there will always be something to show that natural instinct of self-preservation; and in order to keep their heads under shelter of the breastwork from the enemy's shot, they fire their muskets at random up into the air. Seeing this, and to prevent the bad effect of such examples in fighting, I have made a feigned lunge at a man's breast with my drawn sword, and have been obliged to threaten death to any man that should show such a bad example; though it must be allowed to be only a failing and not a fault among new undisciplined landsmen first coming into action, who, at seeing a man shot through the head above the breastwork, may show a little fear, but by practice may prove brave afterwards. Therefore, to remedy this defect which I perceived in fighting the small arms, in fitting out a privateer afterwards, we had a rail, as in common, breast-high on each side the quarter deck, and on the rails were fixed light iron crutches, with the arms about a foot square, and a shoulder to keep the bottom of the crutches about six inches above the rails, and thin boards about six inches broad, laid upon the bottom of the crutches; and netting with large square meshes were formed just to hold a hammock with its bedding longways; and from the gunnel to the rail was boarded up on each side of the stanchions, and filled up with rope shakings, cork shavings, etc., which are found sufficient proof against musket ball, which made so ready and good a fence for the quarter-deck musketeers that the most timorous could point his piece with the utmost confidence between the rail and the netting, and fire right upon the enemy, by having his head, as well as his body, under such secure shelter. For the same reasons, in clearing and preparing the ship for fighting, I used to make the forecastle and top-men lash the hammocks, to shelter them, horizontally on the outside of the fore and topmast shrouds, close to one another, breast high, and then a single hammock above, leaving a little vacancy to point and fire their muskets through, which guards that tender and most important seat of knowledge, the head, as well as the other parts of the body which it governs, from the enemy's small shot."

The commander of a privateer was not free to roam the wide ocean at his own sweet will in search of prey. He usually had his station allotted to him by his owners, and in any case his self-interest would lead him to select a "beat" frequented by passing vessels. Discoursing on a ship cruising on her station, Captain Hutchinson reveals the tactics pursued by himself and Captain Fortunatus Wright in the War with Spain and France (1739-1748). "Cruising," he says, "the war before last, in the employ of that great but unfortunate hero, Fortunatus Wright, in the Mediterranean Sea, where the wind blows generally either easterly or westerly—that is, either up or down the Straits—it was planned, with either of these winds that blew, to steer up or down the common channels, the common course, large or before the wind in the day-time without any sail set, that the enemy's trading ships astern, crowding sail with this fair wind, might come up in sight, or we come in sight of those ships ahead that might be turning to windward; and at sunset, if nothing appeared to an officer at the mast-head, we continued to run five or six leagues as far as could then be seen before we laid the ship to for the night, to prevent the ships astern coming up and passing out of sight before the morning, or our passing those ships that might be turning to windward; and if nothing appeared to an officer at the mast-head at sunrise, we bore away and steered as before. And when the wind blew across the channels that ships could sail their course either up or down, then to keep the ship in a fair way; in the day-time to steer the common course under the courses and lower stay-sails; and in the night, under topsails with the courses in the brails, with all things as ready as possible for action, and to take or leave what we might fall in with in the night.

"Many other advantages attend cruising without any, or but with low sails set. As above mentioned, in the day-time and fine weather, when other ships are crowding with all their lofty sails set, they may be seen at twice the distance that you can, which gives you the opportunity to see them a long time before they can see you, and to take their bearing by the compass, and observe how they alter, by which it may be perceivable how they are steering, and you may consult what is best to be done, if it is too late in the day to give chase, which should always be considered. For, three-mast ships, in fine weather, with all their lofty sails set, may be seen from each other's mast-heads seven leagues distance, which must make a seven-hours chase, at three miles an hour difference in the ships' sailing, which is a great deal with a leading wind; and if the chase happens to be to windward, must make it still longer in proportion of time to come up with her; and when they perceive they are chased, and think themselves in danger of being taken, they will naturally use all possible means to escape out of sight by altering their course in the dark, if they cannot be got near enough for you to keep sight of them in the night. For these reasons, without the time, situations, circumstances and appearance require you immediately to give chase with all your sail at the first sight of a vessel, it often happens that you may stand a much better chance to speak with the vessel by endeavouring to waylay and conceal your design and ship from them; which may be done even in the day-time, with all the sails furled as before mentioned, till within four leagues' distance, when it is computed a ship's hull, in a clear horizon, begins to appear above it. When this concealment can be made, and all is ready prepared to take or leave, and you can fall in with the expected enemy in the night or early next morning, if they are found unprepared for action it must give you a great advantage over them. But when you cannot be concealed from the enemy's vessels in sight that may be coming with a fair wind towards you, then it should be considered whether, instead of giving chase with all your sail set in fine weather, it may not be better to disguise your ship, to appear as an inoffensive neutral ship, by getting your fore and mizen-top-gallant yards down, and the masts struck with only their heads above the caps, and either stand upon the wind with the main-top-gallant sail set, if not noticed, till by tacking you can fetch near the intended chase, or to steer near the same course with them, with stop-waters towed in the water, which I have seen done with success to make the ship sail so comparatively slow as to induce an enemy to come faster up with you, than you could with them, by chasing."

The reader has now been primed with sufficient nautical information to enable him to assume command—in imagination, at least—of the largest privateer afloat. The subject of cartridges was an important one to all engaged in warfare. The old-fashioned cartridges were almost as dangerous to friends as to foes, and at the commencement of the Seven Years War in 1756 we find Mr. Robert Williamson, the printer, publisher, and editor of *Williamson's Advertiser*, announcing in his paper that he sold "prepared cartridges of all sizes for the use of privateers and other ships of war. This preparation," says the advertisement, "prevents any spark from remaining in the gun after its discharge, and thereby not only secures the life of the person who re-loads the gun, but increases its execution, by saving much time; for it may be instantly re-charged without sponging—advantages always experienced in the use of these cartridges, and too important in time of action to be neglected by any sea commander. The great demand for these cartridges in the late war was a sufficient proof of their utility; and no other recommendation of them is necessary than appeal to those commanders who then used them." Some very sad and horrible accidents resulted from the use of the old-fashioned cartridges, and many a fine fellow was blown to eternity, or maimed for life, while in the act of re-loading a gun in which the remains of an old cartridge still smouldered. Mr. Williamson, whose portrait hangs in the Liverpool Free Public Library, William Brown Street, was a most enterprising man, combining with his printing and bookselling business, that of a broker and keeper of an employment registry. He also sold "ransom bills (French and English), and an abridgment of the Articles of War, designed for the use of privateers and vessels that carry letters of marque."

The first newspaper ever published in Liverpool was the *Leverpoole Courant,* printed in 1711-12 by S. Terry, in Dale Street; but to Robert Williamson belongs the credit of publishing the first Liverpool newspaper that attained a venerable age. Started on the 28th of May, 1756, as *Williamson's Liverpool Advertiser and Mercantile Register*, its title was changed on January 6, 1794, to *Billinge's Liverpool Advertiser,* Mr. Billinge being then editor. Its name was again changed to that of the *Liverpool Times*, which it retained until 1856, when it ceased to appear. A second Liverpool newspaper, called the *Chronicle,** was started in 1756-7, but was discontinued in less than three years. In December, 1765, Mr. John Gore published the first number of the *Liverpool General Advertiser,* or the *Commercial Register*, the title of which was afterwards changed to the *General Advertiser.*

* *Williamson's Advertiser* of November 25, 1757, says: "Mr. Robert Fleetwood, bookseller, and Mr. Sadler, printer, have declined being any longer concerned in publishing the *Chronicle*, or any other newspaper." The *Chronicle*, jealous of the advertisements in Williamson's paper, charged him with inserting too many puffs of quack medicines.

The reader will pardon this digression, for it is to the enterprise of these two men, Robert Williamson and John Gore, every historian of Liverpool is indebted for much valuable matter. The name of Gore suggests an employment in connection with privateering which we have not yet noticed, that of linguist or "linguister." In 1780 the following advertisement appeared in the papers:—"Wanted a person who understands perfectly the Dutch language, to go linguist in a good stout privateer. Such a person (if sober and well recommended) will meet with good encouragement by applying to John Gore." In the slave ships the position of linguist was often held by "ladies of colour."

Let us now enquire into the cost of fitting out a privateer. Fortunately, we are not obliged to resort to mere estimates or guess-work, having before us the original accounts of the *Enterprise* privateer, of Liverpool; Captain James Haslam, commander. She sailed on her first cruise in September, 1779, with a crew of 106, composed as follows:—Captain, first, second, and third lieutenants, sailing master, 2 master's mates, 2 prize-masters, surgeon, captain of marines and his mate, carpenter and his mate, boatswain and his 2 mates, gunner and 3 mates (the fourth absconded), cook, cooper and his mate, 4 quartermasters, armourer, captain's clerk, ship's steward, 2 cabin stewards, sailmaker, 20 seamen, 6 "three-quarter" seamen, 13 "half" seamen, 9 "quarter" seamen, 18 landsmen, 3 boys, and 3 apprentices. The amount of wages advanced to the seamen was £645 8s. 5½d., each officer and man receiving two months' pay in advance. The disbursements made to tradesmen and others in connection with the outfit amounted to £1,388 5s. 3d.; making a total expenditure of £2,033 13s. 8½d. This sum was debited to the owners of the privateer in the following proportions, according to the amount of their shares:—

Thomas Earle	-	3/16	- -	£381	6	4
Edgar Corrie	-	2/16	- -	254	4	2½
Francis Ingram	-	2/16	- -	254	4	2¾
William Earle	-	2/16	- -	254	4	2½
Dillon and Leyland	-	2/16	- -	254	4	2½
Peter Freeland	-	1/16	- -	127	2	1¼
Thomas Eagles	-	1/16	- -	127	2	1¼
Edward Chaffers	-	1/16	- -	127	2	1¼
James Carruthers	-	1/16	- -	127	2	1¼
William Denison	-	1/16	- -	127	2	1¼
				£2033	13	8½

For the rate of wages paid to the commander, officers, and crew; the names of the tradesmen supplying the outfit, together with the amount of their respective accounts, and other curious matter connected with the first, second, and third cruise of the *Enterprise*, the reader is referred to the appendix. It is worth noting here, however, that the French and Spanish commissions or letters of marque cost £41 17s. 4d., and that a considerable expenditure appears to have been incurred in bringing seamen from Whitehaven and Chester to Liverpool, due probably to the extraordinary number of privateers despatched from Liverpool about this period, and the consequent difficulty of procuring crews. The amounts which are set down for clothes, etc., for the French prisoners, and the present of £21 to the dispensary, prove that the owners of this privateer were generous and humane men. The name of Mr. Egerton Smith, father of the founder of the *Liverpool Mercury*, appears as supplying stationery to the privateer for each cruise. He was a schoolmaster and printer, at one time in Redcross Street, and afterwards in Pool Lane (now South Castle Street), where the future editor of the *Mercury* was born in 1774.

The total amount expended on the *Enterprise* in her

three cruises, together with the value of the ship, was as follows :—

Outfit for first cruise - - - -	£2033	13	8½
Outfit for second cruise - -	568	17	2
Outfit for third cruise - - -	2413	4	2
	5015	15	0½
Value of the ship - - -	2050	0	0
	£7065	15	0½

The *Enterprise* captured several prizes, the proceeds of one or two of which, probably, more than covered the above outlay. When it is borne in mind that in the year 1779 there belonged to the port of Liverpool a fleet of 120 privateers, whose aggregate tonnage was 31,385, carrying 1,986 guns, and 8,754 men ; some idea of the benefits accruing to the tradesmen of the town, as well as to the merchants and the ship owners, may be formed. The value of the prizes taken by these privateers has been put down at upwards of one million sterling. Judging by the amount expended upon the *Enterprise*, the fitting out of the 120 privateers must have imparted a wonderful activity to all branches of trade. But the mere fitting-out of private ships of war carrying no cargo, could not compensate for the sad falling away of the lucrative slave traffic, owing to the war ; and it is extremely doubtful whether the patriotism of the town would have borne the strain of so many privateers, if more ships could have been profitably employed in the African slave trade. The testimony of the Rev. Gilbert Wakefield is pretty clear on this head : "The principal cause of the multitude of privateers from Liverpool during the French and American War," he says, "was the impediment which this event had put in the way of the African slave trade, whose head-quarters, as I have observed, are fixed at this place."

The following are copies of the instructions given to Captain Haslam by Messrs. Francis Ingram & Co.:—

"LIVERPOOL, 16*th September*, 1779.

"CAPTAIN JAMES HASLAM,

"SIR,—You being appointed commander of our ship *Enterprize*, and being compleatly fitted for a cruise of six months, arc by the first oppertunity to sail from hence and make the best of your way to sea by the North or South Channel, as the wind may offer most favourable, but we prefer the former if to be effected without any extraordinary Risque, as being a path less liable to meet with any of the enemy's Cruizers, and having a chance to meet with American vessels bound to Sweden, etc. In this case don't keep too near the coast of Ireland, and be sure to gain the longitude of 20 West from London before you go to the southward of the latitude of 53, but shoud you go through the South Channel, a true W.S.W. course, 180 or 200 leagues from Tusker, would be the most likely to lead you clear and obtain the longitude of 20, as aforesaid, by the time you would get into latitude of 48; in either case, when the westing is gain'd you are to cross the latitudes under an easy sail to the Island of St. Mary's, then to cruise about five degrees to the westward of it, now and then stretching half a degree to the southward, as vessels may run in that path to see it and yet avoid coming too near for fear of being captured.

"If in the course of three weeks you meet with no success, you are to proceed to the westward of the longitude of Corvo, and stand across north and south from half a degree to the northward of Corvo to half a degree to the southward of St. Mary's. The whole of your cruize in these stations we remit to three months from the time of your being the length of St. Mary's, unless some extraordinary intelligence may be had, in which case it is left to your discretion, hoping that you will at all times weigh every circumstance maturely for and against, aided by the sentiments of such of your officers as may be depended upon.

"Shoud you be so fortunate as to take any prize or prizes in those stations of the value of £10,000 or upwards, you are to see them safe into some good port in Ireland, running down in the latitude 52, gaining that paralell in longitude of 15 west from London at least, then taking the North or South Channel as wind and weather may offer; but if not of that value, dispatch them with a trusty officer, taking care not to put too many of the enemy in proportion to your own men on board, giving the directions for his proceeding as aforesaid, with caution not to trust many of his own people aloft at a time on any account whatever, as many prizes have been retaken by the prisoners for want of such Precaution.

"Shoud you meet with no success, you are then to proceed to the latitude of Ushant, coming no farther to the eastward than 16 west from London, and cruize between that station and Corvo, and shoud you have no success in a reasonable time, finish your cruize between the latitude 37 and 48½, taking care as you increase your latitude to make easting in proportion, and on the contrary as you make southing, to increase your westing, either to the eastward or westward of the Western Islands, likewise as may be thought most eligible; and should you take any prize or prizes of the aforesaid value, you are to act accordingly, and take or send them for the North or South Channel as circumstances may offer, and shoud you loose company with any prize or prizes when conducting them, you are then to regain your station with all convenient speed, and let them take their chance, and for fear such an accident may happen, be sure to give such orders to the prize-master, and put people and necessarys on board as may best insure safety.

"You are strictly order'd not to meddle with any neutral vessel whatever unless you are certain by her papers or other indisputable information (freely given without bribery, promised gratuities, or Force) that she has taken in her loading in North America, therefore you are not to pay any regard to the Giddy solicitations of your Crew, so as to be misled by them, but act upon your own Reason, and for that purpose we desire you will read your printed Instructions from the Admiralty, given with your Commissions, with the utmost attention, and you cannot err.

"In case of your taking a prize, let every Paper, Letter, etc., be immediately secured and sent home with her, all Money and Valuables that can be easily removed to be taken on board your ship, and on you or your prize arrival at any port in Ireland, let an express be sent immediately with a Letter to Mr. Fras. Ingram to the first post Town, by a carefull hand, and repeated a post or two after for fear of Miscarriage, and the greatest care taken not to break Bulk, as the lower class of people in Ireland make use of every scheme to mislead and defraud.

"We order that upon any capture being made that your Lieutenant, with two trusty officers, do, as soon as possible, examine the Trunks, Chests, etc., of the officers, passengers, and crew, and that they take from them all Letters, Invoices, Papers, etc., and other valuables, delivering them to you, with a particular account of the same, signed by them in order to obviate any jealousy or misunderstandings. You will likewise examine the prisoners separately with great attention touching the destination of any ship or ships they may have been in company with, or of the destination of any vessel within their knowledge, and likewise gain all the information as to the Destination of Fleets, etc., and if anything of consequence as to national matters be obtained, communicate it to the first King's ship you meet, taking care at all times to compare the different Informations, so that you may not be deceived, to do which you may be assured every artifice will be used.

"We particularly recommend that the prisoners be not plundered of their Cloths and Bedding, but that they may be used with all tenderness and Humanity consistent with your own safety, which must be strictly attended to; and as true Courage and Humanity are held to be inseparable, we hope your crew will not be wanting in doing that Honour to their Country, the contrary of which is disgracefull to a civilized nation.

"You will take particular care that your crew be treated humanely, that every one be made to do their duty with Good Temper; as Harmony, a good look-out, and steady attention to the main point are all absolutely necessary to be attended to, the success of the Cruise greatly Depending upon it.

"Herewith you have sundry letters of credit, and shoud you have occasion to draw upon London, you must draw upon Messrs. Jos. Denison & Co.

"In case of your Death, which God forbid, your first Lieutenant is to succeed you in the command, and so in succession, and to follow these orders. Wishing you a successfull cruise,

"We remain,

"Your assured Friends,

"F. INGRAM & CO.

"Messrs. HORN & SILL, Lisbon.
"Messrs. PEDDAR & CO., Cork.
"Messrs. SCOTT, PRINGLE & SCOTT, Madeira."

"LIVERPOOL, 13*th Septr,* 1779.

"MESSRS. SCOTT, PRINGLE AND SCOTT,

"GENTLEMEN,—In case Captain Haslam of the *Enterprize* Privateer shoud put into Madeira, you will please to supply him with what necessaries he may want, and for the amount of which you are to value upon Messrs. Joseph Denison & Co., London.—I am Gent[n]

"Your most obd. Servt.

"FRA. INGRAM, FOR SELF & Co."

"Copys of the above wrote to

"MESSRS. JNO. PEDDER & Co., Cork. "MESSRS. HORN & SILL, Lisbon.	"Copied the above 12 June, 1780, and given to Captn. Haslam."

On the 22nd of October, 1779, the *Enterprise* returned to Liverpool, bringing in with her a valuable prize called *L'Aventurier*, of 22 guns and 50 men, bound from Martinico to Bordeaux with a cargo of cotton, tobacco, sugar, coffee, cocoa, and cassia fistula. There appears to have been some insubordination on board the privateer, and from the tenor of the following letter, we gather that the owners were not altogether satisfied with the commander.

"LIVERPOOL, 17 *Nov.*, 1779.

"CAP: HASLAM,

"SIR,—It is our positive orders that in case of your taking another prize, that you do not return to Liverpool on any account or pretence, but that on your taking a capital prize of not [less] than ten thousand pounds, you are to convoy her into the first port in Great Britain or Ireland. And further, you are expressly ordered to continue your cruise for five months from your departure now from the Rock, as by the Custom of the Port the detention in the River is not included in the time allowed for the cruize. We depend on the conduct of you and your officers to carry a proper command on board the vessel and to prevent any Disobedience or further attempts to Mutiny.

"We remain, &c.,

"FRA[S.] INGRAM & COMY."

At the commencement of the second cruise the muster roll had dwindled to 88, notwithstanding the introduction of new blood, and the amount advanced for wages was only £63 4s.

On the 15th of June, 1780, Messrs. Francis Ingram and Co. handed to Captain Haslam the following orders, which, in point of dignity, clearness, and shrewdness, are surely not unworthy of a fine old British merchant, one of the olden time :—

"CAPTAIN HASLAM,

"SIR,—Our ship *Enterprize*, of which you are at present commander, being compleatly equipped and Manned for a six-months cruise, you are by the first favourable oppertunity to sail from hence and make the best of your way for the latitude of Belleisle, and run down that paralel until you make it, as it is the place where all vessels bound for Nantz and Bordeaux take in their Pilots. You are to remain in that station one or two weeks, standing off and on in such direction and at such distance as you may think proper. If no success in that time, you are to stand across the Bay under an easy sail, making free with the land (if the wind and weather will permit), so as to fall in with the Spanish coast about Bilboa, and from thence coast it up by slow movements to Cape Finisterre, and when there stand off N.N.W. by the Compas about 40 or 50 Leagues Distance, and so in again S.S.E., spending two or three weeks in that manner, as it seems to be an Eligible track to catch both outward and homeward-bound French and Spaniards.

"If no success, range back again towards Bilboa, and from thence cross over for the mouth of the River of Bordeaux, taking all prudent libertys with the French and Spanish coasts and so on to the Island of Belleisle. This method we woud have you to pursue for the first three months, and if no material success in that time we suppose your water will be nearly expended, and in that case you are to proceed to the Western Islands and do the needfull there with all

Expedition, and as no profit accrues to a vessel lying long in Harbour, you are to proceed from thence to Cape Finisterre and finish your cruise between that and the latitude of Ushant, as far to the westward as you may think prudent, *and no more*, as the odds is considerably against rambling in the wide ocean, whereas Headlands and Islands usually run down by vessels are the surest places to find prizes.

"Notwithstanding the particular directions we have given above, we woud have it perfectly understood that the Execution of them as to Winds and Weather is left entirely to the good conduct of yourself and the officers (you are desired to consult). Upon all matters of consequence we wish you to consult with Mr. Cotter and him only, and to follow your joint opinions, and as your ship is a prime sailer you may make free with the enemy's coast without danger, shewing at all times a true british spirit to your Crew, with whom we hope you will cultivate the greatest Harmony and treat them with the greatest tenderness and Humanity, at the same time preserving the most strict discipline and command.

"You must by no means detain any Dutch or neutral ship unless bound from an Enemy's port to an Enemy's port with French, Spanish, or America property on board, and that to appear by their regular papers and not from any hearsay Information from the crews, as great Expences have been incurred by such imprudence. But shoud it appear by a clear examination of the Papers as aforesaid that the Goods on board are the property of an Enemy, you are to make a prize of them, in which momentous business take every precaution not to be misled or overawed by the Impropriety [importunity?] of your crew. You are particularly to observe that by an Act of Parliament passed lately the cargoes of any neutral ships bound from the Islands of Grenada, St. Vincents, and Dominica to a neutral port, having certificates on board signed by two English merchants or planters residing there, signifying that the entire cargoe is the produce of that island and was taken on board there, are exempted from capture and therefore are not to be molested.

"If you take a prize or prizes to the value of ten thousand pounds or upwards, put Mr. Cotter in command and see her safe into Milford, Cork, or Kinsale, as the winds and weather may be, and when there put some other trusty officer on board instead of Mr. Cotter, who is then to return to his station in the *Enterprize*. Give notice to us of the arrival of such prize immediately by Post, and order that letter may be repeated three or four times afterwards the following days, adding any new matter that may occur in the intervals. On such return into port keep a strict command over your people, and proceed again with all possible dispatch upon your cruise, as heavy Wages, Provisions, and Premiums of Insurances are constantly going on.

"Shoud you take a vessel belonging to a scattered Fleet, we direct you to pursue them or continue your Cruize for the stragling ships so long as you have a man to board with. If the prizes you may take are of less value than ten thousand pounds, dispatch them with a trusty officer, taking care not to put too many of the enemy in proportion to your own men on board, giving directions for his proceeding as aforesaid, with caution not to trust many of his people aloft at a time on any account whatever.

"Upon taking a prize, secure all the papers immediatly and remove all valuables, as money, etc., into your own ship, or let them remain on board the prize as you may think proper. Mr. Cotter and two other officers to examine the Trunks of the Officers, Passengers, etc., and direct them to deliver all papers found therein to you with a particular acct. signed by them of any Money or valuables which must be delivered to you for safety. Examine the prisoners separately, with great attention as to the destination of any ship, ships or Fleets they may have been in company with, or of which they have knowledge, and compare their informations to prevent you from being deceived by false intelligence.

"We desire you to be carefull to prevent the prisoners from being plunder'd of any article whatever, to prevent any insult to the meanest of them, that you treat them with Humanity and all the Tenderness that is consistent with the Security of your ship, or your prizes, which must be strictly attended to.

"Keep a good look-out on all occasions, and make short work of any action you have by runing close alongside before you open your fire, for depend upon this that by engageing them very close the officers opposed to you will be unable to keep their Men to their Guns. Improve this advantage therefore to the utmost, which the Discipline on board your ship, and the courage of your people, will indisputably give you.

"If you fall in with any British man-of-war and the Captain attempts to impress any of your people, represent to them respectfully the injury you and We sustained by Cap. Phipps interrupting you last cruise, and produce the Memorial to the Admiralty and Mr. Gascoyn's letter upon that subject, which will certainly prevent any worthy british officer from Empressing any of your people a second time at sea.

"Herewith you have Sundry Letters of Credit, and in case of your death (which God forbid) your first Lieutenant is to succeed you in Command, and so on in succession, and to follow these orders.

"To Conclude, this is the last cruize the ship is to make as a privateer, and our motives for fitting her on this cruise have been a dependance on her sailing, with a Confidance in your making the most of that advantage, and of the stations we have pointed out for your cruise, in which you will be well supported by Mr. Cotter. As it is the last cruise, make the most of it, and be assured of this, that you and Mr. Cotter in Consequence of such Conduct, will meet with every degree of regard, favor, and attention to your future Interests from us, Who are very truly

"Your assur'd friends,

"FRA^S. INGRAM & Co."

"To Hoist a White Flag at the main top Gallant mast Head—wch Rich^d. Wilding will answer on his own Pole to the southward of the Lighthouse.

"Capt. Haslam.

"On your coming in (or a prize) you are to make the above signal, wch will be answered by Rich^d. Willding. Success attend you.

"FRA^S. INGRAM."

The gallant commander would doubtless read the following considerate note with mixed feelings:—

"Liverpool, 5*th June*, 1780.

"Capt. Haslam.

"Sir,—In case you shoud be so unfortunate as to be taken, the owners have agreed to allow you six pounds a month during your captivity.

"I am sir your most Hum^e Servt

"FRA^S. INGRAM & C^O."

Owing to their manner of life and reckless habits, captivity, and especially captivity in a French prison, meant certain death to hundreds of seamen. Perhaps none fought more desperately during the war than those men who had once experienced the horrors of a French gaol. The next and last document with which we shall trouble the reader, in connection with this ship, runs as follows:—

"LIVERPOOL, 12*th June,* 1780.

"TO ANY OF HIS MAJESTY'S CONSULS,

"SIR,—In case the *Enterprize* Privateer, Capt. Haslam, shoud put into Port for Provisions, &c., &c., must beg the favor you woud supply him with what he may want to the amount of Five Hundred pounds and your Bill shall be punctually honored for the amount. We are, &c.,

"FRA. INGRAM & Co."*

On her third cruise the *Enterprise* carried 105 men. Her surgeon on the first cruise was Henry Barr; on the third cruise, Edward Lowndes. It is worthy of note that after the lapse of more than a century the names of Barr and Lowndes are honourably represented in the medical profession in Liverpool; but we are not aware that there is any connection between the surgeons of old Liverpool and those of greater Liverpool. In this final cruise the commander, officers, and men appear to have been shareholders, Captain Haslam having 16 shares, first lieutenant 8, second lieutenant, sailing master, surgeon, and carpenter 6 each, petty officers in proportion, seamen 2 shares each, "three-quarter" seamen 1¾ shares each, and so on down to the boys, each of whom had half a share. The total number of shares thus allotted was 212. The second lieutenant, sailing master, surgeon, gunner, and boatswain received £4 10s. per month wages; the carpenter, £5 per month; four sailing mates, £4 5s. each; boatswain's mates, £4 each; quartermasters and the gunner's mate, 75s.; surgeon's mate, cooper, cook, steward, armourer, and full seamen, 70s. per month; "three-quarter" and "half" seamen from 40s. to 65s.; landsmen from 20s. to 46s. per month.

*From the original account books in the possession of T. H. Dixon, Esq., The Clappers, Gresford, and kindly lent by him to the author. The penmanship of the above instructions, and of the accounts, is remarkably neat, but the orthography, as the reader sees, is conceived on a free and easy scale in keeping with the subject and the times.

CHAPTER II.

The Story of Captain Fortunatus Wright and Selim the Armenian Captive.

The "Spacious days of great Elizabeth" were the golden age of privateering, in the sense that the profession was carried on by men cast in the heroic mould, who disdained to draw too nice a distinction between privateering and piracy. Elizabeth was the sailor's friend, "the restorer of the glory of shipping, and the Queen of the North Sea." Camden tells us that "the wealthier inhabitants of the sea-coast, in imitation of their princess, built ships of war, striving who should exceed, insomuch that the Queen's navy, joined with her subjects' shipping, was, in short time, so puissant that it was able to bring forth 20,000 fighting men for sea service." The ships so benevolently provided by the wealthier inhabitants of the sea coast were, of course, privateers, but Liverpool was at that time too insignificant and poor a place to indulge in the romantic and fashionable patriotism of the age.

It is in this reign we find privateering first mentioned in connection with Liverpool. In 1563 a privateer, fitted out by Sir Thomas Stanley, of Hooton, son of the Earl of Derby, brought a prize into the river Mersey "with great rejoicings." Another privateer, fitted out by the licensed victuallers of Chester, brought in a French prize, whereupon the "shipping shot off so noble a peal of guns, so quick and fast one upon another, that the like was never heard in these parts of England and Wales." In the year 1566 two prizes arrived, one of which was subsequently ransomed. It is now impossible to say when the first private armed ship left the port of Liverpool, but as the Tower in Water Street was for many ages the seaside residence and place of embarkation of the Derby family, it is probable that their ships, armed, of course, against corsairs, or for naval warfare, were among the earliest that set out. The ships of the Stanleys, in fact, are mentioned in our old poetry. In the ballad of "Lady Bessie" Lord Stanley promises Elizabeth of York to send her messenger Humphrey Brereton, to Henry VII.

> "I have a gude shippe of mine owne
> Shall carry Humfrey ;
> If any man aske whoes is the shippe?
> Saye it is the Earle's of Derbye.
> Without all doubt at Liverpoole
> He tooke shipping upon the sea."

Nearly five hundred years have flown since Isabel of Lathom gave her hand—and, let us hope, her heart—to the gallant Sir John Stanley, who received from his father-in-law the site upon which he erected the Tower in Water Street. The close connection thus begun between the Stanleys and the citizens of Liverpool has grown and strengthened with the years, and while these lines are being penned, the sixteenth Earl of Derby sits in his official residence as Lord Mayor of Greater Liverpool, within bow-shot of the site of the ancient fortress and town house of his ancestors. Though he may not possess "a gude shippe" of his own to carry Humfrey, he has but to telephone down the street to the neighbourhood of the Tower, and floating palaces, surpassing in splendour the happiest dreams of "Lady Bessie," will be placed at his disposal—for a consideration—to take shipping upon the sea.

About the time of the sailing of the Spanish Armada, the Town Council providentially laid in 300 pounds of gun-powder, and ordered "a gun" to be set up at Nabbe (afterwards Pluckington) Point, above the pool. It was the good fortune of a Liverpool captain and shipowner, however, to render a more important service to his Queen and country at that exciting time. Worthy Master Humfraye Brooke brought to England the first intelligence of the Armada being at sea. He was outward bound from Liverpool to the Canaries when he espied the Biscayan division of the Spanish fleet in the distance, sailing north. Suspecting its errand, he put ship about and made all haste to Plymouth, whence he despatched couriers, or perhaps went himself, to London. He received substantial marks of favour from the Government for his foresight, prudence, and activity. Liverpool was not then able to add much to the fleet of upwards of a hundred merchantmen, which joined the twenty ships of the Royal Navy and took so distinguished a part in baffling, defeating, and dispersing the "invincible" Armada.

In 1634 the memorable levy of ship money took place. The whole county was assessed at the sum of £475, of which Liverpool was required to pay £15, raised in the following year to £25. During the Civil War, the Tower in Water Street was garrisoned by the retainers of Lord Derby, the castle being held by Lord Molyneux. We cannot linger over this period of Liverpool history in which the fiery Prince Rupert found that the men of Liverpool were foemen worthy of his steel, for the "crow's nest" which he despised was not taken without an incessant cannonade carried on for eighteen days, and numerous assaults, in which he lost 1,500 men. It is sufficient for our present purpose to observe that the capture of the town by the Parliamentary forces was a serious blow to the royal cause, as it gave Parliament and its partisans the power of fitting out vessels of war in the Mersey, and of thus interrupting the communications with Ireland, whence the Lord-lieutenant of the King, the Marquis of Ormonde, was preparing to send supplies and reinforcements to the Royal party. Several frigates, or small vessels, were fitted out at Liverpool—one of them by Colonel John Moore. A number of Liverpool frigates, under the command of Captain Danks, cruised in the Irish Channel, sometimes blockading Dublin, and cutting off the supplies of provisions, coal, and other necessaries, which that city previously obtained from England. The cruisers also added much to the difficulty of sending over reinforcements to England. So great was the inconvenience produced by the Liverpool Squadron that the Marquis of Ormonde strongly urged the royalists in Chester to attack Liverpool by sea.

The Marquis, writing to Lord Byron, January 16, 1643, says: "When they (the Royal fleet) are gone, it is too probable the Liverpool ships will look out again, if that town be not in the meantime reduced, which I most earnestly recommend your lordship to think of and attempt as soon as you possibly can, there being no service that, to my apprehension, can at once so much advantage this place (Dublin) and Chester, and make them so useful to each other."

The merchants of Liverpool have always been a shrewd, far-seeing race, and an instance of their readiness to make the most of their opportunities turns up in an unexpected quarter. In the recently published Kenyon MSS. we find, under date 1702, "Reasons humbly offered by Henry Jones, Esquire, for building a mould or harbour in Whitsand Bay, at the Land's End, in Cornwall." The tenth reason adduced is as follows:

"By all the above it is likewise further manifest that even in times of peace there hath not nor can be secure trading 'twixt St. George's and the British Channels, or anywhere to the westward of the Land's End, without this proposed mould, and that for want of it there hath been and may be more ships lost (yearly, besides the men's lives) than three times the value of what would erect the same. Hence, the Leverpoole merchants, during all the last war, possessed those who trade from London that their ships might come safer north about Ireland, unload their effects at Leverpoole, and be at charge of land-carriage from thence to London, rather than run the hazard of having their ships taken by the enemy, or wreckt, by reason of the great dangers of Scilly, the Land's End, Mount's Bay, Lizzard, and all the South Channell to London, which hath proved an unspeakable detriment to all the trading seaport towns that border upon the British Channell ; which evills would effectually be prevented were there an harbour and light-house at the Land's End of England."

In the reign of George II. Liverpool ships, in common with those belonging to other British seaports, were plundered, and their crews maltreated by the Spanish Guarda Costas, whose depredations, carried on with impunity for several years, aroused at length the indignation of the whole country. In 1728, while the fate of Europe continued in suspense, while the English fleet lay inactive and rotting in the West Indies, the sailors perishing miserably without daring to avenge their country's wrongs, the merchants of Liverpool, London, Bristol, and other places petitioned the House of Commons for redress. The House instituted inquiries, and passed resolutions accusing the Spaniards of violating the treaty between the two crowns, and with having treated with inhumanity the masters and crews of British ships. The King, in reply to the address of the Commons, promised to procure satisfaction. The outrages went on and grew in number and daring until, in 1737, the whole nation cried for vengeance, and petitions from merchants in all parts of the country poured into the House of Commons, which, at length, in Grand Committee, proceeded to hear counsel for the merchants, and examine evidence, by which it appeared that amazing acts of wanton cruelty and injustice had been perpetrated by Spaniards on the subjects of Great Britain. In the following year the King informed Parliament that a Convention with Spain had been ratified. When the terms of the Convention became known, many merchants, planters, and others trading to America, the cities of London and Bristol, the merchants of Liverpool, and the owners of ships which had been seized by the Spaniards presented petitions against it. In a great debate in the Commons, Mr. Pitt denounced the Convention as dishonourable to Great Britain, but, in spite of strong opposition, the Convention received the approval of both houses. In 1739, Spain, having failed to pay the money stipulated in the Convention as compensation to those who had suffered by the depredations, letters of marque and reprisal were granted against the Spaniards. The British Minister at Madrid politely explained to the Court of Spain that his master, although he had permitted his subjects to make reprisals, would not be understood to have broken the peace, and that this permission would be recalled as soon as his Catholic Majesty should be disposed to make satisfaction. The King of Spain failed to appreciate the nicety of the distinction perceived by the British monarch, and proceeded to defend himself by vigorous words and actions. A declaration of war on both sides soon followed, and in 1744 France declared war against England. Referring to this period, the author of "Williamson's Liverpool Memorandum Book," published in 1753, advanced the remarkable theory that the town flourished more in war than in peace.

"In the last war, 1739 to 1748," he says, "trade flourished and spread her golden wings so extensively that, if they had possessed it seven years longer, it would have enlarged the size and riches of the town to a prodigious degree. The harbour being situated so near the mouth of the North Channel, between Ireland and Scotland (a passage very little known to or frequented by the enemy) afforded many conveniences to the merchants here, untasted by those of other ports, which invited numbers of strangers from different parts to begin trade and settle here, finding it so advantageous a mart. Trade since the late peace has not been so brisk as formerly, but it appears by the Custom House books to be much revived. The chief manufactures carried on here are blue and white earthenware, which at present almost vie with China (large quantities are exported for the Colonies abroad), and watches, which are not to be excelled in Europe. All the different branches are manufactured in and about the town, to supply the London and foreign markets."

It is true that in this war the commerce of Liverpool suffered much less than that of London, Bristol, and Hull from the privateers of the enemy, but the prosperity had probably more to do with black than golden wings, the number of slave ships having grown from one vessel of 30 tons in 1709, to 72 ships of 7547 tons in 1753. The progress made during the first half of the eighteenth century, long before cotton had been added to tobacco, sugar, rum, and slaves, as the commercial deities of Liverpool, will be seen from the following tables :

A comparative statement of the number of ships that arrived at, or sailed from the Port of Liverpool for six years preceding the year 1751 :

The number of ships that arrived at, or sailed from the Port of Liverpool for six years.					Ships belonging to the Port for the same time.		
	Inwards.		Outwards.				
Years.	Ships.	Tons.	Ships.	Tons.	Ships.	Tons.	Men.
1709	374	14,574	334	12,636	84	5,789	936
1716	370	17,870	409	18,872	113	8,386	1,376
1723	433	18,840	396	18,373	131	8,070	1,114
1730	412	18,070	440	19,058	166	9,766	1,710
1737	402	17,493	435	22,350	171	12,016	1,981
1744	403	22,072	425	20,937	188	13,772	2,621

In 1749 the total tonnage of vessels that entered the port was 28,250 tons. In 1751 the number of ships that entered was 543, with a tonnage of 31,731. For the next hundred years Liverpool went on steadily, doubling her trade about every sixteen years.

In the year 1744, Liverpool appears to have possessed four privateers, namely, the *Old Noll,* of 22 guns and 180 men, Captain Powell; the *Terrible,* of 22 guns and 180 men, Captain Cole; the *Thurloe*, of 12 guns and 100 men, Captain Dugdale; and the *Admiral Blake*, whose armament is not stated, commanded by Captain Edmondson. The *Terrible* recaptured, and sent into Waterford, the *Joseph,* of Bristol, laden with logwood, tar, etc., which had been taken on the homeward passage from Boston by a Bilbao privateer. The *Terrible* also recaptured the *Bromfield*, of Bristol, Captain Sharp, which had been taken by the French on the passage from St. Kitts to Bristol. The *L'Amiable Martha,* from St. Domingo for Bordeaux, was taken and carried into Cork by the *Terrible*. The prize cargo consisted of 370 hogsheads and 44 barrels of sugar, 57 casks of coffee, 11 hogsheads of indigo, one hogshead white sugar, 1,270 pieces of eight, and five cobs of gold. In 1746 the *Terrible* captured a Martinico ship and sent her into Liverpool. In July, 1744, we read that the *Thurloe* had captured a vessel with wine; and about the same time that the *Vulture* privateer, of Bayonne, 14 carriage guns and 118 men, had been taken and carried into Cork by the *Thurloe* and the *Blake* privateers of Liverpool. The *Admiral Blake* took a Martinico ship, and, in company with the *Thurloe*, carried into Cork the *Admiral*, a rich French ship from Martinico for Bordeaux. In the capture list of August, 1744, we read that the *Thurloe* privateer of Liverpool, and her prize, a Martinicoman, were taken by a French privateer, but afterwards retaken by the *Thurloe's* consort, the *Old Noll*, with the Frenchmen on board, and carried into Cork. After a smart engagement, the French privateer, of 36 guns and 300 men, sheered off. The *Old Noll* took the *Providence*, from Bordeaux for Martinico, and carried her into Kinsale ; and recaptured the *Hannah*, Captain Fowler, from Jamaica, which she sent into Cork. The *Old Noll* also took a prize off the Start, and a fishing vessel with 30 men, "three of them Irish," and the *City of Nantz*, a very large ship from St. Domingo for Nantz, which she convoyed to Liverpool. Finally, the *Old Noll* recaptured and carried to Liverpool the *Sarah*, from Carolina for London, which had been taken by a French privateer called the *Count de Maurepas,* who had captured five prizes. In November, 1745, the sad intelligence reached Liverpool that the *Old Noll* had been sunk, with all her crew, by the Brest squadron. In June, 1748, the capture lists recorded that "a Dutch ship, from Bordeaux to Dunkirk, with bale goods and spices, and a French sloop from Cape Francois, coming express with the account of the English taking Port Louis," had been captured by the *Warren* privateer, of Liverpool. Early in the same year a vessel called *L'Amitie*, bound for St. Domingo, was taken and carried to Antigua by a Captain Johnson, of Liverpool. Liverpool commerce suffered heavily from the privateers of the enemy, and the few captures recorded above offer a sad contrast to the long list of Liverpool vessels taken during the war.*

Captain Fortunatus Wright was undoubtedly the most famous British privateer commander of his time, and Liverpool's favourite hero during the first half of the eighteenth century. In the few memorials of his life and character which we have gathered together he strikes the imagination as the ideal and ever-victorious captain, around whose name and fate clings the halo of mystery

* See Appendix No. 1.

and romance. Smollett, the historian, has paid the following tribute to his memory: "Sir Edward Hawke, being disappointed in his hope of encountering la Galissoniere, and relieving the English garrison of St. Philip's, at least asserted the empire of Great Britain in the Mediterranean, by annoying the commerce of the enemy and blocking up their squadron in the harbour of Toulon. Understanding that the Austrian government at Leghorn had detained an English privateer and imprisoned the captain on pretence that he had violated the neutrality of the port, he detached two ships of war to insist in a peremptory manner on the release of the ship, effects, crew, and captain; and they thought proper to comply with his demand, even without waiting for orders from the Court of Vienna. The person in whose behalf the Admiral thus interposed was one Fortunatus Wright, a native of Liverpool, who though a stranger to a sea-life, had in the last war* equipped a privateer, and distinguished himself in such a manner by his uncommon vigilance and valour, that if he had been indulged with a command suitable to his genius, he would have deserved as honourable a place in the annals of the navy as that which the French have bestowed upon their boasted Gue Trouin, Du Bart, and Thurot. An uncommon exertion of spirit was the occasion of his being detained at this juncture. While he lay at anchor in the harbour of Leghorn, commander of the *St. George* Privateer of Liverpool, a small ship of 12 guns and 80 men, a large French xebeque, mounted with 16 cannon and nearly three times the number of his complement, chose her station in view of the harbour, in order to interrupt the British commerce. The gallant Wright could not endure this insult; notwithstanding the enemy's superiority in metal and number of men, he weighed anchor, hoisted his sails, engaged him

*War of the Austrian Succession.

within sight of the shore, and after a very obstinate dispute, in which the Captain, lieutenant, and above three score of the men belonging to the xebeque were killed on the spot, he obliged them to sheer off, and returned to the harbour in triumph. This brave corsair would, no doubt, have signalised himself by many other exploits, had not he, in the sequel, been overtaken in the midst of his career by a dreadful storm, in which the ship foundering, he and all his crew perished."*

Professor Laughton, in his "Studies in Naval History," very properly doubts whether Smollett is entirely correct in his statements regarding Wright's early life. "His father," he says, "who was of Cheshire origin, was a master mariner and shipowner, and I have little doubt that Wright himself followed the sea in his youth probably as his father's apprentice, or afterwards in command of one of his father's ships. The evidence is indeed very strong that he was far from a stranger to a sea life. William Hutchinson, for many years dockmaster at Liverpool, and who, on the title-page of his 'Treatise on Practical Seamanship,' styles himself as distinctively 'Mariner'—the sort of man who, in the last century, would have divided the human race into seamen and land-lubbers—speaks with evident pride of having served under Fortunatus Wright, and frequently refers to the practice of 'that great,' 'that worthy hero,' as illustrating different points of seamanship. He had, however, retired from the sea, and settled down as a merchant and shipowner. Beyond that, little is known, but it is believed that he became involved in a tedious and costly lawsuit on account of one of his ships with letters of marque detaining a vessel in which the Turkey Company had an interest. In this there is possibly some confusion with a later incident, the circumstances of which are before us; but at any rate we may

*Smollett's "History of England," vol. 1, page 337.

accept the statement that, consequent on this lawsuit, and not caring to abide another with which he was threatened, he realised his property and left Liverpool." For these personal details, Professor Laughton was indebted to the kindness of Mr. Fortunatus Evelyn Wright, Consul for Sweden and Norway at Christchurch, New Zealand. Mr. F. E. Wright, or rather his elder brother, Mr. Sydney Evelyn Wright, formerly a paymaster in the navy, is the lineal representative of our privateer captain, as well as of John Evelyn, the author of "Sylva," and the first treasurer of Greenwich Hospital.*

According to Smithers' History of Liverpool, Fortunatus Wright was the son of Captain John Wright, mariner, who died in April, 1717, and who gallantly defended his ship for several hours against two vessels of superior force, as is recorded on a plain tombstone in St. Peter's churchyard; which records also that "Fortunatus Wright, his son, was always victorious, and humane to the vanquished. He was a constant terror to the enemies of his king and country." After giving the substance of Smollett's account, Mr. Smithers adds, "but tradition tells that he became a victim to political interests. The tombstone is silent as to the cause of his death." It is to be regretted that so little is known of the early life of a brave man, of whom Liverpool has reason to be proud.

In June, 1742, Captain Fortunatus Wright was travelling in Italy, where he met with an adventure which is thus related in a letter from Horace Mann, the British Resident, to his friend Horace Walpole:—

*Captain Wright's daughter, Philippa, married Charles, the grandson of John Evelyn, of Wotton, whose daughter, Susanna, married her first cousin once removed, John Ellworthy Fortunatus Wright, who served as a lieutenant in the navy during the war of American Independence, and retired after the peace of 1783. He was subsequently appointed master of the George's Dock, Liverpool, where he was accidentally killed in the year 1798. Some of his descendants, doubtless, still reside in Liverpool, though the elder branch of the hero's family emigrated many years ago to New Zealand.

"For this last week I have had Complaints made to me which were brought by an Express, of an Englishman, one Wright's design to storm the Town and Republick of Lucca, which horrid design was manifested by his obstinate refusal to deliver a couple of Pistols to the Guards at the Gate, and his presenting one of them cocked at the Corporal, and twenty soldiers that demanded them of him, threatening to kill them if they persisted. Much mischief might have ensued had not a Colonel with thirty more soldiers taken this valiant Squire Prisoner. He was conducted with the above attendants to his Inn, where he found another Guard, and two were placed in his bedchamber, till one of the Lucchese noblemen to whom our Countryman had recommendations, found means to persuade the Republick that no mischief should ensue. He was kept three days prisoner, when at four o'clock in the morning, just as his Servant was setting out post to tell me, he received a message from the Gonfaloniere, by an officer who speaks English, 'that since he had been so daring as to endeavour to enter the Town by force of Arms, it was therefore ordered that he should forthwith leave the State—never presume to enter it again without leave from the Republick; and that there were post horses at the door of his house, as well as a Guard of Soldiers, to see him out of the Territories of the Republick!' He answered a great deal not much to the purpose. However, his compliance with the orders put an end to what had made a great noise, and for three days had put their Excellencies in an uproar."*

It is supposed that after this remarkable adventure Captain Wright lived with his wife and family either at Leghorn or Florence for about four years. His connection with John Evelyn, and his letters of introduction to the

* "Mann and Manners at the Court of Florence," vol. 1, pp 72-73.

Lucchese nobleman, show that he was a man of good social position. Professor Laughton, who has seen specimens of his handwriting, pronounces it to be that of a man of culture and education. "The hand," he says, "is not of a commercial character, still less is it the hand of a rude seaman, more familiar with the marling-spike than the pen."

Soon after the outbreak of war with France in 1744, Wright conjointly, probably, with the English merchants in Leghorn, fitted out the brigantine *Fame* "to cruise against the enemies of Great Britain." In the "Gentleman's Magazine" for January, 1744, it is recorded that the *Swallow*, Captain Wright, from Lisbon for London, had been taken by the *Begonia* and ransomed at sea, her former captain, Mr. Hutchinson, being detained as security. We have no means of knowing whether the *Swallow* belonged to our Captain Wright or not, but it is scarcely conceivable that with Fortunatus Wright and William Hutchinson on board either the *Swallow* or any other vessel would have struck. We know, however, that it is to this period of Wright's romantic career that Captain Hutchinson refers in his observations on a ship cruising on her station, which we have quoted in a previous chapter as illustrating the tactics of these two daring and successful commanders.

In the "Gentleman's Magazine" for November, 1746, we read that two French ships from Smyrna for Marseilles were taken "by the *Fame* privateer, Captain Wright, fitted out by the merchants at Leghorn, and carried into Messina;" and a month later the same publication stated that the *Fame* had captured 16 French ships in the Levant, worth £400,000 sterling; also that 18 of our West India and other ships were carried into French ports. The "London Gazette" reported the captures as follows:—"Sixteen French ships, taken by the *Fame* Privateer, Captain Fortunatus Wright, in the Mediterranean; two of them about 200 tons each, brought into Messina on October 13, and the others sent into Leghorn. The largest of the two ships was fitted out by the French factories on the coast of Caramania with 20 guns and 150 men; but after a smart engagement of three hours with the *Fame* off the isle of Cyprus, the Frenchmen ran their ship ashore and escaped, while the English took possession of the ship, and got her afloat again."

On the 19th of December, 1746, the *Fame* captured a French ship, bound from Marseilles for Naples, with the Prince of Campo Florida's baggage on board, and carried her into Leghorn, notwithstanding that the French vessel had a pass from his "Sacred majesty, King George the Second." This was a most irregular, not to say irreverent, action, the only excuse offered being the omission of the vessel's name in the pass. She was sent into Leghorn to be condemned in the usual way; and, no doubt, the Prince of Campo Florida used very sulphurous language when he heard the fate of his equipage and baggage; so did Mr. Goldsworthy, the English Consul at Leghorn, who was aghast at the "insult" offered to his Majesty's pass. We are not sure that Wright himself was in command of the *Fame* when this "outrage" on majesty was committed, but he speedily received a very strongly-worded exhortation from the consul to set the prize at liberty. The captain would not give way to the consul, but afterwards, on the representation of the British Minister, he agreed to refer the affair to the naval commander-in-chief on the station, who decided against him, and the prize was released.

A far more serious international dispute next claimed his attention. Early in 1747 the Ottoman Porte complained that Turkish property on board French ships had been seized by English privateers, and especially by Captain Fortunatus Wright, in the *Fame*. Mr. Goldsworthy, the English Consul at Leghorn, who had been instructed to enquire into the matter, wrote to Captain Wright for an explanation, and received a reply which was the reverse of satisfactory to the Turkish merchants whose property had been confiscated. "The two ships named," wrote Captain Wright, "had each of them a French pass, and both of them belonged to Marseilles. They also hoisted French colours and struck them to me; nay, the latter engaged me for a considerable time under these colours. For these reasons I brought them to Leghorn, and have had them legally condemned in the Admiralty Court, by virtue of which sentence I have disposed of them and distributed the money."*

The fact that the prize money had been realized, distributed, and probably spent by the captors, though grievous to the Turkish mind, was not permitted to end the matter. The influence of the Turkey Company was strong enough to procure from the British Government fresh instructions dated March 30, 1747, for the Privateers and Admiralty Courts in the Mediterranean, to the effect that Turkish property on board even French vessels was not prize. Captain Wright naturally refused to allow the order in his case to be retrospective, and as he positively declined to disgorge, an order was sent from England to have him arrested and sent home. On December 11th, 1747, the Tuscan authorities obligingly clapped him into prison, but refused to deliver him up to Consul Goldsworthy, who vainly argued that as commander of an English private ship he was subject to consular jurisdiction. Captain Wright remained a prisoner in the fortress of Leghorn for about six months; then an order came from Vienna to hand him over to the English Consul. Whilst Goldsworthy was waiting for an opportunity to send the stubborn hero to England, a new command bade him set him at liberty on

*Goldsworthy to the Deputy-Governor of the Turkey Company, Feb. 20th, 1747.

the ground that Wright had "given bail in the High Court of Admiralty to answer the action commenced against him." This was done on or about June 10th, 1748.

The special ground of this action, which ran on in a manner highly pleasing to the legal mind and profitable to the legal pocket, was the seizure of Turkish property on board the *Hermione*, a French ship taken by the *Fame* on February 26th, 1747, the proceeds from which Captain Wright refused to give up. The suit was still pending in June, 1749, a year after the captain's release, for on the 4th of June he sent a long statement of his case to Consul Goldsworthy, concluding in these characteristic words:—

"The cargo was all sold at public auction, for which I have proper vouchers; therefore, I am surprised at the manner the Turkey Company have represented this affair, or that they should trouble his Grace, after they have prosecuted me, after they had caused me to be confined near six months at their instance, and have since found their libel totally rejected, and that I am acquitted from the charge. They attacked me at law; to that law I must appeal; if I have acted contrary to it, to it I must be responsible; for I do not apprehend I am so to any agent of the Grand Signior, to the Grand Signior himself, or to any other power, seeing I am an Englishman and acted under a commission from my prince."

The correspondence about the *Hermione* was still going on in 1750, when Wright entered into partnership with Captain William Hutchinson. It seems that Wright did not disgorge after all, but how the lawsuit ended—whether it was nursed to death by the lawyers, or merged in some diplomatic settlement with the "Grand Signior," is not known. It might be supposed that Wright having in 1746 taken 16 ships, valued at £400,000, was in a position to recoup the losses of the Turkish litigants. Professor Laughton thinks the value of those prizes was a gross exaggeration. "Wright," he says, "was owner as well as captain of the brigantine, and her ship's company must have been small; his share of such a sum would have rendered him wealthy; but he does not come before us in the after years as a wealthy man." It is, however, expressly stated in the "Gentleman's Magazine" for 1746, that the *Fame* was fitted out by the merchants of Leghorn, therefore Wright was not the owner, though he may have had a share in the venture. To capture so many important prizes, and make himself the terror of the French in the Mediterranean, required not only a daring commander, but a considerable crew, both for fighting the enemy and manning the prizes. In a list of British privateers in 1745 we find the *Fame*, fitted out in London, carrying 50 guns and 380 men, and commanded by Captain Comyn. She surpassed all the other privateers—numbering 98—in the number of her men and guns, and yet we can trace none of her exploits. It is very probable that Captain Comyn was succeeded by Fortunatus Wright, who immediately made the vessel justify her name and superior armament. This, however, is purely conjecture.

The *Fame* was not idle while Captain Wright was cooling his heels in the fortress of Leghorn; for in the "Gentleman's Magazine" for January, 1748, it is recorded that a French ship from Alexandria to Marseilles had been carried into Leghorn by the *Fame* privateer.

In 1750 Captain Wright joined with Captain William Hutchinson in purchasing and fitting out as a merchant ship the old 20-gun frigate of war, *Leostoff*, which made several trading voyages to the West Indies, under the command, probably, of Hutchinson, while Wright settled down with his wife and family at Leghorn.

When the speedy renewal of the war between England and France became apparent in 1755 and early in 1756, Captain Fortunatus Wright set about building a small vessel at Leghorn, to cruise against the "hereditary enemy" of Great Britain. This was the *St. George* privateer, destined to be as famous, but not so fortunate for its gallant commander, as the *Fame*. On the declaration of war, the Tuscan government, whose interests were closely bound up with those of France, and whose neutrality was in practice only a thinly-veiled partiality, took measures to prevent the English ships in port from increasing their crews or armament, either for defence as merchant ships or for privateering purposes. Captain Wright was too well-known for the destination of his vessel to be a matter of doubt to the government officials, and he was compelled to resort to stratagem in order to have her properly equipped for her intended cruise. It was with an air "childlike and bland" that he applied to the authorities to know what force they would permit him to carry out of the port as a merchant ship. This was ultimately fixed at four small guns and 25 men, every precaution being taken by the officials to ensure that the limit was not exceeded. Wright gravely urged them to have guard boats rowing round him to make more certain, and so conducted the whole affair that in taking leave of the governor, he obtained from him a written certificate that he had complied with the limitation.

He sailed out of the port of Leghorn on the 25th of July, 1756, in company with three or four merchant vessels homeward bound to England, which, amongst other things, carried an efficient armament and ship's company for the *St. George*. The enemies of England at Leghorn secretly rejoiced, no doubt, thinking that Wright and his convoy were sailing into the lion's mouth, for they must have known that a French privateer had been cruising for the past month off the harbour, expecting to make a rich but easy capture of the poorly armed little *St. George* and her convoy. The captain of the French privateer had asked in Leghorn, "Pray when does Wright intend to come out? He has already made me lose too much time." The French commander had indeed very substantial reasons for desiring a meeting. His vessel, a xebec (carrying lateen sails on three masts) had 280 men on board, and mounted 16 carriage guns, besides swivels and a great number of small arms. She "had been fitted out with a particular view to take Captain Wright, who, having done the French much damage during the last war, had been marked out by the French King, who promised the honour of knighthood, a pension of 3,000 livres per annum for life, and the command of a ship of war, to whoever should bring him into France alive or dead. The merchants of Marseilles had also promised a reward, double the value of Wright's vessel, in a writing pasted up on their Exchange."*

The subsequent proceedings of Wright and the French candidate for knighthood at his expense are given in a letter from Leghorn to a merchant in Liverpool, dated July 30, 1756:—

"Your brave townsman Capt. Fortunatus Wright's late gallant action is at present the topic of conversation here; the heads of which are as follow: Capt. Wright sailed the 25th inst. with three other small vessels under convoy of Capt. Wright, who engaged to see them safe as low as Gibraltar. The Government here would not allow him to carry more than four guns and 25 men, not intending to infringe on the privileges of this neutral port. When he got clear of the harbour, he bought eight guns more from some commanders of vessels and prevailed on 55 of their men to enter on board his ship; so that he had 12 guns and 80 men with him. About 8 o'clock next morning, a French privateer of 16 guns, with above 200 men on board, who had been cruising a month off of our harbour, in order to intercept the English ships, bore down upon them. Capt.

*'Gentleman's Magazine,' August, 1756.

Wright made a signal for his convoy to run and save themselves, whilst he boldly lay by for the enemy; about twelve the engagement began in sight of above ten thousand of the well-wishers to the French, but in three-quarters-of-an-hour he silenced the xebeck, who made off, (ill shattered) with her oars; had there been any wind, Capt. Wright would easily have taken her. Two other privateers appearing in sight and attempting to cut off his convoy, hindered his continuing the chase, he choosing rather to protect them than to run the risque of their being taken. Next morning he brought them safe back into this port. He lost his lieutenant and four men, and had 8 others wounded; but the xebeck suffered very much, a lucky shot having carried away her prow, on which were 30 men attempting to board him; he so maltreated her, that it is generally believed they lost above 80 men, besides their captain and lieutenant.

"There has been an edict published at Marseilles by the French King's order, offering a reward of double the value of Captain Wright's ship, a pension of 3,000 livres per annum, besides being honoured with the Order of St. Lewis and having the command of a king's ship, for any person who will take him.

"Capt. Wright, for his gallant behaviour and protection of the merchantmen agreeable to his promise has had a present given him of £120 sterling, collected by the English Factory; the foreigners are going to make a purse for him, and it is to be hoped his townsmen will not be backward with you, for his gallant behaviour in disabling a French privateer, and to enable him to support himself under some difficulties. This State having (though very imprudent) thought proper to stop him since his return, alledging that his ship was armed out of this place; but the whole Factory can prove to the contrary, he having suffered his ship to be searched by the first and second captains of this port, who went on board by the Governor's order, and two guard-boats attended him to hinder any arms or ammunition coming off shore. The French here daily ship off ammunition for Marseilles, and our States say 'tis no more than common merchandise; though they will not permit any Englishman the same privilege.

"Our Consul here has sent an account of the affair to Sir Horace Mann, the Resident at Florence,* and we are in hopes, through his means and the whole Factory's, who are all hearty in the cause, that the British Government will take notice of him; especially as the French have set so high a price on his head, and think him so dangerous an enemy to them; they having not yet forgot his brave actions last war."†

This was an astonishing victory, gained over an enemy of double his force, who had had ample time to put his crew in efficient order, while Wright's hastily-gathered reinforcement of 55 men, composed of Slavonians, Venetians, Italians, Swiss, and a few English, were called upon to fight at a moment's notice. In the "Gentleman's Magazine" the xebeck is said to have "received much damage, and lost her captain, lieutenant, the lieutenant of marines, and 88 men, 70 more being wounded; she bore away and left Capt. Wright the honour of having preserved four vessels, some richly laden, which had put themselves under his protection for convoy, after having in vain waited for a ship of war."

* 1756.—"Day by day, meanwhile, our Minister at Florence was in extreme agony at the dark hour which had fallen upon old England. His Florentine friends told him that Minorca wou'd be given to Spain, and probably Gibraltar would be restored to her. When he heard that the Genoese had joined France, Mann recognised the old saying of them as people *Senza fede*. 'What an opportunity has been lost' (July 20th); 'at present two privateers of 16 guns and of 24, that are between Corsica and Leghorn, prevent any of our Merchantmen leaving that port.' The partiality of the Florentine Regency for the French enraged him. It is so great, he writes, in August, that there is no bearing it."—Mann and Manners, vol. 1, p. 389.

† *Williamson's Advertiser*, August 20th, 1756.

Instead of a knighthood, a pension for life, and a higher command, the French captain met with defeat, death, and the attendant disgrace of being vanquished by an "inferior force." The Tuscan authorities, exasperated at the Tartar caught by their French friend, soon showed their leaning. Wright had no sooner anchored than the governor ordered him to bring his vessel within the Mole under pain of being brought in by force. As an officer holding his Britannic Majesty's commission he refused to obey; whereupon two snows anchored alongside the *St. George* and took charge of him. This high-handed proceeding roused the indignation of the captains of the English ships in the Mole, who offered to haul out and make common cause with him. Wright, however, chose in this instance a peaceful course, placing himself in the hands of the British Resident at Florence, who immediately demanded satisfaction from the Regency. What likelihood there was of getting it in the then state of public feeling may be gathered from the following extract of a letter received in Liverpool from a merchant residing at Leghorn, dated August 30th, 1756.*

"The loss of Mahon hath exposed us to the most insulting sneers; and it has been very mortifying to see a rabble—though of boys—go about for several nights with white cockades, crying '*Viva Franchia; burn the English;*' which cry has again been renewed on occasion of the holiday of St. Lewis, kept here with great rejoicing. We were in hopes that Captain Wright's (late of Liverpool) gallant behaviour—which we were all spectators of—in defeating a strong French privateer off the port, would have restored our credit a little; but it has served only to exasperate these Italians against us the more, because disappointed of a fresh triumph over us, as they made full account of seeing Captain Wright fall a sacrifice to the

* *Williamson's Advertiser*, September 24, 1756.

enemy, whom they encouraged to cruise off, on purpose, and furnished with intelligence of Captain Wright's motions, which were watched narrowly."

The Regency, in fact, declined to redress the wrong done, and turned the tables by complaining that they were the injured parties, Captain Wright having deceived them by going out with a greater number of men and arms than had been authorised, or seen by the examining officers, who boarded the *St. George* by the governor's orders. They further charged him with violating the neutrality of the port, making improper use of the emperor's colours, and repeatedly disobeying their orders to come within the Mole. The British Resident replied, denying the alleged deception, and pointing out that the men and arms went out of the port on board other vessels; that the engagement had taken place twelve miles off, the Frenchman being the aggressor. As to their orders to Wright to come within the Mole, they had no business to give them. Before sailing he was within their jurisdiction, had complied with their instructions, and received the governor's certificate to that effect; but since he had sailed under the English flag, and now held the King's commission, he owed no obedience to the authorities of Leghorn, whose action was a gross injustice and a breach of neutrality. This polite interchange of views went on for two months, when the affair was unexpectedly taken out of the hands of scribes and diplomatists by a man of action—Sir Edward Hawke, who had just superseded Admiral Byng as Naval Commander-in-Chief in the Mediterranean. In the *Liverpool Advertiser* of October 8th, 1756, we read the following significant extract of a letter from Leghorn:—

"Admiral Hawke has sent the *Jersey* of 60 guns, and the *Isis* of 50 guns, to Leghorn to demand from the Magistrates Capt. Fortunatus Wright, of your port, whom they have detained, and has only given them three days to consider of it." A week later the editor published another letter from Leghorn, dated September 28:—

"Agreeable to my last on the 23rd inst., the men of war arrived from Sir Edward Hawke demanded Capt. Fortunatus Wright. The express sent to the Regency of Florence brought for answer, *that they must submit and deliver up Capt. Wright, for there was no repelling force;* accordingly the guards delivered him. On the 25th the men of war carried him off in triumph, in company with a number of merchantmen that were lying here waiting for a convoy; Capt. Wright has got 150 brave fellows on board his ship, with whom it's presumed, he will revenge himself if opportunity offers. The fort fired by way of disapprobation at parting with him, three guns, but not with any design to do any damage."

Professor Laughton in his "Naval Studies," referring to this affair, states that Sir Edward Hawke sent Sir William Burnaby with the above-named ships "to convoy what merchant ships were waiting, and to bring the *St. George* away, maugre the captain of the port, the governor of Leghorn, the regency, or the Emperor himself. The Governor protested; but Sir William put it, without undue periphrasis, 'that his orders were to take Captain Wright away under his protection; and in case either the barks or the forts fired, he would be sorry to see himself under the indispensable necessity of returning shot for shot.' The governor preferred dealing with the men of the pen, and sought comfort from Mr. Dick, the consul, who, however, had none to give him, and told him he had heard Sir William Burnaby say he would take her away. 'Well then,' said the governor piteously, 'there's an end of it; what can we do? the French will see it's not our fault.' And so on 23 September the *Jersey* and *Isis* departed, the *St. George* accompanying them, and sixteen rich merchant ships, homeward bound."

Our next information regarding this ever-victorious commander is derived from the public prints for November 19th, 1756, which state: "There are letters in town by the last mail, which mention Capt. Fortunatus Wright having been engaged by two French men-of-war, which he fought for several hours, and at last got clear off." And again: "Capt. Fortunatus Wright has taken and sent into Malta two French prizes, viz., the *Immaculate Conception*, Kampbell, from St. John D'Acre, and the *Esperance,* Richards, from Salonica, both bound to Marseilles, reputed to be worth £15,000 at least."

Ere the news of these captures had reached his native town, Captain Wright had put into Malta, where he found partiality for the French as strong as at Leghorn, the English ships in the harbour being kept under close surveillance. Writing from on board the ship *Lark*, at Malta, to Consul Dick, at Leghorn, on November 3rd, Captain Robert Miller feelingly complained that, "Our ships, persons and colours are treated with the utmost scandal, shame and indignity, even to the highest degree, and with such cruel severity that it is almost impossible for anybody to believe it that have not been eye-witnesses of it. . . . Capt. Fortunatus Wright, of the *St. George* privateer, has been used here in a most barbarous manner."

The authorities certainly treated Wright in a most unfriendly and arbitrary fashion, refusing to allow him to buy the slops and bedding which his men sorely needed, and ordering him to send ashore a number of English sailors whom he had received on board the *St. George*. These men had been put ashore there from prizes taken by French privateers. As an officer holding the king's commission, Wright scorned to deliver up British subjects who had taken refuge under the British flag. His contumacy brought a galley royal alongside, whose captain told him his orders were to sink him if he offered to stir an anchor, and if he made any resistance "to board him and cut every soul to pieces." The seamen were accordingly forcibly taken out of the privateer and landed, their visions of rich captures under the famous and fortunate commander shattered, for they could scarcely have gone aboard as simple passengers and non-combatants. The *St. George* put to sea on the 22nd of October without the stores she needed, and twenty-four hours later she was followed by an enemy who had been abiding his opportunity. In the words of Captain Robert Miller, "the large French privateer of thirty-eight guns and upwards of 300 men, commanded by Captain Arnoux, was in this port at the same time, and sailed just twenty-four hours after Wright, to take him, as Wright was still in sight of the port. But when the great beast of a French privateer came out, Wright played with him, by sailing round him and viewing him, &c., just to aggravate him, as Wright sailed twice as fast as him ; and indeed she is a prodigious dull sailer for a privateer, and very crank."

Williamson's Advertiser for December 3rd, 1756, stated : "We have advice by the way of Marseilles that Capt. Fortunatus Wright has taken and sent to Malta another French ship bound from Sydon to Marseilles, esteemed very rich, being laden chiefly with silks, Burdetts, and cottons. Great rewards and honours are promised to any of the French privateers who shall take him. He is a brave, prudent man, and the only scourge the French feel in those seas."

On the 10th December the same journal published the following, dated Florence, November 20 : "On the 10th inst. anchored at Leghorn a French prize, laden with cotton, wool, and other goods from the Levant, valued at about 8,000 dollars, taken by Capt. Wright, of the *St. George* privateer, being the fifth capture he has made since his departure from Leghorn." The losses inflicted by this single privateer upon the commerce of France were so great that the French King resolved to take extreme measures for Wright's destruction. *Williamson's Advertiser* of December 17, 1756, contained the following "extract of a letter from a house at Leghorn to a gentleman concerned in the *St. George* privateer, commanded by Capt. Fortunatus Wright," dated November 22 :—

"The news we have to communicate to you, relating to Capt. Wright, is of his further success in the capture of another prize which he has sent into Cagliari ; we got the notice the day before yesterday, by a vessel from thence, particularising her cargo to consist of 4,000 or 5,000 sacks of wheat, which we compute to be worth £9,000. Pray God continue his prosperity and preserve him from his cruel enemies ; may we use this phrase, as we have advice from Marseilles that two ships of 20 guns, and a settee of equal force, and all well-manned, are there fitting out purposely for him, with orders to give him no quarter, but burn him on board. We are sorry to give you this alarm, but a French gentleman, a friend of ours, is now in our house, and confirms every particular. We have to add, the disgraceful situation we are all in, and the miserable state of our trade, the French Privateers in these seas being innumerable. P.S.—Since writing the above our partner is returned from the Consul, who has acquainted him of the equipment against Capt. Wright with this addition, that the two ships are fitting out by the King of France, and the Settee by the Chamber of Commerce of Marseilles ; and that they have orders to set him on fire in any road where they may find him."

Early in 1757 Wright seems to have had more than one ship under his command. Among the captures mentioned in the "Gentleman's Magazine" for February is a French snow, taken by the *King George*, Wright, letter of marque, and carried to Lisbon. In the Liverpool paper for March 25th, it is said that "a large privateer is fitting out for Captain Fortunatus Wright, which is to be sent to him as soon as ready, and then he will be commodore of three ships."

One of the French vessels, fitted out especially for Wright's capture, or rather for his utter destruction, was the *Hirondelle*, of Toulon. Mr. Tatem, the British Consul at Messina, writing on the 19th of January, 1757, gives the following account of her reception by Wright, then in the *King George*:—

"The *King George*, Captain Fortunatus Wright, has lately had two smart engagements in the Channel of Malta, of three hours each (one in the night, the other by day), with the *Le Hirondelle*, a French polacca of 26 guns and 283 men; but notwithstanding the great inequality in men, guns, and weight of metal, yet Captain Wright obliged him to sheer off, and they both put into Malta the 2nd of January to refit. But poor Wright has met with worse treatment there than he did before, for although he had several shot under water, which made it absolutely necessary to heave down, yet, by the interest of the French faction, he was denied that liberty; and afterwards, upon account of two slaves having taken refuge on board him, he has been sequestered in port, and cut off from all daily provisions, and even water, till he restores them. But as the *Jersey* was hourly expected in Malta, we hope Sir William Burnaby will obtain his release. The *Hirondelle* is one of the vessels fitted out from Toulon expressly to seek him."

On January 22nd Horace Mann, the British Minister at Florence, wrote to Mr. Pitt that the Regency had been lamenting the decay of the Leghorn trade; that he had pointed out that their gross partiality, and their violent action in the matter of Fortunatus Wright, were two of the causes of this decay; that, yielding to these representations, they had assured him of their intention to observe a strict neutrality; and that, on the strength of this, he had written to Captain Wright "that he might send all the French prizes he had made to Leghorn, as, at my request, he had kept them in deposit till he should hear from me that he might do it with safety."*

Two months later he writes again, showing the kind of welcome Wright would meet with if he attempted to enter the port of Leghorn :—

> "The Council sent a gentleman belonging to the secretary's office to me, earnestly to desire that, in order to avoid any further inconveniences with regard to him, I would order Captain Wright to keep at such a distance from the Port as would not oblige the Government to take any notice of his being there. . . . Finding that they thought themselves tied up by the orders they received lately from Vienna with regard to Captain Wright, I thought it my duty, purely for the sake of avoiding any new disputes, to write to the Consul in the manner they desired. The estafette was immediately sent back to Leghorn with my letter, in order that, as soon as Captain Wright's vessel appears in sight of the port, a bark may be sent off to him, with the Consul's directions not to enter into the harbour."

In *Williamson's Advertiser* for April 1st, 1757, we read that "letters from Leghorn, brought by the Flanders mail yesterday, advise that Capt. Fortunatus Wright, who, after a hard engagement with a French ship of superior force followed her to Malta, has been relieved by the *Jersey* man of war, and were both sailed from thence, and were expected daily to arrive in Leghorn. The *Jersey* is to convoy from thence to England, four rich ships that are armed, which have been detained a considerable time on account of a French man of war and a frigate hovering off that port." There is a reference to the detention of these merchantmen, and to Wright and his prizes in an interesting letter, dated March 25th, 1757, from Sir Horace Mann, to Horace

* Naval Studies, p 222.

Walpole. He alludes to the misery and misfortune of Admiral Byng, but he looked on the sentence of death as an act of vigorous justice. Without implying Voltaire's phrase that the Admiral was shot "*pour encourager les autres*," Mann hoped that it *would* give courage to others. He had seen much of our sea captains during his official residence at Florence, and he says:—"Let us hope that the sentence may produce for the future some reformation in the conduct of our sea officers, which was so publickly criticised in the last war. I wish we could see a Fleet in these parts now. Something must be done to recover our maritime reputation. The sea swarms with French Privateers, who daily take all the merchant ships that venture out. I have dissuaded the people at Leghorn from sending many ships away that are laden for above a Million sterling, which, we know, the French have stationed several Privateers and Ships of War to wait for. They have advice boats continually going backwards and forwards, and others are at anchor at Porto (illegible), to be ready to follow Captain Wright and his prizes that had taken refuge at Port Ferrajo, from whence, if they can escape, we daily expect them at Leghorn. A plan has been agreed upon to indemnify the Captains of the Merchant Ships, who are ruined by laying, at a vast expense, in port, by making a small average on the goods they have on board, otherwise they would have ventured out at all hazards."*

But Captain Wright was never more to enter the port of Leghorn. *Williamson's Advertiser*, in its London correspondence, dated May 19, 1757, contained the following intelligence, which must have been received with universal sorrow in the good old town in which the hero was born and bred, and of whose brave and adventurous, yet prudent spirit, he was the shining personification:—

*Mann and Manners, vol. 1, p. 402.

"A private letter from Leghorn brings advice that Captain Fortunatus Wright, of the *King George,* a Letter of Marque ship, having sailed from Malta with a French prize for the said port, met with a great storm on the 16th of March, during which the officer that had charge of the prize went down into the cabin or under the hatches to bring up certain colours to hoist as signals of distress or danger, as there was then a French Privateer in sight; but when he came upon deck again the *King George* was no longer to be seen; so that there is room to fear this gallant officer, with 60 stout fellows, are all gone to the bottom. The prize made the port of Leghorn, and gave there this account."

There was, however, just one ray of hope left. In another corner of the paper was printed a letter from a merchant in Leghorn to the owners of the *Anson* and *Blakeney* privateers, dated May 9th, stating "that five English sailors, belonging to Capt. Fortunatus Wright, who left Cagliari on the 10th ult., and came up in a vessel belonging to Genoa, inform me that the *Blakeney*, Capt. Fowler, and *Anson,* Capt. Speers, were then in Cagliari." Commenting on this, the editor remarks: "his mentioning Capt. Wright's sailors gives us some hopes that the account of the loss of that brave man, mentioned in the first page of this paper, is premature." To cast further doubt on the news of Wright's death on the 16th of March, there came a letter from Leghorn to a merchant in Liverpool, dated May 16th, which ran as follows:—

"I have the pleasure to acquaint you that I have just received from our Consul at Messina an account dated the 26th of April"—nearly a month after the supposed catastrophe—"of Capt. Fortunatus Wright being very well, and has taken another prize since his departure from Malta. And as this so exactly tallies with the account I had from the master of a Maltese vessel arrived here last week (whom I mentioned in my last to have seen him in the Vere of Messina), we have no room to doubt of the truth of his safety, which has given inexpressible pleasure to me, and a general satisfaction to all in this place. A Danish ship, just now arrived here from Tunis in eight days, was visited six days ago, between Sardinia and Sicily, by the *King of Prussia* privateer, of your port, Capt. Maccaffee, all well and in high spirits. We are in great hopes that he and other vessels on the same station will meet with great success, as the Smyrna French fleet, consisting of 16 or 18 ships, only convoyed by a polacca, who was dispatched some time ago in pursuit of Capt. Fortunatus Wright, and engaged him off Malta, but was bravely repulsed, is soon expected to sail for Marseilles, for which place is also bound a French polacca from Alexandretta, valued at twenty-five thousand pounds sterling."

Another Liverpool newspaper, the *Chronicle and Marine Gazetteer,* of June 3rd, 1757, also published a letter almost identical with the above in substance, and of the same date, but apparently emanating from another correspondent at Leghorn, which ran as follows : "I have just now received a letter from the Consul at Messina, of the 26th ult., with the agreeable news that Capt. Fortunatus Wright was arrived in that port and had brought in with him a brig richly laden. A Danish vessel this day arrived from Tunis, was visited six days ago by the *King of Prussia* privateer of your place, betwixt the Islands of Sicily and Sardinia ; all well on board and in high spirits. The *Ambuscade* man of war has taken six prizes bound from the Levant to Marseilles, and sent them to Malta and Messina, from whence they are daily expected here to be sold. She has also taken a French ship, and carried her into Tunis, which the captain sold for £12,000 sterling ; which being arrived here in safety will sell for one third more than she cost. I have letters from Smyrna of a fresh date, which mention 16 or 18 sail of French ships being ready to sail under convoy of the polacca who some time ago attacked and was bravely repulsed by Captain Fortunatus Wright off Malta. There is likewise a polacca on her departure from Alexandretta for Marseilles, deemed worth £30,000 sterling; which I am in hopes will fall into the hands of some of our cruisers in these seas."

"Captain Fortunatus Wright," adds the editor of the *Chronicle*, "a gentleman of this town who in the last and present war, in a small privateer, gained immortal honour, and the universal esteem of his country, by distressing the enemy, and defending himself in a surprising manner against superior force, at sundry times set out on purpose to take him, was lately reported to be lost in a hard gale of wind; but by this day's post we have certain accounts of his being safe in the Bay of Messina with a prize. This joyful news gives every true Briton a sensible pleasure, and must certainly animate every heroic soul with a noble spirit of emulation; that should adverse fortune crush them in the service of their country, their fall may be justly lamented, as his supposed one was—which we are glad to say was premature."

Then follow these lines, which we reproduce more as a curiosity than as a model for future Dibdins and Bennetts:—

"ON THE UNIVERSALLY-ACCEPTED AND AGREEABLE NEWS OF THE ARRIVAL OF THE BRAVE CAPTAIN FORTUNATUS WRIGHT AT MESSINA, IN SICILY.

"He lives, he lives! in spite of all his foes—
Celestial Pow'rs were pleas'd to interpose;
He lives to conquer—lift the Flag on high,
And let the joyful cannon greet the sky.

"Through ev'ry part of Britain, let the joy
Touch ev'ry Briton—ev'ry Gaul annoy;
To ev'ry heart, as on th' electric mass
The quick pervading joy shall sudden pass;

E

All feel at once the permeating stroke,
The pleasing shock, for this their Heart of Oak.

"At the masthead, see ev'ry streamer flies,
To recompense the streamers of our eyes.
Britannia wept! reverse of tears, she smiles;
Her son is safe, the glory of her isles!

"Her tears encreased old Ocean's briny tide;
Her heaving sighs the tempest's breath supply'd;
Her sighs and tears had rais'd the tempest high,
And raging winds had sung his elegy;
When Neptune from the hoary billows rais'd
His awful head—the storm was all appeas'd;
The rocking winds, in deep attention's form
Bend forward—and he thus harangues the storm:

"Britannia is my bride—ye winds obey;
Be still thou tempest—be at rest thou sea:
This is my son—convey him to yon coast
And let Britannia know, He is not lost.
Bid her suspend her tears—her darling Wright,
Her Fortunatus still survives to—Fight.
What, tho' a price on his devoted head
Was set by France, who wish'd, and thought him dead;
For why? His arm was equal to a Fleet!
Tell her no wave shall be his winding-sheet;
That—I'll prevent—If war has doomed his fall,
It must be, shall be—from a Cannon Ball."

Notwithstanding the above statements that Wright was quite well and active, and had been actually seen in the neighbourhood of Messina when he was supposed to be at the bottom of the sea, and in spite of the fact that both "Lloyd's List" and the "Gentleman's Magazine" for June, 1757, state that "the *St. George* Privateer, Capt. Fortunatus Wright, has carried into Messina a French brig, richly loaded," the fate of the hero remains a mystery to this day.

There may have been good grounds for the local tradition mentioned by Smithers, that Wright fell a victim to political interests. He had plenty of powerful enemies ashore, and was always safer on the high seas, whatever might be the odds against him. With British oak beneath his foot, the British flag aloft, and a sprinkling of English seamen among his crew, he was afraid of nothing afloat.

Sir Horace Mann, writing on the 2nd of July, 1757, says: "The trade of Leghorn, upon which the wealth of this whole state chiefly depends, is reduced to the lowest ebb, insomuch that the arrival in the port of a single prize a few days ago, was looked upon as an object of such importance, and exaggerated by the Italians in terms that sufficiently showed that they are now convinced how much their welfare depends upon the navigation of the English merchant ships not being interrupted. The French have many tartans disguised, but well armed, that cruise between Leghorn and Porto Ferrajo, ready on all occasions to intercept such as are of no force, at the [same] time that they can run near the shore when a ship of any strength appears. A few stout privateers, as in the last war, would totally prevent this, and they would enrich themselves by the French vessels from Marseilles that would fall into their hands. Captain Wright, of the *St. George* privateer, did great service of this kind in the beginning of the war; but it is feared by some circumstances, and by his not having been heard of for some months, that he foundered at sea. Several prizes made by him have lain some months at Cagliari in Sardinia, waiting for an opportunity to get with safety to Leghorn."

The English prestige in the Mediterranean had been reduced to a low ebb through the incompetence of the Government at home and the lethargy of the naval commander-in-chief on the station, and the only Englishman whose name was a terror to the French had mysteriously disappeared from the seas. But the power of England was going to be felt in those quarters where it was most despised and hated. A merchant in Leghorn, writing on the 18th of July, 1757, to a house in Liverpool, said, "Last night arrived here Admiral Osborne with seven sail of men of war, who has instructions to demand satisfaction of the Maltese for their cruel behaviour towards the brave Capt. Fortunatus Wright, whom we have great reason to fear is no more, and we are in hopes he will see justice done to the other privateers who have had the misfortune to carry their prizes into their ports."

Again, in *Williamson's Advertiser* of August 27th, we read that "there are letters from Leghorn which mention that Admiral Osborne, who arrived there with seven men of war on the 17th, was fitting for sea with all expedition, having received advice that five French men of war were preparing to sail from Malta for Toulon, whom he expected to meet with in their passage; after which he was going to Malta, to demand satisfaction of the Maltese, for their injurious behaviour to the captains of several of our privateers, particularly to the brave Captain Fortunatus Wright."

We cannot close this account of one whose career has been justly described as "more romantic than any romance," and as "a succession of romances," in a more appropriate manner than by quoting the following characteristic stories, both relating to the period when Wright was cruising in the neighbourhood of Malta. The first is related by the author of "Naval Studies," "on the authority of the first Earl of Charlemont,* who says it came to his knowledge during his residence at Malta, about 1750, and was told to him 'by the most credible eye-witnesses.' No names are mentioned, but there is scarcely room for doubt that the hero of it was Fortunatus Wright. He is described

* Memoirs of James Caulfield, Earl of Charlemont, by Francis Hardy, vol 1, p. 47, etc. Laughton's "Naval Studies," p. 212.

by Lord Charlemont as a captain commanding an English privateer of some force, and 'of such skill and bravery that he reigned paramount in the Mediterranean, daily sending into the port of Malta French prizes of considerable value.' In a society such as then ruled in Valetta, this stirred up much angry feeling, the Austrians and Piedmontese jeering the French or Spaniards, and many duels took place in consequence. At length the French knights, irritated beyond measure by the taunts of their adversaries, and the continued success of the English captain, determined to put a summary stop to both, and sent urgent representations to Marseilles, in consequence of which an armed vessel, of force almost double that of the Englishman, was specially equipped and sent to Malta under the command of 'an officer of the highest character for courage and naval knowledge.' After being duly feted by the French party he sailed out of harbour to look for the Englishman, as to a certain victory. Days passed by ; both parties were aglow with expectation, and the ramparts on the sea front were constantly thronged with anxious crowds. Two ships at last appeared in sight. As they came nearer it was seen that the one was towing the other ; that the one was the French ship for which they were looking ; that the other was much shattered. They hoisted French colours, and who so jubilant as the French knights ! Amid exulting cheers they turned into the harbour, between St. Elmo and Ricasoli. All Valetta, Senglea, and Il Borgo were called to witness the triumph of the French ; when—O cruel disappointment !—the white flag suddenly disappeared, giving place to the victorious flag of England. The Marseilles ship was a prize to the English privateer."

The second story, entitled "The History of Selim, from the Armenian's Letters," represents Captain Wright as acting a very noble part. Though extremely romantic, the incidents are neither impossible nor improbable. The

French privateer mentioned in the story appears to fit in very well with the *Hirondelle*, sent out from Toulon to seek Wright, and whom he fought in the Channel of Malta.

The History of Selim, from the Armenian's Letters.*

Against the inclination, yet not without consent of my parents, I quitted Armenia, and embarked on board a Genoese trading vessel, proposing to study the civil and military discipline of Emanuel Victor, the great Prince of Sardinia. While I was in daily expectation of seeing Genoa, our ship was taken by a Spanish vessel navigated by corsairs. We were soon loaded with irons; and though I was treated more favourably than others on a religious account, yet I was robbed of the money which I had designed for the expenses of travelling, excepting a few sequins that lay concealed in my clothes. As soon as we arrived at Oran we were thrown into a loathsome dungeon, guarded by Spaniards; and the little lenity that appeared was now shown to the Christians. Their clothes were restored, while I was stripped of my outer garment; their allowance of victuals was usually greater, and I was often compelled to labour, while my fellow prisoners were indulged with ease. In this state I continued seven months, and then I was, with five others, sold to a young Moor, and conveyed with my companions to a spacious house two miles distant from Oran, near a little village called Arzew, where the uncle of this young Moor had laid out a plan of spacious gardens, the labour of which was reserved for me and my companions. As soon as we arrived our fetters were removed, for our escape was impossible, the house and intended garden being enclosed

* The "Gentleman's Magazine," for the year 1757, pp 367-9.

in some places by a wall 20 feet high, and in others by a broad trench, and keepers were constantly employed to watch us. Here I continued labouring three months, without any hopes of redemption, sometimes amusing myself with the flowers and fruit trees, and at others conversing in the Arabic tongue, of which, from the knowledge I had before my captivity, and my intercourse with some captives in the prison, I had now attained an easy pronunciation. My country dress being permitted to me, the native slaves were kinder to me than to the Christians; and becoming an interpreter among them, I acquired a sort of pre-eminence that gave me opportunities of doing my fellow captives little offices, which society in distress will extort from the most savage. But the severe labour to which we were daily confined began to waste my strength. Our keepers remitted nothing of their watchfulness over us, nor the young Moor of his care over them. Not an hour of the day passed wherein his eye was not upon our labour. He delighted in seeing us faint beneath our loads; and once when I tottered beneath a heavy burthen he ordered fifty lashes to a Christian who ran to support me.

After three months' toil in the midst of an inclement winter, the spring began to open, and brought with it a sweetness and beauty that would have relieved any but slaves, who had once been happy, and now, by no crime, were condemned to misery. Sometimes I had thoughts of telling the Moor who I was, and exciting his pity by a recital of my misfortunes; but he appeared so avaricious that should he know that I was the son of a Turkish Aga his demands would be greater than my friends could satisfy; wherefore, I resolved to bear my afflictions in silence, and leave the event to God. As soon as the year began to blossom, news was brought me by the native slaves that the uncle of the young Moor and his family were arrived at his country seat, and that in three days the young Moor would set out for Oran to inspect the affairs of his uncle in that city. The joy which I felt for a few moments was little short of what freedom would have given ; but the natives soon informed me that the uncle was more avaricious, cruel, and perfidious than his nephew ; that having no sons, he had preferred his nephew to the inheritance of his large possessions, and that he had one favourite daughter whom he designed for his wife. The hopes conceived from a change of masters now vanished, and I considered myself as one of those unfortunate wretches destined to walk through peril and toil, without any ray of comfort to cheer them in their passage. Two days passed and the uncle had not set his foot in the garden, being troubled with a disorder common in that country to many of his age and sedentary life ; yet he was carried to a window, where, as our keeper said, he constantly observed us ; and indeed the keeper often raised his voice, and exercised the lash, to demonstrate his strict attendance of us. Four days after, the old man's disorder so increased, that being no longer able to approach his window, he was confined to his bed. During this time the severity of our keepers somewhat abated; the daughter of the Moor also, who came at her father's request to oversee the garden, would often bring fruits and other pleasing refreshments to the native slaves, of whom she enquired concerning us, and frequently would recommend to them to treat us tenderly. As the Moors rise early, no morning passed whereon she did not visit the house of the native slaves, and never went unprovided, so that she became their idol. When she had visited the natives, she was often seen to pass through a shady walk into a greenhouse near the dwelling of the captives, where some conjectured she paid her devotions, and others that she watched the labourers. But whatever might be the cause, it was observed that when the natives carried no part of their extraordinary provisions to us unhappy captives, the next day she omitted her kindness to them; thus our captivity was lightened. I once more indulged hopes of escape, and laboured more cheerfully than ever. On the 20th March, just as our labour was begun, our young benefactress surveyed the whole garden, and having passed the Moors, approached where the captives were employed; drawing her veil entirely down, and wrapping herself in a hyke of blue satin, she spoke to them as she passed, and coming near to me, who was last in the lot of ground, and then had a heavy burthen on my shoulders, she turned her face, still covered, towards mine, and laying her right hand on her breast—which is the Moorish salutation,—said, in a gentle tone, "*Holy Alla relieve thee, stranger.*" Many days passed, and some of my fellow captives became so reconciled to captivity, that if the uncle and nephew had been removed they would have been easily persuaded to serve Zaida while they lived. But the indulgence we received only gave me more time to reflect on my hard fortune, and one night, while I was stretched on a grass plot along the side of the Moor's palace, singing a mournful history of my misfortunes, I was surprised by a loud knocking at the gate and the neighing of horses; and instantly a soft, disordered voice from a window above said trembling and hastily: "*To thy apartment, stranger; Morat! Morat! Alla guard thee.*" I fled, blessing the voice that warned me, and spent a tedious night in broken dreams and waking expectations of cruelty from Morat, by whom such expectations were never disappointed. In the morning, long before the sun, he had surveyed the garden, and finding our labour had not equalled his desire, with his first salutation he struck me to the ground, and, before I recovered, three of my companions were lying speechless. While he was proceeding in his cruelty, a slave came pale and breathless from the house, and faltering could only pronounce:

"*Zaida, Zaida no more.*" Morat persevered, and having given each captive his blow, returned to the house. Bruised and dejected we groaned through the day's fatigue; but neither the bruises nor the toil preyed on my mind so much as a fear and desire to know what had befallen our young benefactress. Weariness brought with it no rest. I lay all night sleepless, and before daybreak heard our keepers relating that Zaida, having beheld the first mark of her cousin's cruelty to the captives, had fainted and continued some moments lifeless; that a cry that she was dead had reached Zelebin's—her father's—ear, and so afflicted him, that even her recovery added little to his, the sudden joy rather oppressing him the more; and, lastly, that Morat was gone to Oran, being called thither by sudden business. I rose overjoyed, and informed my fellow prisoners that the storm was over. The next day Zaida walked twice through the garden, carefully observing us through her veil, and as she passed by me twice repeated the Aslemash, pressing her hand more closely to her breast, and saying, "*Alla guard thee.*" Zelebin's disorder increased, and the fright had occasioned a fever, which was likely to prove fatal. On the 28th of March it was my lot to be employed under the greenhouse to which Zaida usually paid her morning visit; nor did she fail that day; for I had scarce taken the spade in my hand when I saw her veiled at the window. When the course of my spade had brought me under the window, she dropped a tulip, with which she had been playing, at my feet; I took it up, and ran round the building to present it to her, but before I could reach the entrance she was gone. I returned, admiring the largeness and colour of the flower, and was struck by characters like letters in the inside. Examining more attentively, I found the tulip lined with two folds of fine paper, which I took out, and hardly had conveyed to my pocket when one of the keepers approached and took the flower from me. With what impatience did I labour through the day! Evening came, and being alone in my cell, I read the following letter :—

"Holy Alla protect thee, stranger; I have enquired much concerning thee, and feel a sharp pain when I see thee treated cruelly. If thou seekest thy freedom, I will contrive to give it, for I am loved by my father's servants, who will not betray me. I have provided for thee a Moorish turban, and a rich hyke, in which thou mayest pass concealed. There is another present which I would give thee, but thou shalt see it first, for it may be burthensome to thee. If thou wilt be early with thy spade at the greenhouse, I will shew thee what I would give thee. Be cheerful, stranger, for if Alla will permit, I will do thee much good."

All the impatience of the day equalled not the restlessness of the night. I was up before the birds, and at day-break the spade was in the earth; Zaida came with the sun, and observing none near but me she threw back her veil, and looking on me with a sweet confusion, dropped another tulip and retired. It was the first time I had seen her face, and some moments passed before I could take my eyes from the window. I conveyed the flower to my pocket-book, and worked through the day in a hurry of joy that was painful to support. The burthen of the tulip was this :—

"Stranger, thou hast now seen what I would give thee; but then I would have thee ask it. I will consent to be thy wife if thou wilt take me with thee to thy own country. There is a French ship now near Arzew, and the French will carry us anywhere for money. But say not thou wilt take me, if thou hatest me. Speak thy mind, for I will do thee good in whatever way thou desirest. Holy Alla watch over thee."

With my pencil I wrote the following answer at the back of her letter :—

"Great Alla reward thee, gentle Moor; I will not only

ask what thou shewedst to me this morning ; but I call our prophet to witness that I will have no other wife but thee. Whatever thou desirest I will do ; but there is one captive who hath been kind to me, and I would free him too."

This she received from the window, and retiring a few minutes, returned and said in her native tongue : "Be thou and thy captive friend at the garden door to-morrow at nine of the night."

The wished-for evening came, and Zaida with her own hands opened the door, attended by a faithful servant, and informed me that her father could not live another night ; that horses and dresses were ready, and she had sent by her servant to a hut on the waterside all the money with which her father had entrusted her ; and that a French privateer was preparing to sail in less than two hours. I urged her immediate departure, and she gave me a turban and a satin hyke, and my fellow captive the coarse dress of a slave, covering herself in the like garment, that all might pass as my servants. Thus prepared, we walked silently from the house before ten, and at a small distance mounting our horses, arrived in a short time at the hut. The captive Swede, whom I had released, immediately went on board the privateer to learn her destination, and was informed that she had orders to cruise near Malta, in order to take a bold Englishman, called Fortunatus Wright ; and if the winds would permit, we should be landed in that island. In a few minutes we sailed, and the next morning were many miles distant from Africa.

Ten days were passed before we obtained a sight of Malta, and we had scarce dreamed of landing there when a signal was made for standing out to sea in pursuit of a ship, which upon a nearer view was found to be the very privateer which the French captain had orders to take. Instantly I ran down, took Zaida in my arms, and supported her courage with all the animating words I was master of. Once she sunk upon my breast, and I had but just recovered her when the signal was made for the engagement. The fire became hot, and the conflict bloody. I continued comforting Zaida till the event became doubtful, when pretending to her we were victorious, I sprung upon the deck, and observing that the English endeavoured to board us ahead, I slew the first who attempted our deck, and beckoning to the French to follow me, leapt on board the enemy's ship, unseconded by any, excepting my Swedish fellow-captive, who seeing me overpowered, leapt back and regained his ship. Thus I was made a prisoner, and my fair Moor left a prey to all the wretchedness of despair. After several vain attempts to board each other, the two ships parted, the French steered towards France, and I was carried into Malta. Good heaven ! how soon was changed the gladsome prospect of happiness to the darkest view of misery ! The good captain, whose prisoner I was, observing my despondence, ordered me to be set free, though I had killed one of his men ; and when I informed him, by a Maltese interpreter, of my unhappy story, and my resolution to go in quest of Zaida, he gave me one hundred guineas, and advised me to sail for England ; "where, though I am unhappily exiled from it," said he, "you will be generously treated, and will hear the fate of the French privateer." He then informed me of her name, and the port from which she was sent ; "when you find that she is landed, you will then be at liberty," said he, "to visit France, and if the French captain be generous as he seems brave, he will restore his passenger with all her possessions." He recommended me to an English captain then at Malta, and having kindly wished me good fortune, we parted.

Two long months I was tossed at sea ; on the 10th of August we arrived at our destined port, where the first object that struck my eyes was the French vessel in which I left the lovely Zaida ; hope and fear almost deprived me of reason ; with difficulty I told the captain all my story, and he, with the readiness of friendship, sent his boat to enquire whether any woman were taken prisoner on board the French prize ; but we received no information, for the sailors who then manned the ship were strangers to her captain. We landed at a fair town,* on the banks of a small river called Avon ; and the captain, who had not drowned his humanity in the rough element on which he traded, conveyed me to the prison, where after searching various apartments, at last I found my fair afflicted Zaida lying on the ground with her head on the lap of her woman, and the Swede sitting near to guard her. As soon as she saw me her voice failed her ; I had almost lost her by an agony of astonishment and joy as soon as I had recovered her. Hours were counted ere she would believe her senses, and even days passed over us, in which she sat with a silent admiration, and even still doubts whether all is real.

* Bristol.

CHAPTER III.

Privateers of the Seven Years' War.

In 1745, Captain Robinson, who had put into the Isle of Skye during his passage from the Baltic, brought the news to Liverpool that the Young Pretender had landed in Scotland. An express was immediately despatched to the Secretary of State with this important intelligence, and vigorous preparations were made to defend the town in case of attack. The sum of £6,000 was quickly contributed to defray expenses ; and a regiment called the "Liverpool Blues," 900 strong, was raised from amongst the inhabitants. The Jacobites passed within 16 miles of Liverpool, but did not dare to risk an encounter with the local forces. When the Duke of Cumberland pursued the retreating insurgents, the Blues, thirsting for glory, joined his army and assisted at the siege of Carlisle.

In 1746, one of the most gallant defences recorded in naval annals was made by Captain Nehemiah Holland and crew, of the Liverpool ship *Ann Galley*, bound for Antigua. Her crew consisted of 14 men, with four guns of one-and-a-half inch bore, six muskets, six pistols, and six cutlasses. When in sight of Antigua, she was attacked by a French privateer, mounting 10 six-pounders, with 100 men. The action was fought in view of the people on the island. The French boarded the *Ann Galley* three several times, but were driven back each time with considerable loss, leaving, ultimately, 18 of their crew dead on the English ship, and 50 to 60 wounded on their own vessel. The *Ann Galley* did not lose a single man. The defence was conducted with considerable skill. Preparations had been made by barricades to protect the crew against boarding; and trains of powder were laid to explode every time the assault was made, which wrought havoc amongst the boarders. The *Ann Galley* took fire twice during the engagement. In a list of Liverpool ships, published in 1753, she appears as a Guineaman, owned by Messrs. Wm. Whalley & Co., carrying 340 slaves, and still commanded by the valiant Nehemiah.

On the ship's return to Liverpool, Captain Holland was presented by his owners with a silver punch bowl, containing two gallons, with the following inscription engraved:—

"The gift of the owners, to Nehemiah Holland, Captain of the *Ann Galley,* who, with inimitable bravery, preserved and defended her against the infinitely superior force of a French enemy, August 21, 1746."*

In 1749, a Liverpool privateer captured a French sloop of war called *Le Lion D'Or*, which was subsequently converted into a whaling vessel. As the *Golden Lion,* she sailed from Liverpool in 1750 for the Greenland fishery, being the first vessel from Liverpool to engage in the trade. In the Mayer Collection in the Public Museum, William Brown Street, is a noble punch bowl, seventeen and a-half inches in diameter, presented in 1753 to Captain Metcalf, of the ship *Golden Lion*, by his employers, on the completion of her second successful voyage in the Greenland Whale Fishery. It is painted in blue, with a representation of the ship inside the bowl. The following is a copy of proposals, in 1749, for the purchase of the vessel in shares, and for fitting her out for the Greenland trade, with the names or firms of the merchants who subscribed to them, and embarked in the concern, and of the shares which they respectively took:—

*In Picton's Memorials of Liverpool the ship is incorrectly called the *Ann Galkey*.

" PROPOSALS from GOORE and BULKELEY to all such persons as shall become Subscribers hereto, for the sale of the ship *Golden Lion*, now belonging to them, and for fitting her out for the GREENLAND WHALE FISHING TRADE for the next season.

" 1st.—That they, the said Goore and Bulkeley, do consent and agree to take the sum of Two Thousand Pounds sterling for the Vessel and her Materials (the Great Guns with their Tackle and Firearms only excepted), the Persons subscribing hereto do oblige themselves respectively to pay his or their proportion according to the amount of the share subscribed for towards the Payment of the said sum of two thousand Pounds in two months from the Date of the Bill of sale.

" 2nd.—That the Joint Concern in the said Vessel shall be divided into twenty or more equal shares, every Person having the liberty of Subscribing one Share more or less, so that none subscribe for less than half a share.

" 3rd.—That twenty shares being subscribed for, the Bargain shall be valid, otherwise void.

" 4th.—That the Subscription being completed, every Subscriber shall and is obliged to pay his or their Proportion of the Outfit, Disbursements, Wages, or other Charges, into the Hands of the Persons appointed Managers, when and as often as by them the said Managers required.

" 5th.—That the Subscribers, or a Majority of them, do immediately after the Completion hereof, appoint two or more of the said Subscribers to be Agents for the directing of the whole Proceedings of the Voyage and equipping the Vessel.

" Lastly, the said Goore and Bulkeley agree on their Part to hold one whole share.—In witness whereof, We the Persons willing to be concerned have Subscribed our Names and Shares this eighteenth Day of December, 1749."

SHARES.

Thos. & John Backhouse.—half a Share
John Nicholson & Co.—half a Share.
David Edie.—half a Share.
Joseph Jackson.—half a Share.
Jo: Manesty.—half a Share.
Richd. Nicholas.—half a Share.
Jas. Gordon.—half a Share.
Thos. Shaw.—half a Share.
John Atherton.—half a Share.
Heywood Benson & Co.—half a Share.
John Parke.—half a Share.
Richd. Golightly.—half a Share.
Owen Prichard.—half a Share.
Tho. Mears for Self, & John Okill.—half a Share.
Richard Savage.—half a Share.
Charles Goore for William Hurst.—half a Share.
Charles Goore for Nathl. Bassnett.—One Share.
Thomas Seel.—One Share.
Foster Cunliffe & Sons.—One Share.
Saml. Ogden.—One Share.
Edwd. Trafford.—One Share.
John Knight.—One Share.
John Brooks.—One Share.
John Hardman.—One Share.
Sam. Shaw.—half a Share.
Jam. Crosbie.—half a Share.
Chas. Lowndes.—half a Share.
Edwd. Cropper.—half a Share.
John Tarleton.—half a Share.
Law$^{ce.}$ Spencer.—half a Share.
Edward Lowndes.—half a Share.
Edward Parr.—half a Share.
Edwd. Roughsedge.—half a Share.
Joseph Bird.—half a Share.
John Seddon.—half a Share.
James Pardoe.—half a Share.
John Entwistle.—half a Share.*

*From the original in the possession of the late Mr. Samuel Staniforth, reprinted in Brooke's History.

So catholic was the spirit of enterprise displayed by most of these gentlemen, that their commercial operations embraced not only whales but negroes, and for one whale's blubber melted by their agency, they might have counted thousands of human hearts either stilled for ever, or crushed by lifelong slavery. Messrs. John Okill & Co., were the only African Merchants not engaged in the slave trade.

The vessel, after being commanded for a long time by Captain Metcalf, was lost, whilst a full ship, as it was termed, in coming out of the ice during one of her voyages. She was accustomed, when in Liverpool, to lie near the south west corner of the Old Dock; which, from that circumstance, was called the Golden Lion berth. She ought not to be confounded with another well-known but more modern vessel, called after the former, also, the *Golden Lion*, which was commanded by Captain Thompson. The latter was employed in the same trade, and belonged to Messrs. T. Staniforth & Sons. She was afterwards withdrawn from the fishery, let out to the Government, and employed in the victualling service, and whilst so employed, in coming home from the Mediterranean, was captured by the French. We shall have occasion to notice her later on. The first ship built at Liverpool, and employed in the trade, was launched in the year 1775, from Mr. Sutton's yard.

The Greenland Fishery was then of importance to Liverpool, and one of the principal merchants concerned in it was Mr. Thomas Staniforth, father of "Sulky Sam." It fluctuated very much, but at one time there were twenty-three ships from Liverpool employed in it. The seamen engaged in it were, as an encouragement to the Greenland trade, protected by Parliamentary enactments against impressment. "Instances," says Brooke, "were not unfrequent during the war, of a body of seamen engaged in that trade, going to the Liverpool Custom-house armed with harpoons and whaling knives to defend themselves against the press-gang, until they could reach the Custom-house, where lists of their names being furnished, on oath, by the owners, the seamen gave security to the satisfaction of the Commissioners of Customs, to proceed in the vessels to which they belonged to the Greenland Seas, or Davis' Straits, in the whale fishery, in the following season ; and they then received a certificate of protection, under the provisions of the Acts of Parliament, 13th George the Second, Chap. 28, Sec. 5 ; 11th George the Third, Chap. 38, Sec. 19 ; 26th George the Third, Chap. 41, Sec. 17, and 31st George the Third, Chap. 43, Sec. 5 ; and they were then privileged from impressment until after the expiration of the season of the fishery, and until the termination of the voyage. Every harpooner, line manager, or boat-steerer who had given such security as above-mentioned, was allowed to sail in the colliery or coasting trade, without being liable to be impressed during the time of the year that they were not employed in the fishery."

A building for extracting the oil from the fat or blubber of whales, and provided with boilers for that purpose, was erected by Mr. Nathan Kershaw at the south end of the Queen's Dock, near the bottom of Greenland Street ; and since the enlargment of the dock, the site of the building now forms a part of it. Mr. Kershaw also endeavoured to establish the manufacture of glue there, from the skin of whales' tails, but the whole works were a failure, and the odour from them was anything but agreeable to the neighbourhood. The Liverpool branch of the Greenland trade gradually declined, until it ceased to exist. One of the last vessels remaining in that trade was the *Lion*, Captain Hawkins, belonging to Mr. Staniforth. He sold her to Mr. Hurry, and she was afterwards lost in the ice in 1817, but the crew were saved.

When the press-gangs came on shore the utmost confusion

and dismay took place among the denizens of Bridge Street, Wapping, Little Bird Street, and thereabout. On the 30th of June, 1755, upon the arrival of the ship *Upton* in the river, from Maryland, the *Winchelsea* man-of-war, then lying at anchor off the town, sent her barge, under the command of a lieutenant, to board her. On the *Upton's* men finding the barge's intention, they seized their captain and chief officer and fastened them in the cabin. As the *Winchelsea's* barge ran alongside, the *Upton's* men swore that the man-of-war's men should not board them, and if they did they would depress their guns and fire upon them. At that time every merchantman was more or less armed, and able to make a stout resistance in case of attack. Seeing matters thus formidable, the *Winchelsea's* barge sheered off to put back for a reinforcement. The *Upton's* men, seeing this, lowered their yawl and pulled to shore. They were, however, followed by the *Winchelsea's* men, when a fierce encounter took place, shots being fired on both sides, the struggle ending by the yawl being upset. Two of the crew swam ashore, 15 others were captured, and two were drowned. The officer commanding the barge was shot in the cheek, the ball passing clean through his mouth. Several seamen on both sides were mortally wounded.*

In 1750, six convicts who had been transported for 14 years and shipped at Liverpool, rose at sea, shot the captain, overcame and confined the seamen, and kept possession of the vessel nineteen days. Coming in sight of Cape Hatteras, they hoisted out the boat to go on shore, when a boy whom they had not confined hailed a passing vessel and attempted to make known the position of affairs, but was prevented. The wretches then drove a spike up through his under and upper jaws, and wound spun yarn

*In Stonehouse's "Streets of Liverpool," the date of this incident is erroneously given as May 30th, 1775.

round the end that came out near his nose, to prevent his getting it out. They then cut away the sails from the yards, left the ship and went on shore. "But," says the *Pennsylvania Gazette* of August 11, "a New England sloop coming by soon after, and seeing a ship driving in the sea in that manner, boarded her, found things as described, and carried her into North Carolina, from whence a hue and cry went after the villains, who had strolled along to Virginia. They were taken at Norfolk, and one of them confessed the fact, upon which they were ordered up to Williamsburg for trial as pirates."

In August, 1753, fourteen Danish ships, laden with timber, were seized at Liverpool; "the reason whereof was," says the paper, "because it was of the growth of Livonia, and Danish ships act contrary to treaty when they bring other wood into England than that of the product of their own country."

In May, 1756, the peaceful pursuits of commerce received a rude shock in Liverpool and other parts of the country by the commencement of another war with France. In the previous wars the trade of Liverpool had suffered much less than that of London, Hull and Bristol from the privateers of the enemy, but at the outset of the Seven Years' War—"the most glorious war in which England had ever been engaged," as Lord Macaulay calls it,—swift and well armed French privateers found their way into the North passage and the Irish sea, and kept Liverpool blockaded for many weeks. Great damage was inflicted on the commerce of the port, and the town itself was threatened with attack by the gallant Thurot. This war continued during the whole of the remainder of the reign of George the Second, and during the first and second years of George the Third. The early part of the contest was marked by very humiliating disasters, both on sea and land, but the latter part of it, carried on under the vigorous administration of the first William

Pitt, was extremely successful, terminating gloriously for England at the Peace of Paris, in 1763. Canada and all the other French possessions in North America were conquered by Wolfe and Amherst; the rich province of Bengal was captured by Clive; and the French fleets, after having been victorious over the unfortunate Byng, were in their turn defeated by Hawke and Boscawen.

The spirited manner in which the French commenced the war, and the superiority and activity of their privateers, caused an immediate and enormous increase in the premium for insurance against sea risks. The rates on vessels from Liverpool to Jamaica rose to twelve guineas per cent.; to North America to ten guineas; from Carolina to Cowes and a market to twenty guineas; from North America to Jamaica to twelve guineas; from Liverpool to Gibraltar to twenty guineas; and from Newfoundland to the Mediterranean twenty-five guineas; rates almost sufficient to put an end to commercial enterprise.

As trade no longer "flourished and spread her golden wings so extensively," but had come almost to a standstill, the Liverpool merchants took a leaf out of the enemy's book, and forthwith began to fit out their ships as privateers and, in some cases, to build new vessels for the specific purpose of cruising against the enemies of Great Britain. The *Revenge, Mandrin*, and *Anson* privateers sailed from the port on the first of July, 1756, and the *Brave Blakeney* followed in August. These vessels were very successful on their first cruise, particularly the *Anson*, which returned in a few weeks with a French West Indiaman worth £20,000; and the *Brave Blakeney*, which brought in two other prizes of great value, named *La Gloire* and *Le Juste*. Then the whole community became mad after privateering, and shares in these ventures were eagerly taken up.* Other

* Soon after war had been declared, Messrs. Edmund Rigby and Sons, ironmongers, advertised that they had contracted with the proprietors of Birsham Foundry for the delivery of a large quantity of good swivel and carriage guns.

privateers were fitted out and sailed in the following order: The *Mercury*, on November 20th; the *Isaac*, on the 29th; the new *Anson*, on December 3rd; and the *Grand Buck*, on the 8th. In the following year two other privateers sailed, the *King of Prussia* on January 29th, and the *Liverpool* on June 10th, 1757. In 1758, the *Resolution* and the *Spy* were sent out to prey on the enemy's commerce. The French, however, played at the same game, with equal, if not greater vigour. It was calculated that at the beginning of March, 1757, the French had at sea no less than 200 privateers, while many more were being built.

On the 1st of July, 1756, the *Anson* privateer, Captain Edward Fryer, a brigantine of 150 tons burthen, 16 carriage guns (four, six, and nine-pounders), 24 swivels, and 100 men, belonging to Mr. George Campbell, a member of the Common Council, sailed on a cruise. A captain who spoke her off the Tuskar, reported that she sailed very fast, and that "the men were in great spirits, giving him three cheers, with their cutlasses brandished over their heads," in a very suggestive manner. Before the 19th of July, the *Anson* had taken four prizes, one being a storeship, bound to Canada with a number of French officers, 300 cannon, and other warlike material, as well as important despatches on board. In the Bay of Biscay, the *Anson* encountered *La Juno*, a Bayonne privateer of 12 guns and 108 men, which she captured and carried into Kinsale, after an

They also sold iron balls, cast in moulds, and "much preferable to those cast in sand." The guns had been proved by the Woolwich proof, but purchasers were at liberty to prove them again. Would Messrs. Armstrong or Herr Krupp have shown more enterprise, had they lived in 1756? Gunpowder was then manufactured by Messrs. Cunliffe, Stanton and Craven, at their mills at Thelwall, the powder being stored at the Black Rock Magazine, in Cheshire. Messrs. Cunliffe & Co., advertised "damaged gunpowder wrought over again fit for service, at 20s. per barrel." Prior to this date there was a powder-house on Brownlow-Hill, opposite the spot which is now the north end of Clarence Street. The building was used for the confinement of prisoners of war, during the American Revolutionary struggle. The records of the mock Corporation of Sephton mention Mr. John Stanton as holding the office of "contractor for gunpowder," while a Mr. Matthews was "hooper, cooper, and powder-keg maker" to the same convivial, but really influential body.

engagement of four hours and a half, in which the French had four men killed and several wounded. The August number of the "Gentleman's Magazine" stated that the *Anson* had already made "above 5,000 per cent. of what was expended in fitting her out."

On the 20th of September, the *Anson* fell in with the frigate *Artabonetta*, 320 tons burthen, 14 guns, from St. Domingo for Nantz, and took her after an engagement of two hours, in which the prize had three men killed and two wounded, her rigging maltreated, and her sails shot to rags. "The *Anson's* people," piously observes the editor of *Williamson's Advertiser*, "have had particular marks of Providence conferred on them this cruise; for though they have taken the *Alexander*, a prize of upwards of 400 tons, outward bound, worth £5,000, and the *Juno*, a privateer of equal force with themselves (besides the forementioned prize), yet not one of their hands has received the least damage." The cargo of the *Artabonetta*, consisting of sugar, coffee, indigo, and "superfine St. Domingo cotton," was valued at £20,000. Both the *Artabonetta* and the *Young Alexander* were sold by the candle at R. Williamson's shop, near the Exchange. Captain Fryer did not sail again in the *Anson*, but took command of the *Hope* (Letter of Marque), 300 tons and 18 guns, besides swivels, which was advertised to sail for Barbadoes, touching at the Cape de Verd Islands, and which offered "good encouragement for seamen and able-bodied landmen wishing to try their fortunes."

The new commander of the *Anson* was Captain Gersham Speers, and on the 28th of January, 1757, she sailed on another cruise, in consort with the *Blakeney*, Captain George Fowler. On March 11th, they passed through the Straits of Gibraltar, and on May 28th boarded a French polacca, freighted with Turks and their effects, from Alexandria for Tripoli, whom they ransomed for 1,000 Barbary dollars. On June 3rd, they fell in with another

French polacca, which they ransomed for 600 Barbary zequins. A few days later they saw four sail, and gave chase to a snow, which hoisted Turkish colours and ran on shore on the Barbary coast. The privateers then stood off and gave chase to a large ship in the offing, which gave them the slip. They pursued two sail to leeward, and captured a French brig from Smyrna for Marseilles. The French captain informed them that the large ship was one of the convoy, and that the snow was his consort. They then stood in shore, sent their boats on board her, soon after got her afloat, and made sail with their two prizes for Cagliari. They appear to have made three other captures on the way to that port, one of which was re-taken by a Maltese man-of-war. From Cagliari, they sailed for Gibraltar, where they arrived safe, after beating off a stout French privateer. Having got two of their prizes condemned there, they convoyed them to Liverpool, but were parted from them in a gale of wind, one of the prizes, the brig *Union*, going on shore on the coast of Ireland. Some of the cargo was saved and sold by auction in Liverpool, as also was the other prize, the "good snow *St. Nicholas*," and in March, 1758, the two fortunate privateers were also brought under the hammer.

The *Brave Blakeney* privateer, Captain William Day, a brigantine of 14 carriage-guns and 20 swivels, also belonging to Mr. George Campbell, sailed from the Mersey in August, 1756, and falling in with the *Hawke* privateer, of Exeter, agreed to cruise in company. On the 6th of October, Cape Finisterre bearing W.S.W. about 25 leagues, they chased and came up with two ships and two snows from St. Domingo, which drew up in a line to engage the privateers. The two ships were the *Robuste*, a French Guineaman of 14 guns and about 40 men, and the *Le Juste*, 450 tons, 22 guns (10 of which proved to be wooden ones), 4 swivels and 27 men. The *Blakeney* being the foremost of the two privateers, shot ahead of the *Robuste*, and attacked the *Le Juste*, who returned his fire very briskly, aided by one of the snows that lay ahead of the privateer, and the other upon his weather quarter. They all fought the *Blakeney* two hours, the *Hawke's* metal not being heavy enough to enable her to assist her consort, and then the *Le Juste* struck. In the meantime the *Hawke* came up and bravely boarded the *Robuste*, which was astern, the Frenchmen running from their quarters immediately on the appearance of the boarders. Their captain was shot in his thigh. When Captain Day had secured the *Le Juste*, he gave chase to the snow, *La Gloire*, which had crowded away while he lay by fishing his wounded masts, which caused a delay of nearly an hour. In endeavouring to escape, the snow threw overboard four guns, a sheet anchor and best bower cable; but the *Blakeney* got alongside of her, fought her, and took her. In the engagement, a crossbar shot broke the shank of one of the *Blakeney's* anchors, and a piece of the bar passing between the thighs of one poor fellow, took most part of the flesh away on each side. Two other men were also wounded. The *Hawke* received considerable damage, and had one man killed, while another man had his arm shot off, and another a leg broken. As soon as Captain Hewston, of the *Hawke*, had secured his prize, he gave chase to the second snow, the *Victoire*, of 10 guns, which escaped under cover of night. When the *Blakeney* engaged the four vessels, she had only 13 guns (viz., 2 nine-pounders, 1 six-pounder, 2 four-pounders, 8 three-pounders), 20 swivels and 67 men and boys on board. All the ships were much damaged in the action, and lay some time to refit. After manning both his prizes, Captain Day had only 45 men left on board his own ship, besides Frenchmen, who were superior in number. The united cargoes of the two prizes, which arrived safe in Liverpool, consisted of the following :—

232	hogsheads	3	tierces	195	barrels	White Sugar.
547	do.	28	do.	27	do.	Muscovada Sugar.
288	do.	—	do.	218	do.	Coffee.
15	do.	—	do.	4	do.	Indigo.

And some hundreds of hides.

In August, 1757, we find Captain Day in command of the *Prussian Hero* (Letter of Marque), 400 tons burthen, 20 guns (six and nine-pounders) and 80 men, belonging to Mr. Richard Savage, and engaged in the American and West India trade. In March, 1758, on his outward voyage, he fell in with five French privateers off Martinico, three of whom engaged him at once, but after a smart fire he got clear of them. Off the east end of Jamaica, he was attacked by a privateer of 16 guns, full of men, who ran his jib-boom into Captain Day's mizen shrouds, where it was immediately lashed, and as fast as the Frenchmen boarded the ship, they were as vigorously repulsed, the Captain animating his men in a surprising manner, and killing ten of the enemy with his own hand. The slaughter was so great that the deck ran with blood. However, Captain Day, finding there was no likelihood of overpowering them, on account of the superiority of their number, cut the lashings, and his mainsail filling, he soon left them. The engagement lasted about two hours, and Captain Day had only one man killed.

The names of two or three Liverpool estates are derived from valuable prizes made by privateers belonging to the port. The St. Domingo estate, in Everton, was so called by Mr. George Campbell, who, in 1755, purchased the first lot of land thereabouts from the Halsall family. In 1758, he frequently added to it, and gave it the name it bears to commemorate the capture of the rich St. Domingo ships by his privateers. The mansion, erected by Mr. Campbell, was a rather eccentric sort of place, resembling an ecclesiastical edifice. The history of this house is rather curious.

After Mr. Campbell's death, the property was purchased from his executors by Mr. Crosbie for £3,800. Mr. Crosbie having paid down £680 deposit, was unable to complete his bargain, and became bankrupt. The estate was then put up for sale at the Pontack, in Water Street, but no sale was effected. Eventually Messrs. Gregson, Bridge and Parke became the purchasers at the price Mr. Crosbie had agreed to pay for it, in addition to his forfeit money. These gentlemen, in 1773, resold the estate to Mr. John Sparling, merchant, who was High Sheriff of Lancashire in 1785, and Mayor of Liverpool in 1790, for £3,470, thus entailing on the three speculators a loss. Mr. Sparling took down the old house and erected a handsome mansion, into which he removed from a large house in Duke Street, the second below York Street on the south side. Mr. Sparling was one of the old school, appearing on 'Change in knee breeches, broad-flapped coat, gold laced waistcoat, broad shoes with gold buckles, and wearing a three-cornered hat. He left a proviso in his will that the St. Domingo estate should be occupied by no other than a "Sparling" by name. Finding it impossible to get a tenant under this condition, the will was set aside, in 1810, by an Act of Parliament to nullify the clause and enable the executors to sell the property. Mr. Sparling was interred in Walton Churchyard, where he erected, in his lifetime, a handsome tomb, which he could see from the windows of his mansion. He was the projector of the Queen's Dock, which he disposed of to the Dock Trustees in 1783. In 1811, his executors sold the St. Domingo estate for £20,295 to Messrs. Ewart and Litt. Mr. Ewart next purchased Mr. Litt's interest, and on the 13th September, 1812, re-sold the estate to the Government, to be converted into barracks. The price was £26,383, subject to 19s. 3d. lord's rent. Prince William of Gloucester, resided at St. Domingo House, when Commandant of the district, in 1803. The

Prince was very affable, and made himself exceedingly popular in the neighbourhood. Scandal said that he was often to be seen turning down Gloucester Place of an evening, to visit a fair lady who dwelt therein. St. Domingo estate was next sold in two lots, one lot, the land, being bought by Mr. Atherton, and the other, the mansion, by Mr. Macgregor. Soon after the purchase had been completed, Mr. Atherton inquired of Mr. Macgregor when he was going to take away his house. Mr. Macgregor said he did not intend to do so. "What do you mean to do with it?" "Why, I think I shall let it, or perhaps live in it myself." "Well, but how will you get to it, because I have bought all the land around it, and you have no right of way?" Mr. Macgregor found that he was at Mr. Atherton's mercy, and, it was said, sold him the mansion at a great loss. After other mutations, St. Domingo House became, as at present, St. Edward's College.

But to return to Mr. Campbell. In 1745, he commanded the "Liverpool Blues," about which a good story is told by Stonehouse. The regiment started one November morning, about three o'clock, to march to Warrington to guard the bridge, and, if need be, destroy it, as the Earl of Cholmondely, the commandant of the district, either from scarcity of workmen or distrust of the many Jacobites in the neighbourhood, felt himself in a position of difficulty. On arriving near Penketh Common, the vanguard of the "Blues" was seen hastily retreating, when the main body came to a halt. The valorous vanguard reported that there was a large body of the enemy ahead, occupying the road and part of the common. A party of skirmishers was then sent forward, when terrific screams and shouts were heard through the darkness of the night. It was then proposed at a council of war, called on the emergency, that the main body should deploy into the fields, and endeavour to take the enemy in flank; however, before the movement could be effected, the skirmishers had come in with each a prisoner in the shape of a goose, whereupon the main body of the gallant "Blues" charged *en masse* and completed the victory their advance guard had commenced; and it was said in Warrington that so many geese were never cooked in one night as there were on the occasion of the arrival of the "Liverpool Blues" in the town.*

Messrs. Henry Hardwar & Co., merchants, fitted out a privateer known as the new *Anson*, a ship of about 200 tons burthen, 12 carriage guns (nine-pounders), 24 swivels, and 130 men. She was commanded by Captain Wm. Cuthbert, who had been first lieutenant of the brig *Anson*, and all the officers and most of the brig's crew sailed with him from the Mersey, on December 3rd, 1756. Four days later they recaptured the *Rebecca*, of Hull, which arrived safe in Liverpool, but was again taken on her passage from Liverpool to Hull, and carried into Boulogne. Mr. Robert Williamson advertised that he had a letter in his hands from part of the crew, then in a French gaol, containing information advantageous to the insurers, but he adds—"no letters answered unless post paid." In June, 1757, the new *Anson* was taken by the French, and the affair is thus described by Capt. Cuthbert in a letter to his owners, written from "St. John the Angelick, 40 miles from Rochelle":—

"On the 12th inst., I fell in with a large fleet, to which I gave chase; fell in about the middle of them, and as there were three large ships to windward, which I took to be the convoy, edged to leeward, attempting to cut some of them out; but there found the *Amitie* frigate, of 24 guns and 250 men, which we engaged from half-an-hour past four to three-quarters past seven o'clock, and as my brave lads behaved like Englishmen, I believe we should have carried her, had not the *Warwick* of 60 guns, and two more frigates, bore down upon

* "The Streets of Liverpool," by James Stonehouse, p. 211.

us. As soon as they steered alongside, they gave me their broadsides both from their lower and upper decks. I gave them one in return, and struck, after receiving three shot between wind and water, and most of my rigging cut to pieces. They have stripped all my people, and only left me two shirts and two pair of stockings. I have at last prevailed on them to permit Mr. Robinson (the first lieutenant) to come out of the common jail in Rochelle, where all the rest of my brave boys are confined, a French merchant having become surety for us; and I hear they are to be marched backwards into the country 100 miles, till a general cartel is settled. I am not permitted to say more." The Captain died soon after.

The *Revenge* privateer, Captain John Gyles, and the *Mandrin*, Commodore Mackaffee, two schooners or wherries, about 35 tons burthen, prime sailers, mounting 2 two-pounders and 20 swivels, sailed on a cruise, in company, on July 1st, 1756. Though small of size and armament, they were great in daring, and soon met with success. In August, the *Mandrin* sent into Crookhaven a brigantine laden with iron and cheese for Brest, and about the same time the two privateers took two Dutch ships out of the river at Bordeaux. On September 1st, the *Revenge* captured the dogger *Maria Esther*, from Rochelle for the Mississippi, with flour, pork, lead shot, Spanish bar iron, cotton, flannel, knives, velvet, linen, "Ozenbriggs," wine, brandy, medicines, beaver hats, silk stockings, candles, linen handkerchiefs, ruffled shirts, shifts, black pepper, and other merchandise, which, together with the vessel, were sold by auction at the Pontacks' and Merchants' Coffee-houses. The *Revenge* returned from her cruise in October, and the privateersmen, when they came on shore, "made a handsome appearance, each man having a clean French ruffled shirt on, which they had taken overboad a bark bound to Bayonne. When the privateers boarded her they found twenty-four Frenchmen hid below, and none but Spaniards upon deck; however, they took care," says the paper, "to ease them of their dollars, silver buckles, private adventures, &c., and have brought in 732 ounces of silver, 13 ounces of gold, five chests of India goods, two tons of coffee, &c." The *Mandrin* having been blown on shore in Bootle Bay, was sold by the candle at the shop of the versatile Williamson.

In an age when charges of cowardice in face of the enemy were freely made against admirals and commanders in the Royal Navy, and courts martial were as common as blackberries, it was not strange that a privateer captain should fall under suspicion. The *Advertiser* of Nov. 5, 1756, contains "a vindication of Captain John Gyles' character." The owners of the *Mandrin* and *Revenge* privateers having appointed a meeting between Captain Mackaffee and Captain Gyles, to hear a true statement of the quarrel between the said captains, and to examine into the cause of the report spread in the town, which had defamed Captain Gyles by branding him with cowardice, the parties met at Pontack's Coffee-house, when Captain Mackaffee voluntarily signed a declaration completely exonerating Captain Gyles, who, far from showing the white feather, had single-handed attacked a French ship before the *Mandrin* could come up.

The method of financing the brave but improvident crews of the privateers was a curious one. In the paper of Nov. 12th, a victualler named Edward Walker gives notice that, having been appointed agent to the majority of the companies of the *Revenge* and *Mandrin* privateers, lately arrived from a cruise against the French, in which cruise they took "several valuable prizes and private plunders from the enemy," and having not only furnished many of the said companies with meat, drink and lodging, but likewise upon proper assignments procured Savil Wilson, merchant, of Liverpool, to furnish them with money and apparel, which they were in great need of, therefore he naturally desired all payments and settlements of the said prizes and private plunders, which concerned his clients, to be made to him; and no doubt Messrs. Savil Wilson and Edward Walker made their own little private plunder out of the necessities and recklessness of "poor Jack."

Prior to the publication of the "vindication" of his character, Captain Gyles had been appointed commander of the *Mercury* privateer, of 16 guns (six and nine-pounders), 24 swivels, and 130 men, belonging to Messrs. John Hulton and Co., which sailed on a cruise the end of November. On December 21st, Captain Gyles wrote to his owners, from Castlehaven, as follows:—

"On the 16th of December, we gave chase to a French ship of 22 guns, in latitude 43° 40′, longitude 11° 10′, which we came up with about 12 o'clock at noon (after having fired 10 rounds of our bow-chase guns, nine-pounders, which they answered with their stern-chase), and engaged broadside and broadside for five glasses.* They shot away all our standing rigging, wounded both our lower masts, and carried away our Troysail mast; hulled us in several places between wind and water, and an unlucky shot struck us four feet under water. We very soon had seven feet water in the hold, and expected to sink every minute, the water being level with our platform, and all our water casks afloat in the hold, which hindered us from plugging the hole in the inside; upon which we struck, and called out for quarter, but the enemy kept a continued fire into us, which determined us to throw all our guns overboard, whilst part of our people were baling the water out of the hatchway. Soon after the Frenchmen hove out a signal of distress, but we could not assist one another, and I believe never two ships were in a more shattered condition, for they appeared to be as much disabled as ourselves. If it had not been for that unfortunate shot, I believe we should have taken her. Four of my men are killed and thirteen wounded. I have received a shot in both legs, and have not been able to turn

* Two hours and a-half.

myself in my hammock since. I am more concerned for the loss of my cruise than my own wounds; and if it please God to spare my life, and one leg, I will have the other knock at the French. As soon as the ship is in condition I shall return to Liverpool."

A merchant at Kinsale, writing five days later, informs the owners that the *Mercury* was in a very shattered condition. The crew, in their distress, had thrown overboard 12 carriage guns and most of the swivels, two anchors and cables, and other articles. Their powder, bread, and most of their stores were "ill damaged," three more of the wounded men were dead, and two, besides the captain, were dangerously ill on shore. Captain Gyles had been wounded in both legs, two of the four wounds he received in his left leg being very dangerous. The gallant behaviour of Captain Gyles and his ship's company had recommended them to Colonel Townshend and the gentry of the neighbourhood, who were extremely kind to them. It was supposed that the ship which engaged the *Mercury** was the *Bristol*, of Bordeaux, 22 guns, which on her arrival at Rochelle, reported having fought an English privateer of 16 guns, and left her sinking.

Captain Gyles arrived in Liverpool on January 20th, 1757, and in February a notice was inserted in the papers

*Some question of marine insurance appears to have arisen in connection with this privateer, as the following letter on the subject was addressed to the publisher of the *Advertiser*:—"Some disputes that have lately happened between the owners and insurers of the *Mercury* privateer, if they have no other good effect, are at least sufficient to show us that our present method of insurance upon privateers is greatly defective; and that though the insurers have, upon account of certain exceptions, been induced to run the risk of the whole for a very small premium, yet the property of the adventurers has not thereby been truly secured. It is therefore submitted to the consideration both of the adventurers and insurers, whether it would not be more eligible to fix some premium, which should be sufficient to pay all averages and losses whatsoever, except powder and shot expended in attacking or defending; and also to insure for the whole cruise, for such time as the crew are engaged, without any exception as to their coming into or going out of port during such time; and also to make a clause in the policy, that in case the ship shall not be certainly seen in safety after the expiration of the limited time, she shall after the expiration of months, be esteemed a lost ship within the limits of the insurance and paid for accordingly."

calling upon the crew of the *Mercury* to repair on board to finish her six months' cruise, on pain of forfeiting their share of prizes, and of being prosecuted for the advance money by them received. She sailed from the Mersey on the 11th of March, and on the 25th recaptured, off Cape Peñas, the ship *Liverpool*, from Jamaica for London, laden with 247 hogsheads of sugar, 26 puncheons of rum, 8 hogsheads of ginger, 18 casks and 32 bags of pimento, 1 bag of cotton, and 38 logs of mahogany, all of which were duly sold by auction at the famous Merchants' Coffee-house. On May 12th, they took possession of the brigantine *John*, of Greenock, laden with pickled salmon and iron hoops, which was lying troy in the sea, without a soul on board, and which they sent to Liverpool.

Little occurred to the *Mercury*, except daily speaking neutral bottoms, and now and then an English privateer, until May 25th, when they gave chase to a sail, which, as soon as they came up with him, saluted them with a broadside, which they "returned freely." During the skirmish he carried away the *Mercury's* main top-sail yard, and damaged her rigging. Night coming on, they parted, but fell to it again at three in the morning. At five, Captain Gyles determined to board, steered alongside, and received the enemy's whole fire. The *Mercury's* top-men called out that he was well provided with close quarters, and had double their number at small arms, whereupon the experiment of boarding was abandoned as too dangerous. "We engaged him an hour-and-a-half close alongside," says Captain Gyles, in his journal, "and they answered our fire briskly, carrying away our stays, braces, topmast and futlock shrouds, great part of our rigging, and riddled our sails. At eight, set the men to work, to splice and knot our rigging, etc. Finding them so well provided, and double the number of our people on board, we agreed to leave him, and I apprehend he is an outward bound Angolaman. None of my people were hurt, except the gunner, who received a musket ball in his right breast."

On Sunday, July 10th, while the *Mercury* was lying in Fayal harbour, they saw a large ship in the offing, and immediately gave chase, only to find, however, that they had "got the wrong sow by the ear," the ship being a French privateer of 18 guns and 200 men. They stood away from him, but he followed and soon gave them the contents of his four bow chase guns. They then hove their broadside to him, and at four o'clock an engagement began, which was warmly maintained by both sides till half-past eight. The *Mercury* received three shot between wind and water, and was otherwise much damaged, but no one on board was hurt. Night coming on, they lay till morning, expecting a renewal of the engagement, but as soon as daylight appeared, they saw the enemy six leagues off. They were obliged to keep one pump going. At twelve o'clock, they felt a great shock, like an earthquake, and returned to Fayal harbour. On the 14th July, they left Fayal, and off Port Pine, where the French privateer had put in, "gave Monsieur three cheers, which he returned," but durst not follow them. On the 7th of August, in company with the *Bellona* privateer, of London, they captured a Spanish snow, laden with French goods, and sent her to England.

The snow *Mary*, Captain Richmond, in her passage from Liverpool to America, was taken by the *Le Roche* privateer, of 22 guns, and, nine days later, retaken by His Majesty's ship *Torbay,* the captain of which, having taken all the Frenchmen but two out of the prize, put eight Englishmen on board to carry her into port. On the second night, after they parted with the man-of-war, the two Frenchmen broke into the cabin, where the master was sleeping, and killed him, wounded most of the men, and confined them below in the steerage for eight hours. One of the Englishmen, by the glimpse of daylight, finding loopholes in the after bulkhead, luckily met with a musket, knocked a plug out, and shot one of the Frenchmen dead. The other Frenchman immediately jumped overboard, and clung to the rudder ring. The Englishmen, having got him on board again, had the humanity to spare his life, and carried him prisoner to Dale, near Milfordhaven. About the same time, the *Landovery*, Captain Johnson, from Liverpool for Jamaica, in company with two other ships, two days after leaving Cork, fell in with a large French privateer, which chased the *Landovery*, and took her after an engagement of an hour-and-a-half. The *John*, Captain Peter Gibson, on her passage to Virginia, was taken by a French frigate and scuttled. The crew were carried to Dinan and close confined, except the Captain and mate, who had liberty to walk the town within the walls.

One Sunday afternoon in October, 1756, the impressed men confined on board the *Bolton* tender at Hoylake mutinied, and, after knocking down the sentinels and securing their arms, took possession of the vessel. In attempting to recover the ship, the mate was knocked down with the butt end of a musket by one of the mutineers, and while he was down, two others struck him with an iron bar and a handspike, "though entreated to the contrary by several who begged for his life." He was then forced into a boat and put on shore, where he died of his wounds in two hours. Several people were "ill hurt in the scuffle," and about forty of the impressed men made their escape to Liverpool. On the following Wednesday afternoon, as Lieut. Siddal was taking one of the captured mutineers down to a boat, the man was rescued by a mob. In the evening, having doubtless partaken liberally of refreshments, they assembled again, broke open the watch-house, where another of the deserters was confined, "used ill the master of the watch, broke several of his ribs, and took off the man in triumph."

The brig *Jenny*, Captain Brown (Letter of Marque), belonging to Messrs. John Tarleton & Co., on her passage to the Leeward Islands, took the *Legere*, 300 tons burthen, 10 guns and 30 men, from St. Domingo for Bordeaux, laden with a valuable cargo of sugar, coffee, and indigo, which was sold by the candle at the Bath Coffee-house, in Liverpool.

On the 9th of October, 1756, in the latitude of Tobago, the *Catherine* (Letter of Marque), Captain Augustine Gwyn, had a very close and sharp engagement for eight hours with a large French snow, which struck to the Liverpool vessel. Both ships had their rigging and sails shot to pieces. The *Catherine* had only one man wounded, while the enemy had three killed. The prize was subsequently retaken, and run on shore, through the gross carelessness of the prize-master. Soon after this affair, the *Catherine* was chased by a French privateer of 10 guns, and full of men, "who came up with us," says Captain Gwyn, "and fought us three glasses, but my people behaved gallantly and beat them off. They made attempts to board; we raked them with our stern chase, which made them glad to sheer off. In this engagement none of our people were hurt, but almost every rope was shot away, and our sails, &c., greatly damaged." In 1757, while in command of the *Fame* frigate, 350 tons burthen, 20 carriage guns (twelve, nine, and six-pounders), and 80 men, a Letter of Marque, belonging to Messrs. John Tarleton & Co., Captain Gwyn, carried into Kingston, Jamaica, a French privateer of 8 guns and 80 men, which he took off Antigua. He also fell in with three other privateers, at one of which he fired 30 shot, and in all probability would have taken her, if night had not prevented him. He likewise brought in a Dutch sloop, laden with French sugars; "but," says the correspondent, "as the Dutch are artful traders, probably they may evade our laws and escape with impunity, which too many of them have done this war, notwithstanding their being notorious carriers of contraband goods to our natural enemy." On August 31st, 1758, the *Fame* retook the brig *Truelove,* of Lancaster, and the brig *Jane*, of Sligo, which had been taken by the famous *Marshal Belleisle* privateer, of St. Malo, commanded by the gallant Captain Thurot.

The paper of November 5th, 1756, contains the following advertisement :—

> "All gentlemen seamen, and able-bodied landmen that are willing to fight the French and make their fortunes, may meet with suitable encouragement by entering on board the *The Grand Buck Privateer*, Captain John Coppell, Commander. A ship of 300 tons burthen, frigate built, 6 feet between decks fore and aft, mounting 20 carriage guns, twelve, nine and six-pounders, 20 swivels, and 200 men. N.B.—The ship will be ready for sea in a fortnight, and now lies in the South Dock. Apply to the Captain ; or to Messrs. Robert Clay & Compy., Merchants."

Alas for the vanity of human intentions! Notwithstanding this brave invitation, there is nothing to record of the doings of this privateer, except that she sailed on her cruise on December 8th. The ship *Cunliffe*, which arrived from North Carolina, reported having passed a derelict rolling in the Atlantic, which some of the *Cunliffe's* crew recognised as the *The Grand Buck*.

The *Isaac* privateer, 16 guns, Captain David Clatworthy, sailed from Liverpool on November 29th, 1756, and on the 8th of January, 1757, carried into Kinsale, the *Le Victoire,* of Havre, bound for St. Domingo, with bale goods, gunpowder, etc., valued at £6,000. She had on board 9 carriage guns, 6 swivels, and 50 men, who fired two broadsides before they struck. Having returned to Liverpool to refit, the *Isaac* sailed on another cruise in the following June, and on July 15th, Captain Clatworthy wrote to his owners, from Plymouth, as follows :—

"On July 1st, in latitude 43°, longitude 10° 57′ from London, at six in the morning, we, in company with the *Shark* privateer, Captain Abraham Harman, chased a ship and a snow, which proved to be Spaniards from Cadiz; one bound to Ferrol, the other to the Groyne. As soon as we had discharged them we discovered a smoke right ahead, and in a short time heard the firing of cannon, upon which we both stretched that way and soon saw a large French ship engaged with three English privateers. At half-an-hour past ten we gave her a gun, and hoisted our English colours, which she answered with her broadside. We returned the compliment, wore ship and berthed ourselves upon her quarter, where we lay for two hours and a half. She then struck. During the engagement, we fired our bow chase 43 times, and broadsides as fast as they could be repeated. The last shot fired was one of our nine-pounders, which went in at her larboard quarter, and killed a relation of the Captain's; upon which they instantly struck, and gave the victory to the *Isaac*, and have since declared that had it not been for us, they would not have been taken. I sent Mr. Valens (first lieutenant) and 20 men on board her. She proved to be the *Prince of Conti*, from L'Orient in Old France, bound to the East Indies, Capt. De La Motte, Commander, her burthen 800 tons, mounts 50 guns, (18 twenty-four-pounders, the rest twelve and nine-pounders), and had 195 men; but as they threw their papers overboard, with most of their small arms, we can give no other account of her cargo, but that she has stores on board, and by all the intelligence we can get, cash, and have reason to think it will prove no inconsiderable sum. The privateers engaged with her were the *St. George*, Robinson, *Black Prince*, Creighton, and *Boscawen*, Harden, all of London; the two first of 22 guns each, the last of 16. After we had settled our affairs on board, as well as the hurry and confusion would admit, it was agreed to make for the first port we could reach; as I had it not in my power, against so many voices, to bring her to Liverpool. Could I have done it, nothing would have given me so much pleasure as the shewing you one of the finest vessels you perhaps ever saw there. We had the misfortune to burst one of our six-pounders, which killed one man, and wounded another, who is since recovered. As soon as we get in, I shall enquire out some gentleman of undoubted character to act for me in case I should sail before I have the favour of your answer. I make no doubt of your acting to our mutual interest. You may depend on my using all diligence to serve my worthy owners. Our vessel sails so incomparably well that they are all courting me for a Consort; nor would I have you think I compliment myself when I inform you we have had the thanks of the whole fleet for our behaviour in the action. Our officers all acted with courage and discretion, and our men with the greatest bravery; and I believe that had we, in company with a vessel equal to us seen the prize first, we should have needed no farther assistance. If I should sail before I hear from you, I shall leave an exact inventory with a proper officer, but should be glad if you were here to act for me and crew, as the concern is too considerable for any but trusty hands. I am, &c., David Clatworthy. P.S.—The first day I was on my station, I fell in with 8 sail of French Martinico ships, and two frigates, which we lost in the night."

The *Prince de Conti* was reported to be worth £100,000, exclusive of the cash on board, which must have been a very large amount, as she was bound to Bengal to purchase English merchandise.

This capture of a rich prize was not effected without some heat and jealousy arising between the gallant commanders. Capt. Harden, of the *Boscawen*, felt aggrieved, and wrote to one of his owners in these forcible terms:—

"Notwithstanding the many and villainous reports you have heard of my being astern and out of gun shot when the Frenchman struck, you yourself may judge of the truth of it, as our boat was on board, brought off the captain and several of his principal officers, and returned again on board long before any of the other boats were there. It would have been impossible for us to board her first had we been out of gunshot or at a greater distance than the rest. All the men on board our ship are ready to swear we were nearest when she struck, and those that have spread this infamous report have not souls to stand to it like men, for when they were charged with it by myself and challenged, they meanly denied it and begged pardon, and in everything relinquish those great feats they boasted of in the papers."

In 1758, the *Isaac*, on her passage to Barbadoes, took the *L'Aimable Marie*, from Nantz for St. Domingo, and beat off a French privateer of 12 guns, after an engagement of an hour and a half. Twenty years after, the *Isaac* appeared in the Channel as the American privateer *General Mifflin*, and played sad havoc with the commerce of her former friends.

The following very remarkable letter, dated December 11th, 1756, was received from on board the *Hibernia*, Capt. Watson, "off Rogipore, a little to the southward of Bombay":—

"This day about noon, we saw several calevats, or rather, gallevats, or war-boats armed with swivel guns and doubly manned. They were at a considerable distance, and crowded about a ketch, which they seemed to tow along. The captain and chief mate, who were both well acquainted with the Malabar coast, immediately declared it was the Meelwan fleet, which had made a prize of this ketch, and was towing her in shore, to get out of our way. These Meelwans, or Kemasants, as the Portuguese call them, are a nest of pirates, a little to the southward of Gary, and formerly subjects to, or allies of Angria, the grand pirate, on the Malabar Coast. The chief mate was positive that it was Capt. Scott, of Blay's ketch, and that it would be a piece of good service to retake her from the pirates, whose calevats were twelve in number. Accordingly we bore away likewise in shore, and endeavoured by all means to come up with them, but there being little wind, and we having a luggage boat of 70 or 80 tons to tow after us, they lugged the ketch along, and kept at a considerable distance from us all the afternoon. However, about half an hour after eight o'clock at night we came within gun-shot of them, when firing only two nine-pounders among them, the calevats abandoned their prey, and ran in shore into shallow water, whither we could not follow them. Our chief mate and twelve men went armed cap-a-pie, with their muskets and cutlasses, and took possession of the ketch, which the pirates had robbed of part of the cargo. It belonged to some merchants of Calicut, and was bound for Muscat with Malabar goods, as cassia, pepper, bottlenuts, cardamoms, sandal-wood, &c.

"The vessel itself, with what is left of the cargo, I judge will amount to 8,000 rupees, or £1,000 sterling, which we shall divide among us, according to the rates of the navy. The affecting part, however, of this affair is what follows:—Our chief mate had orders to send all the prisoners on board the *Hibernia*, and to keep possession of the ketch with our men. But, good God! when they came on board, what a moving sight! Out of 25 men, hardly any could walk, or even stand, without being supported. Thrice had they sustained the attack of 12 calevats, and as many times repulsed them, partly with their swivel guns, and partly with stones, spears, and cutlasses. This hot action lasted about two hours, during which the ketch's people behaved exceeding well, and the captain or Nokedy, as the country people here call him, killed five of the assailants with his own hand. However, as the pirates were twelve times more numerous than the ketch's people, they got on board her a fourth time; when the Nokedy asking his men which of them would stand by him, two of them only offered themselves, and were in a manner cut to pieces along with their Captain, who fell fighting heroically, if I may be allowed such a term, for his liberty and property. He was the only man that was killed outright, but almost all the rest were wounded in a most frightful manner, particularly the two men who stood by their Captain to the last. Some of them must have been stabbed as they retired, the wounds being in the hinder parts of the body; but the two brave men already mentioned received all theirs in the fore parts. One had a piece as large as the palm of the hand almost cut off from the forehead, and a deep cut on the crown of his head, which we imagine will prove mortal, as the skull is fractured. Good God! what a gash it is! These two wounds, it seems, laid him flat, sprawling upon the deck, and indeed any one of them, especially the last, was sufficient to stun the most stout-hearted. The other was cut and slashed all over the body. He had received a frightful wound on the right side of his face, which had cut off the lower half of his ear, and laid open the jawbone quite to the chin, and even the integument of the neck so deep, that the jugular vessels appeared. The patella, or small bone of his left knee was divided in two by another slash that reached four or five inches in length. Another gash across the outside of the left thigh penetrated to the bone, dividing asunder a large nerve as big as a man's finger. He had received another wound between the elbow and wrist of his left hand, which had cut asunder the nerves which serve to move the fingers, and penetrated quite to the bone. All this time he stood fighting the enemy with his right hand, till at last a wound received athwart the fingers of his right hand, whereby one finger was cut off and two others deeply wounded, proved a finishing stroke, so that, no longer able to hold his cutlass, he fell down upon the deck, bleeding at numerous and also very deep wounds. And, indeed, it is surprising he could have stood so long, considering the vast discharge of blood from his wounds. Capt. Watson, whose humanity on this occasion deserves particular praise, acting as Surgeon's assistant, preparing bandages, tents, plegets, plaisters. &c. He took a great deal of pains in washing, cleaning, and dressing their wounds; and, besides the plaisters put up in the medicine chest, made use of Balm of Gilead,* which he poured plentifully

*The celebrated "Balm of Gilead" was prepared in Liverpool by two Hebrew quacks, named Solomon, father and son. Dr. Solomon (the younger) made a large fortune out of the Balm; and died about 1819. He resided in his later years at Gilead House, Kensington, and was a curious political and social character. In 1803, he started the first daily newspaper published out of London.—See "Historic Notes on Medicine, Surgery and Quackery"; by the present writer, in the *Lancet*, April-May, 1897, or the "Streets of Liverpool," by Stonehouse.

in the wounds, securing all with tents and plegets dipt in the same balsam, which he had purchased at Judah the last summer. And the better to see them taken care of, they were all brought on board the *Hibernia* till we should arrive at Tillecherry, where the captain intends to put them all under the care of Dr. Gill, the Company's Surgeon."

Williamson's Advertiser of December 17th, 1756, contained the following spirited description of another private ship of war:—

"Now fitting out for a cruise, and will be ready to sail next week against the Enemy of Great Britain, the ship *King of Prussia* privateer, under the command of William Mackaffee. Burthen 250 tons, mounts 16 carriage guns (all nine-pounders) 20 swivels, and 154 men. All gentlemen seamen and ablebodied landmen, that are willing to imitate the brave King whose name the ship bears, in curbing the insolence of the French, and making their fortunes immediately, will meet with suitable encouragement by applying to Messrs. Thomas Parke & Stanhope Mason, Merchants, or the Commander. N.B. This ship carried a commission the last war, met with great success in taking many prizes, and is a remarkable prime sailor."

She sailed from Liverpool "to curb the insolence of the French," by capturing their property at sea, on January 29th, 1757. Captain Mackaffee, writing from Gibraltar on the 22nd of April, gives the following account of his movements:—

"After a long and tedious cruise, we arrived at Gibraltar. The day after our arrival there was an engagement between five English and four French men-of-war. Our ships were superior by one gun. On hearing their fire we slipped and made for the Gut, where we fell in with the outward bound fleet and engaged them. Five sail struck to us, but the four French men-of-war, which were their convoy, bearing down upon us, I was obliged to quit, but soon joined the fleet, and it being dark they could not discover us. I came alongside the French Commodore, and boarded one of the fleet, without the loss of a man. When the prisoners came on board, we gave chase and fell in with a French privateer and one merchant ship. We drove them both into Cadiz, and then returned to Gibraltar. I have lost three men. Our ship's company is in great spirits, we being extremely well manned. The *King of Prussia* is a fine ship, and carries her metal well. The prize, having the French King's Commission on board, was easily condemned."

The prize so daringly captured was the snow *La Favourite*, whose cargo was invoiced at 30,000 livres. The French prisoners reported that she had 20,000 dollars on board, and was the richest vessel* in the fleet. Admiral Saunders and his squadron, having heard the firing of the privateer, were entitled to a share of the capture, "but the noble spirited Admiral gave up his claim in favour of the captors, and the rest of the captains followed his example."

Writing from Candia in July, 1757, Captain Mackaffee tells his owners that after leaving Gibraltar he was obliged to abandon his proposed cruising station, being pestered by English privateers. He proceeded up the Straits as far as the Channel of Malta, where he took a Swedish ship of 22 guns, laden with French property, from Smyrna. She had 370 bags of cotton, and her hold full of wheat. Captain Mackaffee instructed his first lieutenant to proceed with her to Gibraltar, but the heating of the wheat, the number of enemies swarming around, and the unruly conduct of the Swedes on board, compelled the lieutenant to put into Malta, where the cargo was condemned and the vessel discharged. The captain then relates his own doings in the *King of Prussia*, as follows :—

"I then proceeded farther up the Arches, having intelli-

*She was sold by auction in Liverpool on March 27th, 1758, with her entire cargo, which consisted of the following curious assortment : Castile soap, tallow, wax candles, sweetmeats, capers, bitter and sweet almonds, flour, cheese, cordial drams, lavender and Hungary waters, caplier, shoes, "sallet oyl," kidney beans, earthenware, nails, perfumed poma, etc., raisins, paving tiles, anchovies, writing paper, claret (120 hogsheads), wood hoops and medicines.

gence of many French vessels passing, where we fell in with three English privateers. They informed me that the *Anson* and *Blakeney*, of Liverpool, were not twelve leagues off us. According to orders, would not engage to keep company with any of them. I put into Zinda in Candia, watered and victualled the ship at easy expense, having met with so much money which belonged to the French merchants on board the Swedish ship, which defrayed expenses. I stayed in there six days at the first time, and put out, giving out I was bound to Constantinople, immediately went round Candia, where we had the good fortune to meet with a French ship with Turkish passengers, and cargo on board. I ransomed her for three hundred Turkish chequins, which was paid by the said merchants before we parted with her. Four days after, I had the good fortune to fall in with a French ship called *La Murice*, mounting 12 guns, from Zinda, bound to Marseilles. After seven hours' chase, she struck, without firing a gun. Her cargo by the French Captain is valued at 225,000 French Livres. Meeting with contrary winds, and having many prisoners on board, was obliged to put in a second time into Candia. During the time since I left this place, orders was sent from the Grand Seignior, that all vessels, whether English or French, that brought in any prizes, should not be condemned until he was acquainted. I am very much afraid the condemnation will be very expensive. The Governor of Candia this day is to give me security for the prize, until he hears from the Grand Seignior, for which reason have made him a very rich present of a Turkish carpet, as there is nothing to be done here unless by presents. I put into Zinda in Candia, being the best port, and was detained at Candia as prisoner at large, until the prize came in, for fear I should put out with my prize, by their fort, but as the prize is now in their possession, hope things will be easy. I have desired to go out on a cruise, until the prize is condemned, and then call and bring her along with me, having now remaining on board the brave *King of Prussia* 100 men in perfect health, besides those with the prize at Malta and Candia."

On the 30th of August, Captain Mackaffee writes again to his owners, from Syracuse, in Sicily, in the following pleasant vein :—

"I acquaint you with pleasure that we have taken another large French ship, employed as a caravan to carry Turks and their effects from one port to another, whom I ransomed for 1,000 chequins ; there were about 100 Turkish gentlemen and passengers on board, who immediately advanced the money. We have a brave ship's company, and expect to eat a Christmas dinner in Liverpool, having not less than 30 laced hat gentry, and not one sick man on board. My expenses have been so trifling that I expect to trouble the owners with few bills ; and I shall do my endeavour to gain the respect of my King, Country, and Owners.

"What we have had from neutral bottoms we have paid for with French money ; and you will hear of no complaints, except using the enemy too well, and not plundering their clothes. I propose sailing on the remainder of our cruise the first fair wind ; and as we met with six carriage guns and four swivels, which the enemy was so civil to give us, I have lent them to Capt. Benn, who mounts now 18 guns, &c., &c."

In November, the *King of Prussia* carried into Syracuse a French polacca, richly laden, which they had taken off the island of Candia. In February, 1758, the privateer arrived at Gibraltar with a prize from Malta, and on the 7th of August the successful cruiser was sold by auction at the *Golden Fleece*, in Liverpool. In March, 1759, we read that the *Thames*, laden with the valuable cargo of the prize belonging to the *King of Prussia* had arrived in the Downs from the Mediterranean. The reader may possibly envy the owners of this gallant vessel, and wonder what they did with all the plunder. One of them, at any rate, did not prosper, for in the paper of December 8th, 1758, we find the following :—

"On Tuesday night died Mrs. Dorothy Parke, widow of the late Capt. Parke, formerly a commander in the West India trade. It is said that the misfortune of her sons, Messrs. John Parke, of London, merchant, and Thomas Parke, ironmonger, here, being both bankrupts, affected her so much that she immediately took to her bed, and appeared to be broken-hearted."

In the year 1757, the activity of the French privateers was phenomenal. It was computed from the number of Carolina ships taken that the French had got the year's whole produce of indigo from that colony, excepting about 60,000 lbs. brought in one or two ships that escaped the enemy. "Much to the honour of a nation possessed of above 200 men of war!" observes the newspaper. "Happy if the trade from Carolina had put into Ireland and waited for convoy; but though we have been unfortunate with respect to several merchant ships, we have taken above fifty French privateers, whereas we can't learn that they have taken more than three of ours." This, of course, refers to the whole kingdom. A month later, the number of French privateers taken had risen to 78, but we read that the account "of the increase of the French privateers upon all our coasts are most shocking and alarming, and unless timely dispersed must terminate in the ruin of our commerce." The French privateers swarmed in every sea, many of them cruising from 100 to 180 leagues to the westward of Cape Clear, in lat. 48° and 49.° They cruised so thick round the island of Antigua that it was next to a miracle for an English vessel to get in there, except under convoy. It was stated that from August, 1756, to February, 1757, the French privateers had taken 70 English vessels, chiefly owing to the small number of 20-gun ships and sloops stationed in that quarter of the world for the protection of British commerce. A ransomer who came over from Dunkirk stated that from January 27th, 1756, to March 5th, 1758, the privateers belonging to that place had taken 136 British ships, 78 of which they ransomed.

From the commencement of the war up to July 12th, 1757, the French had taken 637 British vessels; the ships taken from the French during the same period were 681 merchantmen and 91 privateers, making a total of 772. It was calculated that the English had profited by captures upwards of two millions. But, in spite of Brown's "Estimate," a book published at this time, in which the author, as Macaulay observes, "fully convinced his readers that they were a race of cowards and scoundrels; that nothing could save them; that they were on the point of being enslaved by their enemies, and that they richly deserved their fate;"—in spite of all this, the pluck of the British race was not a bit cowed. In the paper of May 13th, 1757, we read that "the spirit of privateering had diffused itself amongst all our colonies abroad in so extensive a manner that even many of the Quakers breathed revenge against our perfidious enemies. The *Sprye* a privateer belonging to Philadelphia, of 22 nine-pounders, and 208 men, commanded by the brave Obadiah Bold (a Quaker), sailed for Tobago," on a cruise. It must have been a rich treat to see the gentle Obadiah in action, "thee"-ing and "thou"-ing his brave fellows, while directing their fire to the vitals of the enemy. We picture him to ourselves pacing the quarter deck, a harmless-looking, little man, placid amongst the crashing of cannon-balls, the rattle of small arms, the shrieks of the wounded, and all the attendant horrors of a tough seafight when the Bloody Flag is flying. He is the coolest, yet most determined, man on deck, and at his silvery voice, raised in command, men are hurled headlong into eternity. But let us leave the realms of fancy, and stick, as we have hitherto done, to strict matters of fact. The spirit of privateering had diffused itself in Wales also, for the paper of April 7th, 1757, tells us that "Last Saturday sailed out of Beaumaris Bay, in Wales, the *St. David* privateer, of 20 carriage guns and 16 swivels, commanded by Captain Reeves Jones, fitted out by a Society of Ancient Britons." After being out a few days, the *St. David* returned to Beaumaris bringing in with her a new French privateer of 12 carriage and 16 swivel guns, which she had taken after a very smart engagement of two hours and-a-half. The French had 29 men killed and the Welsh five, which argues that some of the valour of Fluellen and Glendower still remained in the land.

On the 5th of December, 1757, the ship *Trafford*, Captain Marshall, in her passage from Virginia to Liverpool, fell in with a privateer from Louisbourg, and engaged her closely four hours and-a-half, but was obliged at last to strike, having received considerable damage, and her ammunition being nearly expended.

The commanders of the King's ships appear to have been shamefully relax in the unpleasant duty of convoying merchant vessels, and in pursuing the privateers of the enemy, during the early part of this war. It was customary to announce the date of sailing as follows:—

> "The captain of his Majesty's sloop *Otter* gives this public notice to the merchants, freighters, and owners of such trading vessels now in this port, and the ports of Chester and Whitehaven, bound up the English Channel as far as Plimouth, that he proposes sailing by the 28th instant, when all vessels who are ready may have the benefit of his convoy.—Dated in Hylelake, July 20, 1757."

In March, 1757, the Lords of the Admiralty signified to the merchants, that "on its being made to appear to them that any commanders of men-of-war had been defective in their duty of convoying or protecting their ships, or pursuing privateers, on notice being given, they should be dealt with according to their deserts."

That it was necessary to make an example of somebody, is clear from a letter written by Captain Isaac Winn, of the

Dolly & Nancy, who had been blamed by the underwriters for sailing from Dartmouth to Liverpool without convoy, which resulted in his ship being taken by a privateer of St. Malo, and his imprisonment at Dieppe. After describing his efforts to join some men-of-war which were convoying a fleet of merchant ships, and the commodore's conduct in ignoring his signals of distress, etc., the captain relates what befell him one morning, when a calm had succeeded the gale :—

> "At daylight we saw several sail all around us, one of which we took for a man-of-war, and not far from us, to windward, was a sloop, which we took for a tender. She was so like a Folkstone cutter that the people I had for those that were impressed at Dartmouth, took her for the tender belonging to their ship. It being quite calm, she rowed up with us, nor did we perceive our mistake till we heard them talk French on board the privateer (as she proved), which was not till she was within half a musket shot, for before they talked English. They immediately boarded us, the consequence of which was (as we had nothing to defend ourselves) our being taken, being the eleventh prize taken by the said privateer this winter, or rather the last, by the shameful neglect of our cruisers, which is so flagrant that the French themselves laugh at it. When I told the Captain I did not doubt of meeting some of them before he got me to France, "*Well*," says he, "*if we do, they will not chase us if we don't hoist French colours*." We were not above a league off Beachy Head when taken, and it continued calm till midnight. Thus, gentlemen, you have lost a good ship and a good freight, and we all that we had on board, and our liberty. I am in the common prison, without so much as a shirt to shift myself with, having nothing but what I had on when taken. They allow me fourpence per day to live upon, out of which I pay for a bed for myself and mate, otherwise I must lie amongst straw and filth, with the rest of my poor fellows. This is hard usage. Notwithstanding I can sooner forgive the authors of it than

the villains who commanded the above-mentioned two men-of-war, who, if they had suffered us to join them, might have prevented our being taken. Rage and vexation hinder me from adding any more, than that I am, with the greatest respect, and sorrow for your loss, gentlemen, Your friend and servant, ISAAC WINN.

P.S. All who have any relations at Liverpool are well, and give their love to said relations. I write this with the help of a wood pen, and soot and water. Are the French prisoners so used in England? I hope not."

In April, 1757, Captain Walter Barber, in bringing his ship the *Resolution* to Liverpool to be re-fitted as a privateer, undertook to convoy 33 sail from the Downs to Spithead. "We were for forty hours visited by three French privateers, till Captain Barber beat them off," writes one of the captains to his owners. "He is the most honourable commander I ever was under." After striking on a rock three times and losing her rudder, the *Resolution* arrived in Liverpool. A romantic affair in connection with this ship, is thus reported in the paper of May 20th, 1757:—

"A young person, five feet high, aged about nineteen, who entered in January last on board the *Resolution* privateer, Capt. Barber, under the name of Arthur Douglas, proceeded with the ship from London to this port, went aloft to furl the sails, &c., when called upon, was frequently mustered amongst the marines at the time they exercised the small arms, and in short executed the office of a landsman in all shapes with alacrity, was on Saturday last discovered to be a woman by one of her mess-mates. 'Tis said that he found out her sex on the passage, and that she, to prevent a discovery, then promised to permit him to keep her company when they arrived here; but as soon as they came into port refused his addresses. The officers in general give her a very modest character, and say by her behaviour that she must have had a genteel education. She has changed her clothes, but will not satisfy any of them with her name or quality; only that she left home on account of a breach of promise of her lover. 'Tis remarkable that during their passage down, on the appearance of a sail, she was eager to be fighting, and no ways affected with fear or sea sickness."

The genius of our novelists can expand this crude outline into a stirring sea novel of the orthodox size, as expeditiously as certain bacilli are said to transform barley meal into the richest port wine.

The *Resolution* cruised in company with the *Spy* privateer, Captain Pierce, of Liverpool, "a brave vessel under foot, which could either speak with or leave" any cruiser on the sea, but beyond a smart brush with a French frigate of 36 guns, which they engaged several hours and ultimately beat off, the joint cruise does not seem to have been eventful or profitable, both vessels being forced into Cork by loss of masts, etc. Captain Barber afterwards commanded the *Shawe* (Letter of Marque), 200 tons burthen, 12 guns, belonging to Mr. Edward Deane, which was captured while employed in the Jamaica trade, by the French frigate *Gronyard*, of 26 guns and 130 men. The British frigate *Favourite* afterwards took the *Gronyard*, which was said to be the richest prize taken during the war, and the best sailer the French ever possessed.

On December 22nd, 1757, about 30 leagues to the westward of Vigo, the *Spy*, in company with the *Mercury*, took the *Mutiny* privateer, of St. Jean de Luz, a brig of about 40 tons burthen, 2 carriage guns, 6 swivels, and 58 men, commanded by Dominique Cannonier, who two days after leaving port, in order to intercept English vessels in the Portugal trade, had the mortification of voyaging to Liverpool a prisoner in the *Spy*.*

* The following curious advertisement appeared in *Williamson's Advertiser*, January, 1758:—"Notice is hereby given to all gentlemen, seamen, and brave landmen, that have courage to face Monsieur, that the *Spy* Privateer, Thomas Pierce, Commander (who the last cruize took the *Mutiny* Privateer of Bayonne, the Captain of which says he saw forty English Privateers before, and tho' chased, was in no danger of being taken by them; but depending still upon his Heels had

The Muse of Poetry, in a rather distressed condition, visited Liverpool in 1758, and produced the following lines, which appeared in the *Chronicle*:—

ON THE RESOLUTION AND SPYE PRIVATEERS.

As poor Britannia pensive stood, depress with grief and pain,
Her tears encreas'd the briny flood, and swell'd the curling main;
"O where are now those hearts," she said, "those sons of ancient praise,
Whose look would strike each foe with dread, and endless trophies raise!
But see! distrest and drooping now, I can no longer hold;"
She sigh'd and moan'd, then fainting bow'd, struck with a death-like cold.
Then flew two lovers of the maid, rais'd up her failing arms;
Offered their lives her cause to aid, and guard her from alarms.
Britannia kind, as always wont, admir'd their noble mind,
And bade them think, in danger's front, that she would still be kind.
Take you my spear, my shield take you, as proofs of my regard;
And think each glorious deed you do, you've valour's just reward.
Unarm'd I'm safe, protected so, on you I will rely;
Command my *Resolution* you, and you my fav'rite *Spye*;
Hence then, my heroes, scourge my foes; acquire a glorious name;
Return with laurels on your brows—in death I'll sound your fame.

From the journal of Captain Robert Grimshaw, of the *Spy* privateer, we find that he sailed from Liverpool, on the 16th of March, 1758, and on April 10th, in company with the *Resolution* privateer, Captain McKee, also of Liverpool,

made bold to look at the *Spy*, who after a long chase, shew'd him the Way to Liverpool.) Mounting twenty-two carriage guns, besides swivels, to carry about one hundred and fifty men, and will be completely fitted and ready to sail in ten days on a six months' cruize."

took a Spaniard, from Marseilles for Nantz, laden with Castile soap, Brazilletta dyers' wood, olive oil, etc. On the 17th, they recaptured the ship *Marlborough,* from Jamaica for London, laden with 176 casks of sugar, 12 puncheons of rum, 27 casks and 85 bags of pimento, 82 mahogany planks and 152 hides; the ship and cargo being ultimately sold by auction, at the Merchants' Coffee-house. On May 14th, the *Spy* sailed from Beaumaris to finish her cruise; on the 19th, anchored in seven fathom water, below the Spit, at the Cove of Cork; the 22nd, they got a boat down from Passage with 15 new candidates for fortune's favours, several of whom, the Captain tells us, "wanted to go ashore, but not being allowed, two or three came in a riotous manner upon the gangway, with clubs, threatening to knock down the first lieutenant; upon which he fired a loaded pistol at them, which dropped one; then we put 16 others into irons, and afterwards had a quiet ship." They left Cork on May 26th, and on the 28th, gave chase and came up with the *St. Philip* and *St. Jago,* from Dublin to Cadiz, laden with beef, butter, hides, linen handkerchiefs, buckskin breeches, etc. They detained her till next morning, and finding, on examination, that her cargo and bills of lading differed from each other, as likewise her clearances from the Custom-house, they felt constrained to make a prize of her, and sent her to Liverpool to be examined. On June 16th, they fell in with the *Princess Carolina*, with 236 hogsheads of French sugar, 57 bags of cotton, and 252 bags of coffee; and the *Eendracht*, with 859,790 lbs. of French sugar, 25,030 lbs. of cotton, 256,036 lbs. of coffee, 2,058 lbs. of indigo, and 150 hides, both from St. Eustatia, for Amsterdam. From the earnestness of the commanders to secure their own goods, etc., Captain Grimshaw and his officers suspected that the cargoes were French property, therefore felt it their duty "to carry them in," which they did in safety, after burying Henry Roberts, "who had catched the smallpox."

The *Resolution*, Captain McKee, having assisted her consort the *Spy* to recapture the *Marlborough*, gave chase to the *Machault* privateer, of 24 guns and 230 men, which had taken that vessel, but the Frenchman got clear in a squall. On May 17th, they gave chase to three sail, which they boarded the following day, and felt justified, on the inspection of their papers, to deem two of them legal prizes. One was the snow *St. Jacob*, from St. Eustatia for Amsterdam, with indigo, sugar, etc., and the other the *Catherine Maria* galley, from Curacoa for the same port, with coffee, sugar, indigo, 10 chests and 1 cask of silver, etc. On the 29th September, the *Resolution*, in company with the *Nelly's Resolution*,* of London, took the *Smyrna Galley*, a Dutch ship from St. Eustatia, laden with coffee, indigo, cotton, and 400 hogsheads of sugar, which they despatched for Liverpool. In November, the *Christopher*, from St. Croix, another prize taken by the *Resolution,* was lost on Spanish Island. The *Spy,* 160 tons burthen, and the *Resolution*, 400 tons, were sold by auction, February 2nd, 1759.

The *Tartar* frigate, Captain Hugh MacQuoid, 320 tons burthen, 22 guns and 70 men, belonging to Messrs. Halliday and Dunbar, on her passage to New York, in company with the *Union*, Captain Smith, took a Dutch bottom, homeward bound with sugar, etc., but the prize had to be released. The *Tartar* was afterwards stranded on the coast of Scotland, and only £1,000 worth of her cargo saved. The *Philadelphia*, of 10 guns, owned by the same firm, was "esteemed one of the fastest sailing ships belonging to America."

The *Johnson*, Captain Gawith, on her passage to Virginia, took a French brig privateer, and the *Betty* (Letter of Marque), Captain Rimmer, took a ship bound from

* Probably the *Ladies' Resolution*, which was the name of a privateer fitted out by the ladies of London.

Martinico to Marseilles, which she carried into Barbadoes. A large Dutch ship from St. Domingo was taken and sent into Liverpool by the *General Blakeney*, Captain Loy (a Letter of Marque), bound for Jamaica. Captain Francis Lowndes,* of the *Baltimore*, from Liverpool to Maryland, fell in with three large French ships off the Island of St. Mary, but could not bring them to an engagement. A little later, however, he took the *Resolute*, a French vessel from Curacoa for Amsterdam, "with a pretended bill of sale to the Dutch," having on board the following goods:—

240	Casks of	Sugar =	231,901lbs.	1320 Hides.
209	ditto	Coffee =	35,803 ,,	41 Packs Tobacco.
28	ditto	Indigo =	5,937 ,,	208 Sticks Wood.
26	ditto	Cocoa =	6,229 ,,	

On May 30th, 1757, Captain Salisbury, in the *Ottway*, on his passage from Liverpool to Virginia, took, after a chase of three hours, and without firing a gun, a brigantine from St. Domingo for Bordeaux, laden with sugars, coffee, and indigo, valued at £6000.

In the same month the *Marlborough*, Captain Ward, on her passage from Liverpool to Virginia, met a large French ship of 16 guns, which she fought two hours, when night put an end to hostilities. About five in the morning they fell to work again, and continued a warm engagement till noon, and then parted by mutual consent. The *Marlborough's* sails and rigging were much shattered, and she had one man killed and four or five wounded. "My people," writes Capt. Ward, "behaved well. The French captain called out to us several times to strike, but we answered him with three cheers."

On the morning of the 7th June, 1757, the *Thistle*, Captain George Foster, a small ship of about 150 tons burthen,

* The paper of November 10th, 1794, chronicles the death of "Francis Lowndes, aged 69, formerly master of a vessel, etc., and since many years Clerk to the Pilots' Committee in this town."

belonging to Mr. John M'Cullough, merchant, of Liverpool, and carrying only 2 four-pounders, 12 three-pounders (mostly for sale), 8 swivels, and 20 men, saw a sail edging towards her, which afterwards proved to be the *La Jeune Anna* from Bordeaux to Martinico, burthen 350 tons, laden with wine, provisions, iron and dry goods, mounting 8 nine-pounders, 2 four-pounders, and carrying 49 men. The *Thistle* hauled up for her, and came within gunshot about four that afternoon, when a smart engagement was fought for about an hour, "and then Monsieur took to his heels." The *Thistle* crowded after him all night, and at four in the morning gave him a few broadsides, upon which he struck, having had three men killed and 14 wounded, while the people on board the *Thistle* escaped scatheless. A passenger named Blythe, from Manchester, distinguished himself by his conduct and bravery in the action. The French officers were almost distracted when they stepped on board the *Thistle*, and informed the captain that their adventures cost in France 400,000 livres. Captain Haffey, of the *Polly*, who brought the news of the capture to Liverpool, reported that the Frenchmen were so enraged to find themselves on board so small a vessel as the *Thistle*, that they attempted three times to retake the prize after Captain Foster had sent them off with the boats and provisions sufficient to carry them into Dominica.

In a letter dated Antigua, July 24th, 1757, Captain Thomas Onslow, of the snow *Hesketh*, describes his experiences on the outward passage as follows :—

"On Monday the 13th of June, (being then in lat. 18.20 running for Anguilla, and bound for Jamaica) at break of day we saw a sail off our starboard quarter, finding her to stand towards us, about half-past eight, being then very nigh, perceived her to be a French Privateer, we prepared ourselves in readiness for their reception, and at nine began to engage, which lasted till half-past eleven, when they thought proper to sheer without gunshot of us, but as they continued following, and at a particular distance, I apprehended her consort was not far off. My fears soon after proved too true, another privateer appearing on our larboard bow, and in ten minutes we had them both alongside us, which obliged me to strike. The first was a sloop from Martinico, of six guns, all three-pounders, 10 swivels, and 70 men on board, of whom we killed two, wounded five, broke down his gunnelling on the larboard bow, burst one of their guns, with one of our six-pound shot taking the muzzle off it, and carried away his jibb stay. As she was in such a shattered condition, I am surprised his men escaped so well, they being obliged to keep both pumps working to keep her above water. We received no other damage than that of a few blocks being split, and some rigging cut by their small shot. The other was a sloop from Guadaloupe, of 12 six-pounders, 18 swivels, and 135 men, called the *Invincible,* Joseph Lizard, commander, on board of whom I was ordered, and after leaving 22 Frenchmen in the vessel with my people, they steered for Guadaloupe. At daylight next morning we were close in with Antigua, and at eight o'clock we fell in with the *Duke of Cumberland* privateer brig, of 14 six-pounders, 20 swivels, and 135 men, Joseph Thomas, commander, belonging to this island; the small privateer perceiving what she was, and being ill shattered, they made the best of their way off. Whilst a bloody engagement ensued between the large sloop and English brig, they ordered me down into the hold, where I had not been long before there were company enough, some without legs, and others wanting arms, in all 19 wounded, the number killed unknown to me, and after an hour-and-half engagement, the brig left the sloop, and run up for our snow, received a few shot, the brig only firing one gun loaded with langrell (which killed two, wounded three, and in half-an-hour after, one of the three expired) and immediately struck. My poor fellows were relieved whilst I was carried to Guadaloupe. During my whole confinement on board, and whilst on shore, I was treated much better than any prisoner could expect, and they kept me only five days before I was sent on board a flag of truce, with 17 commanders of vessels, and upwards of 100 sailors.

"They have taken into this place since the commencement of the war, 124 sail of English vessels, 73 of them square rigged, and have 14 privateers out from this island, mounting from 12 to 6 carriage guns. On the 27th ult., I arrived at this island, and on the 8th, the admonition days being expired, my vessel was advertised for sale, and purchased by some gentlemen here, who gave me the command of her, and in three days I loaded her with rum, for Dublin, in order to proceed from thence to Liverpool. I am now under way with a convoy of four of his Majesty's ships, and upwards of 100 merchantmen, having the same ship's company I brought out. Capt. Carruthers of the *Elizabeth and Mary*, belonging to our place, and Capt. Dan. Baines of the *Black Prince*, of Whitehaven, who was taken by the French men of war on the coast of Africa, are coming passengers home with me."

The following letter was written from St. Eustatia on November 30th, 1757, by Captain Richard Venables of the *Cæsar* (Letter of Marque), a frigate of 400 tons burthen, 22 guns (twelve, six, and four-pounders), 70 men, bound from Liverpool to Cork and Jamaica:—

"I am at last got safe here, and find that within these 10 days, the Dutch have brought in 14,000 barrels of beef in their own vessels, which has entirely supplied this market. Our vessel behaved extremely well, and sails fast. We had not the good luck to meet with anything but neutral bottoms till we got within 20 leagues of Antigua; about daylight fell in with a sloop. At half-past seven she began to fire at me; we reserved our fire till we came near, then gave him our bow chase (twelve-pounders) and as many guns as we could bring to bear on him. He fired nine shots only, hauled his wind for about an hour, and then bore down upon us again, but finding our metal heavy, left us. Our ship being deep laden could not come up with him. He mounted 14 guns, and carried 140 men, I understand by a Dutch ship arrived since we came here. The next night we fell in with a second, who only kept us under arms all night, and as soon as daylight appeared hauled off. Whilst I am writing, Mr. Thomas Eaton, mate of Capt. Potter of the *Quester*, is come on board; they were taken last Sunday, about 20 leagues to the eastward of St Bartholomew. He had 87 slaves on board, whom the privateer took out of the *Quester*; and as they could not get the brig to windward, they bore away for this port. The mate, the boatswain, and a boy came with her, and attempted to rise upon the five Frenchmen, but were overcome. The mate is now under our doctor's care, and is likely to do well. The *Cavendish*, with 170 slaves, is carried into Guadaloupe. The *Pickering* and two other Liverpool snows carried into the same island. I can't see how a vessel of small force can well escape, the privateers are so numerous. Capt. Jones, of Liverpool, is taken on the coast of Guinea, and I am informed by gentlemen who have been in the French islands, that on some days 10 or 12 English ships are carried in there."

The owners of the *Cæsar*, Messrs. Gregson and Bridge, also owned the ship *Alexander* (Letter of Marque), 16 guns and 50 men, commanded by Capt. John Ross.

One of the finest privateers belonging to this period was the *Liverpool*, 22 guns (18 of them twelve-pounders) and 200 men, commanded by Captain William Hutchinson, the companion of Fortunatus Wright in some of his cruises. The privateer was fitted out by Mr. Henry Hardwar, and others, including the captain. Mr. Hardwar, who at one time was collector of customs at Liverpool, had the good luck to win, in December, 1758, a prize of £1,000 in the lottery. In 1762 the land about the Everton Beacon was let to him for 2s. 6d. per annum, and he afterwards bought it for a few pounds.

The *Liverpool* sailed from the Mersey on June 10th, 1757, and in going down Formby Channel lost one landsman, who was drowned. "On Saturday, June 18th, 1757," writes Captain Hutchinson in his journal, "in lat. 48·0

18 mins. long., from London, made a sail from the mast-head bearing S. from us, called all hands to quarters, and gave chase with all sails set. At 8, the ship hauled up her courses, and by appearance seemed to prepare for action. At 10, they threw out a French ensign and fired a gun. We answered them only with French colours, but they, not trusting us, began to fire their stern chase pretty briskly, upon which we gave them two of our bow chase. The ship yawed and gave us her larboard broadside. Several of their shot went through our sails, and one of the crossbar shots (a six-pounder) struck the fore topmast and fell upon our deck. We immediately gave her both our broadsides, upon which she struck. Sent our boats on board the prize for the prisoners. On examination she appears to be the *Grand Marquis de Tournay*, Francis Dellmar, commander, from St. Domingo for Bordeaux; is pierced for 24 guns (20 upon the upper deck and 4 upon the lower deck), but has only 12 six-pounders mounted. She came out of St. Domingo with 31 sail, under convoy of six men-of-war, one of 80 guns, four of 74 guns, and a frigate of 36 guns, who saw them through the windward passage and then left them. Found on board the prize, Captain John Mackay, and his crew, of the *Sarah*, brig, bound from Bristol for Boston, whom they had taken on the 3rd ult. The English prisoners report that the brig was retaken on the 15th by two men-of-war, and that the Frenchmen had behaved extremely civil to them."

The cargo of the *Le Grand Marquis de Tournay*, valued at upwards of £20,000, as advertised to be sold by the candle at the Bath Coffee-house, consisted of 494 hogsheads, 13 tierces, and 4 barrels of sugar; 19 butts, 35 hogsheads, 30 tierces, and 83 barrels of coffee; 2 butts, 7 hogsheads, 24 tierces, 31 barrels, and 4 ankers of indigo; 22 whole, and 117 half hides; and 8½ tons of logwood. The vessel, also sold by auction, was described as "a firm, good ship of about 450 tons burthen, pierced for 22 guns, prime sailer and very fit for a privateer or merchantman." Referring to this capture, the Liverpool paper tells us that "all the officers and the whole ship's company gave Captain Hutchinson the best of characters, both as to conduct, courage and humanity. He would not permit the least article to be taken from any of the French prisoners, and to the honour of the whole crew, each man behaved well in his station. Some of the landsmen, who had not been at sea before, could scarcely be kept within bounds, they were so eager to come to action. Several who had entered themselves for seamen, on trial proved to be incapable of their duty, and have been since they came into this port discharged."

In a few days the *Liverpool* sailed on the remainder of her cruise fully manned, and on the 26th July, gave chase to a sail which she came up with a mile from Ushant. The Frenchmen on board, guessing that the *Liverpool* was an English cruiser, escaped from their vessel in the long boat. The prize, which Captain Hutchinson took possession of without firing a gun, for fear of alarming the fort, proved to be the *Sampson*, 200 tons burthen, from Antigua for Bristol, laden with 248 hogsheads, 25 tierces, and 9 barrels of sugar, 20 puncheons of rum, and 33 bags of ginger. She had been taken six days before by a French privateer. "The people arrived here in the *Sampson*," says the *Advertiser*, "give the ship *Liverpool* a very great character, and say that she sails remarkable fast. They fell in with six sail of French men-of-war and wronged them, and had not seen any vessel but they could either leave or speak with. All hands were well and in great spirits."

Having despatched the *Sampson* to Liverpool, Captain Hutchinson went in futile quest of a large French merchantman, of whom a vessel had given him intelligence, and meeting with a 17-gun French privateer, chased her on shore on the coast of France. He also destroyed a fishing schooner, after stripping her and taking off the crew prisoners. Falling in about this time with the *Fame* privateer, of 10 guns and 70 men, belonging to Guernsey, the two privateers made an agreement to cruise in consort, to share what they should capture in proportion to their guns and men, until they arrived at Kinsale, the place of rendezvous. The Guernsey captain being "extremely well acquainted with the French coast," a pretty little scheme was arranged between him and Hutchinson. They cruised close in shore, for the purpose of entering Bordeaux river and cutting out some of the ships in that harbour, the little Guernsey man appearing as a French privateer, with a prize—the *Liverpool*—in company, but as soon as they got into twenty fathom water, they fell in with the ship *Turbot*, and a brig and a snow in company with her, all of which they captured, the Guernsey privateer convoying the three prizes to Kinsale, while the *Liverpool* gave chase to three other vessels in sight. The *Turbot* was described as a ship of about 200 tons burthen, laden with 500 barrels of flour, 400 barrels of wine, 200 barrels pork, 100 barrels beef, 100 ankers brandy, 4,000 gold and silver laced hats, 3,000 pairs shoes, slops, &c.* One of the three prizes, the brig *La Muette*, laden with bale goods, small arms, wines, stores, etc., was entirely lost in St. Bride's Bay, near Milford Haven, where the natives plundered all that was saved of the cargo. Another of the prizes, the brig *Six Brothers*, about 100 tons burthen, arrived safe in Liverpool, and was sold by auction with all her cargo† at the Merchants' Coffee-house,

* When the *Turbot* was advertised to be sold by auction with all her materials and cargo, her burthen was given as about 220 tons, and her cargo as consisting of 110 tuns of red and white wine, 7 cases of sweet wines, 15 tuns, 2 ankers and 20 casks of brandy, 2 casks of loaf sugar, 138 cases of soap (quantity about 50 pound weight each), 4 barrels of prunes (quantity about 684 pounds weight), 28 casks of vinegar, 6 tons of bay salt, 200 barrels and 34 ankers of pork, 104 cases of sweet oil, 37 cheeses, 21 casks of shoes, 2 bales of coarse jackets, 10 bales of coarse cloth containing 30 pieces each, 1 bale containing 9 quilts, 50 cases of drams, 4 barrels of artichokes, 34 barrels of rice, 150 casks of flour.

† The cargo was described as follows: 57 tuns of red and white wine, 250 barrels of flour, 73 casks of pork, 100 cases quantity about 40 lbs. of soap each, 120

The *Liverpool* arrived in the Mersey from her cruise on the 24th of November, 1757. The following is an extract from Captain Hutchinson's journal:—

"On September 11th, left Kinsale; little occurred, only speaking neutral bottoms and English privateers, till Thursday, October 7th, gave chase to a snow; little winds and calm, obliged to ply our oars; spoke a Spaniard, who informed us that the chase was a privateer that had only been 13 days from Dunkirk, and had met with no success. We continued the chase till Wednesday, the 2nd inst., and then saw several sail, particularly two vessels engage; from the inequality of the fire, we judged the larger to be a French ship privateer, and the other a Bristol snow, whom the *Duke of Cornwall* privateer had that day told us of. Night coming on, about three quarters past seven, in lat. 47, long. 12.30, came up with the ship, standing and stemming for her quarter, and hailed him in French by mistake. Without answering he made us feel the weight of his broadside, and carried away our foretop-gallant mast, part of the head of our foremast, fired a shot through the middle of our main-mast, carried away our lower steering sail boom and fore chain plate, three of our lower shrouds and bobstay, and gave us a shot which went through our bends near the water's edge. He ill damaged our sails and running rigging, and wounded 28 of our men. We soon found our mistake, the vessel proving to be his Majesty's ship the *Antelope*, in company with her prize, a French privateer, taken in sight of us. We lay by all night repairing our rigging, &c., and a fleet in the morning appearing in sight, immediately crowded after, and soon found them to be Sir Edward Hawke and Admiral Boscawen's fleet, 14 in

firkins of butter, 100 cases of candles, each about 30 lbs. weight, 200 cases of sweet oyl, 100 Dutch cheeses, and 2 casks of cheese, 1.500 lbs. weight of nails, 2 casks of twine, 10 anchors of lamp oyl, 5 chests containing 100 fuzees, 5 chests containing bayonets for fuzees and other hardware, 2 casks of shoes, 50 bundles of woodhoops, 6 pairs of boots, 2 bales of light canvas, and 1 bale of shirts for negroes. The flour was stored at Mr. Trafford's warehouse in Trafford's Weint; the pork and butter at Mr. Earle's cellar in Strand street, and the soap, candles, and cheese at Mr Earle's warehouse in Hanover Street. The Traffords have long since vanished from Liverpool life, but the Earles still assist in making history at home and abroad.

number. The *Royal William* brought us to, and we kept them company. On Thursday, the 10th, nine more sail of men-of-war joined us, in the whole 23, but several of the ships parted company, owing to thick, hazy weather. We continued with them till Monday, the 21st, being then in lat. 47, longit. 12.30. Were obliged to leave the fleet, consisting of 18 sail, on account of a fever and flux raging amongst our ship's company, owing, it is presumed, to the unlucky accident of wounding our men. We buried six, and had 103 sick when we left the fleet, having not quite finished our cruise."

"On Wednesday," says the *Advertiser*, "were committed to the waves, universally lamented, the remains of Mr. James Holt,* a young volunteer on board the *Liverpool* privateer, whose personal merit and bravery gained him the general respect of the commander, officers and whole crew. He was son to an eminent manufacturer in Rochdale, Lancashire."

The *Liverpool* having been new masted and completely fitted for another cruise, was ready for sea at the end of January, 1758, but we do not learn anything more of her movements until April 30th, when she sailed into the Mole of Leghorn with three prizes—the tartan *St. Lewis*, laden with hemp, sugar, marble, copper, etc., the tartan *Jesus, Mary and Joseph*, laden with corn and linen rags, and the tartan *Joseph, Mary and Joseph*, with timber for the King's yard, all from Marseilles for Toulon. These vessels formed

* The Holts are numerous in the neighbourhood of Rochdale, and claim to be off-shoots of the Holts of Grizlehurst. One of Liverpool's merchant princes, Mr. George Holt, the founder of the firm of George Holt & Co., cotton brokers, India Buildings, Liverpool, was born on Midsummer Day, 1790, at Town Mill, Rochdale. The event took place at six o'clock in the morning, and an old servant remarked—knowing well the characteristics of the family—that he had "just been born in time to begin a day's work." At the age of 17, and with a guinea, the parting gift of his father, in his pocket, he came to Liverpool as an apprentice to Mr. Samuel Hope, a cotton broker. In order to eke out his slender resources during the years of apprenticeship he carried on upon his own account a small business in coarse canvas for mending cotton bags. At the age of 22, when his apprenticeship expired, he was offered a partnership by his employer, and this he accepted. He married Miss Emma Durning, eldest daughter of Mr. Robert Durning, in 1822, and became a "numerous father." His career was marked by unusual versatility and energy, and of such are the makers of great seaports and large cities.

part of a small fleet of coasting craft which Captain Hutchinson had fallen in with, and had not M. de la Clue in his return to Toulon appeared in sight, the *Liverpool* probably would have taken the whole fleet. Captain Hutchinson's plan was to capture a fishing boat, which he sent close in shore to cut the enemy off from the land. He had taken a fourth vessel, but being a heavy sailer he was obliged to let her go as soon as de la Clue's squadron appeared. He had a narrow escape from a French fleet off the coast of Portugal, and was actually reported in Lloyd's List as taken and carried to Toulon. The captured vessel, however, proved to be the *Enterprise*, of Bristol, Captain Lewis. The safety and continued activity of the *Liverpool*, was demonstrated by her sending into Cagliari a French privateer of 24 guns and 200 men, which was said to be "worth 50,000 dollars, exclusive of head and gun money as a privateer." On the 23rd of August, the *Liverpool* arrived in the Mersey, bringing in with her the ship *Roy Gaspard*,* a French privateer of 22 guns, burthen about 350 tons, bound from Messina to Marseilles, which she had taken and carried to Gibraltar. The *Liverpool* had previously sent home two Dutch vessels named the *Sarah and Margaretta* and the *Jong Barbara*, laden with sugar, coffee and indigo, which she had taken on their passage from St. Eustatia.

Captain Hutchinson, being greatly interested in his scheme for supplying the town with live fish, relinquished the command of the *Liverpool* to his first lieutenant and

* The *Roy Gaspard* was sold by auction at the Merchants' Coffee-house, a tavern at the south-west corner of St. Nicholas's Churchyard, with a doorway opening upon the churchyard. It was erected about the middle of the eighteenth century, and was for many years the favourite resort of the commercial community. The large room entering from the churchyard commanded a fine view of the river. Here, during the latter half of the eighteenth century were held the principal auction sales of ships and property. It was the boisterous conduct of the sea captains at this tavern that led to the erection of the Athenæum in Church Street, a haven in which Mr. Roscoe, Dr. Currie, and other men of literary tastes could rest undisturbed by slave captains and privateer commanders.

relation, Captain Ward, and on September 1st, the following advertisement appeared in the newspapers :—

"For a third cruise against the Enemies of Great Britain, the fortunate ship *Liverpool* privateer, under the command of Capt. John Ward, and will be ready for sea as soon as she comes out of the Graving Dock. She carries 22 guns (18 of which are twelve-pounders), and 160 men. All gentlemen Seamen and others who are willing to try their fortunes, may apply to the Commander, or Mr. Henry Hardware, Merchant."

Either the change of commanders, or some mysterious underhand work, raised difficulties, which led to the insertion of the following notice in the *Advertiser* of September 15th :—

"Whereas the seamen who have entered to go the cruise in the ship *Liverpool* privateer, agreed and were warned by the public cryer to go on board the said ship on Monday Evening, and are not yet gone on board ; This is to give Notice that the gentlemen who had subscribed for the outset of the said ship, to send her in quest of the *Marshal Belleisle*, think it now too late ; therefore, all seamen who are inclined to go the six months' cruise, as was at first intended, may apply to Capt. Ward, near the Old Dock Gates."

The editor, commenting on the above in the same issue of the paper, says :—

"On Saturday last, Capt. Wm. Hutchinson, late commander (and part-owner) of the *Liverpool* privateer (notwithstanding he had appointed his lieutenant to the command of the ship, intending to stay at home in order to forward his scheme of supplying this market with live fish), proposed to undertake the command of her once more, and attempt to curb the insolence of Monsieur Thurot, of the *Marshal Belleisle* privateer, cruising in the North Channel, to intercept the trade of this neighbourhood. Upon which the principal Merchants generously opened a subscription, to indemnify the owners of the privateer, and to advance each seaman five guineas in hand, for one month's (31 days) cruise, exclusive of their right to the customary shares of prize money. Notwithstanding 207 seamen had signed the articles, yet as soon as the ship was ready for sea, on Tuesday, only 28 appeared, which obliged the subscribers to drop the cruise, knowing that unless she got out immediately, it would be impossible to execute the proposed expedition in time. We can't avoid remarking that the intended scheme was the most generous one ever offered in these parts, and that probably no seamen ever had before such great encouragement offered them for so short a cruise. Whoever were the obstacles in preventing the scheme being put into execution, will always be deemed enemies to the trade of this port, especially when the public are acquainted that upwards of 700 pounds was generously subscribed to the outset, exclusive of insuring the value of the vessel to the owners, and the gentlemen had undertaken to procure several hundred pounds more from their neighbouring friends."

The *Liverpool* sailed on another cruise on Sunday, October 15th, 1758, and in the following January arrived at Falmouth, having captured some Dutch ships and sent them to Ireland. On March 1st, she arrived in the Mersey, though her cruise was not fully expired, and on April 12th, 1759, she was sold by auction at the Merchants' Coffee-house, having in her capacity as a privateer proved herself worthy of the name she bore.*

In February, 1759, Captain Hutchinson was appointed by the magistrates and common council, principal water bailiff, and dockmaster of Liverpool, a position he held for about forty years. About three months later, a man named Murphy, one of the *New Anson* privateersmen, presented a loaded pistol at Captain Hutchinson, saying, "D——you, you are a villain," an act and sentiment which the captain promptly reciprocated by seizing the man by the collar and

*In the paper of Sept. 28th, she is advertised in a new character:—"For New York, and will be clear to sail in three weeks, the ship *Liverpool*, burthen 250 tons, a remarkable fast sailing vessel, with good accommodations for passengers. For freight, redemptioners, indented servants, or passengers, apply to Messrs. Trafford & Bird, Merchants, or James Chambers, Commander."

wrenching the pistol—which luckily missed fire—from his grasp. Mr. Murphy was secured, tried at Lancaster, and sentenced to serve in the navy for life.

Captain William Hutchinson was a remarkable man. His work on seamanship and naval architecture, and the variety of pursuits in which he was engaged during a long and busy life, his charities and his hobbies, all go to prove that he was of a higher type than the generality of men in his calling at that period. Judging by Sir Horace Mann's description of certain English admirals and sea captains with whom he had dealings in his official capacity as English Resident at Florence, the two Liverpool privateer commanders, Fortunatus Wright and William Hutchinson, compare very favourably in education, intelligence, professional skill and daring, with many officers of rank in the King's navy. Mr. Bryan Blundell, who was well acquainted with Hutchinson, said "that his whole life was one unwearied scene of industrious usefulness," and this is confirmed by the closing words of the preface to the "Practical Seaman,"* where the author says, "as my best endeavours have hitherto been exerted for the public good, without any other motive, so will they be continued by the public's humble servant, William Hutchinson." In the same preface he refers to the unexpected difficulties he found "in being a new writer, venturing to lead the way on so important and extensive a subject in this learned, criticising age;" but, he says, "for my imperfections as a scholar, I hope the critics will make allowance for my having been early in life at sea as cook of a small collier; and having since then gone through all the most active enterprising employments I could meet with as a seaman, who has done his best, and

"Principal Dock Master of Liverpool, Captain William Hutchinson at No. 1. on the north side of the Old Dock Gates. One whose great knowledge and ingenuity has proved of infinite service to this port, and to whom the British mariner stands indebted for a learned and curious Treatise on Practical Seamanship, &c."
—Prestwich's MS. History of Liverpool, p. 239.

who, as an author, would be glad of any remarks candidly pointing out how to improve his defects, if there should be a demand for another edition." As a native of Newcastle-on-Tyne, he was naturally proud of the seamanship of those among whom his early life had been passed, and mentions that "the best lessons for tacking and working to windward, in a little room, are in the colliers bound to London, where many great ships are constantly employed, and where wages are paid by the voyage, so that interest makes them dexterous and industrious to manage their ships with few men in a complete manner in narrow channels, more so than, perhaps, in any other trade by sea in the world." He tells us that the sight of a fleet of 200 or 300 colliers sailing out of the harbour of Newcastle for London in one tide, and their dexterous navigation in passing and crossing each other in so little room, made a "travelling French gentleman of rank to hold up his hands and exclaim, that it was there France was conquered." While expressing his belief, based on long experience in different trades, that the seamen engaged in the coal and coasting trade to London, "are the most perfect in working their ships in narrow, intricate and difficult channels, and in tide ways," he admits that "those in the East India trade are so on the open seas." He believed that the custom of heaving the hand-lead and singing out the soundings, "which is peculiar to our seamen," originated in the coasting trade to London, where their success and safety depended greatly upon it, and quotes a saying attributed to Dr. Halley, that the system of navigation in his time depended upon three L's, meaning, Lead, Latitude and Look-out. There was some kind of Ship Club in Liverpool in his time, for he says:—

"A late great mathematician at Liverpool, Mr. Richard Holden, who found Theory from the Attractive Powers of Nature to agree with my observations on the tides, and made a most excellent tide table from them, used often to say, at what we called a Ship Club, that there was no hidden or unknown principle concerned in the art of building, sailing, working, and managing of ships, but the laws of motion, the pressure of fluids, and the properties of the lever, which are all well known to British Philosophers, and that nothing was more deserving their attention and pursuit, in order to bring these arts to their utmost perfection."

It is much to be regretted that Captain Hutchinson did not write a regular narrative of his own adventurous and useful life. It is only by stray paragraphs scattered through his voluminous printed work to illustrate various points in the argument, that we are able to form an imperfect sketch of his career, as a supplement to what has gone before. Having risen from the position of cook's cabin boy and beer drawer for the men in a small collier, to the dignity of a forecastle man, he made his first voyage to Madras and China in 1738, "when our East India ships had open waists." "Not having water to go over the Flatts in turning to windward down the Swin, the common track for our deep-loaded colliers, our vessel," he says, "shipped and leaked so much water, that it took all the pumps to keep her free, so that when we got into the Downs, the crew protested against going the voyage, without her being lightened, but a 50-gun ship of war being near, a signal was made, and they came and took the principal ringleaders out, and we proceeded on the voyage." There was no Mr. Plimsoll to fight for poor Jack's rights in those days, and, indeed, had he miraculously turned up, they would have bundled him on board a tender and made an excellent man-of-war's man of him, as no doubt they did of every "collier" they impressed. On this voyage to the East Indies he was three months terribly ill of the scurvy, and found himself benefited by the use of tea, a habit confirmed by what he saw of the Chinese style of living at Canton, and for the rest of his life he, like Dr. Johnson, became a confirmed lover of tea. The doctor is said to have taken forty cups a day, hence, probably, his morbid fear of death; but Hutchinson only took tea twice a day, and his method of making it on board ship, where there were no tea utensils, was by putting the leaves into a quart bottle filled with fresh water, corked up, and boiled in the ship's kettle along with the salt beef. This mode of brewing was a great success, especially in stormy weather, when teapots, cups and saucers and such like could not have kept their sea-legs.

He acted as mate of a bomb's tender in Hieres Bay with the fleet under Matthews and Lestock, and shortly after sailed in "a fine frigate-built ship for the Leghorn trade, that carried 20 six-pounders on her main-deck, and went a-cruising in the Mediterranean." It was probably at this time he made the acquaintance of Fortunatus Wright. The following incident may have occurred on board Wright's ship, when Hutchinson, as Professor Laughton suggests, was officer of the watch, or he may have been in independent command. He was at any rate cruising in the Mediterranean during the war of 1747, with the prisoners of three French prizes on board, at their entire liberty on deck. He had just sent nearly all his own men aloft, to execute an order, when he providentially noticed one of the French captains about to give the alarm for the Frenchmen to rise and take the ship. Hutchinson immediately ran up to the Frenchman, pistol in hand, "and told him coolly that he should be the first that should die by the attempt, which stopped his proceeding." This affair taught him two lessons—that prisoners for the future should be sent up aloft to assist in the work; and that ceremonious professions are not to be depended on, for the French captain in question, when first brought on board, was the pink of politeness. He "made many apologies and begged that he might not be ill-treated for the resistance he made in defending his ship, and was answered that he should be treated rather better than worse for doing his duty like a brave and honest man."

In 1750, he was concerned with "that worthy hero, Captain Fortunatus Wright," in purchasing and fitting out the *Leostoff*, 20 gun frigate of war, with lighter guns and materials than she formerly carried, and loaded her with a general cargo for the West Indies. During one of his voyages, he slept on the bare ground in the Bay of Honduras, fell very ill of the flux, and was suddenly cured in a more surprising and original manner than if he had taken a modern patent universal healer. He was then acting as commodore of a fleet of ships in the Bay, and being told that some strange vessels were entering without first sending in their boats, as usual, to make known who they were, he gave orders to fire at them. As the ships still came on, heedless of the warning, he got alarmed, was roused to action, and immediately recovered.

Captain Hutchinson, in conjunction with his partner, Mr. Ward, made a plucky endeavour to perform for Liverpool, in 1757, the service which Frank Buckland rendered London in the nineteenth century. In the paper of June 10th, of that year, we read that Messrs. Hutchinson and Ward had fixed a large store-well-vessel in the river, near the Woodside-house, in which they fed their fish as the codsmacks brought them in, and for the conveniency of the Cheshire markets they sold fish on board. The enterprise was not successful, and probably swallowed up some of the money made by the Captain in privateering and other "active enterprising employments," as well as a subsidy granted by the corporation in aid of the scheme. The curious reason given in the paper of February 15th, 1760, for disposing of the *Resolution*, a codsmack employed in the industry, is "the prejudice that prevails here against fish brought in smacks, though the best in kind, by which its consumption is so hindered that the proprietors cannot with prudence support the vessel longer, though they have had handsome allowance from the Corporation for the support thereof."

His zeal for the interests of Liverpool was conspicuous on many occasions, and could not have been greater, had he been born within sound of St. Nicholas' bells, and entitled to call himself "a genuine Dicky Sam." When Thurot and his squadron entered the Irish Sea, and threw Liverpool into a state of wild excitement, Captain Hutchinson's daring spirit was shown. With one associate only, he raised a number of volunteers to man a few armed vessels then in the river, with which he determined to attack and conquer the enemy, or perish in the attempt. The news of the gallant Frenchman's defeat and death arrived, and rendered the enterprise unnecessary.

He was the inventor of reflecting mirrors for lighthouses, and, in 1763, he erected at Bidston the first mirror of that kind ever used, consisting of small reflectors of tinned plates, soldered together; and he also made larger ones, as far as 12 feet diameter, formed of wood and lined with numerous plates of looking-glass. A ridge of rock and gravel, lying between the Rock Perch and the south point of the Brazile sandbank, was named after him, because he removed some obstructions which had been placed there by "designing villains," and opened a passage by cutting away the rock and deepening the channel.

From the 1st of January, 1768, to the 18th of August, 1793, Capt. Hutchinson continued a series of observations on the tides, barometer, the weather, and the winds, the MSS. of which he presented to the Liverpool Library. From these were obtained the data by which the Holdens, father and son, calculated the tide-tables.

On a blank leaf, at the commencement, the following memoranda are written in his own hand :—

"These five years' observations from 1768 to 1773, upon the tide were made from solar time, and the winds from the true meridian, and their velocity judged according to Mr. Smeaton's rule, our great storms going at the rate of sixty miles an hour. The thermometer, kept indoors, at the head of a staircase, four stories high, by Wm. Hutchinson, at the Old Dock Gates, Liverpool. The first sheets were cut out to give Mr. Richard Holden, and aided him to make out the 3000 observations mentioned in his preface of his Tide Table, by which he founded a theory, from natural causes, to agree therewith."

It is to be regretted that Mr. Hutchinson did not suffer copies to be taken rather than break such an uniform series of observations, made with a punctuality and accuracy that do infinite credit to his perseverance and talent. They form most invaluable documents for reference and comparison.

His meteorological tables were kept in the following manner :—

MORNING.

1768 January 1 Friday	M's age 11	Moon's distance in miles 284,384	Moon's declin. N. 25 29	Moon's South E. 9 59	Time of High Water H. M. 8 45	Height Ft. in. 14 3	Winds, their velocity in miles 1-60 S.E. 35	Weather. Hazy and a hard frost

EVENING.

Time of High Water H. M. 9 10	Height Ft. in. 14 2	Winds and velocity E. 30	Weather. Cloudy and a keen frost.	Tide's daily difference. One tide M. 55	Barom. 29.2	Ther. 31

In May, 1775, he added to these a rain gauge.

On the 4th of June, 1777, it was resolved: "That the Corporation do make a compliment of Ten Guineas to Captain William Hutchinson for his late ingenious publication of a book entitled the 'Practical Seaman,' being deemed a book of great utility to commercial places." This work elicited the following tribute from a competent authority:—

> "Sir Thomas Frankland presents his compliments to Mr. Hutchinson, and hopes he will approve their Institution. He makes their Superintendent read over, with the eldest of the boys, his Treatise on Seamanship*; which he thinks seems as if written for the instruction of their Maritime School at Chelsea. November 30th, 1781. N.B. He wishes the officers of the Navy would study it also."

Scattered through Captain Hutchinson's work is a vast amount of matter which enables us to realize the difficulties of the old navigators, who, previous to its publication "were left entirely to learn their duty by their own and other people's misfortunes." The captain was the pioneer, not only of scientific seamanship, but of scientific shipbuilding, for the marine architects, as well as the mariners of those days, were either too conservative to adopt new methods, or attempted impossibilities in defiance of the laws of Nature. The annals of the eighteenth century teem with terrible catastrophes arising from the crass ignorance of shipbuilders. Vessels, with hundreds of people on board, suddenly capsized before a puff of wind, simply owing to a radical defect in the

* In 1791, a new and enlarged edition of the work was issued with the following formidable sub-title:—

"A Treatise founded upon Philosophical and Rational Principles, towards establishing fixed rules for the best form and proportional dimensions in length, breadth, and depth of Merchant's Ships in general; and also the Management of them to the greatest Advantage, by Practical Seamanship; with important hints and remarks relating thereto; from long approved experience. By William Hutchinson, Mariner, and Dock Master at Liverpool. Liverpool: Printed by Thomas Billinge, Castle Street, 1791."

"This book is most humbly dedicated to His Royal Highness William Henry Duke of Clarence, President of that most patriotic Society, instituted at London, for the improvement of Naval Architecture, by His Royal Highness's Most Humble Servant, William Hutchinson."

principle of construction. Many Liverpool ships were lost in this manner, the *Pelican* privateer, which overset opposite Seacombe, with the loss of about 70 lives, being comparatively a minor catastrophe. Captain Hutchinson, speaking of ships being built too high, gives the following instance:—"We had a late fatal loss of a large new frigate on her first voyage, which had overset with upwards of five-hundred slaves, and her crew all drowned except two seamen and three slaves; which added to the many other such instances, proves the necessity to endeavour to get such general rules fixed to prevent as much as possible such dreadful losses." Two or three Liverpool vessels were built on lines suggested by Captain Hutchinson, notably the *Hall* and the *Elizabeth* for the Jamaica trade, both of which proved veritable "greyhounds of the Atlantic."

He tells an anecdote of Mr. Bryan Blundell, the noble-hearted founder of the Blue Coat Hospital:—

> "Being appointed from our Pilots' Committee with Mr. Bryan Blundell, Merchant, who had been a great and successful shallop-racer in the West Indies, to go with two of our pilot sloops and pilots to survey our neighbouring ports, to fix rules to examine our pilots by, the sloop we happened to be in was the worst sailer of the two. Mr. Blundell said he would make it sail better than the other without meddling with the mast, sails, or rigging, or trimming her more by the head or stern; which he did by getting the Pilots to move the heaviest loose materials from fore and aft into the main body amidships, which answered the designed purpose, and made her beat the other sloop as much as they beat us before."

He was never happier than when making experiments and observations, afloat or ashore. At one time he is an eye witness of some curious experiments made in a close room by the "ingenious Mr. Smeaton," for the purpose of discovering the fixed standard of velocity for windmill sails, prior to the framing of his table of winds; at another time, he stands in a boat's stern sheets, going to attack a ship, and takes particular notice that cannon shot will rebound and rise about a man's height out of the water. Having had a narrow escape, he observes: "I saw the shot first graze the water right ahead of us, and then rise and go directly over our heads, and make ducks and drakes right astern of us." So keen was this scientific instinct and habit of observation in him, that even in action he made mental notes of such facts as this: "I have seen a bombshell turn round in the air, by the centre of gravity being near the middle," etc. Most philosophers in similar circumstances would doubtless feel more solicitous regarding their own centre of gravity. He experimented with a model of a ship in a cistern of water to test the statement of his friend Mr. Henry Bird, "a great shipbuilder at the Greenland Dock, London," that 33 degrees, or three points of the compass was the best angle "for sailing vessels' rudders to be fixed to traverse to"; and not content with the cistern he "having the management of our three long graving docks at Liverpool, where we have in common ten or twelve ships at a time repairing and cleaning," with a bevel tried the traverse of many ships' rudders, and found that Mr. Bird's rule was right. It appears that the Parkgate method of hanging the rudders was heterodox, and caused the loss of ships. He gives a curious account of the elaborate experiments he made with models of ships, to find out their centre of gravity and motion. The spectacle of the "old sea dog," who had peppered and been peppered by "the enemies of Great Britain" (and even by Great Britain herself, as in the unfortunate affair of the *Antelope* man-of-war), being thus engaged with his miniature ships, reminds us of Uncle Toby and Corporal Trim conducting imaginary sieges and campaigns in the kitchen garden, with this difference, that Captain Hutchinson's hobbies had the merit of being useful.

In 1789, he founded the Liverpool Marine Society, for the benefit of masters of vessels, their widows, and children, the first president being Mr. Thomas Staniforth. Mr. Hutchinson subscribed one hundred guineas to it, and other benevolent institutions of the town were liberally supported by him. This is the more remarkable, because his stipend was never more than one hundred guineas per annum, although his duties at one time comprised both those now exercised by the harbour master and those of a dock master.

It is said that he was accustomed to observe a particular day, in each year, as one of strict devotion, in commemoration of his providential deliverance at one period of his life, when, after the loss of the vessel in which he sailed, he, and others of the crew being without food, had drawn lots to ascertain which of them should be put to death, in order to furnish a horrible and revolting meal to the survivors. The lot fell upon Mr. Hutchinson, but he and his fellow sufferers were saved by another vessel which hove in sight.

Captain Hutchinson died at a ripe old age, in February, 1801, and was interred in St. Thomas' Churchyard, close to the Old Dock and the office in which a great portion of his life was passed. Upon the site of that dock now stands a vast and gloomy pile of buildings, in a wing of which the rulers of our modern docks meet and deliberate; but to most of them the name of Captain Hutchinson is scarcely known, and his deeds and personality are to them vague and shadowy as those of the heroes of the Iliad.

He was evidently a kindly though firm commander. "I once," he says, "had the pleasure of taking up one of my seamen from under water, and to all appearance drowned, but by our exertions recovered him, and the first words he was able to speak (perceiving me busy about him) were, 'my dear Captain, pray for me.' To which I replied, that as he was now in a fair way of recovery, I hoped he would be able to pray for himself, and be thankful to Providence for his narrow escape."

In February, 1798, when a voluntary subscription, which produced over £17,000, was entered into in Liverpool, to assist the government to meet the enormous expenses of the war, Captain Hutchinson pledged himself to contribute £20 per annum as long as the war lasted.

Captain Hutchinson was a religious man, and held that, "since many a fruitful and flourishing land has been made barren for the wickedness of its inhabitants, every impious and profane man ought to be treated as the greatest enemy to his country." He quotes Archbishop Tillotson's saying that "no man can plead that he was born with a swearing constitution," and recommends all commanders of ships to have a reasonable part of Divine worship publicly performed on board every day, "which," he observes, "to our shame, be it spoken, is often, even in our large East India ships, scandalously neglected. This, I can say from profitable experience, contributes greatly to produce good order, harmony, and piety on board, and check disorder, vice, and immorality of every kind, even amongst the most dissolute and ignorant in privateers, as well as merchants' ships." We are not accustomed to associate privateering with Divine worship, but here we have a privateer commander as devout in his way as John Newton, the slave captain. For the first fifteen years of his sea life in different trades, he never saw any religious duty publicly performed on board, except that in an East India ship for two or three Sundays, when they drew near the Cape of Good Hope, they had prayers, which ceased when danger passed away. He blames the East India Company for "shamefully rating their large ships only at 499 tons, in order to evade the expense of a clergyman, and the penalty of the law for not carrying one."

It would be impossible to form a true estimate of the character of this fine old privateer captain without reading the concluding words of his "Practical Seamanship." Referring to the "grand atoning sacrifice," he characteristically observes :—

> "And how devoutly should we implore the promised assistance of his aiding and sanctifying grace to conduct us safe through this transitory voyage of life to a blessed and happy eternity. Let us then, under the direction and guidance of this great Author and Captain of our Salvation, our all glorious Redeemer, Christ Jesus, pursue our course with steadiness and resolution, and fight manfully under his banner; looking up to him for succour in all our distresses and difficulties, who is powerful in heaven and earth, and will never forsake or reject those who sincerely love and trust in him. To whom be glory for ever. Amen."

This was the spirit that animated Cromwell and his "Ironsides," and no wonder the Liverpool privateersmen fought so well, when men like William Hutchinson commanded them.

The *Windsor,* Captain Joseph Clarke (a Letter of Marque), about 300 tons burthen, of 12 six-pounders and 40 men, belonging to Messrs. Edward Trafford & Sons,* was taken on her passage from Liverpool to Philadelphia by a French privateer of 18 guns, after engaging some time, and carried to Bayonne. One of the crew of the *Windsor*, writing home from Bayonne prison, which was "very sickly," tells how Captain Clarke, Captain Grubb, and a Mr. Berry of Liverpool, made their escape from a French country town, where they were at large on their "parole of honour." They were soon retaken on their way to St. Sebastian and re-secured.

* Mr. Henry Trafford, who died in 1740, during his mayoralty, had expressed a wish that his body should lie in state, and that an oval glass plate should be inserted in the lid of his coffin, so that the spectators who knew him might take "a last, lingering look." The wish was carried out to the letter, and even children were held up to see the show. The Trafford's were a notable family in Liverpool during the eighteenth century, and became connected with the Leighs, of Oughtrington. In 1761, Mr. Edward Trafford (who had been mayor in 1742) and his sons, Mr. Richard Trafford (bailiff in 1755), and Mr. Wm. Trafford, all lived in King Street, and Trafford's Weint in that locality still commemorates the family.

The three prisoners lay all in one bed. Captain Clarke in the dead of night, observing the guards to be asleep, made his escape through the window, got clear off, and ultimately arrived safe in Spain. Captain Grubb and Mr. Berry, when they awoke and found their friend missing, attempted to follow his example, but were seized when in the window, and sent to Bayonne Castle. Captain Clarke at one time commanded the *Trafford,* 200 tons burthen, 10 guns (six and four-pounders), belonging to the same owners.

On the 25th of February, 1758, about 25 leagues S.S.E. from Cape Tiburon, the *Adventure*, Captain George Washington, a ship belonging to Mr. Joseph Manesty, merchant—the friend and employer of John Newton—was attacked by the French brigantine *St. Louis*, of 10 six-pounders, 18 swivels, and 120 men, which she fought five hours, most of the time yard-arm and yard-arm. Captain Washington, writing from Kingston, Jamaica, gives the following details:—

> "During the engagement I had one man killed (got at Cork) and one wounded. The brig had two killed and 19 wounded. We received four shot between wind and water, several in the upper works; gaft shot away, mainyard, fore-topmast, and top-gallantmast disabled, two guns dismounted, topmast stays, shrouds, backstays, futtock shrouds, shot away, and not a lift or brace standing, but one strand of the main-topsail brace. We had scarce any running rigging but what was shot away, sails in such a shattered condition that they will not be fit to bend any more. Our powder being all expended, to my great mortification, we were obliged to haul down the colours. They saw our powder chests out of their tops, or they would have boarded us. We must inevitably have been most part of us killed had it not been for Matrosses and Kendal cottons we got out of the hold, and put upon the inside of the filling up plank in the waist, for they had 60 men at small arms. They stripped us of our clothes and instruments, and carried us into Port St. Louis. On the 23rd of March our ship's company arrived here, and on the 5th inst.

we buried Samuel Chatterton, apprentice. Capt. Boats was pleased to make me an offer of the vessel this comes by, but I chose rather to keep your servants together, and to go to Savannah La Mar, and take in your interest there. Large vessels sell high, as it begins to be late in the year, and willing to get in your debts and sail first convoy. I have with Mr. Richd. Watt bought a brigantine of about 80 tons, which we have called the *Mary*. She is a good vessel, well found, and hope may get money. I assure you, Sir, your outstanding debts gives me no small concern, but hope to be more careful for the future, as I see the many evils attending it. You may depend that I'll do my utmost endeavours to bring matters to a conclusion. I have ten tons of logwood from Mr. Roper, and will sail for Savannah La Mar in two days with a vessel of force. Your servants are all with me. We shall certainly sail first convoy, which will be about the 10th of June, and am, with gratitude for all favours conferred on me, &c."

This letter throws some light on the affairs of Mr. Manesty, by whose subsequent failure the Rev. John Newton lost all his savings, which he had entrusted to the keeping of his generous benefactor and former employer. Captain Boats was the celebrated merchant, "Billy Boats," or Boates, of whom, and Mr. Richard Watt, we shall have occasion to speak later on.

On the 15th of April, 1758, in latitude 46.20 N., longitude 12, west from London, the ship *Pemberton*, Captain Walter Kirkpatrick, having outsailed her consorts, had the misfortune to fall in with the *Machault* privateer, of Bayonne, 26 guns, and 320 men, which she mistook for a homeward merchantman. On discovering the Frenchman's force, the *Pemberton* made sail, and kept up a running fight with her stern chase guns two hours and-a-half. In a letter to his owners, written from a French prison, Captain Kirkpatrick thus describes what ensued :—

"She soon gained on us, and when within pistol shot, we fired broadside for broadside an hour and an half, and had it not been for the continual fire from her small arms, whose balls were like showers of hail and obliged my men to run from their quarters, perhaps we might have got clear, notwithstanding her superior force. Thus overpowered, we were obliged to strike. Our rigging, masts, yards, and sails were very ill shattered, though our people were tolerably well sheltered. Four of our people were wounded; George Godsall (since dead). Mr. Woolley Maisterson had his leg shot away by a 12 pound ball, which dismounted the gun he was quartered at, went through the dog's body, and split in two on the capson. He is now in a fair way of recovery; Edward Langshaw was ill hurt, but since recovered. All the rest in good health. The Captain and officers behaved very well to us, the former complimented me with my hanger, saying I deserved one for fighting so long, and ordered me all my clothes, watch, books, and instruments, of which I got part, the remainder being plundered during my being on board the privateer, which I think is as near the model of the *Liverpool* man of war (now in Liverpool) as possible. This day we are ordered all into close confinement, and those who can find bail for £150 are allowed to go on parole about ten miles into the country. Your letters of credit will be extremely acceptable, &c."

On May 29th, 1758, the *Ellen* (Letter of Marque), Captain Kirby, 14 carriage guns (8 four-pounders, 4 six-pounders, and 2 two-pounders), in latitude 48°, 150 leagues W. by S. from Cape Clear, met with a large French ship, mounting 18 guns (six-pounders) and full of men, "whom he engaged very warmly for near three hours, till dark. Captain Kirby received three shots in his hull, which went through him, five through his mainsail, six through his fore top sail, and one in the head of his foremast. In the morning, Captain Kirby gave him a broadside; Monsieur returned the compliment and took to his heels, which surprised them, as she appeared to be full of soldiers, and of so much superior force." In June of the same year, Messrs. Joseph and

Jonathan Brooks, the owners of the *Ellen*, while searching the snow *Prince William*, a Dutch bottom, captured by that vessel, found concealed in a barrel of coffee, a large packet of French letters, several of which were advices to merchants in France, particularising great part of her cargo to be French property, and shipped under cover. Other letters in this important find mentioned large quantities of goods shipped in different Dutch bottoms, etc. In the summer of 1759, on her passage from Liverpool to Jamaica, the *Ellen* fell in with three French privateers, whom she engaged several hours, but a fourth privateer joining the others, they all boarded the *Ellen*, took her, and carried her into Martinico. The gallant Captain Kirby and seven of his men were wounded in the action.

Captain Spears, of the *Granville*, who arrived in Liverpool from Edenton, North Carolina, with a cargo of tar and tobacco, had the ill-fortune to meet with the *Jupiter* privateer, of Bayonne, 22 guns, and 250 men, Captain Jean Maubeaule, and agreed to pay the said captain £500, as ransom money for his ship and cargo on his arrival at Liverpool. The French captain, a Frenchman of the old school, treated Captain Spears very politely; offered him bread, water, and anything his ship afforded, but begged to carry off Mr. Alexander Scott, the chief mate, as "ransomer," or security for the due payment of the ransom money. The *Jupiter* also captured the *Knutsford*, Captain Sefton, from Liverpool for St. Kitts, and ransomed her for fifteen hundred guineas.

The practice of ransoming vessels for large sums of money continued during the whole of this war and during part of the American war, but it was then declared illegal. The late Sir John Tobin, when a boy, on his first voyage narrowly escaped being carried off as a "ransomer," along with the mate and one of the able seamen of the ship. Fortunately for him, the captain of the privateer, who was an Irish

Frenchman, Captain Kelly by name, had known his father at Douglas, and on finding whose son he was, sent him away rejoicing.

But a ransom, however desirable, could not always be arranged, as will be seen from the following letter, written by Captain Josiah Wilson, of the *Aurora*, to his owners in Liverpool, from St. Andero in Spain, and dated June 12th, 1758:—

> "We were unfortunately taken by the *Jupiter* Privateer, belonging to Bayonne, on the 24th of May last, in lat. 45 30 N. and 33 40 W. Long. from London, which is farther to the westward than anyone could imagine an European Privateer would cruize. She has taken six prizes this cruize, exclusive of our vessel, three of which belong to Glasgow. One of the prizes she took was got as far to the westward as 40 degrees from London. We fell in with her in the night, but never saw her till the morning, when she was about a league to windward of us, and steering the same course we did. I took her for some English merchantman, as her guns were all housed. We were well prepared for an engagement, but as soon as she came alongside, they ran out their guns, and fired into us; two of their shot struck us but killed none of our men. There was no contending with a ship of her force, for she mounted 22 guns, 12 nine-pounders, and 10 six-pounders, with 280 men, and frigate built. I could not ransom her upon any account, for as the ship's cruize was just out, they determined to return and convoy their prize to Bayonne. We were busking in the Bay of Biscay ten days, where I was in great hopes of being retaken by some English cruizer, but am now out of all hopes, the prize as well as the privateer being both at anchor in this port, which is about 35 leagues from Bayonne."

In November, 1758, Captain Wm. Part, formerly a commander in the Virginia trade, died at Prescot, and his remains were brought to Liverpool for interment. His funeral was attended by ten of the oldest seamen's widows, to whom he left each thirty shillings for a gown, handkerchief and hood ; and by the Blue Coat Hospital boys, to whom he had been a great benefactor. He left £50 to the Infirmary ; £200 to a school which he built at Hale, and loaves of bread to be given to the poor of that township every Sunday. He had, it seems, intended £600 more for the school, but withdrew it on account of some disagreement with the lord of the manor. To a number of his poor relations he left legacies, but would not allow them to attend his funeral. He also built four almshouses in Prescot, in which parish this curious old "sea dog," who performed other good deeds, had resided some time—probably to be inaccessible to his poor relations !*

During the year 1758, the French had decidedly the best of the privateering. In March, no ships of any sort sailed from Liverpool, or arrived in the port, for some weeks, owing to the boldness of the Frenchmen "which laid an effectual embargo on the coast." Yet the paper of December 1st, was able to speak in the following comfortable strain:—"It has been remarked by those who have access to know the truth, that England never carried on a greater trade, not only in any time of war, but even in any time of peace than at this period, and this chiefly at the expense of our enemies' commerce; so that the nation is thereby a double gainer. And never in the memory perhaps of any now alive were Great Britain's power and reputation abroad higher than at present."

The reduction of Cape Breton was a fatal blow to the French trade, and most beneficial to the British, for the rates of insurance to America, etc., fell from 25 and even 30 per cent. to no more than 12, while the enemies' rates rose in proportion to the falling of ours.

* Baines, in his "History of Lancashire," states that over the porch of the Grammar School at Childwall, founded and endowed by William Part, and affording instruction to about twelve boys, is this inscription :—" M. S. Hoc Ædificium Gulielmus Part a longa Majorum hujus Pagi Indigenarum oriundus suo solius Impendio extruxit Censuque Donavit Anno S. H. MDCCXXXIX."

On the 13th of February, 1759, forty-five merchants and shipowners of Liverpool addressed a letter to Mr. Robert Williamson, printer of the *Liverpool Advertiser*, requesting him to suppress the list of vessels sailing from the port, as they had "too much reason to apprehend" that it had "been of very bad consequence this war." The signatures attached to the document are those of the principal ship-owners of Liverpool at that stirring period :—

Matthew Stronge	Richard Savage	Robert Cunliffe
Robert Cheshire	John Bridge	John Tarleton
John White	George Drinkwater	James Gildart
R. Armitage	William Williamson	John Backhouse
George Campbell	Robert Hesketh	John Welch
James Clemens	John Maine	Arthur and Benjamin Heywood
John Stronge	John Ashton	Halliday & Dunbar
William Gregson	Thomas Mears	John Gorell
James Brown	Henry Hardwar	George Campbell & Sons
John Parr	John Hughes	Ralph Earle
Thomas Rumbold	Edward Parr	John Crosbie
John Stanton	William Crosby	Scroop Colquitt
John Hammer	John Ansdell	Charles Goore
William Fleetwood	Samuel Woodward	William Earle
William Trafford	William Reid	James Clegg

On the 22nd of February, 1759, the *Catherine* (Letter of Marque), 12 guns and 35 men, Captain Seth Houghton, on her passage from Liverpool to Montserrat, fell in with a French privateer of 16 guns and 145 men, with whom, after exchanging a few shot, about seven in the morning, they came to a general and close engagement, which for the most part was within pistol shot, till four in the afternoon, when the *Catherine*, overpowered by numbers, was obliged to strike. During the action, the privateer sheered off twice, having seven of her men killed and seven wounded, and mounted four more guns, which she had been obliged to dismount a few days before when chased by an English man-of-war. The crew of the *Catherine* had the mis-fortune of killing one of their own men, and hurting the third mate and two others, when firing their first broadside. The Captain was wounded by a musket ball, and narrowly escaped a four-pound shot, which carried away part of his waistcoat. After they had struck, the French captain complimented Captain Houghton on his gallant behaviour, and would not allow him to be plundered, like the rest of the prisoners. While they were exchanging prisoners, the second mate of the *Catherine,* and a boy from Cheshire were drowned, through too many privateersmen jumping into the boat in their eagerness for plunder. The shrouds, masts and hull of the *Catherine* suffered severely in the action. The privateer had taken 18 prizes.

The *Upton* (Letter of Marque), Captain Birch, arrived at Gambia, from Liverpool, on May 9th, 1759, with a prize, taken off the Canary Islands, the cargo of which was valued at £5,000.

The *Prince Frederick*, Captain Frierson, on her passage from Liverpool to Guadaloupe, had a smart engagement of three hours with a privateer of 10 guns, whom they obliged to sheer off.

Captain William Lethwayte, of the *Wheel of Fortune*, from Liverpool for Tortola, writing from Antigua, says:—

"In the evening of the 24th of May, 1759, in lat. 17°, and about 25 leagues to the eastward of this island, we fell in with a French privateer, and at half past six o'clock next morning she attacked us, and continued till eight, when she sheered off to stop her leaks and repair other damages she had received in the action, it being very smart and within half pistol shot. At eleven she renewed the attack as brisk as ever, till twelve, at which time, being little wind, she got her oars and rowed from us, a second time to stop her leaks, &c. This being done, she hauled to the northward out of gun shot, but kept hovering in sight all afternoon and night. Next morning, being the 26th, we saw another sloop of twelve carriage guns, 22 swivels, and 120 men, who spoke with the one we had engaged, then astern of us. Immediately they bore down and both began to engage us very warmly, but to our greater surprise and mortification, before we had fired above 30 shot, we perceived two other vessels bearing down, which proved to be French privateers; one a sloop of ten carriage guns, 16 swivels, and 90 men, and the other a schooner of the same force. At the same time we saw a fifth privateer stretching for us from the S.E. so that we thought it prudent to strike (though against all our inclinations if could possibly be avoided) rather than to risk our lives and not the least probability of getting clear. The *Eward*, Capt. Kevish, the *Swan*, Capt. Slazer, of Dumfries, both last from Liverpool, and the *Cork Packet*, Capt. Champion, of Cork, were in company when taken, and being defenceless, shared the same fate. We were all carried into Martinico, where I was taken ill with the flux, but am now perfectly recovered, and expect to sail in the first Antigua fleet."

In July, 1759, the *Vengeance* man-of-war, of 26 guns, formerly the celebrated French privateer of that name, arrived in the estuary of the Mersey, and about a week later, the *Golden Lyon*, Captain Thompson, returned from the Greenland fishery. The whaler, in stretching in with the buoys laid in the mouth of the Mersey, fell in with two cutter tenders, one of which kept company with her till within gunshot of the man-of-war, and then hoisted a signal for four boats, which boarded the *Golden Lyon*. The lieutenant in command of the man-of-war's men, hailing the crew of the whaler, declared that he would impress all of them except the officers, unless they entered as volunteers, whereupon the men of the *Golden Lyon*, 60 strong, answered that as they belonged to the Greenland Fishery, they would not be impressed, and to enforce their words, brandished their long knives and harpoons, vowing vengeance on the man that attempted it. This demonstration terrified the man-of-war's men, who jumped into their boats, while the lieutenant got on the quarter deck of the *Golden Lyon*, and ordered the *Vengeance* and her tenders to fire at the whaler, which was within pistol shot of them. Part of the Greenlandman's crew then forced Captain Thompson and his officers into the cabin, standing sentry over them, and keeping the lieutenant of the *Vengeance* on deck, to run the same chance of being shot with themselves; whilst the remainder filled the sails, and crowded away from the *Vengeance*, which slipped her cables, and fired her bow chase into the *Golden Lyon* "as quick as possible." Several of the nine-pound shot struck different parts of the town, but luckily did no other damage than destroying a boat in a builder's yard, though many hundred spectators were very near it. Other shots carried away the rigging, sails, and mizenstay of the *Golden Lyon*, whose crew, however, carried her safe into the dock clear of the man-of-war's people.

On the following day, the whaler's crew proceeded to the Custom-house, to give bond, and to renew their protections, according to Act of Parliament. Immediately after they had done, a large party of the press gang forced themselves into the Custom-house, fired several pistols, and committed other outrages, crowning the whole by impressing Captain Thompson and five of his crew. The rest escaped by various ways, some risking life and limb by jumping through the windows; others climbed on the house tops and over the walls. Whilst the press gang were taking the impressed men down to the water side, they were hooted by some women, one of whom "was shot through the legs with a brace of balls." The paper of August 3rd, announced that Captain Thompson had been discharged from the *Vengeance* man-of-war; that several bullets fired by the press gangs in the Custom-house had been found, and that the magistrates and merchants were determined to prosecute the ruffians for their insolence, "one of the magistrates being then in the Custom-house, and very ill-treated for commanding the peace, etc." This was bad enough, but the commander of the *Vengeance* was capable of inflicting even greater injustice on the wretched seamen who fell into his hands. About the same time, the slave ship *Ingram* (Letter of Marque), returned to port from one of her usually pleasant and profitable little trips to Africa and Jamaica. The crew, having secured the captain, attempted to get clear of the man-of-war and four tenders, but, "the tide being spent," says the paper, "the ship's company and officers were all impressed, except the chief mate and commander. On their being brought on board the man-of-war, Captain ——, ordered each man to be tied up, stripped, and whipped. This needs no comment, for had the seamen committed any offence against the laws of this realm, they were entitled to an Englishman's right." It is no exaggeration to say that in some respects, the British sailor at this time, and for long afterwards, was worse off than the negroes he assisted to oppress. His freedom was a sham, and the law which made it so has never been repealed, though it may never be enforced again. Unfortunately there is a probability that a few years hence a genuine British seaman will be a greater curiosity than that "animal exceedingly rare," whose fossil bones puzzled "the Society upon the Stanislaus."

The brig *Providence*, Captain Parke, on her passage from Liverpool to Tortola, was attacked by a French privateer of 12 guns, 18 swivels and 80 men, which got clear off, much shattered, by dint of superior sailing, after a smart engagement of two hours, during which the French had six killed and seven wounded, while only one man was wounded on board the *Providence*.

Captain Quirk, of the *Prussian Hero* (Letter of Marque), of 18 guns, and 60 men, writing from Guadaloupe in December, 1759, says:—

"I arrived here the 8th inst. from Barbadoes. On my passage from thence, I fell in with three French privateers, viz: one a sloop of 10 guns, a sloop of eight guns, and a schooner of 6 guns, all whom I engaged very briskly two hours. The two sloops rowed up in order to board us, having their bowsprits crowded with people, stink-pots, &c., on which I ordered the guns to be double loaded with round and grape shot, and gave them such a warm reception, as obliged them immediately to sheer off, much damaged, and undoubtedly with the loss of many men. They were no sooner got a little distance off than joined by two other privateers (in all five) to whom I gave chase, a breeze springing up, as fast as possible, till they ran close in shore off Martinico. I then steered my course for this place. Had wind favoured me at the beginning of the action, should have taken at least one of the three."

Captain Quirk formerly commanded the snow *Betty* (Letter of Marque), of 10 guns, belonging to Mr. Peter Holme, merchant. She was taken on her passage from Jamaica, by the *Count de St. Florentine* privateer, and retaken by the *Royal Hunter* privateer, of New York, and sent into Rhode Island. The French privateer was taken herself soon after.

Some of the privateersmen, having received their bounty money in advance, decided to fight another day; and from the following advertisements offering rewards for the apprehension of such gentry, we are enabled to form some idea of the personal appearance and dress of the very mixed specimens of humanity who composed the fighting crews:—

"Ran away from the ship *Liverpool* privateer . . . John Coulston, a middleaged man, about 5 foot 7 inches high, wears his own hair, brown complexion, and very much marked with the small pox. Had on, when he went away, a cheque shirt and 2 waistcoats, one made of white flannel, trimmed round with black tape and black buttons, and the other a blue frize; wore a brown pair of fustian breeches, dark blue stockings, and round pewter buckles. Any person who will secure the said Coulston, by applying to Charles Williams, at the sign of the Whale Fishery, on Sea-Brow, shall receive a handsome reward."

Two seamen, who had run away from the ship *Pember-*

ton, after receiving four guineas each advance as their bounty money for proceeding in the said vessel, were thus described :—

"William Toutcher, seaman, aged 22 years, served his time out of Whitehaven in the coal trade, just arrived from a French prison, (and has procured a pass from the Worshipful William Goodwin Esq : Mayor of this town to proceed to Dover, the place of his residence, being born there) about 5 feet 8 inches high, wore a green jacket, a white flannel waistcoat, a pair of trowsers, and a wig ; is of a middling fair complexion, and a stout able young fellow.

"John Melody, seaman, born at Winchester, served his time in the navy, and lately belonged to the *Fame* privateer of Guernsey ; about 5 feet high, aged 30, wore a blue serge waistcoat, with a row of white buttons down each side, an old blue waistcoat, a wig, and sometimes trowsers. Whoever apprehends either of the above seamen, so that they may be brought to justice shall receive two guineas for each man, by applying to Charles Magee, Boatswain of the ship *Pemberton*, Walter Kirkpatrick, commander, in Redcross-st.

"N.B. They came here in the ship *Liverpool* Privateer, Captain Hutchinson."

Four men, who ran away from the *Spy* privateer, were described as follows :—

"Daniel Lindsay, a full-faced man, about 20 years of age, 5 feet 4 inches high, had on when he went off a blue jacket, a white waistcoat, a check pair of trousers, and wore a cap or wig. Henry M'Cormick, of a fair complexion, about 19 years old, 5 ft. 9 inches high, wore a black wig, a blue jacket, a white waistcoat trimmed with black. John Smith had on a blue rug great coat, a brown frize coat under it, a curled light coloured wig, and a slouched hat. Robert Maxwell had on a snuff coloured fustian coat. He was very much pitted with the small pox, had a brown complexion and sometimes wore a wig over his hair. Both very much addicted to gaming."

The treatment of the English prisoners of war in France during this war appears to have been excessively severe. Captain James Settle, of the *Annabella*, a ship laden with 1400 barrels of tar, deerskins, reeds, etc., from Cape Fear, was taken in November, 1756, by the *Luce* privateer, of Brest, who stripped him and his people almost naked. "We have pleasure to inform our readers," says the Liverpool paper, "that the French prisoners brought into this port have met with more humanity from our privateers' brave crews." But the Frenchmen in Liverpool were not happy, though lodged in the ancient fortress of the Stanleys, for, on April 22nd, 1757, we read that one Monday night, between 11 and 12 o'clock, the prisoners took out a window, and by the help of a rope, four of them got down into the street and made their escape. The noise they made alarmed the neighbourhood, and the rest were immediately secured. A reward was offered for their capture, but without success. In 1759, several French prisoners got away at one time. Many of these came back to the Tower of their own accord, while others were captured in a state of starvation.* In December, 1756, the Lords of the Admiralty took the dancing room and buildings adjacent, at the bottom of Water Street, and fitted them up for the French prisoners "in a very commodious manner, there being a handsome kitchen, with furnaces, &c., for cooking their provisions, and good lodging rooms, both above and below stairs." "Their lordships," says the paper of December 31st, "have ordered a hammock and bedding (same as used on board our men-of-war) for each prisoner, which it's to be hoped will be a means of procuring our countrymen, who have fallen into their hands, better usage

* In December, 1759. James Seabrook, silversmith, was committed to Lancaster for assisting one Jaques L'Uleur, a French prisoner of war, to make his escape from prison, and "the honourable the Commissioners for prisoners of war" sent positive orders to their agent in Liverpool, to prosecute with the utmost rigour all persons that should "mediately or immediately" assist any prisoner of war to make his escape.

than hitherto, many of them having been treated with great inhumanity."

The Tower of Liverpool, in which the prisoners of war were confined at this period, stood at the bottom of Water Street, on the north side, on the site of the present Tower Buildings. Viewed from the river, the Tower was a picturesque and venerable object. It was built of red sandstone, in the Norman style; at one time battlemented, but afterwards crenelated. It would appear that the original structure consisted of a large square, embattled tower, with subordinate towers and buildings, forming three sides of an interior quadrangle, which were altered from time to time. Including its gardens, it occupied an area of 3,700 square yards. Between the Tower and the river there was, at one time, a passage leading into St. Nicholas' Churchyard, and eventually this passage became the street called "Prison Weint." Two houses then skirted the river side, one of them the "Ferry House" tavern. The inhabitants of the town used to walk and show off their finery in the Tower gardens. After being for centuries the town house of the Earls of Derby—the theatre of stirring events, stately functions, and feudal jollifications—the tower, in 1737, passed out of the hands of the Stanleys, who sold it to the Clayton family, by whom it was let to the Corporation for the borough gaol. For years after this transformation and lapse of dignity, the utmost disorder reigned within it, and scenes of the grossest depravity were frequent. There was a large, open space in the interior, in which the prisoners took exercise, and here both debtors and criminals—men and women—were allowed to meet promiscuously. The debtors' room was made use of, amongst other purposes, as a chapel, and also as a general assembly room. It is said that the ladies went there from their houses in blue cloaks and pattens, coaches not being then in general use. On these occasions, the sounds of the music were so plainly heard throughout the building, that the prisoners used to "jig" it as well as the free merry-makers. In 1775, John Howard, the philanthropist, visited the tower gaol, of which he gives a deplorable account. The place, which had just been purchased by the Corporation for the sum of £1535 10s., was insufferably dirty, grimy and wretched. There were two large yards, in one of which poultry was kept, and in the middle of it was a great dunghill. The cells were seven in number, 6 ft. 7 in. in length, 5 ft. 9 in. in breadth, and 6 ft. high. In each cell three persons were locked up nightly. There was a large dungeon, with an iron grated window looking on the street, in which as many as twenty and thirty prisoners were confined at a time. There was no infirmary, nor accommodation for the sick. The women debtors were lodged over the Pilot-office, in Water Street. Mr. Howard made strong representations to the authorities with regard to the disgraceful state of the prison, but nothing seems to have been done in the way of improvement, except some whitewashing and cleaning. The philanthropist received the freedom of the borough, and was lionised for his investigations. In 1803, when Mr. Neild, another philanthropist, visited the gaol, its condition was rather worse than better. The whole prison was then filthy in the extreme, the dirt in some of the passages being three to four inches thick, while the large dunghill, ducks, poultry, etc., shared the courtyard with the herd of male and female felons and debtors. Spirits and malt liquors were freely circulated through the prison, without restriction. A low typhoid fever was constantly prevalent among the prisoners, and the most shameless extortion and robbery also prevailed, the strong over-coming and tyrannising over the weak. The debtors, whose rooms overlooked Prison Weint, used to hang out bags or gloves by a string, with a label attached, "Pity the poor debtors." When any money was placed in the bag, it was drawn up and spent in drink.

On the 12th of March, 1796, died Mrs. Lyons, wife of the keeper of the gaol, and on the following day Mr. Lyons died. The two were conveyed to the churchyard of St. Peter, in two hearses abreast; then followed one mourning coach, next two coaches abreast, and then two more coaches abreast. Thousands of persons gathered in the streets to witness the unusual procession. The cause of death in both cases was said to be gaol fever.

During the Seven Years' War, the celebrated surgeon, Harry Parke, then a very young man, attended the French prisoners in the Tower gaol, and in after years he performed the first operation in conservative surgery, at the Liverpool Infirmary, on a poor sailor, who was subseqently able to follow his arduous calling, whereas, if he had fallen into the hands of an ordinary "sawbones" of the period, he would have lost a leg, and probably his life. That pioneer operation conferred countless boons on humanity, and the name of Harry Parke stands high in the annals of surgery. In excavating the foundations of the first Exchange, the remains of a secret subterranean passage were discovered. It was explored for a considerable distance, and stated to be a communication between the Tower and an old house near the White Cross, which stood at the top of Chapel Street, opposite the end of Old Hall Street. Although the discipline of the prison was so lax that some of the French prisoners occasionally made their escape, it does not appear that they ever discovered this passage, which reminds us of the one described by G. P. R. James in his romance of "Heidelberg." Several of the Jacobites implicated in the Rebellion of 1715 were confined in the Tower, and four of them were executed at Gallow's Mill, near London Road. In 1788, two men were hung on the top of the Tower, for a desperate robbery at a house on Rose-hill. The old Tower continued to be occupied both by felons and debtors down to July, 1811. It remained unoccupied until 1819, when the building was pulled down, and the materials sold by auction for £200. In the Derby Museum, William Brown Street, there is a remnant of one of the old doors of the Tower.

Having formed some idea of the kind of "hospitality" extended to the French captives in Liverpool, we now turn our attention to the English prisoners of war in France. On November 30th, 1756, the master of a merchant vessel wrote from Bayonne, as follows:—

"I am still close confined in this prison, as are all our masters and men without distinction; our usage differs nothing from that of the worst criminals in England, irons only excepted. No one is permitted to speak to us without the commandant's leave; our letters are all opened and read before they are delivered to us, and we are not allowed to purchase any provisions or necessaries from the town's people, but must take every thing from the commandant's mistress, who charges us at the rate of two shillings for what she buys in the town for sixpence. The French commanders, who are prisoners in England, write to their friends in France that they are close confined there, which is the reason of our confinement here; but you informed me in your last that they were all at liberty at Petersfield and other places upon their parole of honour, and that two of them, with a surgeon, had been advertised in the papers for running away. It is evident that they have no honour at all, or they would not have deserted, nor have propagated such a palpable falsehood, which injures us here extremely, for we humbly conceive we are entitled by the law of nations to the same good usage here as the French partake of in England, and as this is a national concern, it ought to be truly represented. There have been built and fitted out in this port within these three months, no less than ten privateers, carrying from 16 to 24 guns upon one deck; and if there is not a cartel of exchange settled soon, I am afraid that many of our common sailors, who are now about 200 prisoners in this castle, will be induced by threats or promises to take on in the enemy's service, where they are offered great encouragement."

In January, 1759, nine English captains, who had been prisoners in France, arrived at Plymouth, and their treatment by "the polite nation" was thus described in the newspapers :—

"Monsieur's behaviour was most barbarous and cruel; the most brutal savage would have shown more compassion. On their first entrance on board their ships they stript them of everything, even to their shirts; as to the common people M. Bompart insisted that they should do the same duty as on board our ships of war; upon refusal, to undergo the same discipline and live on bread and water; but as they did the ship's duty, they were allowed per day four ounces of salt meat, and what they call soup, made of horse beans with common oil. The several captains before mentioned were treated in the same manner. On their arrival at Brest, they were all put down in a dungeon 40 feet under ground, and not permitted fire or candle, though they often petitioned for it, but to no purpose; they had straw to lie upon, but were obliged to pay dear for it. As to the provisions allowed them per day, it was three ounces of poor beef, such if brought to our markets would be burnt. Several of the gentlemen have brought over the allowance with them of every species. They were indulged with three half pints of sour white wine per day, but debarred from water, which if sweet, was much better; but to do them some justice, they had bread sufficient. What was most singular is that they were debarred of laying out their own money, or drawing bills, no person being permitted to come near them; in short, by the report that several of the gentlemen give, they were treated worse than we treat dogs, of which they highly complained and telling them how the French prisoners were used in England, they answered 'that we were afraid to use them otherwise.' At their arrival at Vannes they were put amongst common felons, who were condemned to die, in a most nauseous gaol. The case of poor

Capt. Gordon and his ship's company is a most deplorable one; the whole crew perished in the French ship they were taken in, she being lost on some rocks near the shore; the crew, who were confined in irons, were by the French captain called English dogs, and told they should perish as such, and would not suffer a man to let them out. Their behaviour to Capt. Turner was likewise very cruel, and to the English prisoners in general, forcing them to enter into their service. This can never go unnoticed by those in power."

A gentleman who arrived in Liverpool from Dinan, where he had been confined prisoner of war several months, stated that on the 7th of June, 1758, news arrived there that the English had landed. All the English prisoners, numbering 1300, were immediately marched from thence up into the country, in the night time, guarded by a troop of dragoons and about 500 militiamen. In passing through the villages, the officers and soldiers of the escort showed the prisoners to the country people as a parcel of English vagabonds caught at St. Malo, where they had attempted to land for the purpose of plundering the country. The prisoners were compelled to travel 22 hours without any refreshment, save a little dry bread and small cider at one of the villages. At the small town of Le Mene they were all driven into a church, without distinction of rank, and given some hay to lie on. "Then a strange thing happened," as the novelists say. St. Vierge's image fixed up in the church, tumbled down and broke its valuable neck, upon which some of the English prisoners were clapped into a dungeon, "the priests suspecting that they had done the act; however, on farther enquiry, and a full hearing, they were discharged, being proved innocent." Here, they were joined by upwards of 100 more prisoners from Lorient, and four boys, taken with some horses belonging to the English train of artillery. After a rest of two days, they were marched farther—to Ploermel, "a royal town or city," from whence, after a day's stay, they were driven like sheep to Josselin, where they were all confined in an old palace belonging to the Duke of Rohan, and it was some days before the officers among the prisoners had the liberty of boarding in the town. The usage of the governor who guarded them was intolerable. Two of the English seamen were killed and several others wounded during their march. The French, believing the rumour that the town of St. Malo had been ransomed for a sum of money, were in a state of consternation.

Advertisements like the following were common during war time :—

> "With or without convoy for Jamaica, and will sail in May from Liverpool, the new ship *Nancy*, Benjamin Holland, commander, burthen 500 tons ; carries 22 carriage guns of nine and six-pounders, 10 swivels, and 70 men, and will carry a Letter of Marque."

The *Nancy* was ultimately captured by the French on her voyage to Jamaica. The name of Holland is still associated with the commerce of Liverpool. Here is another advertisement, which one would suppose, at first sight, reassuring to poor Jack :—

> "Merchants and Commanders of ships may be furnished with commissions for private ships of war, and Letters of Marque, on proper security given not to molest any vessel but of the nation at war with us. Also protections for Seamen, from being imprest by any of His Majesty's ships, on the shortest notice. By G. Parker, in Castle street."

But things were not what they seemed in the sailor's "Psalm of Life," for we read in the paper of June 29th, 1759, that there had been a very smart press that week, without any regard either to outward or homeward bound protections. It was said to be the hottest press throughout the nation that had been known since the commencement of the war.

The consequences of taking the best seamen off the merchant-ships were often most disastrous to the owners, who, however, had their remedy at law. In an action tried in London this year, before Chief Justice Willes and a special jury, a Mr. Nickelson, of Poole, was awarded £1,000 damages and costs, against Captain Fortescue, of the *Prince Edward* man-of-war, for impressing the men out of the *Thomas and Elizabeth,* from Newfoundland to Poole, in consequence of which the said ship was lost.

The *Austin*, Captain Holme, on her passage from Liverpool to Barbadoes, was taken and carried into Martinico, after a running fight of eight hours, by a schooner privateer of 6 carriage guns and 10 swivels, and a sloop of the same armament. The *Austin* was condemned at Martinico, and her cargo, etc., sold. So great was the demand for guns at that place, that the only carriage guns she carried (2 three-pounders) were sold for £100. The French had then fitted out of Martinico, 74 privateers, the largest mounting 10 guns, and the smallest only two. Some of the owners of the privateers had entered into an agreement to allow all English captains taken by their ships two dollars per day for the first three weeks after being brought into port, and afterwards to consign them to the king's allowance, which was very scanty, owing to the dearness of all provisions. They drew their chief supply from St. Eustatia, from whence several vessels arrived with Irish provisions, which sold for 20 dollars per barrel. Owing to the number of American ships captured, Indian corn was so plentiful that the French would scarcely hire people to land it. After a stay of eight days at Martinico, Captain Holme was sent up to Barbadoes in a cartel ship, and returned home passenger in the *Merrimack*.

The *Tyger*, Captain Burrowes, on the passage from Liverpool to Jamaica, re-captured the *Speedwell*, from Virginia for London, which had been taken by a Bayonne privateer.

The prize had on board, 315 hogsheads of tobacco, 7 barrels and 19 kegs of indigo, 15 tons of pig iron, and 4000 staves.

On January 8th, 1760, the *George and Betty*, Captain Edward McGill, from Liverpool for Jamaica, was taken by a French privateer of 10 guns and 90 men, after a chase of six hours, and carried into St. Pierre, Martinico.

In the following advertisement we have a description of a small armed merchantman of the period, the kind of ship which, when manned by Liverpool seamen, was generally found to be a very "ugly customer" if interfered with:—

"For Sale by the Candle, at R. Williamson's shop near the Exchange, in Liverpool. On Monday, March 10th, 1760, the sale to begin at 1 o'clock at noon precisely, the Ship *Planter;* burthen about 200 tons, square sterned, Lyon-head, takes the ground well, mounts two six-pounders on slides in the cabin, three new four-pounders on deck, four swivels, and is pierced for 16 carriage guns, being deep waisted with iron stanchions and double netting fore and aft, and suitable for the African or American trade; being 10 feet deep in the hold, 4 feet 9 inches between decks, from the mainmast forward, and from the mainmast aft 6 feet 2 inches, with all her materials, 2 new cables, one new anchor, and all her stores as she arrived lately from London, and now lies at the upper end of the South Dock. Inventories to be had of Mr. David Kenyon, merchant, or Robert Williamson, Broker."

In May, 1760, the old *Eagle* snow, the oldest ship belonging to the port of Liverpool, was wrecked near the Point of Ayre, Isle of Man, on her passage to Guadaloupe.

When George the Third ascended the throne in 1760, Liverpool had surpassed Bristol in tonnage, and had, therefore, become the second port in the kingdom. In this year, Samuel Derrick, master of the ceremonies at Bath, visited the town. Writing to his friend, the Earl of Cork, he says:—

"When the famous Thurot was in the Channel, this town expected that he would honour them with a visit, and they made good preparation to receive him. The ear of a bastion was run out at the main dock head; the walls of the Old Churchyard, under which he must have passed before he came abreast of the town, were strengthened with stone buttresses and mounds of earth; and the whole furnished with some very fine eighteen pounders, which were so disposed as fully to command the river. The merchants were regimented under the command of the Mayor, as colonel, divided into four independent companies,* uniformly clothed and armed, each man at his own expense. Besides, Lord Scarborough and Major Dashwood marched from Manchester, at the head of the Lincolnshire militia, upon the first notice of danger, without waiting for orders from above; so that had this bold adventurer presented himself, there is no doubt but he would have been opposed with a true British spirit of resolution and gallantry."

In another letter, he says:—

"I need not inform your lordship that the principal exports of Liverpool are all kinds of woollen and worsted goods, with other manufactures of Manchester and Sheffield and Birmingham wares, &c. These they barter on the coast of Guinea for slaves, gold-dust, and elephants' teeth. The slaves they dispose of at Jamaica, Barbadoes, and the other West India islands, for rum and sugar, for which they are sure of a quick sale at home. This port is admirably suited for trade, being almost central in the Channel, so that, in war time, by coming north-about, their ships have a good chance of escaping the many privateers belonging to the enemy, which cruize to the southward. Thus, their insurance being less, they are able to

* In the paper of March 14th, 1760, we read that:—"On Tuesday last Col. Spencer's (the Mayor,) Capt. William Ingram's, and Capt. John Tarleton's independent companies of this town were reviewed by the Right Hon. the Earl of Scarborough, in Price's (now Cleveland) Square, and went through the manual exercise, platoon and street firing, etc. The companies were all clothed in their new uniforms at their own private expense; the Colonel's company in blue, lapelled and faced with buff; Capt. Ingram's in scarlet coats and breeches, lapelled and faced with green; green waistcoats, gold laced hats and queue wigs; and Captain Tarleton's in blue, with gold vellum button holes; Capt Thomas Johnson's company of the train of artillery wear the uniform of the navy, blue and buff, with gold laced hats."

undersell their neighbours; and since I have been here, I have seen enter the port, in one morning, seven West India ships, whereof five were not insured."

Whether Liverpool was M. Thurot's object is uncertain, for on the 28th February, 1760, his "best laid schemes" were put an end to for ever. On that day his squadron of frigates was brought to action a few leagues south of the Isle of Man, by a squadron of English frigates under the command of Captain Elliott. After a sanguinary battle, in which the French fought with desperate valour, the whole of the French frigates were taken. Captain Thurot fell covered with wounds on his own deck, and nearly 300 of his officers and men were killed or wounded. By this victory Liverpool was again rendered perfectly safe, but the volunteers remained embodied till the close of the war. The threatened attack was doubtless beneficial to the community, as the emergency brought out the true men, and aroused to action the finest qualities of the Briton.* On the very day Thurot was slain,† the French prisoners confined in the Tower of Liverpool were marched under a guard of "Invalids" for Chester Castle. They were brought back on March 6th.

The following interesting account of Thurot's descent upon Islay was written on the spot, on February 19th, 1760, by Mr. David Simpson, an eye-witness, and forwarded

* When Thurot's expedition was expected, in 1760, it was said that Everton Hill was alive with people from the town, waiting the free-booter's approach. A party of soldiers was then encamped on the hill, and I have been told the men had orders, on Thurot's appearance, to make signals if by day, and to light up the Beacon if at night, to communicate the intelligence of the French fleet being off the coast to the other beacons at Ashurst and Billinge, Rivington Pike and elsewhere, and so spread the news into the north; while signals would also be taken up at Halton, Beeston, Wrekin, and thence to the southward. The most perfect arrangements for the transmission of this intelligence are said to have been made, and I knew an old man at Everton who told me that he had on that occasion carted several loads of pitch-barrels and turpentine, and stored them in the upper chamber of the Beacon, to be ready in case of emergency. He said that during the French war, at the close of the reign of George II., the Beacon was filled with combustibles, and that there was a guard always kept therein.—"Recollections of a Nonagenarian."

† Sir James Picton gives the 4th of March as the date of Thurot's death—a curious mistake, when even Gore's Directory sets forth the true date so conspicuously.

express to a Liverpool merchant, by the codsmack belonging to Captain Hutchinson and his partner, Mr. Ward:—

"Saturday last Commodore Thurot, with three French ships, viz., one of 54 guns, one of 36, and one of 20, came in here from the westward, and betwixt this island and Cantyre they were hovering for five or six hours; at length came close to this land and hoisted an English ensign, which made us imagine they wanted a pilot. Your friends Archibald and Hugh Macdonald went out with a boat and five men, and brought them to anchor at the entry of the Sound of Islay, in Clagin Bay. I was there on Sunday last, where they landed about 600 men in order to plunder the country, and surrounded a parcel of cattle belonging to a gentleman of the place, which they carried off, and they said would be paid for by bill on the French Ambassador at the Hague. Our sloop lay in a harbour close by them loaded with kelp bound to Liverpool, and had 21 bags of flour on board, which Thurot likewise took away, but did no other prejudice to the vessel. They have about 1,500 land forces on board, with a great number of officers, mostly gentlemen, with whom I was in company. They are almost starved for want of provisions, being at allowance of four ounces of bread per day. The land officers and Thurot have disagreed on account of his coming into these channels, &c., and they want him to proceed immediately to France. Thurot's vessel the *Bellisle* is very leaky. I send you now by the bearer one of the swords they left on board my sloop, which I suppose is all the payment Mr. MacDonald and I shall get for our flour. On the sword is struck the words *Volontaire de Bellisle.* You'll please to return it when the bearer comes back this way. Five days before the French put in here they parted with one of their comrades off Barrahead, which they imagine is foundered at sea, or driven into some of the Highland islands. The *Bellisle* broke her rudder, which he told me forced him into these channels. I have been these two days last past ranging the coast, in hopes of meeting with the cod-smack before, in order to dispatch her express to England, and having now met with her, immediately send her, and I hope you and the rest of the merchants of your place will satisfy the owners of the cod-smack for their trouble. One of our 50 gun ships would take Thurot's three vessels, the *Bellisle, Blanque,* and *Thurot*, for they are crowded with men so much, that they are scarce able to fight their guns; but Mons. Thurot says that if he once gets half gunshot from the best ship in England, he could clear himself by his fast sailing. The season here is very rough; but Thurot will go either through St. George's Channel, or round Ireland, as best suits him, being determined to execute his original scheme. There are a number of English and Irish amongst his crew. We have sent an express to Edinburgh; however, hope the cod-smack will bring the first intelligence to you. We are deprived of the use of arms here, or should have been able to have defended our country from being plundered. The ships lay close inshore between Mr. Arthur's head and Ardmore point; and you may depend on this relation, as I was eyewitness to the facts here."

We have now arrived at the end of the Seven Years' War, and naturally pause to ask ourselves if privateering paid? On this point Sir James Picton observes:—

"It has been sometimes asserted that the merchants of Liverpool greatly enriched themselves in the last century by the practice of privateering. At a subsequent period there were a few exceptional instances of this, but during the Seven Years' War the results to the Liverpool merchants were most disastrous. From a list published in July, 1760, it appears that in four years from the commencement of the war there had been taken by the French, of vessels *belonging to Liverpool alone*, the number of 143, or 36 in each year. The tonnage is not given; but as they were all sea-going vessels, principally in the West India and American trades, the losses must have been enormous."

We have searched diligently for the list in question, but failed to find it. Its discovery would have saved us the great trouble of compiling an independent list, which will be found in the appendix, and which is incomplete, owing to certain issues of the newspapers being missing. It tends to confirm the statement referred to by Sir James Picton as to the number of vessels taken, but he might have added that probably one-third of the captured ships were slavers, a fact which added enormously to the losses, each slave ship representing three distinct sources of profit in a single round voyage. But we must bear in mind that but for the activity of the Liverpool privateers, these losses would have been greater. Every prize they made rendered the enemy poorer, and reduced the aggregate loss to the port. To put an extreme case: if all the merchant vessels of Liverpool had been captured by the enemy, and a single Liverpool privateer had been fitted out, sent to sea, and returned with a single prize of more than her own value, privateering in that case must have been held to pay, for without that prize the port would have been the poorer. Assuming the prize to be an enemy's privateer, the gain would be even greater, for the destructive power of the enemy was thereby reduced, and consequently a certain number of British ships saved from capture by that privateer, whose guns might at once be turned against the commerce of the enemy. If the commerce of Liverpool suffered so heavily during this war, while she had a gallant little fleet of privateers scouring the seas, harassing the enemy, and bringing in valuable prizes, how much greater would have been the losses if the privateers had not been sent out at all! But those who concede that privateering benefited the town materially, may contend that it damaged the people morally. On this head, Sir James Picton, whose moral sentiments are always admirable, says:—

"The pursuits of the Liverpool Merchants during a great part of the eighteenth century, will not bear very severe scrutiny in a moral point of view, taking the standpoint of the present day. The practice of privateering could not but blunt the feelings of humanity of those engaged in it, combining, as it did the greed of the gambler with the ferocity of the pirate. War is hateful in any form, but undertaken by a nation with the discipline and courtesies of a regular force, it assumes an amount of dignity which hides to some extent its harsher features; whilst marauding expeditions undertaken by private parties combine all the evils without any of the heroism of war; greed is the motive power, and robbery and murder the means of its gratification. Its influence on the community which encourages it cannot but be deleterious."

With these sentiments theoretically considered, we must coincide in the words of Artemus Ward, "too true, too true," and President Kruger might even endorse the last clause as a prophetic utterance, but the fact is, the moral condition of Liverpool in the eighteenth century was such that privateering, as carried on by Hutchinson and the other commanders of whom we have spoken, was more likely to elevate than lower the people. It is clear from the writings of John Newton, Gilbert Wakefield, Goronwy Owen and others, that the standard of morality was so low in Liverpool, that even the introduction of piracy itself into the Mersey, as a fine art, would not have perceptibly altered the manners and morals of the masses during the first half of the eighteenth century; and Mr. Clarkson's experiences in the town, at a much later period, prove that there was more room for improvement than for deterioration. On the other hand, there were sentiments and qualities evoked and developed in connection with privateering, that tended to raise those who had fallen to the lowest depths. To be fired with enthusiasm, to cruise about the seas in "great spirits," replying to the enemy's remarks with "three cheers" and hot broadsides, to face death manfully "for the honour of Liverpool," and even for pelf, if not for King and country, must have done good to many a bankrupt soul and pocket, and could scarcely deteriorate the men who embarked upon such work—

> "As full of peril and adventurous spirit
> As to o'er-walk a current roaring loud
> On the unsteadfast footing of a spear."

As to the moral effect of privateering upon the merchants themselves, we fail to discern any signs of the greed of the gambler, or the ferocity of the pirate, in the instructions given by Mr. Francis Ingram to Captain Haslam—which instructions, be it remembered, were private, and are now first made public.* Whatever atrocities may have been committed by the privateers of other nations, or of other ports, or by pirates in the name of privateers, we cannot call to mind a single action committed by any Liverpool privateer unworthy of its character as a private ship of war carrying the King's commission.

See Chapter I., pp. 21-30.

CHAPTER IV.

PRIVATEERS OF THE AMERICAN WAR OF INDEPENDENCE.

DURING the twelve years of peace which intervened between the Seven Years' War and the first American war, the commerce and wealth of Liverpool increased more rapidly than they had ever done before. Liverpool had taken the lead of all the seaports of the empire in the American and the African trades, and also possessed a large share in the trade of the West Indies; the two latter branches of commerce being too frequently cemented with the blood of slaves. As the pressure of the wars with France and the continental powers fell with greater severity on the commerce of London, Hull, and even Bristol, than on that of Liverpool, owing to their geographical position and their greater commercial intercourse with Europe, so the commercial ruin caused by the first and second war with America fell more severely on Liverpool than on any other port, owing to the extent of its American and West Indian connections.

The American War of Independence opened a new chapter in the world's history. The obstinacy and imbecility of George III., and the despotic instigations of his consort, Queen Charlotte, forced the descendants of the "Men of the *Mayflower*" to teach tyrants for all time the lesson that the subject as well as the king had a divine right. It was the privilege of the freedom-loving British colonists in North America, to fix the attention of the whole civilised world upon a maxim which it had taken ages of social misery and oppression to evolve—"Taxation without Representation is Tyranny." The world has never ceased to wonder at the crass stupidity which converted loyal colonists into rebels, but it is not necessary here to review the causes that led up to hostilities. Suffice it to say that the first blood in the fratricidal conflict was spilt at Lexington, in April, 1775. On the 4th of July, 1776, thirteen of the colonies declared themselves independent of Great Britain. In 1778, France acknowledged the independence of America, and declared war against England. War with Spain commenced on the 17th of April, 1780, and with Holland on the 21st of December, 1780. This desperate and wide spreading contest with America, France, Spain and Holland, continued to the year 1783, when Great Britain finding the attempt to subdue her late colonies hopeless—the people at home being by this time disgusted with the folly of their rulers—abandoned the attempt, acknowledged their independence, and made peace with their allies.

In January, 1775, prior to the affair at Lexington, an influential meeting of merchants from all parts of the kingdom trading with America, was held at the King's Tavern, Cornhill, London, to protest against the violent proceedings of the government towards the colonists, and to petition for the repeal of all the acts which interfered with their friendly relations towards the mother country. The West India merchants from Liverpool and other towns, also assembled at the London Tavern for the same purpose, when strong resolutions were carried by a large majority. These remonstrances proved fruitless, and we search in vain for one conciliatory sentence from the British Government at this time. Within a month afterwards, 8000 tons of shipping had to return from America without cargoes, the blockade not allowing them to land. Whatever the great body of merchants thought of the turn affairs had taken, the wiseacres of the Liverpool Common Council, on the 11th of September, 1775, presented a loyal address to the King, expressing their "abhorrence of all traitorous and rebellious disturbers of his Majesty's peace and government, and hoping that the rebellious Americans might soon be sensible of their error, and return to an acknowledgment of the power of the British Legislature"—a very pretty and appropriate sentiment coming from such a quarter. Liverpool soon began to feel the effects of the war. A writer in the *Liverpool General Advertiser*, of the 29th September, 1775, says—"Our once extensive trade to Africa is at a stand; all commerce with America is at an end. Peace, harmony, and mutual confidence must constitute the balm that can again restore to health the body politic. Survey our docks; count there the gallant ships laid up and useless. When will they be again refitted? What become of the sailor, the tradesman, the poor labourer, during the approaching winter?" London also suffered heavily, for in November, 1775, it was announced that 600 vessels formerly employed in the trade with America, were lying idle in the Thames. As early as February, 1776, only seven vessels entered at the London Custom House in a whole week; a circumstance not known before for 40 years.

In the Autumn of 1775, the Americans began to fit out privateers at Philadelphia and other ports. In January 1776, it was announced that there were American privateers in all parts of the Atlantic, and very soon they swarmed round every one of the West India islands. Meanwhile the King's cruisers were not idle. In the first half of the year 1776, they captured seventy-two American vessels. Thus the energies of the two nations were turned to the destruction of commerce with terrible effect. The foreign trade of Liverpool rapidly declined, until it sank to a small part of what it had been before the war. There were at that time 170 American cruisers at sea. Amongst other prizes, they took 23 valuable

West Indiamen in the summer of 1776.* In the paper of September 6th, 1776, we find the following remarkable statement put forth with the evident intention of minimising the importance of the American successes :—

> "As we daily read of many rich vessels taken by the American privateers, it may not be disagreeable to acquaint the friends of Old England with two essential circumstances ; first that some of the rich vessels taken from us by the rebels have no existence whatever but in the newspapers ; and secondly, that the principal part of the rest go out with the professed view of falling into the hands of the enemy. To elucidate this point we must observe that the Saints are in great distress for numberless articles which they cannot procure openly from England, as all commerce with them is prohibited by Parliament, and punishable as high treason. Under these circumstances, therefore, directions are privately issued to their adherents in such of the British ports as are most conveniently situated, to fit out large ships with the commodities particularly wanted. This is accordingly done, and the vessels sail to a given latitude under the plausible pretext of bearing to some well affected part of America. When they arrive at the given latitude, however, provincial privateers are in readiness to seize them, and they strike without a blow, well knowing that their owners are to be amply indemnified for the utmost loss which they may seemingly sustain in this imaginary capture."

It is a remarkable fact that during the early part of the war, while hostilities were confined to the two principal belligerents—the mother country and her rebellious children—the merchants of Liverpool did not enter into privateering with the spirit that had distinguished them in former wars.†

* *The Virginia Gazette* of June 21st, 1776, gives the following statement of the cargoes of certain West Indiamen taken by the American privateers, viz., 22,420 dollars, 187 oz. of plate, 1,052 hogsheads of sugar, 213 puncheons of rum, 70 pipes of old Madeira, 246 bags of pimento, 396 bags of ginger, 568 hides, 25 tons of cocoa, 41 tons of fustic, one cask of tortoise shell. The owners of the privateers are said to have shared £5,000 each, and each sailor £500.

† The British Government appears to have contributed to this remarkable forbearance by its tardiness in issuing Letters of Marque. "In last Tuesday's *Gazette*," says the paper of April 11th, 1777, "the Lords of the Admiralty give notice that they are ready to issue commissions to private ships for cruising against the Americans."

Perhaps they felt reluctant to fight and plunder their former customers, men of their own race and speech, and in many cases, most likely, their own personal friends and correspondents; or they may have considered the game not worth the candle, the sea-borne commerce of the colonies being too insignificant to supply remunerative prizes for the King's cruisers and privateers combined. As soon as the French and Spaniards joined the Americans, however, Liverpool enterprise awoke like a giant from a dream, and put forth its strength as the fathers had done, and on a much vaster scale, as became the second seaport of the empire. Once more visions of valuable French East Indiamen, and treasure-laden argosies of Spain, dazzled the imagination of those who coveted easily acquired fortunes. The slave ships were lying idle in the docks, the war having almost ruined the man traffic, to the great grief and pecuniary loss of many excellent citizens of Liverpool and their friends—certain native chiefs on the coast of Africa. Those vessels that could not be profitably employed in the slave trade were easily converted into privateers, and so great was the energy displayed in their equipment, that, between the end of August, 1778, and April, 1779, no less than 120 private ships of war were fitted out.* Their total tonnage was 30,787, carrying 1986 guns, and 8754 men. The largest of these ships was a frigate of 30 nine-pounders; that of the heaviest metal carried 16 eighteen-pounders; the other vessels were mostly armed with six and nine-pounders. The number of men forming the crews varied considerably in the different vessels, a ship of 250 tons burthen carrying 140 men, while a ship of 1200 or 1400 tons carried only 100 men. "This formidable armament," says Troughton, "proved a considerable annoyance to the hostile powers, and captured several French ships from the East and West

*A list of the Liverpool privateers engaged in this war is supplied in the appendix.

Indies, of such immense value, as enabled the merchants of Liverpool not only to restore their credit and extend their commerce, but to trade upon real capital." Dr. Aikin, in his description of the country round Manchester, says, "Liverpool has in different wars distinguished itself by the spirit with which it has fitted out armed ships for the purpose of annoying the trade of the enemy. How far this is a useful spirit to a trading town, and in what degree the prizes made have exceeded or fallen short of the expenses of the outfits, we shall not inquire. Some of the prizes taken by the Liverpool privateers were of very great value ; and their effect in cutting off the resources of the hostile powers were very considerable." "The undaunted courage and gallantry of the crews of both the privateers and armed merchant vessels of Liverpool," says Brooke, "command our applause, and on numerous occasions excited the admiration of the enemy." Sir James Picton, referring to this period, observes, "there were two blots on the fair fame of Liverpool commerce which could not but have a demoralising tendency on society generally. I allude to privateering and the slave trade. . . Privateering, though practised to a considerable extent, was in private hands, and did not come within the purview of the Corporation, hence there is no allusion to it in the records." Although it is generally held that a corporation has neither a soul to be lost nor a corporeal presence sufficiently tangible and "get-at-able" to receive castigation, one feels that it would have been morally to its advantage if the Corporation had had more to do with privateering and less with slave trading.

The following description of the launching of the *Mary Ellen*, a ship which played the double part of slaver and privateer, is from the pen of "A Nonagenarian":—

"My father was owner and commander of the *Mary Ellen*. She was launched on the 4th of June, my birthday, and also the anniversary of our revered sovereign, George III. We used to keep his majesty's birthday in great style. The bells were set ringing, cannon fired, colours waved in the wind, and all the schools had holiday. We don't love the Gracious Lady who presides over our destinies less than we did her august grandfather, but I am sure we do not keep her birthday as we did his. The *Mary Ellen* was launched on the 4th of June, 1775. She was named after and by my mother. The launch of this ship is about the first thing I can remember. The day's proceedings are indelibly fixed upon my memory. We went down to the place where the ship was built, accompanied by our friends. We made quite a little procession, headed by a drum and fife. My father and mother walked first, leading me by the hand. I had new clothes on, and I firmly believed that the joy bells were ringing solely because *our* ship was to be launched. The *Mary Ellen* was launched from a piece of open ground just beyond the present Salthouse Dock, then called 'the South Dock.' I suppose the exact place would be somewhere about the middle of the present King's Dock. The bank on which the ship was built sloped down to the river. There was a slight boarding round her. There were several other ships and smaller vessels building near her; amongst others, a frigate which afterwards did great damage to the enemy during the French war. The government frequently gave orders for ships to be built at Liverpool. The view up the river was very fine. There were few houses to be seen southward. The mills on the Aigburth road were the principal objects.

"It was a pretty sight to see the *Mary Ellen* launched. There were crowds of people present, for my father was well-known and very popular. When the ship moved off there was a great cheer raised. I was so excited at the great 'splash' which was made, that I cried, and was for a time inconsolable, because they would not launch the ship again, so that I might witness another great 'splash.' I can, in my mind's eye, see the 'splash' of the *Mary Ellen* even now. I really believe the displacement of the water on that occasion opened the doors of observation in my mind. After the launch there was great festivity and hilarity. I believe I made myself very ill with the quantity of fruit and good things I became possessed of. While the *Mary Ellen* was fitting up for sea, I was often taken on board. In her hold were long shelves, with ring bolts in rows in several places. I used to run along these shelves little thinking what dreadful scenes would be enacted upon them. The fact is that the *Mary Ellen* was destined for the African trade, in which she made many very successful voyages. In 1779, however, she was converted into a privateer. My father, at the present time, would not perhaps be thought very respectable ; but I assure you he was so considered in those days. So many people in Liverpool were, to use an old and trite sea phrase, 'tarred with the same brush,' that these occupations were scarcely, indeed were not at all, regarded as derogatory to a man's character. In fact, during the privateering time, there was scarcely a man, woman, or child in Liverpool, of any standing, that did not hold a share in one of these ships. Although a slave captain, and afterwards a privateer, my father was a kind and just man—a good father, husband, and friend. His purse and advice were always ready to help and save, and he was, consequently, much respected by the merchants with whom he had intercourse. I have been told that he was quite a different man at sea, that there he was harsh, unbending, and stern, but still just. How he used to rule the turbulent spirits of his crews I don't know, but certain it is that he never wanted men when other Liverpool shipowners were short of hands. Many of his seamen sailed voyage after voyage with him. It was these old hands that were attached to him who I suspect kept the others in subjection. The men used to make much of me. They made me little sea toys, and always brought my mother and myself presents from Africa, such as parrots, monkeys, shells, and articles of the natives' workmanship. I recollect very well, after the *Mary Ellen* had been converted into a privateer, that, on her return from a successful West Indian cruise, the mate of the ship, a great big fellow, named Blake, and who was one of the roughest and most ungainly men ever seen, would insist upon my mother accepting a beautiful chain, of Indian workmanship, to which was attached the miniature of a very lovely woman. I doubt the rascal did not come by it very honestly, neither was a costly bracelet that one of my father's best hands (once a Northwich salt-flatman) brought home for my baby sister. This man would insist upon putting it on the baby somewhere, in spite of all my mother and the nurse could say; so, as its thigh was the nearest approach to the bracelet in size of any of its little limbs, there the bracelet was clasped. It fitted tightly and baby evidently did not approve of the ornament. My mother took it off when the man left. I have it now. This man used to tell queer stories about the salt trade, and the fortunes made therein, and how they used to land salt on stormy and dark nights on the Cheshire or Lancashire borders, or into boats alongside, substituting the same weight of water as the salt taken out, so that the cargo should pass muster at the Liverpool Custom House. The duty was payable at the works, and the cargo was re-weighed in Liverpool. If found over-weight the merchant had to pay extra duty; and if short weight, he had to make up the deficiency in salt. The trade required a large capital and was therefore in few hands. One house is known to have paid as much as £30,000 for duty in six weeks. . . . To return to the launch. After the feasting was over, my father treated our friends to the White House and Ranelagh Tea Gardens, which stood at the top of Ranelagh-St. The gardens extended a long way back. Warren-St. is formed out of them.

"As a young boy and an old man I have seen my native town under two very diverse aspects. As a boy, I have seen it ranked only as a third-rate seaport. Its streets tortuous and narrow, with pavements in the middle, skirted by mud or dirt as the season happened. The sidewalks rough with sharp-pointed stones, that made it misery to walk upon them. I have seen houses, with little low rooms, suffice for the dwelling of the merchant or well-to-do trader —the first being content to live in Water-St. or Oldhall-

St., while the latter had no idea of leaving his little shop, with its bay or square window, to take care of itself at night. I have seen Liverpool streets with scarcely a coach or vehicle in them, save such as trade required, and the most enlightened of its inhabitants, at that time, could not boast of much intelligence, while those who constituted its lower orders were plunged in the deepest vice, ignorance, and brutality.

"But we should not judge too harshly of those who have gone before us. Of the sea savouring greatly were the friends and acquaintances of my youth. Scarcely a town by the margin of the ocean could be more salt in its people than the men of Liverpool of the last century ; so barbarous were they in their amusements, bullbaitings, and cock and dog fightings, and pugilistic encounters. What could we expect when we opened no book to the young, and employed no means of imparting knowledge to the old?—deriving our prosperity from two great sources—the slave trade and privateering. What could we expect but the results we have witnessed? Swarming with sailormen flushed with prize money, was it not likely that the inhabitants generally would take a tone from what they daily beheld and quietly countenanced? Have we not seen the father investing small sums in some gallant ship fitting out for the West Indies or the Spanish Main in the names of each of his children, girls and boys? Was it not natural that they should go down to the "Old Dock" or the "Salthouse," or the "New Dock" and there be gratified with a sight of a ship of which they—little folks as they were—were still part owners? We took them on deck and showed them where a bloody battle had been fought—on the very deck and spot on which their little feet pattered about. And did we not show them the very guns, and the muskets, the pistols and the cutlasses, the shot-lockers and magazines, and tell them how the lad, scrubbing a brass kettle in the caboose, had been occupied as a powder-monkey and seen blood shed in earnest? And did we not moreover tell them that if the forthcoming voyage was only successful, and if the ships of the enemy were taken—no matter about the streams of blood that might run through the scuppers—how their little ventures would be raised in value many hundred-fold—would not young imaginations be excited and the greed for gain be potent in their young hearts? No matter what woman might be widowed—parent made childless, or child left without protector—if the gallant privateer was successful that was all they were taught to look for. And must not such teaching have had effect in after life? I have seen these things, and know them to be true; but I have seen them, I am glad to say, fade away, while other and better prospects have, step by step, presented themselves to view.

"As a man I have seen the old narrow streets widening—the old houses crumbling—and the salty savouring of society evaporate and the sea influence recede before improvement, education and enlightenment of all sorts. Step by step has that sea element in my townsmen declined. The three-bottle and punch-drinking man is the exception now, and not the rule of the table."*

In November, 1776, the merchants of Liverpool gave public notice that they would discourage the future employment of any persons, who, being masters of vessels, should separate from their respective convoys, or otherwise wilfully disobey the orders received from the commanders thereof. In the same month, the Corporation, to avert the inconveniences and hardships which the impress brought on the freemen and other inhabitants of the town, and also to the trade, business, and commerce thereof; and at the same time to assist Government in manning his Majesty's ships, offered a bounty of two guineas to every able seaman volunteering to enter the navy, and to every volunteer ordinary seaman a bounty of one guinea, over and above the King's bounty. A committee of the Town Council sat in the Mayor's office, within the Exchange, to examine and enter such volunteers.

Captain Wilson, of the *Union*, on his passage was boarded by an American privateer, of 10 six-pounders, and

* "Recollections of a Nonagenarian," by the late Mr. James Stonehouse.

103 men, called the *Sally*, Captain Munro, of Rhode Island, who took ut his cargo of ivory, and Malageta pepper, clearances, etc., and all the letters. The privateer had just taken a ship from Bristol, and another from London, and put on board the *Union* 24 prisoners, and some provisions. Two days later, the *Union* was boarded by another privateer of six guns and 34 men, commanded by Captain Field, of Rhode Island, and again by a third privateer, called the *Cabot*, belonging to the Congress, of 14 guns and 130 men, commanded by Captain Hinman, who searched the *Union* and ordered her to stand to the N.W., which they did until a fleet of ships came in sight, when one of them gave chase to the privateer. The *Cabot* subsequently captured the brig *Watson*, Captain Brison, from Jamaica for Liverpool, which had on board the owner, Mr. James Bier, a man of resolution and resource. The following letter written by him from Dundalk, on December 3rd, 1776, tells how he got his own again :—

"On the 2nd of October, in lat. 41, long. 45, I was taken by an American man-of-war (as they call themselves) called the *Cabot*, a brig of 14 guns, commanded by Captain Elisha Hinman, who had just before taken five large vessels. I was carried on board the privateer, where I applied to the Captain for leave to continue on board my own vessel, which he positively refused; however, after some conversation about privateering, he consulted his officers, and then told me I might go on board; this gave me great pleasure. Had he kept me, I should have taken their man-of-war, they having only about 40 of their own people, and upwards of 80 prisoners. They took all my men, except the captain, one boy, and a passenger, putting eight of their own people on board. We were to proceed to New London or Rhode Island. In about three weeks we got into soundings off Boston, but that night I had determined to re-take her, having brought over to my party two of their people, by promising them £100. Accordingly, at 8 o'clock, they sent me a pistol by the boy, on which I immediately jumped upon deck, clap'd it to the prize-master's breast, and demanded him to surrender the vessel, which he instantly complied with, at the same time the captain and boy secured the lieutenant of marines in the cabin. We then secured the hatches, till I got all the arms, which compleated the business. I bore away for Halifax, but the wind being fair, stood on for Newfoundland. The wind still continuing favourable, stood on for Ireland, where I struck soundings in 27 days. We had but two barrels of beef and three of bread when I bore away, but fortunately had two turtles about 600 lb. weight, which served us three weeks. We ran in here in a hard gale of wind, where we lie in safety, having (thank God) received no damage, except one boat washed overboard, with studden-sails and some spare ropes. Our fire and candles were intirely exhausted. I hope this will be agreeable news, and remain &c., J. B."

The following are copies of the commission granted by Admiral Hopkins, the American naval commander-in-chief, to Captain Hinman, and of the latter's orders to the prize-master put on board the *Watson* :—

"By the power given me by the Delegates of the United Colonies of New Hampshire, Massachusetts-bay, Rhode Island, Connecticut, New York, New Jersey, Pensylvania, the counties of Newcastle, Kent, and Sussex, on Delaware, Maryland, Virginia, North Carolina, South Carolina, and Georgia,

"To ELISHA HINMAN, Esquire,

I, reposing especial trust and confidence in your patriotism, valour, conduct, and fidelity, do by these presents constitute and appoint you to be first Lieutenant of the armed ships in the service of the Thirteen United Colonies of North America, fitted out for the defence of American Liberty, and for repelling every hostile invasion thereof. You are therefore, carefully and diligently to discharge the duty of First Lieutenant, by doing and performing all manner of things thereunto belonging. And I do hereby strictly charge and require all officers, mariners, and seamen, under your command, to be obedient to your orders as First Lieutenant. And you are to observe and follow such orders and directions, from time to time, as you shall receive from the Congress of the United Colonies, or Committee of Congress for that purpose appointed, or Commander in Chief, for the time being, of the navy of the United Colonies, or any other your superior officer, according to the rules and discipline of war, the usage of the sea, and the instructions herewith given you, in pursuance of the trust reposed in you. This Commission to continue in force until revoked by the Congress.

"Ship *Alfred,* Sept. 29, 1776.

"Signed EZEK. HOPKINS, Commander in Chief.

"The above is a true copy of the commission given me, from under my hand 2d Oct., 1776.

"ELISHA HINMAN.

"N.B. Since I received my commission, I received orders from the Commander in Chief to take command of the *Cabot.*

"SIR,—You are to take charge or command of the brig *Watson*, Francis Brison, master, from Jamaica, bound Liverpool, and proceed with her to New London, Rhode Island, or any convenient port in North America. On your arrival apply to the continental agent; at same time advise Admiral Hopkins of your safe arrival.—On board the *Cabot*, in latitude 40.36 N. longitude 43.30 W. Oct. 3rd, 1776.

"ELISHA HINMAN."

The ship, *Leghorn Galley,* Captain Alexander M'Daniel, belonging to Mr. Thomas Earle, was taken on her passage from Liverpool to Jamaica, by an American privateer and carried into Philadelphia. About the end of the year 1772, Mr. Thomas Earle established a line of packets, sailing at regular times from Liverpool to Leghorn, hence probably the name of the vessel. This was the first line of foreign packets established in Liverpool. Captain M'Daniel wrote the following letter to Mr. Earle, from Nantz, on December 19th, 1776 :—

"I have nothing new to acquaint you with from Philadelphia, since the taking of New York, which you no doubt have heard of, only that a decisive battle was expected to be fought when we left Philadelphia. Here is the *Enterprize*, Capt. Weeks, at this place, belonging to the Congress, mounting 16 six-pounders, 24 swivels, and 130 men. She brought over Dr. Franklin, one of the Congress, who is gone to the court of Versailles. She took a brig belonging to Cork from Bordeaux, one Cod master, and a brig from Rochelle bound to Hull, one Fetchett master, about fifteen leagues from this place, and has sold both vessels and cargoes to the French. Here is also a privateer belonging to Charles Town, South Carolina, commanded by one Cockran, mounting 12 four-pounders, and 80 men, besides four other vessels belonging to the Congress, all taking in naval and military stores, and are arming them all."

It is worth noting here, that Dr. Franklin has left upon record some very severe and disrespectful remarks, regarding the picturesque and patriotic profession of privateering, in which we—and it is hoped the reader—are at present highly interested, but we are, on the other hand, morally fortified by the equally vigorous utterance of another distinguished American statesman—Jefferson, who held that privateering is a national blessing, when a country like America is at war with a commercial nation.

In November, 1776, the ship *Sam*, Captain Richardson, on her passage from Barbadoes for Liverpool, with about 20,000 dollars, and 52 cwt. of ivory on board, was taken in latitude 20°, by the *Independence* privateer, of 10 guns and 45 men, John Young, commander, belonging to the Congress. The ship was sent away to Philadelphia, with the ivory and silver ; the captain and boatswain were landed at St. Pierre's, Martinico ; the doctor, mate, and two servants, and four of the crew were left on board the *Sam*, and all the rest of the hands entered on board the privateer. "If the French Governors suffer prize cargoes, without condemnation, to be landed in their islands," observes a Barbadoes

N

correspondent, "our trade must most certainly be quite ruined very soon."

The manners of the lower orders in Liverpool at this period, are exemplified in the following incident. "Tuesday afternoon," says the paper of January 17th, 1777, "some riotous people assembled before a house in Frederick Street, and dragged from thence a poor woman, whom they stripped and inhumanly ducked a number of times in the dock, and otherwise ill-treated, so that she now lies very ill at the Infirmary. Proper steps are taking to discover and punish the offenders. Their resentment was owing to her having given information to the press-gang against a sailor, who had lately married her in the north, had brought her here, where he had a former wife, and refused to give her two shillings to carry her home again." A subscription was raised on her behalf in the town.

During the scare caused by the incendiary, John the Painter, in January, 1777, "at a very great and most respectable meeting of the mayor, magistrates, merchants, and traders," held in Liverpool, it was resolved that a strong and efficient watch be set every night from five o'clock in the evening, till seven o'clock in the morning, to patrol round the docks and through the town. Owners, masters, and others interested, were recommended to have their ships carefully watched, the persons in charge not to be allowed any candle-light or fires aboard during their watch. A committee was appointed to enforce the recommendations, and a great number of gentlemen voluntarily offered themselves to be upon guard, by rotation, each night. Many special constables were sworn, and the magistrates ordered all disorderly, idle and suspected persons having no visible means of getting their livelihood, to be taken up. A strict lookout was kept on all loitering persons being in or coming into the town, and the inhabitants who had lodgers whom they eyed with suspicion, were invited to impart those suspicions to the authorities—an excellent opportunity to settle old scores. The committee met daily in the Council chamber to receive the report of the preceding night's watch, and a justice of the peace was at hand to deal with offenders. "All riotous, disorderly, and idle persons," says the official order, "are hereby cautioned to forbear their wicked courses, and to be early in their houses or places of abode at nights; and all strangers are desired to keep in their inns in due time, and not be strolling about the town at unseasonable hours, to prevent the inconvenience of being taken up by the constables on the watch, the mayor and magistrates being determined rigorously to put the laws in force against all offenders. And the gentlemen, merchants, traders, and inhabitants in general of this great commercial town and port will heartily concur, and diligently assist in their guard, care, and watching for the safety and preservation thereof." Circumstances over which he had no control prevented John the Painter from visiting Liverpool and firing its shipping, and in the paper of March 28th, his ghost is made to sing a doleful warning to other "poor, deluded, guilty souls," to behold their fate in him.

"On the 20th instant," says a letter from Barbadoes, dated the 25th January, 1777, "the *Thomas*, Captain Collison, and the *Sarah*, Captain Frith, both from Liverpool, fell in with a 10 gun sloop, which soon boarded, and, sword in hand, took the former; she and the prize then fired upon the latter ship, which by having the heels of them, got off, but she had the misfortune of being attacked again in the morning of the 22nd inst, near the land, by a schooner of considerable force, with which she had a long and smart action. The enemy attempted sundry times to board her, but were prevented by booms rigged out on each side of the *Sarah*, whose mainmast has a six-pounder through it, the sails, blocks, and rigging all cut to pieces, and his chief mate wounded in the arm. He, poor gentleman, received two musquet balls in his body, and though all imaginable care was taken of him, immediately on his arrival in Carlisle-bay, last Wednesday evening, he died the night before last, and was buried yesterday morning. The engagement was seen from the shore, and it is allowed that poor Frith behaved gallantly. At the bottom of the invitations to his funeral were these words—

*"*Dulce et Decorum pro Patria Mori.*"

"The people on board the *Sarah* imagine they dropped the Captain of the schooner, and several of the crew, who appeared to them to be chiefly French, Mulattoes, and Negroes. The privateer was a good deal shattered in her sails and rigging, and received several shots in her hull. Several of the Independent Gentry are cruising to the east-ward of this island, and some of them well fortified and manned. One Fish, a Salem man, in the brig *Tyrannicide*, of 14 guns and 120 men, has been very successful; the last he took, that we know of, was a brig called the *Three Friends*, one Helme, commander, from London, with a valuable cargo of dry goods and provisions. Said capture was last Saturday. The long boat was given to six of the hands, who arrived here that afternoon, and reported that their Captain and mate entered on board the privateer, carrying with them a sum of money and upwards of 2000 letters for this place. Such swarms of them are to the windward that 'tis to be feared they will do much mischief."

Mr. James Barton, second mate of the *Thomas*, writing to his owners in Liverpool, from Newport, Rhode Island, on February 23rd, 1777, says:—

"I suppose by this time you may have heard of our being taken; but as it is uncertain, I shall just mention a few particulars, and leave the rest to a superior officer. Suffice it, therefore, to say, that on the 21st of January last, in lat. 14 N.

* It is sweet and glorious to die for one's country.

and longit. 56 W. from London, we were met with, engaged, and took by an American Privateer, called the *Revenge*, whereof Joseph Sheffield was the commander, after having two of our people killed, and the Captain, Mr. Harper, and the Boatswain wounded.

"Most of the crew, along with myself, were immediately put on board the privateer; but on the 23rd of January we fell in with a French ship bound to Martinico, whom the privateer obliged to take immediately on board her our Captain, Chief Mate, and all the crew (excepting the Boatswain, who was very ill wounded, a boy and myself, whom they detained and would not suffer to quit the ship). We parted from the French ship the same day, and stood for America with the privateer in company. We had very bad weather after leaving the privateer, and on the 14th of February made land, which proved to be Rhode Island, and not having heard it was taken, we stood in for the harbour, which we were very near, when the privateer, our consort, ahead saw a sail and immediately crowded all he could and stood from her; we followed his example, but it blowing very fresh at N.W. in about two hours the sail came alongside of us, which proved to be his Majesty's ship *Unicorn,* of 20 guns, commanded by John Ford, Esq., who took all the prisoners on board his own ship, and sent two officers and twelve of his men on board of us to take charge of the vessel. We then stood for Newport with the *Unicorn* in company with us till morning, when she left us and went in chase of a sloop, and on the 16th of February we got safely to anchor in the harbour. The privateer took all our letters and papers from on board us. As soon as we arrived here I put the Boatswain on shore, intending to get him into the King's Hospital, which I was advised not to do by the doctor, as he had no disorder but his wounds, and that if he went there he might contract some distemper that would retard his cure. I have since procured him private lodgings and believe his leg will be amputated in a few days.

"Captain Wise and his boy, who were taken by the same privateer in her last cruize, are on board, and I believe they will come home with us. When our ship was taken we had very little water, and the privateer was in the same case; they therefore thought proper to unstow our hold, and take our wine, upwards of 180 firkins of butter, two hogsheads of bread, two barrels of flour, four barrels of pork, a few bags of barley, pease, &c., five or six barrels of gunpowder, some small arms and sundry other articles on board of them, for fear of our vessel being retaken."

In 1779, the *Thomas*, Captain Barton, and the *Sarah*, Captain Hooton, took a prize on their passage to Grenada. The *Sarah* was subsequently lost at Anguilla. The *Thomas* also captured a schooner laden with tobacco, and sent her into St. Kitts.

A letter from Waterford, dated May 1st, says: "We have a certain account of five rebel privateers off Cape Clear and Kinsale, waiting, it is imagined, for the Newfoundland fleet from hence, and transports from Cork. The convoy is only a sloop of 14 guns. They take our ships in our channel. Two of this port are taken. Where it will end I know not." In the same month it was calculated that the value of the West India ships that had fallen into the hands of the Americans was upwards of £400,000 sterling. The following was given as "a perfect list of the naval force belonging to the Congress," exclusive of which there were upwards of 100 sail of privateers:—

SHIPS.	GUNS.	COMMANDERS.
The Virginia	28	James Nicholson.
Hancock	32	John Manly.
Boston	24	Hector M'Neal.
Trumbull	28	Dudley Saltonstall.
Randolph	32	Nicholas Biddle.
Rawleigh	32	John Thompson.
Effingham	28	John Berry.
Washington	32	Thomas Reed.
Congress	28	Thomas Guinall.

Ships.	Guns.	Commanders.
Delaware	24	Charles Alexander
Reprisal	16	Lambert Weeks.
Providence	28	Abram Whipple.
Warren	32	John Hicks.
Montgomery	24	John Hodge.
Lexington	16	John Hallock.
Hampden	24	Hoysted Hacker.
Andre Doria	14	Isaiah Robinson.
Providence	12	John Paul Jones.
Defence	20	James Josiah.
Alfred	28	Elisha Hinman.
Cabot	16	Joshua Olney.
Sachem	10	James Robinson.
Independence	10	John Young.
Fly	10	Elisha Warner.
Columbus	—	—— Cook.
Wasp	8	Lieut. Baldwin.
Musqueto	4	Lieut. Abberton.

In June, 1777, the *Marlborough*, Captain Dawson, on her passage to New York, took the *Three Brothers*, Bentley, from South Carolina, with rice, staves, and indigo, and sent her to Liverpool. The night before they made Cape Clear, Captain Bentley attempted to kill the prize-master, but was prevented by the vigilance of the cabin boy. In 1778, Captain Dawson brought into the Mersey a brig from Boston, laden with boards, cider, etc., which he had taken on his passage from Philadelphia. The *Marlborough* was captured and carried into Bordeaux in August, 1780.

His Majesty's ship, *Ariadne*, took the *Musquito* privateer, 16 guns and 72 men, commanded by Captain Harris, and fitted out in Virginia, by Captain Younghusband, formerly of Liverpool. The *Mary Ann*, Captain Leigh, of Liverpool, and her consort, a sloop, captured thirteen prizes, valued at upwards of £10,000, and carried them into Tortola. The *Mary Ann*, on her homeward passage, was lost on the

Tuskar, but all the crew were saved, as well as the indigo portion of her cargo. The *Hero,* Captain Woodville, a Letter of Marque, 16 guns, took a double-decked schooner, 150 tons, with boards, staves, etc., in the West Indies, and the *Valiant*, Captain Naylor, took an American sloop, with lumber, etc., from Boston. The *Laurel*, Captain White, on her passage from Dominica, took a schooner bound from Bilbao to Philadelphia, with bale goods.

Captain Wm. Buddecome, of Liverpool, was presented with a silver cup, value £18, by the merchants and masters of 18 vessels belonging to New York, for his care in convoying their fleet between the West Indies and New York, after the *Falcon* sloop of war had separated from them. Captain George Ross, of London, was also presented with a silver cup, value £12, for assisting Captain Buddecome.

Letters from Dumfries mentioned that two American privateers had anchored in the Solway Frith. One letter says they had taken 14 prizes; another 9; another account stated that two American vessels had appeared off Kintyre; that they had made 15 prizes, three of which they burnt, and sent the rest to some port in France. A letter from Jersey complained that the American privateers grew daily bolder, having the effrontery to cruise even between Jersey and Guernsey. The largest of them were only 10 gun vessels, and on being pursued, they immediately made for St. Malo.

The following curious extract is from a letter received by a Liverpool merchant from St. Vincent, and dated May 5th, 1777:—

"I had the pleasure of writing to you a few days since by the *Bess*, which ship we are in hopes is got clear of the islands, as we find she has not been carried into that infernal place Martinique, the nest of damned piratical scoundrels. The great frigate, *Oliver Cromwell*, took three prizes in last week, one a Guineaman with 300 slaves, one a ship from London; have not yet heard what Guineaman she is, nor what the other vessel is. The ship *Champion*, belonging to Bristol, is taken at Tobago; she was going from one bay to another, and had 150 hogsheads sugar, and 22 bales of cotton on board. They are now discharging her at St. Lucia. We do not find that the Americans are so much protected any where as at Martinique and St. Lucia, which is under the same government. Was their trade and communication cut off there, which two frigates would in a measure do, no step the ministry could take would distress them more; it is from thence they are supplied with every thing they stand in need of, not only arms and ammunition, but men. A great many experienced officers are gone to the continent; and we have numbers of privateers that are manned with French; some have only an American, and that perhaps a landsman, just to cloak their piratical proceedings. It is said (and I believe founded on truth) that Mons. le Compte D'Argout, present governor of Martinique, is concerned with Bingham, the agent to the Congress, in nine privateers. There are now about 20 sail of English ships in Martinique. Negroes are cheaper there than in Africa, and provisions than in Ireland."

The *Sisters,* Captain Graham, a Liverpool slave ship, was taken on her passage from Africa to the West Indies, and carried into Martinico with 163 slaves on board. The *Lydia*, Captain Dean, from Jamaica to Liverpool, was taken, and sold in Maryland, with the cargo, for £20,400 currency. The *Grace*, Captain Wardley, taken by the *Lexington* privateer in the Irish Channel, was ordered to France, but was recaptured by the prize master and some of the people on board, carried into Torbay, and thence to Liverpool. On the 2nd of June, 1777, the *Elizabeth*, Captain Byrne, on her passage from Liverpool to Jamaica, fell in with the *Fly*, an American sloop privateer, of 14 guns and 104 men, whom she fought for an hour and-a-half, when the American ran alongside and boarded the *Elizabeth*, sword in hand. The captain and crew of the *Elizabeth* were cut and mangled in a shocking manner; three of them were killed and thirteen wounded. The *Johnson*, Captain Jones, from Liverpool to New York, made a gallant fight of three hours against three American vessels (a brig of 14 guns, and two schooners of 10 and 12 guns), but in the end was forced to accompany his captors to Boston. The *Johnson*, re-named the *Marquis de la Fayette,* was subsequently captured by Sir George Collier.

Although all is said to be fair in love and in war, we cannot but feel that the inducements offered to American sailors by the British government, for the treacherous capture of American ships, were unworthy of the British name. In *Williamson's Advertiser*, of July 18th, 1777, we read as follows:—

> "As his Majesty has declared that the sailors on board the American ships, who shall take the same from the person or persons having the command thereof, and bring them into any English ports, shall have two-thirds of what such ships and cargoes shall be sold for (and for three years be exempted from being impressed), it is apprehended that when the same shall be properly known among the sailors, it will be the means of bringing many a valuable ship and cargo into England, which would otherwise go to the French ports."

That this bait was intended for American sailors, as well as for British seamen in the rebel service, is very clearly seen from a case which happened in Liverpool earlier in the same year. In January, there entered the Mersey, "in the presence of thousands of rejoicing spectators," the ship *Oxford*, from York River, in Virginia, laden with 412 hogsheads of tobacco and staves. She had been captured on Sunday, the 11th of January, on her passage to Nantz, by four of the ship's company, who overcame the rest of the crew, eight in number, besides the supercargo and a passenger. It appears that these "four resolute, brave men, two of whom were Liverpool and the other two Lancashire lads," had, in conjunction with others of their comrades, who were put on board other vessels, combined together in Virginia to attempt, on their passage, to take the respective vessels they belonged to. "These brave sailors," observes the paper of January 23rd, "say that a brig loaded with tobacco, from the same place, with some men, part of their association, may be daily expected to arrive here. The *Oxford* originally belonged to Glasgow, and was in the transport service when took by the Americans. As this vessel was not taken by the officers and seamen of his Majesty's ships of war, she becomes, by virtue of an Act of the last session, and also the cargo, forfeited to his Majesty, who will, no doubt, reward the brave captors with the whole, or the greatest part of this valuable prize; which it is hoped will encourage all seamen, that may be engaged in the Rebel Service, to imitate these brave fellows."

Another ship called the *Aurora*, was captured under similar circumstances. She had sailed from America for Nantz, with about 416 hogsheads of tobacco from York River, and was manned by Americans, Frenchmen, and four or five Englishmen. On the passage the Englishmen contrived to make prisoners the captain and the rest of the crew, when, putting her head to the wind, they brought the vessel safely into the Mersey. She was condemned as a droit of Admiralty, and the tobacco was sold "duty free and for inland consumption only" at Messrs. Backhouse's warehouse in Church Street.

"The Lords of the Admiralty," says the paper of July 4th, 1777, "have rewarded the brave seamen who took and brought in here the *Aurora*, loaded with tobacco, in January last, with two-thirds of the cargo, which will be to each of the eight English seamen, who made the seizure, £1828 2s. 9d., and to each of the five American seamen, who assisted to bring the ship into this port, £914 1s. 4½d.*

* From this it would appear that Sir James Picton, in his "Memorials of Liverpool," and the compiler of the "Annals of Liverpool" in Gore's Directory, are

The names of the five Americans were Jesse Jenkinson, Jesse Topping, Gilbert Welsh, Joseph Walker, and Hugh Johnson. On their arrival here, they were impressed on board his Majesty's tender; but, on their applying to the Admiralty, they will, without doubt, each receive their respective shares."

Something of a similar nature occurred in 1862, when a ship called the *Emilie St. Pierre* arrived in the Mersey, commanded by Captain Wilson, who had recaptured her from a Federal American prize crew. Sir James Picton, referring to a certain villa in Everton Terrace, built by Mr. James Parke about the beginning of the present century, and occupied by him for many years, says, "its last occupant, Captain William Wilson, was rather a noticeable man. In 1862, when in command of the *Emilie St. Pierre*, he endeavoured to run the blockade into Charleston, South Carolina, but was intercepted and captured by the Federal war steamer *James Adger*. A prize crew was put on board to carry the vessel to Philadelphia, Captain Wilson, with his steward and cook, being alone retained. By an amazing combination of stratagem and daring, the whole crew were made prisoners and put in irons by Wilson and his two assistants, who, unaided, navigated the ship and brought her in safety across the Atlantic into the Mersey, where she arrived on April 21st. As might naturally be expected, Wilson received quite an ovation.* By a subscription

incorrect in stating that "thirteen seamen received each £1,828 2s. 9d. as share of prize money, being only one-third of the value of the prizes taken." Stonehouse, in his "Streets of Liverpool," states that the cargo of the *Aurora* sold for upwards of £30,000. The proportion received by the seamen was, therefore, two-thirds (=£19,195 8s. 10½d.) as graciously ordained by that wise monarch, who lost us thirteen colonies. We suspect that the *Oxford* was really the *Aurora*, as the latter name does not appear in the arrival lists, although her cargo was advertised and sold, whereas there is no further mention of the *Oxford* and her cargo. Other ships, however, captured in the same manner, arrived in Glasgow and other ports.

* The present writer, then a boy, had the pleasure of being in the company of Captain Wilson a day or two after his arrival in Liverpool, and the romantic impression made by seeing and hearing one's first live hero remains undimmed after the lapse of 35 years

amongst the merchants, he was presented with a gold chronometer, and a tea and coffee service. From the Mercantile Marine Association he received a gold medal, and from the owners the sum of £2,000. The cook and steward received £320 each. Captain Wilson died in September, 1868, and the house has since been removed.*

The following is a copy of a letter from Captain Nehemiah Holland, probably the same who so bravely defended the *Ann Galley*, as already recorded :—

> "At sea, ship *Sarah Goulburn*, lat. 44.0 N. long. 39.00 W., 19th July, 1777 :—
>
> "GENTLEMEN,—I congratulate you upon a prize I have taken this day, named the *Sally*, Thomas Tracy, master, from Charles Town, South Carolina, bound to Nantz, loaded with 470 whole, and 120 half barrels rice, and betwixt twenty and thirty casks of indigo. I have put in Mr. Smith as prize master, who will acquaint you of every particular since our sailing. Am in a hurry to dispatch the prize, as I am informed there were 30 sail more to sail from Charles Town the day after them, and am anxious to be amongst them. You'll please remember me to my friends, as I have not time to write them. I remain, Gentlemen, Your much obligeed humble Servant, N. HOLLAND.
>
> "P.S. Mr. Smith has behaved very well with me, and executed his office as I could wish. Would be much oblished to you to assist him in another birth."

Mr. Smith brought the valuable prize safe into port in August.

"On Sunday last," says the Liverpool paper, of July 25th, 1777, "arrived here, the ship *Pole*, Captain Maddock, in twenty-four days from New York. On the 12th inst at p.m., in lat. 50°, long. 20°, she fell in with the *Tartar*, a rebel privateer, mounting 20 nine-pounders on the main-deck, 8 four-pounders on the quarter-deck, and 4 four-pounders on the forecastle, full of men, supposed two hundred at

* "Memorials of Liverpool," vol 2, pp. 357-8.

least; had an image head, and quarter galleries. All her guns on the main-deck were painted black; those on the quarter-deck and forecastle red. The ship was painted black and yellow, with tarred sides, and short topgallant mastheads. She bore down on the *Pole* under English colours, enquired from whence she came, and whether she was a King's ship. Being answered in the affirmative, the captain gave orders to hoist the Thirteen Stripes, and fire away, on which the engagement began, and continued from five until about twenty minutes past eight, when the privateer sheered off. Captain Maddock had two mates and a passenger wounded, and supposes that near one-half of the people belonging to the privateer must be killed or wounded, he having cleared their forecastle of men three different times, and says he heard dreadful cries among them. The *Pole* had 16 six-pounders, and only forty people, passengers included. Both officers and men behaved gallantly, and to Captain Maddock's entire satisfaction. One of the passengers, an elderly woman belonging to Liverpool, but who had been twenty-seven years in America, handed the cartridges to the men. The ships were within hail of each other during the whole engagement. The word "*Tartar*" was observed on the privateer's stern, and by a list handed about at New York, Captain Maddock finds she was commanded by one Davies, a Welshman, and mentioned there to have 32 guns."

On her passage from Liverpool to New York, in September, 1777, the *Pole* took the *Friendship*, from Bordeaux for Boston, but the prize was retaken by an American privateer. In 1778, the *Pole* took the *Hannah* schooner, and sent her to Jamaica, and the *Prince and Liberty*, an American brig laden with wine, rum, molasses, and dry goods, which she carried to New York. In 1779, she captured the *Salisbury*, from Maryland for Nantz, with 140 hogsheads of tobacco, and the *Hector* from Martinico for

France with sugar, coffee, and cotton. The *Pole* was herself taken the same year, on her passage to Jamaica, by the *Boston* and *Confederacy*, "Continental frigates," and carried into the Delaware.

According to the *London Gazette*, of the 11th of July, 1777, the English cruisers on the coast of America, captured between the 1st of January, 1777, and the 22nd of May following, 203 American vessels, besides recapturing fifteen British vessels taken by the Americans. Thus the work of destruction and the ruin of commerce proceeded with equal vigour on both sides, for the American privateers wrought havoc in the Channel.

The *Gregson*, Captain Wotherspoon, was attacked by two privateers, but beat them both off.

The *Fancy*, Captain Allanson, on the passage from Jamaica to Liverpool, had an engagement in the Gulf with an American privateer of 10 guns and 50 men, killed three of her men, and obliged the captain to produce his papers, which were French, "and then let him go about his business."

The *John*, Captain Watkins, from Liverpool for Halifax, was taken by an American privateer, and retaken by the *Milford* man-of-war, who put a midshipman and two seamen on board to carry her to Halifax, but in her passage, she was again taken by an American privateer.

The paper of July 11th, 1777, contains an abstract of the Act authorising the carrying of the captures therein mentioned into any part of his Majesty's dominions in North America; and for ascertaining the value of such parts of ships and goods as belonged to the recaptors. After reciting the Act of George III. (for prohibiting all trade and intercourse with the rebellious colonies; and two other Acts for restraining the trade and commerce of the said colonies; and to enable persons appointed by the King to grant pardons, issue proclamations, etc.), it enacts:—

"1. That any persons authorised by the King to grant pardons, &c., may by licence or warrant authorise captors, or other persons in their behalf, to carry their captures into any port, &c., in any of his Majesty's dominions in America :

"2. That all captures already carried into New York, or which, before August 1st, 1777, shall be carried in there, or into any of his Majesty's dominions, with such licence aforesaid, shall be deemed to have been lawfully carried into such port ; and after condemnation, may be brought into this kingdom, or any of his Majesty's dominions, upon payment of the same duties, and subject to the same regulations as they now are :

"3. That when any ship, &c., taken by virtue of the above Act, or any goods therein, shall be proved in any of the Admiralty or Vice-Admiralty Courts in America or the West Indies, to have belonged to the subjects of Great Britain or Ireland, or any dominions in allegiance to the King, and to have been taken by any of the inhabitants of the rebellious colonies, and to be in the possession of such unlawful captors, when retaken; such ship, &c., shall be restored to the owners, they paying one eighth of the value to the recaptors, or giving sufficient security to do so, for salvage ; and the judge of the court wherein such ship shall be decreed to be restored shall cause the same to be appraised by persons named by the claimant and recaptors, or they not agreeing, by the Court, such persons being sworn truly to appraise the same ; and no retaken ship, &c., shall be sold for payment of salvage, or on any other account, unless with the owner's consent, except where there shall be no claim for such retaken ships, &c., in which case, the said judge shall order as much of the cargo to be sold as will pay the said 1/8th and the expenses of appraisement, &c., and if the cargo be not sufficient, the ship, &c., to be sold, and the remainder, after paying the salvage, to be deposited in the registry of the Court, for the owners, who may afterwards claim the same ; and except also any part of the cargo appear in a perishing condition, when the same may be sold for the benefit of the concerned."

The *Fanny*, Captain Wignall, on her passage from Liverpool to Halifax, had an engagement with an American privateer, of 16 or 18 guns, for two hours, when the "rebel" sheered off. Captain Wignall believed that she sunk, as he lost sight of her about two hours after she left him, although the weather was clear and but little wind. Captain Wignall expended in the engagement—"89 rounds of shot, 18 lb. double-headed ditto, six canisters of copper dross, 295 lb. of grape shot, in number 1295, making in all 1420 shot and about 250 lb. of gunpowder, beside musquet shot, which was a great many." His officers and men, he tells us, "behaved like true British tars."

In July, 1777, the Lords Commissioners of the Admiralty informed the merchants of Liverpool that they had stationed his Majesty's ships the *Albion*, *Exeter*, *Arethusa*, and *Ceres* to cruise between the coasts of Great Britain and Ireland in quest of American privateers, and for the protection of the trade in those ports. The commanders were directed to enquire for intelligence respecting such privateers, as follows:—The *Albion* and *Ceres* at Dublin and at Campbeltown; the *Exeter* at Milford Haven and Cork alternately; the *Arethusa* at Whitehaven, in her way up channel, and afterwards at Campbeltown and Carrickfergus. Other cruisers were stationed between Scilly, the coasts of Ireland, and Milford Haven for the like purpose. It was certainly high time for their Lordships at the Admiralty to bestir themselves, for in this very month of July the American privateer, *General Mifflin*, of 20 six-pounders, fitted out at Boston, and commanded by Walter Day, made her appearance in the Irish Channel, and captured many prizes, including the *James*, from Glasgow to Oporto, taken the day after she left Glasgow; the *Rebecca*, from Liverpool to Limerick; the *Mary and Betty*, from Liverpool to Ballyshannon; and the *Priscilla*, from Sligo to Liverpool, with linen yarn. Most of these vessels, and of the other prizes taken by the

American privateers, were sent to France to be sold. The *Mary and Betty* was given back to the crew, after being plundered of the most valuable part of her cargo. The *General Mifflin* was originally a Liverpool vessel called the *Isaac*, engaged in the West India trade, and commanded by Captain Ashburner, and most likely the very ship commanded by Captain Clatworthy during the Seven Years' War, both as a privateer and as a slaver. The Liverpool paper gives the following account of the ill-treatment of Captain Richard Cassedy, of the *Priscilla*, who was taken by the *General Mifflin* :—

> "These sons of freedom seized all the captain's clothes that were worth anything, and £88 in cash ; every one of his men they took on board the privateer, plundered the vessel of spare rigging, stores, &c., and one bale of linen, part of the cargo, which chiefly consisted of yarn ; and after leaving several of the crew on board, ordered the captain to be bound hand and foot, and put into confinement. In this miserable situation he remained until the 19th of July, when his vessel was retook by the *Union*, letter of marque, of London, within ten leagues of Bordeaux, and carried into Fowey. The privateer's people split all the sails, except the foresail, by carrying, whenever they saw any vessel. Captain Cassedy was in a very poor state of health when they arrived at Fowey, and not able to stand, through the cruel treatment he received. His remaining so long bound occasioned his flesh to swell to a shocking degree. All his prayers and intreaties were in vain ; the inhuman tyrants had no compassion. Surely the fear of a single man retaking the vessel, could not induce them to this barbarity. She was a constant Irish trader, had not a single gun on board, nor ammunition, or warlike weapons of any kind."

Captain Edward Forbes, of the ship *Sparling*, 300 tons burthen (carrying 10 six-pounders, 4 four-pounders, and 8 swivels), belonging to Mr. John Sparling, writing from Kingston, Jamaica, on July 23rd, 1777, says :—

"I arrived here safe, after a passage of six weeks. In crossing the Bay, I saw several ships, but passed none without bringing them to, boarding and examining them. Off the Western Islands saw a brig privateer, who, on my giving chase, thought proper to alter his course and make the best of his way. I saw nothing more until I was within fifteen leagues of this island, where I was attacked by a large privateer sloop of 12 guns, a number of swivels, blunderbusses, and full of men. They attacked me at four o'clock in the afternoon, with a great deal of vigour, for an hour and-a-half, and then sheered off. I attempted to give chace, but soon found she could sail two feet for my one. She then got her graplins out for boarding when dark, and attempted it three times, but perceiving his intentions, disappointed him, and threw him off his guard, by which means I got our guns to bear, which made him sheer off to refit; this I was not sorry for, as it gave us an opportunity to do the same, for our braces and running rigging were often shot away. The engagement lasted six hours. Getting all my guns to bear, in less than a quarter of an hour I lost sight of him. The ship is a good deal damaged in her rigging, sails, and hull, but no lives lost, which is owing to the good shelter we had on deck, as she constantly fired small arms. I found the sloop had great advantage over us, we being square rigged, she always kept on my quarters. The *Sparling's* sides are hard, but the yankies found means to shew daylight through her in several places, but hope to repair her at a small expence. There is three feet of the starboard quarter entirely knocked out, and some shot in the bends. My men behaved during the action with the greatest courage, and very attentive to command."

In 1778, the *Sparling* took a prize named the *Isaac*, which was recaptured by a privateer, which in its turn was taken by a King's ship.

Captain James Collinson, of the ship *Will*, writing to his owners from Dominica, on October 13th, 1777, says:—

"I congratulate you on our arrival here, on the 8th of

October. On the 7th, we fell in with a rebel privateer sloop with 10 guns and 16 swivels, which we defended ourselves from for full five hours. She boarded us on the larboard quarter with twelve men, which we killed, and made them cry out for quarter several times, but still kept a brisk fire upon them, and paid no regard to their crying for quarter, as they still had their colours up. They were half an hour under our quarter, where we made them fast to us. By their cutting their ropes they cleared from our quarter; then we stood to the northward, and cleared ship ready for them again; when clear, gave chase, and came up with them, gave them three broadsides and three cheers, and left them, as we should run a risk of losing the ship if we had taken them all, though I imagine we killed 40 or 50 of their men; and by information, I find she had on board 120.

"At noon the same day, we fell in with a schooner of 14 carriage guns, which we fought for six hours, and gave them the same as above, but are not certain what number of men we killed, as she did not board us, but was prepared with stinkpots on her bowsprit end. We should have sailed for Jamaica in three days after our arrival, but for the damage we got by the sloop; she carried all the iron stantions away on the larboard quarter. We have not one man killed or wounded, and all behaved like true Britons. We fired the small arms three times for their twice, and every man obeyed his orders. I will write you more particulars from Jamaica, as I strained my forefinger on my right hand in the engagement, but came to no more hurt, although there was a swivel ball came through the speaking trumpet in my right hand. We boarded five sail coming out, but could not condemn them."

Immediately on the *Will's* arrival at Dominica, the following paper was subscribed by a number of gentlemen, and £72 6s. collected.

"For the encouragement of the twenty-five brave fellows belonging to the ship *Will*, Captain Collinson, who, on the 7th inst., gallantly defended the said ship for five hours against a rebel privateer of fourteen carriage guns and about 120 men, and obliged the privateer to sheer off; likewise a schooner privateer of the same force. We subscribers hereto, have given the sum opposite to our respective names."

At this period, on the arrival of privateers, slave ships,* and other armed vessels in the river, it was customary to salute the town with a discharge of cannon, which, from negligence, were sometimes loaded with ball. For the prevention of accidents, the following order was issued by the magistrates:—

"The late alarming circumstances of vessels coming from sea, and those lying in the river, frequently firing balls from their cannon, to the great and imminent danger of the lives and property of the inhabitants of this town, we, the grand jury, having a power invested in us for that purpose, do order, that every captain, or any other commanding officer for the time being, of any ship or vessel, suffering any cannon loaded with ball, or any other shot whatsoever, to be fired from on board such ship or vessel, after such ship or vessel has come round the Black Rock, shall pay twenty pounds for every and each gun so fired, loaded with ball or other shot. And we farther request, that this order may be publicly printed in the newspaper, that no person may plead ignorance thereof; and we recommend that the penalty hereby inflicted may be appropriated to the fund of the Seamen's Hospital."

In a letter from Captain James Wiseman, of the *Isabella*, of Liverpool, dated St. Vincent's, January 20th, 1778, is the following description of a hot engagement between her and American war vessels fitted out by the Congress:—

* The paper of November 14th, 1777, gives an instance of this dangerous practice of firing guns when entering the port, adopted by the slave ships:— "Wednesday before last, a Guineaman coming in and firing, a shot from a six-pounder passed very near a servant of Richard Parry Price, Esq.; and broke a tree near to his pleasure ground at Berkhead. Care certainly ought always to be taken to draw the shot before the guns are fired in the river. Yet, tho' nothing can be more absurd, dangerous, and deserving of punishment than thus firing with shot, this is the third instance which has happened lately. One of the shot passed thro' Water-Street and Dale-Street, and another over the Old Churchyard. So imprudent a practice should be checked by the merchant no less than the magistrate." Mr. Price was lord of the manor of Birkenhead.

"We sailed from Cork 15th December, and had a fine passage of four weeks; and on the 8th January met with an American brig privateer of 16 guns, and fought her for two hours and a half, yard arm and yard arm. We gave her the first and last broadside. I believe she is sunk. We had killed Mr. Godwin, passenger, and John Taylor, seaman, and wounded John Manesty; third mate shot in the hand, which is since amputated, and he is likely to do well; Rowland Evans shot in the leg, since amputated and he is dead; John Jones shot twice through the knee; we expect he will recover. John Webster received a shot in the thigh and another in the arm, but likely to do well; six or seven others slightly wounded. We received 132 shots in our hull and masts, a six pounder went through our mainmast, six foot above deck, and four others higher up, and our main top mast almost shattered to pieces, three shots in our mizen mast, one of them about six foot above deck, and numberless in our hull, most of them betwixt wind and water, and all our rigging entirely shot away. Our Ensign halyard being shot away, and the Ensign falling down, the privateer thought I had struck and gave a huzza, which was answered by a broadside from us. The Captain hailed me to strike, telling me he would never leave me, which I believe were his last words, for I never saw or heard him afterwards; in short, the engagement was hot, and I believe fatal to them, for we could see them falling out of the tops, and hear their shrieks and groans. It falling dark and our rigging being cut to pieces, we could not work our ship, and so lost our prize. The next day, we were chased by a sloop, but when she came within view of our guns, she hauled her wind and run away; our rigging being gone, we were in no condition to follow her. On the 11th we were chased by a brig and a sloop, who soon came up with us (the brig first, and hailed from Halifax, bound on a cruize to the westward) and then dropped astern to his consort, when we got our stern chaces to bear on them and began to fire away, our people still in good spirits; the third shot we carried away the brig's cross jackyard, sent several shots into her bows and rigging, and beat them both off. The brig stayed along side of us for two hours, and told me that General Burgoyne and his army were defeated by the rebels. The next day, being the 12th, we arrived at St. Vincent's and were received by every one with great applause."

The following additional particulars appeared in the *St. Vincent Gazette* of March 7th, 1778:—

"The brig that first engaged the *Isabella*, Captain Wiseman, was the *General Sullivan*, Capt. Darling, of 14 guns, 4 and 6 pounders, fitted out by the Congress; and had at that time, by the Captain's account, 135 men on board, most of whom were able seamen. She arrived at Martinico a few days after the engagement in a most shattered condition, her mainmast so much wounded that Capt. Darling was obliged to get another; the bowsprit carried away, and the hull, rigging, etc., greatly damaged. Captain Darling says he had eleven men killed and twenty-three wounded, many of them very dangerously,—and gives Capt. Wiseman and the crew great credit for their spirited behaviour and good conduct. He expressed great surprise when he found the *Isabella* had only fifty men; acknowledged he was obliged to sheer off, and that it was the second drubbing he had got from Liverpool men, and wished not to meet with any more armed vessels belonging to that port. There is certainly a great deal of propriety in his remark, as the merchants of Liverpool have entered more into the spirit of arming ships than any others in England, in the present contest. Being a mercantile people, they choose to bear the additional expence, rather than have their trade annihilated, which has raised their town, in the last century, from an obscure fishing place to that of being the second commercial port in Great Britain. The second brig that engaged the *Isabella* was the *Resistance*, Captain Tue, of 14 guns and 100 men, and the sloop that was in company was the *Rambler*, Capt. Stanton, of 10 guns and 70 men.

"The zeal and loyalty of the Merchants of Liverpool, in favour of Government," says *Williamson's Advertiser*, of

February 27th, 1778, "is eminently evinced by the number of vessels they have already armed and stationed for the annoyance of the commerce and communication with the natural enemies of Great Britain. The following are now cruising in the American seas, on the coasts of Carolina and Virginia, the *Sarah Goulburn*, Captain Holland, of 20 six and nine-pounders; the *Brilliant*, Captain Priestman, of 20 six and nine-pounders; the *Belcour*, Captain Moore, of 18 six and nine-pounders; the *Pole*, Captain Maddock, of 18 six and nine-pounders; and the *Active* sloop, Captain Powell, of 12 four-pounders."

In February, 1778, an enquiry took place before the House of Lords, as to the amount of injury done to British commerce from the beginning of the war, in which it was stated that the number of vessels destroyed or taken since the commencement of the war was 773, or, after allowing for those retaken, 559; that their value, at a very moderate computation, was £1,800,000; that of the ships thus taken, 247 were engaged in the West India trade; that all articles imported from America had risen enormously in price; tobacco from 7½d. to 2s. 6d. per lb.; pitch from 8s. to 35s. per barrel, and tar, turpentine, oil, and pig iron in the same proportion.* It was considered a sufficient answer to this statement to show that the English cruisers had taken 904 American vessels, of the value of £1,808,000. "It was forgotten," says Baines, "that the enormous sums taken from the merchants of England were not transferred to the merchants of America; nor those taken from the merchants of America transferred to those of England; but that the whole were taken from commerce and turned into prize-money."

In January, 1778, Captain Jolly, in the *Ellis*, on the passage from New York to Liverpool, took the *Endeavour*,

* In 1777, sugar sold in Carolina at upwards of £5 sterling per cwt.

from North Carolina, with flour, staves, and 43 hogsheads of tobacco, and the *Nancy*, from Essequibo, with coffee, cotton, and 115 hogsheads of rum. Later in the year, Captain Jolly, in the *Gregson*, and Captain Washington, in the *Ellis*, cruised in consort, and took the *La Ville du Cap*, from St. Domingo to Nantz, with 224 hogsheads, 6 tierces, and 12 barrels of sugar, 392 casks and 275 boxes of coffee, 16 bales of cotton, 45 barrels of rum, and 6 barrels of indigo. Their next prize was *L'Aigle*, "a large new snow, 70 feet keel and 24 feet beam, pierced for 16 guns," from Port-au-Prince for Nantz, with sugar, cotton, indigo, coffee, etc. In October, the *Gregson* took a privateer sloop of 10 guns, eight swivels, and 64 men, threw the guns overboard, dismantled her by taking on board her swivels, small arms, spare sails, cables, etc., and sent her home. On October 23rd, the *Gregson* boarded a snow with passengers, belonging to St. Pierre and Miquelon, but they had a pass from Admiral Montague, to go unmolested to France. On the following day, the *Gregson* captured the snow *La Genevieve*, from Nantz to St. Domingo, laden with wines, flour, etc. The *Ellis* also took the snow *Josephine*, bound for Dunkirk, loaded with oil, soap, brimstone, casks of straw hats, and boxes of lemon. "Yesterday," says the paper of November 27th, "arrived here the *Gregson*, Captain Jolly, from a cruize, and brought in with her a large ship bound from St. Domingo to Nantz. This is the sixth prize, the *Gregson* and *Ellis*, who sailed in consort, have taken; three of them fine ships from St. Domingo to France." In May, 1779, we read that the *Ellis*, Captain Washington, "who had sent into Liverpool five valuable prizes," had been captured, and carried into Martinico, and in March, 1780, it is stated that the *Ellis* had been re-taken by Admiral Parker. She was a vessel of 340 tons burthen, carrying 28 guns and 130 men; the *Gregson*, of 250 tons, carried 24 guns and 120 men. Both vessels belonged to Messrs. Boats and Gregson.

The *Clarendon*, Captain Amery, arrived at Jamaica from Liverpool, with a fine brig, called the *Defence*, from New London for Martinico, with staves, lumber, etc., which she had captured on her passage.

A Liverpool vessel, commanded by Thomas White, fell in with an American privateer, of 14 guns, and upwards of 100 men, to windward of Antigua. For some reason, Captain White's people refused to fight, whereupon the brave commander blew his vessel up, and only ten of the crew were saved, and put on board a Dutch vessel.

The *Tom,* of 12 six-pounders, Captain Lee, arrived at New York, after a passage of eight weeks, from Liverpool, during which he captured two vessels loaded with fish and lumber, and a schooner of 10 guns and upwards of 40 men. As he could not spare a prize crew to take the schooner home, he took out her guns, disabled her mainmast, and providing "the rebels" with bread and water, turned her adrift. Soon afterwards, he met another privateer of 12 guns, which he fought for over an hour, and would have taken her had she not greatly outsailed him. In 1779, the *Tom* arrived at Antigua with a prize laden with fish and oil, and in February, 1781, he captured the *De Koningin Esther*, laden with 200 hogsheads of sugar, 1,000 bags of coffee, 300 bags of tobacco, 100 bags of cocoa, 300 hides, and 24 casks of indigo; and also the *Jacobus*, with 140 casks of sugar, 1,400 bags of coffee, 48 casks of indigo, 100 bags of tobacco, 200 bags of cocoa, and 1,200 hides. A few months later, the *Tom*, in company with the *Greyhound*, captured a French cutter of 16 guns and 120 men, and carried her into Londonderry. The cutter had four ransomers for 150 guineas on board.

"The *Liberty*, Wardlaw, and the *Prince George*, Gardner, both from Martinico, are safe arrived here," says the newspaper of March 20th, 1778. "They being both defenceless and arriving safe without molestation, we may reasonably presume the privateers are not so numerous, at least in the homeward bound track."

The *John*, Captain Watkins, and the *Suffolk,* Captain Bower, both missing ships, from Liverpool bound for New York, arrived safe at Antigua, having captured a large schooner laden with tobacco, etc.

The *Sally*, Captain Smith, upon a cruise in the West Indies, took an American vessel loaded with timber, and sunk a privateer of 12 guns ; and another ship, belonging to Liverpool, took three prizes on the coast of Carolina. The *Toms*, Captain Houghton, also cruising in the West Indies, took a schooner laden with spermaceti, candles, etc.

The *Lydia*, Captain Evans, on her passage from New York to Barbadoes, took a very valuable ship loaded with masts, etc.

The *Richard*, Captain Lyon, arrived at New York from Liverpool, with a prize worth £4,000, taken on the passage.

The *Sparling*, Captain Denny, arrived at Philadelphia, with a vessel from Charleston for Amsterdam, loaded with rice, etc., which she had captured.

The *Mersey*, Captain Gibbons, on her passage to Philadelphia, took a schooner bound to Boston, loaded with coffee, molasses, etc.

In February, 1778, the *Fanny*, Captain James Wignall, arrived in Liverpool from Philadelphia, after a passage of thirty-one days, bringing in 35 American prisoners taken out of an American privateer, captured by her on the passage, and ordered for Liverpool. The *Fanny* was herself taken in the following July near Sandy Hook, on her passage from Liverpool to New York, by one of the French fleet under Count d'Estaign. The following account of the usage of the prisoners by the French was written by Captain Wignall :—

"About seven in the evening the *Langezant* frigate of 36 guns took us. As soon as the boat came alongside, we were hurried into it, without clothes or bedding, the officer promising us all our clothes, &c., should be safe, and that we should have them in the morning. We were carried on board the frigate, and remained there about an hour, then part of us were sent on board the *Languedoc*, Count d'Estaign's ship. They put us down in the forehold altogether, and about five in the morning ordered us up on the forecastle, where an officer came and searched us all one by one, all our pockets, shoes and stockings, &c., and took from us all our money, watches, papers, &c., then ordered us down into the hold again, where we were almost smothered, and not so much as a drink of water. About noon, I made a motion to the sentinel to permit me to go upon deck on a necessary occasion, which was granted. I then went upon the quarter deck, where I found an officer that could understand English. I desired that we might have permission to come on deck to have air, and that we might have some provisions and water; the answer was we had been forgot. We were then ordered upon the poop, and had served us some bread and stinking cheese, half a pint of wine and half a pint of water per man. About two in the afternoon, an officer came to me, and asked for the key of the strong box, as he called it, (that was the chest where the money was); I told him the key was on board the prize, and if he would permit me to go there, I would get it. A boat was immediately manned, and we went on board, but when I was on board, he would not permit me to bring my clothes or bedding with me.

"The next morning I saw the Admiral, Count d'Estaign. I went to him and begged the favour of him to permit us to go on board the prize for our clothes, which was granted for four of us to go, and was immediately done; but when we came on board, found our clothes and bedding all gone. What chests were left were all empty, and we all returned as we went. We then made out a petition, and I presented it to the Count, setting forth our distresses. He then gave me an order to go on board the frigate, and have her searched for our clothes. I went on board, and the captain ordered one of his officers to go down and search for them, but all to no purpose, as they would not find any of them, and that was not the worst, for several of my men had part of their clothes stripped off their backs.

"August 2, about noon, we landed on Point Judith, 245 in number; at four in the afternoon began our march towards Providence; at eight arrived at North Kingston; about nine had a little raw salt pork and bread served us, that being the first victuals we had tasted that day. The ground that night was our bed, the clouds our covering.

"August 3, at seven in the morning, we began our march; at eleven arrived at a town called Greenwich. We halted four hours, it being very hot, but nothing to be got to eat. About three in the afternoon we moved forward, at nine arrived at Providence, and were put into the market-house for that night; we had walked thirty miles that day, but nothing to eat. At six in the morning we were turned out, and marched through the streets into a yard to be viewed and mustered. At about nine we were marched down to the river near the bridge, embarked in boats and sent down the river and put on board a ship called the *Aurora*, lying at Fox's Point, where we were confined altogether, in number 225. About four in the evening we had some beef and bread sent down to us, having then been forty-five hours without victuals or drink, except water; and when we complained, they comforted us with saying we were too well used, with several other speeches to the same purpose and mortification. The ship had no covering for her hatchways, and when it rained there was scarce a dry place to be found under deck, and nothing to lie on but the bare planks. We then lived middling for three weeks, then were put to one quarter allowance for twelve days; after that to half allowance for sixteen days. We then petitioned General Sullivan, telling him we were nearly starving for want. He then ordered us full allowance, which was continued as long as we stayed there, but the provisions very often were so bad, that nothing but hunger could make us eat them.

"October 7, we were put on shore out of the prison ship,

and marched about five miles on our way towards New London, where we arrived the 12th. On the 15th, were marched back again to Norwig; the 22nd, were marched back to New London, and on the 23rd were embarked on board the *Carlisle,* and on the 28th arrived at New York."

In 1778, the *Belcour*, Captain Moore, on her passage to Jamaica took a schooner valued at £1,200, and some time later a French brig with salt and bale goods, worth £2,500. A letter dated Old Harbour, May 9th, 1779, gives the following account of a terrible catastrophe which attended the capture of this ship:—

"In our passage from Halifax to Jamaica, we unfortunately fell in with a French frigate off Coycas, one of the Bahama Islands, called the *Minerva*, taken from the English about five months before. We engaged her full two hours and an half, the furthest distance she was off was not more than pistol shot, a great part of the time yard arm and yard arm, as we term it, but that you may better understand it, her sides and ours touched each other, so that sometimes we could not draw our rammers. The French, I assure you, we drove twice from their quarters, but unluckily their wadds set us on fire in several places, and then we were obliged to strike. You may consider our condition, our ship on fire, our sails, masts and rigging being all cut to pieces, several of our men severely mangled. The French seeing our ship on fire, would not come to our assistance for fear of the ship blowing up, as soon as the fire reached the magazine, which it did five minutes after I was out of her. The sight was dreadful, as there was many poor souls on board. You will be anxious to know how we that were saved got out of her. We hove the small boat overboard in a shattered condition, being almost shot to pieces, and made two or three trips on board the frigate before she blew up. The next morning, we picked up four men that were on pieces of the wreck. The following is a list of the unfortunate men who lost their lives: William Ion, second mate; Peter Thompson, third mate; Daniel Gibson, surgeon;

John Lyon, surgeon's mate; Thomas Anderson, Wm. Wood, Absolom Crippin, Lawlens Madget, John Kelly, William Crowby, James Carey, James Wrinkle, Edward Mahoney, Richd. Wellsted, George Cample, seamen; three negroes and a child, passengers."

In April, 1778, the famous corsair, John Paul Jones, cruised in the Irish Channel in the *Ranger* privateer, committing much havoc. He sailed boldly into the harbour of Whitehaven, and set fire to the shipping. He then sailed northward, and afterwards landed in the Scottish Isles, remaining on the coast for a considerable time, but occasionally taking refuge in the French and Dutch ports when hard pressed by English cruisers, or when short of supplies. More fortunate than his predecessor, M. Thurot, Paul Jones escaped all attempts at capture, and retired safe to America with his booty. Liverpool was well prepared to give the daring adventurer a fitting reception. "We have the pleasure to inform the public," says the paper of May 1st, 1778, "that there are two grand batteries here of 27 eighteen-pounders, in excellent order for the reception of any mad invader whose rashness may prompt him to attempt to disturb the tranquility of this town. George's battery is commanded by the Mayor, and the Queen's by Captain Hutchinson, both of them accustomed to the thunder of cannon, as are also the several captains and assistants stationed to each gun, which are shoted, etc. Centeries are fixt, and all the requisites so regulated as to be ready for action at the shortest notice. The King's battery for thirty-two-pounders is preparing with expedition, under the direction of Lieutenant-Colonel Gordon, an experienced engineer. To these securities will be added the *Hyæna* frigate, a King's ship built here, which in a day or two will be fit for sea, and will be moor'd in the river; for our security by land we have two companies of Veterans, and four companies of the Liverpool Blues, commanded by General

Calcraft, who resides in Liverpool." The following vessels were afterwards sent to cruise in St. George's Channel :—the *Thetis*, 32 guns ; the *Stag*, 28 ; the *Boston*, 28; the *Heart of Oak*, the *Satisfaction*, and the *Three Brothers*, all of 20 guns.

Early in 1778, the *Active*, of 14 guns and 80 men, Captain John Powell, upon a cruise on the coast of America, took no less than fourteen prizes, mostly loaded with tobacco and rice, and sent them into St. Augustine. In September of the same year, the Liverpool paper stated that the *Active* had been very successful on her second cruise, having carried into St. Augustine a large brig loaded with 1100 tierces of rice and indigo, and a schooner with 72 hogsheads of tobacco, and it was supposed that she had had further success.

"Last week," says the *Liverpool Advertiser*, of September 25th, 1778, "arrived here the *Santa Maria*, from Barcelona for Honfleur, loaded with brandy, etc., taken by the *Wasp* privateer, belonging to this port ; and on Friday, the *Minerva*, from Hispaniola for Dunkirk, taken by the *Sarah Goulburn*, in the Bay of Biscay, with upwards of 118 hogsheads of tobacco, etc., on board ; and yesterday the *Sarah Goulburn* arrived here with another prize called the *Amiable Magdalaine*, from Guadaloupe for Nantz ; her cargo consists of 595 hhds. of sugar, 119 casks of coffee, 145 bales of cotton, 1600 lb. of ivory, and some dollars ; also the *Lady Granby* privateer, with a fine brig, called the *Lady Louisa*, that she had taken, from Newfoundland to Bordeaux, laden with fish."

In 1779, the *Sarah Goulburn*, Captain Lewtas, fell in with a tobacco ship, and chased her eight hours, but by heaving a great part of her cargo overboard she got away. In 1780, she took a shallop, loaded with coffee, and in 1781, on a cruise from New York, she captured two vessels, one from France, and the other from Holland ; and soon after arrived at Jamaica with a Dutchman of 400 tons, laden with sugar, etc., which she had captured.

In September, 1778, the *Lady Granby* privateer, brought into the Mersey, a French snow, called *Le Bon Chretien*, loaded with fish and oil from Newfoundland. On the 2nd of October, we are told that "several ladies of the first rank are about following the patriotic plan of the Marchioness of Granby, by opening subscriptions for fitting out privateers, and it is expected, in a very little time, several will be manned and sent to sea against our perfidious enemies, merely by British pin-money." Lady Granby was very popular in Liverpool. The privateer named after her, the "*Marchioness of Granby*," of 260 tons, 20 guns, and 130 men, was owned by the Marquis of Granby and Mr. Nicholas Ashton. There was a smaller privateer, called the "*Lady Granby*," of 45 tons, 10 guns, and 60 men, owned by Messrs. Ashton & Co., in which the Granbys doubtless had an interest. On June 18th, 1779, we read that the sloop *Lady Granby*, Capt. Powell, took and carried into Antigua, a French snow, laden with salt, dry goods, wine, and brandy.

In December, 1778, the *Marchioness of Granby*, Capt. Rogers, captured a Dutch ship with a French cargo on board, and sent her for Liverpool, but the prize was lost on the coast of Ireland, all but two of a crew of fourteen perishing. On the 20th of January, 1779, the *Marchioness of Granby*, cruising in latitude 44.34 N., took the *Le Labour*, from St. Malo to New England, with sundry merchandize, also a Dutch snow for Cadiz, with a cargo of brandy. The *Marchioness of Granby* was captured, after an obstinate engagement, by the French frigate *Sensible*, 26 guns, and carried into Brest.

In *Williamson's Advertiser*, of October 23rd, 1778, appeared the following tribute to the charms of Lady Granby, who in the spring of the following year "was safely delivered of a son at the Marquis's house in Harrington Street."

TO THE MARCHIONESS OF GRANBY.

In beauteous Granby nature sure design'd
The fairest form to clothe the sweetest mind:
She joins in Granby, courteous, great and good,
The brightest virtues with the noblest blood;
And from her heart, so generous and humane,
Impels benevolence through every vein.
What graceful affability and ease!
How great the power, how kind the wish to please!
Go on, fair excellence, to charm mankind,
Your beauteous face the transcript of your mind;
Teach them what praise to virtue's charms belongs,
And live a lesson to admiring throngs;
That wondering courts may this great truth relate,
Virtue adds lustre to the noblest state.

In the same month, "An Old Seaman" issued the following "Invitation":—

Rouse, British Tars! Old England's boast,
Drive all her foes on either coast—
Let Lewis know and feel that we
Can yet avenge his perfidy,
While Keppel has the Key of Brest.

Be Ready, Lads—slip out in quest
Of riches bound to faithless France,
And bravely take another chance.
Revenge and Riches both invite
A Foe insidious to requite;
May your attempts, my willing Boys,
Be crowned with Honor, Wealth and Joys!

On the 25th of September, 1778, *Williamson's Advertiser* published a list of 18 privateers fitted out in Liverpool, in addition to which there were many others preparing for sea, besides several "Letters of Marque," carrying from 16 to 28 guns. It was computed that at that date—not five weeks since letters of marque were issued—prizes worth upwards of £100,000 had been brought into the port.

In September, the *Jenny*, Captain Ashton, in her passage from Liverpool to Tortola, in company with the *Betsey* and the *Buckingham*, of Lancaster, captured a very valuable prize called *Le Marquis de Brancas*, bound from St. Domingo for Nantz, with a cargo of sugar, coffee, indigo, etc.; also a brig from Newfoundland for Havre, with fish, and sent both of them into Cork. The *Jenny* carried 12 guns and 30 men, and was owned by Messrs. Ashton & Co., of the Island of Tortola.

In the paper of December 4th, we read that the *Bellona*, Captain Fairweather, a vessel of 250 tons burthen, 24 guns and 140 men, belonging to Messrs. Bolden & Co., on a cruise from Liverpool, took a schooner loaded with 75 hogsheads of tobacco, and sent her for Liverpool, but unluckily she got ashore in Carnarvon Bay and bulged. On the 18th of the same month, it is stated that the *Bellona* had taken a sloop called the *Canister*, bound from Virginia to France, with 58 hogsheads of tobacco. On January 31st, 1779, the *Bellona* carried into Lisbon the snow *L'Amitie*, of 18 guns, 10 swivels, and 54 men, bound from St. Ubes to South Carolina, with a cargo of salt, wine, oil, fruit, soap, and several chests, supposed to be arms. In April, 1779, she took a ship called the *Necessary*, and in May, 1780, it was stated that she had arrived at Jamaica with a prize valued at £4000.

The following is extracted from the log of the *Bellona*, on her passage from Lisbon to Liverpool, in April, 1779:—

"April 25th. Gave chace and spoke the snow *Vrow Theadora*, from Barcelona to Rotterdam; spoke a Dutch ship from Barcelona bound to Amsterdam; 27th, saw two sail, one astern chasing us, the other under the lee bow; away to the ship under the lee, he laying by under English colours, fired two guns to leeward, which he answered with two to windward. Made sail and stood from him, he still firing shot and standing after us under English colours. The said ship mounted 24 guns upon one deck, and six on the quarter deck, being a long frigate-built ship, with a plain stern and gilt image head. 29th, fell in and spoke the *Vrow Theadora* that we spoke the 25th. 30th, saw a sail standing towards us; down steering sails and stood towards her; came near, when she put about and stood from us, making and shortening sail, as she out-sailed us, and never would come near; hove about but she did not follow. Seeing that, kept the ship her course, and in a short time she came down towards us; shortened sail, when she hoisted American colours and fired a gun, which was returned several times with the nine pounders. As they scarcely reached her then, she made sail, when we crowded all after her, but the wind being light, she got away. By the description of the vessel in Lisbon, she is called the *Vengeance*, fitted out at Vigo, in Spain, mounting 20 guns, very low, with an image head. May 1st, spoke the ship *Tartar*, Captain Lloyd, from Bristol; out five weeks, had taken nothing. 5th, at four a.m. saw a sail one mile and half under our lee quarter. Perceiving her to be a very large frigate-built ship, in chace of us, thought it prudent to run, as having many people ill; in nine hours near out of sight, when she left off chace. 6th, spoke the *St. Bees* bark, belonging to Whitehaven, Capt. Williamson, from Cork, bound to New York; sailed in a convoy of 24 sail, out 18 days; lost the fleet 29th April. 7th, at seven a.m. saw a sail bearing down upon us, shortened sail to wait for her, proved to be the *Tyger*, Capt. Shaw, from Bristol, mounting 22 guns, out eleven weeks, and taken one St. Domingo man."

On the 28th of September, 1778, a large ship, bound from Archangel for Marseilles, laden with tar, hemp, and iron, arrived in the Mersey, a prize to the *Delight*, Capt. Dawson, a vessel of 120 tons, 12 guns, and 39 men, belonging to Messrs. Rawlinson & Co.

The *St. Peter*, Capt. Holland, upon a cruise from Liverpool, fell in with and captured a French East Indiaman, called the *Aquilone*, valued at upwards of £200,000, but most unfortunately, Capt. Holland and his prize afterwards met with a French man-of-war of 74 guns, and a frigate, who took him and his prize, and carried them into Port L'Orient. The *St. Peter*, 320 tons, 22 guns, and 147 men, was owned by Messrs. Holme, Bowyer, & Kennion.

The *Thomas Hall*, Capt. Beard, one of the fleet from Jamaica for Liverpool, having parted with the convoy in a gale of wind, had the good fortune to capture a ship loaded with rum, which she carried safely to Cork.

The *Rumbold*, Captain Fayrer, a vessel of 250 tons burthen, 20 guns, and 57 men, owned by Messrs. Caruthers & Co., captured a large ship from Alicant, with a cargo of brandy. In 1779, on her passage from Liverpool to Africa, she took a French Guineaman called the *Ulysses,* with 302 slaves, and about two tons and-a-half of ivory on board. In April, 1781, the *Fortuna of Flensburg*, laden with fruit and wine, another prize to the *Rumbold*, arrived in the Mersey.

On the 14th of October, 1778, the cooper of the *Brilliant*, 20 guns, Captain Priestman, wrote from New York, as follows :—

"We arrived here on the 26th of September, after a passage of ten weeks. On the 16th of Sept. in lat. 38 N. long. 65 W. we had a very hard engagement with an American privateer of 28 guns, which lasted for nine glasses, when the privateer ran away; and being a faster sailing vessel than ours, we could not come up with her, having great part of our rigging shot away, and our masts wounded. I believe she was much worse shattered than us. We had three people wounded. I was shot in the hand by a piece of cross-bar shot, but am mending fast. On our passage, we saw several privateers, but none durst engage us but the one mentioned before. We were dogged two days by two privateers and sloops, but imagine they did not like our appearance, as they would not come near us. Our ships of war and cruisers are bringing in here French and American prizes daily."

As security against privateers, all vessels were ordered to

sail under convoy, and in large fleets. In the third week in September, 1778, it was announced that all the principal fleets had arrived safely, namely, the Jamaica fleet at Liverpool and Bristol; the Leeward Islands fleet at Plymouth, and the Lisbon and Spanish fleets in the Downs. The arrivals that week were the largest that had been known for many years. In October, the London underwriters calculated that the losses sustained by the French since the proclamation of reprisals already amounted to upwards of £1,200,000.

The *Two Brothers*, Captain Ralph Fisher, a vessel of 150 tons, 16 guns and 39 men, sailed for a cruise in company with the *Young Henry*, Captain Currie, of 270 tons, 18 guns and 60 men, belonging to Messrs. Hartley & Co. The following letter was received from Captain Fisher by Messrs. Roberts & Co., the owners of the *Two Brothers*:—

"Ship *Two Brothers*, at Spithead, 3rd October, 1778.

"GENTLEMEN,—I have the pleasure to acquaint you we arrived safe here this day with a French East Indiaman of 500 tons, deeply laden, from Bengal, which we took on Tuesday the 29th of September, in lat. 47.28 N. long. 10.30 W. At six in the morning we discovered two sail in the N.W. quarter, wind at S.S.W. upon which the *Henry* and me gave chace to the northwardmost of them. At eight, I found she wronged us much, and was afraid she would pass to windward. I left the *Henry* chacing, and hauled for the westermost ship. At ten I just weathered her, at a long shot, (she was standing to the Eastward) notwithstanding her formidable appearance, I wore round and gave her a broadside, which was well directed. She still stood on, thinking to outsail us and get clear, which I believe she would, had not the second broadside of grape and round shot, which we poured into him, immediately cleared his decks, and he struck to the *Brothers*. The *Henry* had given over the other chace, and passed us a league to leeward; and when she struck he was a long way astern. As she is such a valuable prize, we thought it most prudent for both of us to convoy her into port. I put two mates (Mr. Callow and Mr. Pugmore) and twelve men on board; Currie the same number. She is called the *La Gaston* (her first voyage). She is 90 feet keel and 28½ feet beam, frigate built, and would carry 32 guns. She has only 6 nine-pounders on board and 60 men, including passengers, amongst which is a French General called Nardierre, a Chevalier of the order of St. Lewis. By the General's account she is worth 2,000,000 of livres; I think she is worth more. I beg you will write me by return of post, to the care of the postmaster at Portsmouth; but I think one of you coming yourselves post would be requisite; I wish you would. We have sent off express to Mr. Hartley. As dull as the *Brothers* sails we have stumbled upon a noble acquisition. I had near lost her for want of a fast sailing ship. Had they stood another broadside by keeping all crowded as they had, they would have got clear. The *Henry* was too far to leeward to have come up with her. At three o'clock the same day we took the prize, we fell in with the *Ellis* and *Gregson** who informed us that the day before they had been chased by three French frigates, and that they saw a French 60 gun ship with a prize, a large black ship, standing to the S.E. Inclosed you have the French Captain's account of the cargo, but by information of some of the passengers, there is four trunks of valuable merchandise, and other packages of value. Am, Gentlemen, your most humble Servant, RALPH FISHER.

"400 Bales of Muslin and White Bafts, 150 Tons Saltpetre, 190 Bales Cotton, 11 Pipes, 138 Half Pipes, 34 Bags Sago, 4 Casks Tortoise Shell, 40 Barrels Coffee, 50,000 Billets Ebony, besides other packages of value."

On Sunday, October 4th, 1778, the *Mary*, Captain Bonsall, a vessel of 130 tons, 16 guns, and 40 men, belonging to Messrs. Drinkwater & Co., arrived in the river

* The Editor patriotically remarks:—"The *Gregson* and *Ellis* both belong to this port, and as they are two fine ships and manned with brave seamen, if they got up with the other East Indiaman, we doubt not her being soon brought to England."

Mersey with a ship she had captured in the Bay of Biscay in her passage from Liverpool to Dominica. The prize was called *Le Grand Athanase*, from Port de Paix for Nantz, loaded with about 200,000 lbs. of tobacco and other goods. In December, the *Mary* arrived from a cruise, bringing in with her another prize, called *L'Equité*, bound from St. Domingo for Bordeaux. Her cargo consisted of

239 hhds.,	6 tierces,	9 barrels first white Sugars.	
67 ,,	58 ,,	184 ,,	118 bags of Coffee.
2 ,,	3 ,,	1 ,,	3 ankers Indigo.

16 Bales, 10 bags Cotton. 2 Barrels Cocoa.

1 barrel tortoise shell.

The paper of December 11th, describes how the prize, coming into port, was run aground, on the Dove, "and coming round the Rock, beat off her rudder, which was washed away; when very bad weather coming on, no boats could be got to tow her into the dock. A temporary rudder was sent off to her, but after cutting three cables, being three times ashore, and losing three boats, she at length was run ashore near the New Ferry, with one anchor and cable at her bow, which were never let go. 'Tis to be feared the ship will not be got off, but the materials and greatest part of her cargo are saved." Such were some of the difficulties of navigation prior to the advent of those powerful steam tug-boats which have rendered incalculable services to the shipping of the port. But *L'Equité's* dramatic career was not yet closed. What followed is related by Troughton as an instance of the daring depravity of the inhabitants on the Cheshire coast. A number of lighters were employed to take out the cargo. On the second day, the people from the country, assembling in their hundreds, swooped down upon the vessel, threatening destruction to all who opposed them, forcibly seized and carried off great quantities of the cargo, in consequence of which lawlessness, it was found necessary to call in the aid of the military. Application was accordingly made to the Mayor of Liverpool, and the commanding officer of the Leicestershire Militia stationed there, both of whom declined interfering, the transaction being in another county. The owners of the privateer then sent over arms to their people, for the defence of the vessel. On the following night a numerous mob again assembled, and in spite of the entreaties of the four men who guarded the property, proceeded to renew their depredations. The guard then fired several times over the heads of the most desperate of the plunderers, and at last, for the preservation of their own lives, fired directly upon them, killing one man. This resistance, however, only exasperated the mob, and in the end, to prevent farther bloodshed, the men upon guard took to their boat, and left the prize to the robbers.

In April, 1779, the *Mary* fell in with the rebel privateer *Vengeance*, Captain Wingaze Newman, of 22 six-pounders and 90 men. The *Mary* had only 16 four-pounders, and 48 men and boys, having a short time before sent fifty-five people away with a prize. In the engagement which ensued, lasting one hour and-a-half, the *Mary* had three men killed and twelve wounded ; her main topmast was carried away, the mainmast cut two-thirds through ; she received thirty shots in her hull, five of which were between wind and water, three of her guns were dismounted, and the rigging all cut to pieces. In this condition, the brave Captain Bonsall was obliged to strike his colours, and the *Mary* was carried into the port of Newbury. She did not remain long, however, in the hands of the Americans, being retaken and carried into Antigua.

In August, 1778, the *Molly,* Captain Kendall, from Liverpool for Africa, took the *La Verturane*, from Port-au-Prince for Havre, with sugar, coffee, cotton, indigo, etc. Being a valuable prize, Captain Kendall returned back to convoy her safe to Liverpool, but, meeting with the *Stag* man-of-war between Holyhead and Tuskar, the prize was taken from him, and arrived in Liverpool on December 17th.

On her passage from Africa to Jamaica, with 412 slaves, the *Molly* took and carried in with her, a valuable prize laden with provisions, etc. In March, 1781, we read that she had arrived at Jamaica from Africa, with a cargo of 514 slaves, which was an immense success, considering that she had buried 106 poor wretches on the middle passage. The *Molly* belonged to Messrs. Gregson & Co. She was 260 tons burthen, and carried 16 guns and 70 men. The *Nancy*, Captain Nelson, a slaver of 150 tons burthen, 16 guns and 50 men, owned by Messrs. Pringle & Co., managed at about the same time to convey to the same elysium, 610 slaves, doubtless to the great disappointment of the sharks, which followed these floating dungeons with expectant eyes. In December, 1782, we read of the *Molly*, Captain Kendall, being well on the coast of Africa with 650 slaves, and in April, 1783, it was announced that *La Joletta*, prize to the *Molly*, had arrived in the river Thames. Happy was the merchant who possessed brave and skilful captains, who, not only secured large cargoes of eligible and healthy negroes, but picked up rich argosies on the passage.

In October, 1778, Captain Robert Bostock, of the *Little Ben*, 110 tons, 14 guns, and 50 men, wrote from Exeter to his owners, Messrs. Radcliffe & Co., in Liverpool, as follows:—

> "I am just arrived here, in company with the *Molly* privateer, belonging to this port, with the snow *Le Mallie*, Capt. Mouroy, from Port Prince, bound to Bordeaux, laden with sugars, coffee, indigo, and cotton; they calculate the value here to be £20,000, and a Dutch ship loaded at Marseilles, and bound to St. Villeroy, with brandy, oil, soap, &c. I spoke the *Gregson* on Monday last, who had parted with the *Ellis*. She was chased by three French men of war, and he saw them take an English ship, which he was afraid was a Liverpool vessel. I was very near being taken by a French

40-gun ship; she gave us three broadsides, but we being the best sailor, got clear of her."

The paper of October 30th, mentions that "the *Little Ben*, from Liverpool for Africa, and the *Molly*, belonging to Exeter, had taken a snow bound from Martinico for Nantz, laden with 178 hogsheads of Muscovado sugar, 50 hogsheads, 74 tierces and barrels of white sugar, 5000 lbs. of indigo, 6,500 lbs. of cotton, and 103,896 lbs. of coffee. In July, 1779, the *Hope*, Captain Potter, arrived from Africa, and reported that the *Little Ben* had left the coast full slaved, and that the *Rose* and the *Spy* had likewise left the coast to escape the French frigates, but intended to return to finish their trade, after the French had passed." The *Spy*, Captain Rigmaiden, and the *Rose*, Captain Jackson, were both vessels of 120 tons, belonging to Messrs. J. Zuill & Co., and each carried 14 guns and forty men.

At the end of October, 1778, the *Knight*, a vessel of 220 tons burthen, 18 guns, and 80 men, belonging to Messrs. Hindley, Leigh & Co., Captain Wilson, brought safe to Hoylake a large ship from St. Domingo, called *La Plaine du Cap*, which she had captured. On the 27th of November, another prize, taken by the *Knight*, arrived in the Mersey, the *Catharina*, from Cadiz for Havre de Grace. The cargo consisted of 418 packs of wool, 174 tanned hides, 25 barrels of wine, 16 barrels of cochineal, 28 bales of indigo, 2 bales chocolate, 6 chests gum copal, 5 chests of medicines, 4 chests vanigla, 26 pipes sweet oil, one case of books, and one bale of camlet; the whole valued at £25,000. In the Liverpool paper of January 8th, 1779, we read how the *Knight*, upon a cruise, captured a fine East Indiaman called the *Deux Amis*, from China, with a valuable cargo on board, and was convoying her into Liverpool, but in a gale the prize was forced on shore upon the point of Ayr, while the *Knight* went on shore in Conway Bay, with the loss of her masts only. The crew of the prize, when she filled with water, endeavoured to save themselves by getting upon the shrouds and masts, but the sea breaking over them, and the night being intensely cold, only nine Englishmen and one Frenchman survived it; the rest, thirty-two in number, were starved to death, and the ship broke to pieces. A great part of the cargo was saved. In July, 1779, the *Knight* was sunk by a French frigate. Her crew were saved and landed at Oporto.

The following quotations from a letter dated Baltimore, U.S., December 12th, 1778, show the distress which prevailed in that country during the war:—"Dry goods 1500 per cent. advance. Exchange 600 per cent. few good bills. Hard money 500 advance on current cash. Gunpowder, lead, shot, earthenware, sells at 2500-3000 per cent."

In October, 1778, the *Viper,* Captain Cowell, 160 tons burthen, 18 guns, and 80 men, belonging to Messrs. Birch & Co., captured a valuable prize called *La Judicieux*, from Port-au-Prince for Nantz, laden with sugar, coffee, indigo, cotton, sweetmeats, copper, and lignum vitæ. On the 17th November, in the Bay of Biscay, the *Viper* took the snow *La Amiable Annette*, bound from Cape François to Nantz, with sugar, coffee, cotton, and indigo. On the 8th of March, 1780, the *Viper,* in company with the *Dick*, Captain Hewin, took the *Uriah,* from Newberry to Hispaniola, with lumber, etc., and on the 17th of the same month, they captured the *Count d'Estaign*, of 18 guns, from Cadiz to Virginia, laden with wine and salt. They carried both prizes to St. Kitts. The *Viper* also took a schooner from Salem for Grenada, laden with wine and lumber, and carried her into Antigua. In 1782, she recaptured the *Parnassus*, from Liverpool for London.

At a council held on the 7th of October, 1778, the Corporation resolved to give a bounty of four guineas per man to able seamen, not exceeding 40 in number, who should enter as volunteers on board his Majesty's ship *Penelope*, built in Liverpool, "they being first fairly proved to be volunteers and not impressed men, before the committee of the council formerly appointed for the purpose."

The *Juno*, Captain Beavër, a small vessel of 90 tons burthen, 14 guns, and 40 men, belonging to Messrs. Hartley & Co., in her passage from Liverpool to Africa, took a large Dutchman with a French cargo, which she sent to Liverpool.

In October, 1778, the *Tartar*, Captain Allanson, captured the *Le Concorde*, of 500 tons burthen, from Bordeaux, with 2,500 barrels of flour, 800 barrels of beef, 200 hogsheads of wine, and above 20 bales of dry goods, in which were 600 ounces of silver. On February 26th, 1779, the *Tartar* carried into Antigua, a large New England brig, laden with 380 hogsheads of tobacco. In the paper of May 28th, 1779, we read that Captain Allanson on his passage to Jamaica, captured a French slaver, from Angola with 692 slaves, who were sold in Jamaica for £25,560, currency; a stroke of business which must have caused great satisfaction in the office of the owners, Messrs. J. Backhouse & Co. Captain Allanson also took the schooner *Victory*, from Nantucket, with lumber, fish, and oil; and on his passage home, the sloop *Hazard* from Providence, with lumber. The *Tartar*, though worthy of her name, was not a formidable craft to look at, being only 90 tons burthen, carrying 18 guns, and 80 men.

It may be easily imagined from the nature of their employment afloat, that the privateersmen, when they came into port, were not exemplary pillars of law and order. As a matter of fact, they were a rough and lawless set, the terror of the town, committing many outrages, and breaking open the guard-house to release the impressed seamen there confined. Their riotous behaviour became so alarming as to demand magisterial notice, insomuch that, towards the close of the year 1778, during the mayoralty of William Pole, the following caution was published under his authority:—

"Great complaints having been made to me, as your chief magistrate, that of late, numbers of seamen and others, engaged and entered on board the several privateers, and letters of marque vessels equipping at this port, to cruize against his Majesty's enemies, do frequently assemble themselves, and go armed in a riotous and unlawful manner through the town, and its environs, as well in the day as in the night time, without any commission or other officer being in company, or to command them, to the great annoyance of the inhabitants and others, and who have committed several outrages thereby against his Majesty's peace, and the laws of our country in particular, in forcibly breaking open, and rescuing several impressed seamen out of the houses for the reception of them. These are, therefore, in his Majesty's name, for the future, to caution all such persons assembling themselves in such an unlawful manner and mode, and from committing such unlawful breaches of the peace, and violations of the laws, otherwise, I shall be under the most disagreeable necessity of calling unto my assistance, for the preservation of the lives and property of his Majesty's peaceable subjects in this town, the military stationed here, of which, I hereby require all such persons to take notice, at their peril."

This remonstrance had its proper effect in deterring a lawless banditti from the perpetration of outrages.

The *Bess*, 270 tons, 18 guns, and 100 men, Captain Perry, belonging to Messrs. Slater & Co., on her passage from Liverpool to St. Vincent's, captured in the Bay of Biscay an American snow called the *St. Croix*, with 49 hogsheads (or 40,000 lbs.) of tobacco; 102 hogsheads, 6 half-hogsheads, and two barrels white sugar; 24 hogsheads, 3 half-hogsheads, and one barrel of brown sugar; 116 hogsheads, 15 half-hogsheads, and 28 barrels of coffee; and 826 dollars. In January, 1779, the *Bess*, in company with the *Saville*, of Bristol, took a rich snow, named the *Proteus*, bound from Philadelphia to

France, with 281 hogsheads of flax seed, 166 hogsheads of tobacco, 52 casks of pot-ashes, 15,000 staves, and 140 beaver skins.

The *Arethusa*, Captain Jones, a vessel of 150 tons burthen, 18 guns, and 92 men, belonging to Messrs. Nelson & Co., took a brig from Bayonne, loaded with wine and flour, and sent her for Liverpool, but she was stranded near Waterford. The crew and part of the cargo were saved. The *Arethusa* soon afterwards captured a schooner loaded with tobacco, and sent her for Liverpool. In November of the following year, we read that the *Arethusa* had taken and carried into the Bermudas, two prizes, one a sloop laden with 59 hogsheads of tobacco from Bird's warehouse, James River, the other from Salem for Cape François, laden with fish and lumber. In company with another privateer, the *Arethusa* captured a prize which was sent to the Bermudas, and another from Carolina, which she sent to New York.

On the 28th of October, the *Carnatic*, East Indiaman, was taken by the *Mentor*, Captain John Dawson. This was said to be the richest prize ever taken and brought safe into port by a Liverpool adventurer, being of the value of £135,000. "A box of diamonds of an immense value," says the Liverpool paper, of November 27th, "was discovered on Friday on board the *Carnatic*, French East Indiaman, which is arrived in the river, to the no small satisfaction of the captors." This lucky hit was due in a great measure to the sagacity of Captain Dawson. War with France had commenced in April, but Admiral Keppel only put to sea to look for the French fleet on June 17th. Captain Dawson boldly sailed southwards to intercept the French East Indiamen, which had put to sea before the declaration of war, and met with the reward which his pluck and foresight deserved. The *Mentor*, 400 tons burthen, 28 guns, and 102 men, belonged to Messrs. Baker & Co. Mr. Baker became Captain Dawson's father-in-law, and partner in the noted shipbuilding firm of Messrs. Baker & Dawson, on the Estuary Bank. Mr. Baker was mayor in 1795, and died during his year of office. Messrs. Baker & Dawson bought the Mossley Hill Estate, and erected thereon the mansion which, in joke, was called by the wags, "Carnatic-hall."* In 1778, the firm bought the manor of Garston of the Corporation of Liverpool, and in January, 1791, they conveyed their undivided moieties to Mrs. Elizabeth Kent, widow of Mr. Richard Kent, of Duke Street.

In the Binns' Collection, at the Brown's Library, is the following undated newspaper cutting :—

"A RELIC OF OLD LIVERPOOL.

"The fire which took place on Monday last at Carnatic-hall, situate on Mossley-hill, promises to complete the destruction of a mansion not a little associated with the fortunes and history of Liverpool and its leading citizens during the last 100 years. It is well known what a prominent part Liverpool took in fitting out privateers against the French and other enemies at the commencement of our American troubles in 1775. But it was in November 1778, that the richest prize ever taken by a Liverpool adventurer fell into the hands of Captain John Baker of the *Mentor*, a ship of 400 tons, 28 guns, 102 men, and belonging to the Bakers and Pudsey Dawsons—names then of repute in commercial circles. The prize, a French East Indiaman, the *Carnatic*, turned out to have on board a box of diamonds, valued at £135,000. With the proceeds of the prize and after it Carnatic-hall was built and

* Carnatic-hall, re-built in princely style, is now the residence of Mr. Walter Holland. Towards the close of the eighteenth century, Mr. Samuel Holland, grandfather of Mr. Walter Holland, the present owner, came to Liverpool from Knutsford, in Cheshire, from which place the present Henry Holland takes his title of Viscount Knutsford. Mr. Samuel Holland was no doubt largely engaged in privateering, like other Liverpool shipowners in the long war with France. He is said to have been on board one of his own privateers, and commenced an engagement with either a British man-of-war or another privateer, in the Mediterranean (each vessel having hoisted a wrong colour to deceive an enemy) before the mistake was found out. The family of Holland is an old one, and was resident in the neighbourhood of Liverpool farther back than the days of the valiant Nehemiah, who commanded the *Ann Galley* and the *St. Peter*.

appropriately named. Though the name of its architect is not known, the proportions of its rooms and their decoration denoted a superior hand—possibly that of Adams. It is not known for certain whether there remain any of the family of the builders of the mansion, though about twenty years ago a woman claiming to be so descended appealed for alms to the then occupants of the house. Anyhow, the latter after remaining in the family of its builders till about 1830, came into the possession of the Ewarts, one of whom was sometime member of Parliament for Liverpool, and in 1838, shorn of part of its acreage, it was purchased by the late Charles Lawrence, grandfather of Mr. W. F. Lawrence, M.P. In 1886, after the death of Mrs. G. H. Lawrence, the property was sold to a syndicate, and lately came into the hands of the present owner. In 1847, during the mayoralty of Mr. G. H. Lawrence, Sir Robert Peel visited Liverpool, and slept at Carnatic-hall, as the guest of Mr. Charles Lawrence. In 1886, an old and interesting print, which always hung at Carnatic-hall through its many vicissitudes, representing the capture of the *Carnatic* East Indiaman, was forwarded by the representatives of the then owners to the authorities at the Brown Library. It is probably lost under the accumulation of similar curios, but it would seem worth while for the authorities to provide some room specially devoted to commemorate Liverpool in the olden days, and to set out a selection of prints and curios of special interest. It is probable the old picture depicting the capture of the *Carnatic*, however poor as a work of art, might be not the least attractive exhibit, and stimulate many a youth to deeds of daring and laudable ambition." *

In August, 1779, we read that the *Mentor* had captured two prizes on her passage to Jamaica. In that year she

* The writer of the above interesting article is incorrect in saying that Captain Baker was in command of the *Mentor*, instead of Captain Dawson. The author of the present work, having had his attention called to the old picture in question, by Mr. W. F. Lawrence, M.P., made enquiries, and, with the kindly aid of Mr. Cowell, the Chief Librarian, found it hung in a dark corridor in the private portion of the Brown's Library. "More light" only revealed the sad fact that the picture had been in battle on its own account, and that it was too sorely wounded for reproduction in this volume.

was commanded by Captain John Whiteside, who, on the 4th of November, wrote to Messrs. Baker & Dawson, from Cork, as follows:—

"On Wednesday, the 27th ult., in lat. 47, long. 11., at A.M. daylight, saw four sail bearing S.E.; bore away for them. As we came near, found two of them to be ships, one having Dutch colours hoisted, the other English; the other two sail being a sloop and a schooner. At nine A.M. came so near one of the ships as to perceive she was a frigate, on which we hauled upon a wind to the southward. She immediately hoisted French colours, fired several shot, and gave chace to us. We, finding she came up fast, kept away with the wind abeam, and set every studding sail and small sail in the ship to the best advantage. At noon the frigate was about two miles astern in chace. The 28th October moderate winds and clear weather; the frigate in chace all these twenty-four hours, sometimes coming near us, other times dropping, according as the breeze freshened and died away. We took every possible method to get away from him, but could not get distance enough from him to alter our course in the night. At ten A.M. saw three sail astern, steering after us. At meridian, the frigate about a mile and a half astern, coming up fast with us. The 29th October, at half past meridian, the wind dying away, the frigate came up with us fast, in consequence of which, we in studding sails, up courses, and got all clear for engaging. At one P.M. came to action, which continued very warm till ten minutes past two, when she made all the sail she could, and stood away from us to the southward.

"She was a frigate of 36 guns, carrying 28 twelve-pounders on one deck. We weighed one of their shot, and found it 15 lb. weight. We had our main-top-mast shot away, a great deal of our rigging and sails, and one shot through our main-mast head, just below the trussle trees, which splintered all the larboard cheek of the mast and all the bolts. At five P.M. two of the vessels we saw astern came up, and spoke us, proved to be the *Lyon* and *Tyger* privateers, belonging to

Bristol. We all three made sail after the frigate. At A.M. daylight, had a survey on our mainmast, by our carpenters, found it so much shattered, and impossible to get any fishes on it, concluded putting into Ireland, in order to get a new one. At meridian in company with the privateers, saw two sail, one upon a wind, which we knew to be the frigate, being in the same disguise as when we first saw her, with his fore-top gallant mast down, and a small jigger abaft, and the other a brig standing to the S.E., to which the *Lyon* gave chace I suppose, not knowing the other vessel to be the frigate by being in that disguise, and she on the wind to the N.W. The *Tyger* gave chace to her, we being disabled so much, with our main-top-mast down, and a great way astern of them both, and not consulting with me, thought it most prudent to haul my wind. As the *Tyger* drew near her chace, she perceived her to be the frigate, whereupon she immediately hauled her wind, and stood for the *Lyon*, which I suppose would have been the case had I gone down with her. Our officers and men behaved exceeding well."

In 1782, on her passage from Jamaica to Liverpool, the *Mentor* foundered in a gale, off the Banks of Newfoundland, and 31 of the crew were lost. Captain Whiteside, his second mate, and a boy were saved. The *Sarah Goulburn*, Captain Orr, also foundered.

The *Dragon* privateer, 112 tons, 14 guns, and 75 men, Captain Briggs, belonging to Messrs. Warren & Co., took the *La Bonne Foi*, from Martinico for Dunkirk, laden with 201 hogsheads, 7 tierces, and three barrels of sugar, 165 bags of cocoa, 119 bags of cotton, 22 hogsheads, and 48 bags of coffee; also two ships from Newfoundland laden with fish, one of which was lost off Cork. The prize-master of the other stated that he had left the *Dragon* in pursuit of fifteen more vessels when he parted from her. On the 16th of February, 1779, the *Dragon* took the *La Modeste* (a French Letter of Marque), from St. Domingo for Nantz. When she struck, the sea ran so high that it was impracticable to board her, upon which she was ordered to steer towards Ireland, and carry a light, the *Dragon* keeping close on her quarter. When the weather became more moderate, an attempt was made to man the prize, in which all the boats belonging to each ship were stove. The impatience of the *Dragon's* crew was now raised to the utmost pitch, and, regardless of all danger, five seamen stripped themselves, leaped into the sea, swam to the prize, and took possession. This unparalleled instance of British courage so astonished the French, that they declared none but Englishmen would have thought of such an expedient, and much less have carried it into effect. The prize was worth swimming for, the cargo, which arrived safely in Liverpool, consisting of 517 hogsheads of sugar, 42 hogsheads, 31 barrels, two tierces, 116 quarter casks, and 181 bags of coffee, 39 bales and 29 pockets of cotton, four hogsheads, one barrel, and eight quarter casks of indigo. In February, 1780, we read that the *Dragon* privateer, Captain Reed, had taken "another schooner, bound from Martinico to Boston, laden with sugar, etc., and sent her into Bermuda, where both vessel and cargo were sold." In the paper of September 13th, 1781, the capture was announced of the *Dragon*, Captain Gardner, on a cruise from Liverpool, by a French frigate, which carried her into Brest.

The *Nanny*, Captain Beynon, belonging to Messrs. Hindley, Leigh & Co., a vessel of 220 tons, 14 guns, and 50 men, took a large Swedish brig bound from Lisbon to Rouen, with wine, fruit, and wool. In January, 1779, on her passage to Oporto, she recaptured a brig loaded with provisions, from Limerick for Gibraltar. In the following March, on the homeward passage from Oporto, she had a three hours' engagement with a privateer of 16 guns, and beat her off. The particulars of an engagement between the *Nanny* and the American privateer, *General Arnold*, Captain Brown, of 18 six-pounders, and 100 men, are given in the following letter from Captain Beynon to his owners, dated Cadiz, June 2nd, 1779:—

> "On the 20th of May, off Cape Finisterre, saw a ship in chace of us. Being resolved to know the weight of his metal before I gave up your property, I prepared to make the best defence I could. Between eight and nine o'clock, he came alongside, with American colours, hailed, and told me to haul my colours down. I desired him to begin and blaze away, for I was determined to know his force, before I gave up to him. The engagement began, and lasted about two hours, our ships being close together, having only room to keep clear of each other. Our guns told well on both sides. We were soon left destitute of rigging and sails, as I engaged him under my topsails and jib. We were sadly shattered below and aloft. I got the *Nanny* before the wind, and fought an hour that way, one pump going, till we had upwards of seven feet water in the hold. I thought it then almost time to give up the battle, as our ship began to be water-logged. We were so close, that I told him that I had struck, and hauled my colours down. The privateer was in a sad shattered condition. By the time we were all overboard the *Nanny,* the water was up to the lower deck. When Captain Brown heard the number of men I had, he asked me what I meant by engaging him so long? I told him, as I was then his prisoner, I hoped he would not call me to any account for what I had done, before the colours were hauled down. He said he approved of all I had done, and treated my officers and myself like gentlemen."

One man went down in the *Nanny*. The *General Arnold* was herself taken soon after, by his Majesty's ship *Experiment*, Commodore Sir James Wallace.

The slave ship *Diana*, Captain Colley, from Liverpool and Africa for America, was captured 30 leagues to windward of Tobago, by the ship *General Moutrey*, Captain Sullivan, 18 guns and 200 men, and the brig *Fair American*, Captain Morgan, 14 guns and 90 men. The prize was carried to Curacoa, where her cargo, consisting of 378 slaves, 30 tons camwood, and about three tons of ivory, was sold; the slaves at ten "Joes" per head. The *Greenwood*, Captain Reid, a slaver of 250 tons burthen, 16 guns, and 50 men, belonging to Messrs. Crosbie & Greenwood, was taken on her passage to Africa by the *Vengeance*, an old French frigate of 24 twelve-pounders and 260 men, and carried to Cadiz. The French commander was killed at the first broadside from the *Greenwood*, but the great weight of metal, number of guns and men against him, made it impossible for Captain Reid to continue the action.

In November, 1778, the *Catcher*, Captain Fletcher, a vessel of 110 tons, 14 guns, and 80 men, owned by Messrs. Salisbury & Co., brought into the Mersey a French ship from Cape François for Nantz, with 130,300 lbs. of sugar, 115 barrels of coffee, 7 barrels of indigo, and 12 bales of cotton.

On the 10th of December, 1778, the *Atalanta*, Captain Collinson, 180 tons burthen, 16 guns, and 54 men, owned by Messrs. Fowden & Berry, recaptured the brig *Eagle*, from Newfoundland to Cadiz, with 3429 quintals of fish, and sent her to Lisbon. On the 21st of the same month, the *Townside*, Captain Watmough, 130 tons, 16 guns, and 90 men, belonging to Messrs. Mitton & Co., captured an East Indiaman, laden with coffee, dry goods, etc., but the prize was lost near Beaumaris, the crew and part of the cargo and materials being saved. The *Townside* was captured a few months later, and re-taken by the *Sybil* man-of-war.

"The attention shewn to our trade by the Admiralty," observes the paper of December 18th, 1778, "cannot be too generally known; but it may not be improper to mention that 120 trading vessels are preparing to sail from Domingo to France: a promising harvest for our spirited privateers." From this we perceive that the editorial mind was tainted with the prevailing "iniquity." In the same month it was stated that since the commencement of hostilities with France, the *Kite* cutter had taken, in the English Channel, prizes to the amount of £400,000.

In the beginning of 1779—a year memorable for the numerous instances of gallantry displayed by the seamen of Liverpool, both in privateers and merchantmen—the *Ellen*, Captain Fell, 200 tons, 20 guns, and 70 men, owned by Messrs. France & Co., on her passage to Jamaica, took the *Three Friends*, from Boston to L'Orient, with furs, lignum vitæ, etc.; and a little later she captured the *Fantasie*, 700 tons, from Port-au-Prince for Bordeaux, with 482 hogsheads, 2 tierces and 7 barrels of sugar, 100 bags of cocoa, 10 bags of cotton, 59 hogsheads, 119 tierces and 165 barrels of coffee, 10 hogsheads, 5 tierces and 1 barrel of indigo.

The *Retaliation*, Captain Townsend, 160 tons, 16 guns, and 100 men, belonging to Messrs. Syers & Co., took a brig of 200 tons, laden with tobacco, flour, lumber, etc., and a schooner laden with fish, both from America for the West Indies, and carried them into Antigua. She also captured a large French ship, of 16 guns, laden with bale goods and provisions, which she carried into St. Kitts.

The *Friendship*, Captain Fisher, on her passage from Liverpool to Jamaica, took a Dutch ship, bound from St. Eustatia for Amsterdam, laden with tobacco, etc.; also a vessel from Charleston, with a cargo of rice, indigo, etc.

In February, 1779, the *Betsey*, Captain Fisher, returned from an unsuccessful cruise of three months, but on the 20th of June following, she brought into the Mersey, the *Favourite*, an East India ship, 450 tons burthen, bound for L'Orient, with 1,054 billets red wood, 484 sacks of cowries, 171 bales of cotton, 2 bales dimity, 1,570 bags of pepper, 500 rattans, 27 bags of Mocha coffee, 8 bags of Bourbon coffee, 1,090 bales of coffee, 44½ bales, 2 trunks and 1 box sundries. On thc 29th of August, the *Betsey* was taken on her passage to New York, three days after leaving

Liverpool, by the *Alliance* frigate, of 44 guns, and three other ships in company, cruising off the coast of Scotland, under Commodore Paul Jones.

On the 4th of March, 1779, the *Success* (Letter of Marque), 120 tons, 12 guns, and 30 men, owned by Messrs. Crosbie & Greenwood, and commanded by Captain Niven, arrived in Hoylake from a cruise, having captured *La Probite* from St. Domingo, with a valuable cargo consisting of 94 hogsheads and 4 barrels of tobacco, 43 hogsheads, 3 tierces and 2 barrels of sugar, 14 tierces, 49 barrels and 103 bags of coffee, some cocoa, and hides. On the 14th of May, in her passage to Bermuda, the *Success* fell in with the *Pilgrim* privateer, Captain Hugh Hill, from Beverley, New England, of 16 nine-pounders and 150 men, which she engaged. The sailing-master of the *Success* was killed at the first broad-side, and others of her principal officers dangerously wounded before she struck. The *Pilgrim* had two of her crew wounded, three of her mizen shrouds carried away, and her sails and rigging greatly damaged. The *Pilgrim* cruised afterwards in lat. 50° and long. 13° to 15°, and took the *John*, from New York to Liverpool, and the *Anna and Eliza*, from New York for London, both laden with tobacco, etc. After despatching them, she stood for Sligo Bay, landed all her prisoners, and stood off again to pick up some of the linen and yarn ships. She had taken eight prizes in six weeks.

On the 19th of February, 1779, the *Enterprize*, 250 tons, 20 guns, and 70 men, belonging to Messrs. Brooks & Co., and commanded by Captain Pearce, took the *Paulina*, 450 tons burthen, from Cape François to Bordeaux, with upwards of 500 hogsheads of sugar, besides indigo, coffee, etc. She was pierced for 22 guns. On the 23rd of the same month, Captain Pearce captured *L'Hostilité*, bound from Bordeaux to Port-au-Prince, laden with provisions, etc. Both prizes arrived safe in Liverpool. On the 22nd of October, the *Enterprize,* having changed owners and commander, and having been out about a month cruising on behalf of Messrs. Francis Ingram & Co., whose instructions to Captain Haslam we gave in a former chapter, brought into the Mersey *L'Aventurier*, of 22 guns, and 50 men, from Martinico to Bordeaux, with 105 bales of cotton, 28 hogsheads of tobacco, 600 hogsheads of clayed sugar, 38 hogsheads of Muscovado sugar, 14 tierces and 23 barrels of sugar, 164 hogsheads, 49 tierces and 115 barrels of coffee, 6 tierces, 235 bags and 1 barrel of cocoa, and 2000 lbs. of cassia fistula. On the 14th of September, 1780, the *Courier*, 200 tons, captured on the passage from Bordeaux to St. Sebastian, by the *Enterprize* and the *Stag*, of Jersey, arrived in Liverpool. The prize cargo, as advertised to be sold at the St. George's Coffee-house, consisted of 141 casks of sugar, 82 bales of hemp, 7 hogsheads of claret, 1 hogshead of Virginia tobacco, paint, copper pans, marble slabs, looking glass frames, 12 new chairs, 41 new guns, and 8 new carriages. The brig *Le Vaillant*, another prize taken by Captain Haslam, laden with wine, flour, sugar, etc., was lost on the Burbo, on September 12th, 1780, and only one man saved. About 140 casks of claret and 74 barrels of flour, etc., were recovered from the wreck. In quick succession there arrived the following prizes, captured by the same successful privateer:—The *San Pedro*, 150 tons, taken on her passage from Bayonne to Bordeaux, with 900 barrels of flour, 11 casks of brandy, etc.; the *St. Joseph*, 40 tons, from Bordeaux to St. Sebastian, with 484 casks of resin, etc.; and the brig *Le Moineau*, from Nantz to St. Domingo, with a miscellaneous cargo that would have delighted Robinson Crusoe—flour, claret, nails, canvas, sail cloth, Castile soap, tallow and wax candles, butter, tallow, bread, cheese, pork, salad oil, linseed oil, linens, drugs, Epsom salts, thread, handkerchiefs, thread stockings, men and women's shoes, hair powder, snuff, earthenware, flint glass, barrel staves, etc., coopers' twiggs, copper pans, scales, garden seeds, swivel balls, etc., all sold by auction at St. George's Coffee-house.

The *Molly*, Captain Woods, 240 tons, 14 guns, and 40 men, belonging to Messrs. Rawlinson & Co.; the *Wasp*, Captain Byrne, 220 tons, 14 guns, and 95 men, owned by Messrs. Kennion & Co.; and the *Bess*, Captain Perrey, all belonging to Liverpool, and cruising in company, took a schooner bound from Bordeaux to Philadelphia, loaded with tea, silks, etc., which arrived in the Mersey on February 25th, 1779. They also captured a brig bound from France to the West Indies, laden with provisions. The *Wasp* and the *Bess*, in company, took a ship from St. Domingo for France, with a cargo of coffee, indigo and ivory, which likewise entered the Mersey on February 25th. The *Molly* soon afterwards captured the *St. Augustine*, a three-decked ship, from Port-au-Prince to Nantz, mounting 10 nine-pounders and carrying 40 men. She was laden with 536 hogsheads and 5 tierces of Muscovado sugar, 8 hogsheads, 61 tierces and 97 barrels of indigo, and 300 hides. The *Molly* was captured on September 4th, 1782, on her passage from Liverpool to St. Lucia, by two frigates bound to Marseilles.

The West India fleet sailed from Cork on the 1st of March, under convoy of two 74-gun ships, one of 50 guns, and two frigates. In the same month, a gentleman in Manchester, received the following testimonial to the activity of our privateers, from his correspondent at Bordeaux:—"Very many rich and respectable merchants here, have been already ruined by the great success of your privateers and cruizers. Many more must fall soon. May God, of his mercy to us, put an end speedily to this destructive and ridiculous war."

Early in the year 1779, Captain Ash, of the *Terrible*,

250 tons, 20 guns, and 130 men, belonging to Messrs. Nottingham & Co., took a large snow called *La Victoire*, from St. Domingo to Bordeaux, and on the same day, a large ship from Port-au-Prince, called *L'Age D'Or*, both of which arrived safe in Liverpool. Their united cargoes consisted of 774 hogsheads and 22 tierces of sugar, 29 barrels, 29 half-barrels, 65 quarter-casks, and 40,000 pounds of coffee; 120 bags of ginger, 120 bales, 11 bags and 3 pockets of cotton, and 2 quarter-barrels indigo. Captain Ash also recaptured and sent into Cork, the *Leinster Packet*, from Bristol to Galway, which had been taken the previous day by the *Rocket*, of 16 guns and 110 men.

On Sunday, February 28th, 1779, the *Griffin*, Captain Grimshaw, 130 tons, 14 guns, and 90 men, owned by Messrs. Hall & Co., brought into the Mersey the *Count de St. Germain*, a large St. Domingoman, of 14 guns and 33 men, bound to Nantz with two passengers, and a valuable cargo of sugar, molasses, coffee, cotton, indigo, cocoa and tortoise shell. The Frenchman fought for eight hours before he struck. The *Griffin* had two men severely wounded, one of whom died of his wounds. The arrival of this and several other prizes in the same week "appeared to occasion very general satisfaction" in Liverpool. In April, 1780—more than a year after the capture—an advertisement appeared in the papers desiring the officers, seamen, and others having legal demands against the late owners of the privateer *Griffin*, or her prize *Le Compte de St. Germain* to attend at 33, Edmund Street (the street where John Newton formerly resided), to receive their due. Thus, "Poor Jack," after risking his life and losing a limb in his desperate employment, had long to wait for his share of the plunder, and was too frequently the prey of land sharks. Many of those interested in the *Griffin's* prize had probably gone elsewhere to receive their due, before the invitation to Oldhall Street was issued.

The *Rawlinson* and the *Clarendon*, of Liverpool,

recaptured off the Land's End, the *Weymouth Packet*, which had sailed from Jamaica without convoy and been taken by the *General Sullivan* privateer, of Portsmouth, New England.

The *Dreadnought,* Captain Cooper, took a vessel laden with salt, and sent her into Mount Bay, in February, 1779. On the 5th of March, the *Dreadnought*, Captain Taylor, 200 tons, 20 guns, and 120 men, belonging to Messrs. Wagner & Co., returned from a cruise, bringing in with her the *L'Aimable Agatha* from St. Domingo, with "237 hogsheads of sugar, 80 bales of cotton, 150,000 weight of coffee, and 600 weight of indigo." The head of the firm of Messrs. Wagner & Co., was Mr. Benedict Paul Wagner, the maternal grandfather of Felicia Dorothea Hemans, some of whose earliest poems were written at what is now the "Loggerheads Revived" tavern, in Richmond Row.

On the 30th of March, 1779, five leagues off Cape Clear, the *Polly*, of and for Liverpool, was taken by the French privateer *Monsieur*, of 40 guns, and 450 men. After being ransomed for 1250 guineas, the *Polly* proceeded on her voyage, but on the following day, another French vessel of 36 guns, a consort of the *Monsieur*, fired four guns at the *Polly*, but the latter luckily made the port of Skibbereen before the Frenchman could come up with her.

The *Tom*, Captain Davis, a slaver of 100 tons, 12 guns, and 36 men, belonging to Mr. Clement, was taken on her passage from Africa by two French 36-gun ships, a brig and several armed boats. The *Tom* was purchased from the French to bring the people home. The *Hereford*, Captain Harrison, the *Providence*, Captain Colley, and the *Juno*, Captain Beaver, all slave ships belonging to Liverpool, were taken by the French on the coast of Africa. The *Juno,* a vessel of 90 tons burthen, 14 guns, and 40 men, belonged to Messrs. Hartley & Co.

The *Hunter*, Captain Ashburn, was taken on her passage to New York by a rebel privateer, of 16 guns, called the

Pallas, after an engagement of "five glasses," wherein the *Hunter* had four men killed and sixteen wounded. The American had many men killed and wounded, and was so torn to pieces in her hull and rigging that she had to put into Newbury.

The *Nancy*, Captain Adams, on her passage from Tortola to Liverpool, had a very smart engagement with an American privateer, of 18 guns, and beat her off. Captain Adams's men in the tops with small arms, made great slaughter amongst the privateer's people.

The *Sturdy Beggar*, Captain Cooper, 160 tons, 16 guns, and 160 men, belonging to Messrs. Davenport & Co., took the *St. Michael*, from Cape François for Nantz, with 325 hogsheads, 14 tierces and 4 barrels of sugar, 147 casks and 201 bags of coffee, 22 casks of indigo, 12 bags of cocoa, 1 bag of cotton, and 246 hides, which arrived in Liverpool in May, 1779. On the 4th of September, the *Le Moissonier*, from Cayenne to St. Malo, laden with cotton, cocoa, mahogany, etc., a prize to the *Sturdy Beggar*, Captain Humphrey, arrived in the Mersey; and on the 8th of the same month, the privateer returned from her cruise, bringing in *La Salta Nostra Senora del Rosario,* Captain Buenaventura Prana, from Buenos Ayres to Cadiz, laden with dollars, skins, wool, etc. On the 29th of October, 1779, the *Sturdy Beggar*, then in Fayal Road, parted both cables in a gale of wind, drove on shore, and in ten minutes went entirely to pieces, four of her crew being drowned. Five other vessels were totally lost in the same manner, and at the same time and place.

On the 20th of April, 1779, the *Vulture*, Captain Allanson, took the *St. Cyprian*, 400 tons burthen, from Martinico to Bordeaux; and in August, a large Spanish snow called the *San Esteven*, from Orinoco to Cadiz, with 14,000 rolls of "the Genuine and Fine Oronoque Vorcena or Cannastre Tobacco," 23 tons of cocoa, 400 hides, 370 dollars, and some chests of medicine. Early in 1782, the *Vulture*, on her passage from Jamaica to Liverpool, captured a brig and a snow, one of which foundered on the coast of Ireland.

The *Will*, Captain Lewtas, cruising in company with another vessel, took the *La Meredale*, from Virginia for Cadiz, with 240 hogsheads of tobacco, and 70 barrels of tar and turpentine. The *Nanny*, Captain Harrison, 14 guns and 70 men, belonging to Messrs. Watts & Rawson, on her passage to Jamaica, took a Swede from St. Domingo; and the *Jamaica*, 350 tons, 18 guns, and 110 men, owned by Messrs. Birch & Co., captured two prizes.

The *Ashton*, Captain Thompson, on her passage to the Baltic, had a severe engagement with two French privateers, one a snow, of 18 guns, and the other a brig, of 16 guns. When at last the Liverpool vessel was obliged to strike to superior force, she had only her foremast standing. The mate died of his wounds. The boatswain of the *Ashton* was carried off as ransomer. The same privateers captured and ransomed the *Hannah*, belonging to Messrs. Heywood, and shortly after were themselves taken by the *Fairy* sloop of war and the *Griffin* cutter.

The *Delight* (Letter of Marque), Captain Dawson, 120 tons, 12 guns, and 39 men, belonging to Messrs. Rawlinson & Co., was lost upon Cape May in a fog, and her crew made prisoners.

The barque *Swift*, Captain W. Brighouse, belonging to Messrs. W. Davenport & Co., having lost sight of the Jamaica fleet and convoy, off the west-end of Cuba, proceeded on her passage alone, and was captured by the *General Arnold* privateer, 20 guns and 85 men, Captain James M'Gee, of Boston. Most of the captain's letters, papers, and clothes were taken from him, and he was nearly stripped of everything he had.

The *Tyger* (Letter of Marque), Captain Amery, 300 tons, 16 guns, and 60 men, belonging to Messrs. James France & Co., was taken on her passage from Liverpool to

Jamaica, by a French frigate, and carried into Hispaniola; and the *Adventure*, Captain Hyatt, a privateer of 160 tons, 14 guns, and 80 men, belonging to Messrs. Newby & Co., was taken, in company with a Glasgow privateer, of 14 guns, by an American privateer, of 20 guns.

The privateer *Spitfire*, 200 tons, 16 guns, and 100 men, Captain Thomas Bell, was captured by a French frigate, which also took the *Intrepid*, Captain Buddicome. The *Intrepid* was retaken by the *Dublin* (Letter of Marque), Captain Harding, and carried into New York. The owners of the *Spitfire*, Messrs. J. Zuill & Co., received the following letter from Captain Bell, dated L'Orient, May 2nd, 1779:—

> "I am sorry to acquaint you with my misfortune. On the 19th of April, in lat. 46. 20. long. 5. 10. we met with the *La* ——— frigate, of 32 guns, 26 twelves and 6 six-pounders, removable to either side, which made 19 guns on the side they engaged, and three brass swivels of one pound each, with 246 men. We began to engage about fifty or sixty yards distance, and from that to thirty. The fire was brisk on both sides and well kept up for one hour and fifteen minutes, having then only two guns on the starboard side, and five on the larboard fit for action. We were at last obliged to strike our colours, or rather the Union thereof, as the other part was shot away. In all our sails, from the royals downwards, there was scarce one piece left the size of a sheet. Our standing and running rigging cut entirely to pieces; two ports made into one, and the ship's side like a rabbit warren. You may rest assured I have discharged the duty of a man, both in courage and conduct. Four killed and six wounded."

On the 20th of June, 1779, the *Commerce*, Captain Woods, on her passage from Liverpool to Halifax, fell in with an American privateer, of 16 six-pounders and 75 men, which she beat off after an engagement of three hours and-a-half, with the loss of one man killed and four wounded. On the 24th of June, she fell in with another privateer, of 14 six-pounders, and full of men, which she also beat off, after a very obstinate engagement of four hours, in which she had one man killed, and the Captain and four more men wounded. The *Commerce* had but forty-two people on board, and 14 carriage guns, six and four-pounders. In December, 1780, we read of the *Commerce*, Captain Curwin, being retaken and carried into New York.

The paper of July 2nd, 1779, announced that orders had been despatched to all the seaports to lay an embargo on all ships, that none might put to sea until all the men-of-war which were wanted for immediate service, had got their full complement of men.

The Corporation of Liverpool were at this time—and, indeed, at all times—as effusively loyal in words and deeds as the merchants of Liverpool were far-seeing and warlike in their enterprises. At a special Council, held on June 26th, 1779, it was resolved that an address, under the Common Seal, should be presented to the King; "as a testimony of our duty and affection for your Majesty's royal person, and of our attachment to the welfare and prosperity of your kingdoms at the present alarming juncture; when from the perfidious alliances of our natural and combined enemies, the House of Bourbon, with your Majesty's revolted colonies in America, to succour rebellion against the parent state, this nation and the most formidable powers in Europe must be unavoidably involved in all the calamities of war." Their loyalty did not evaporate in mere professions, for, at the same council, it was ordered "that a bounty of ten guineas for every able seaman, and five guineas for every ordinary seaman, should be offered, and be paid by the Corporation treasurer to every volunteer who should enter on board any of the King's ships of war at Liverpool."

"According to accounts from the Admiralty," says the Liverpool paper of July 9th, 1779, "upwards of 250 Warrants have been granted for making out Letters of Marque in Doctors Commons, since the Spanish

Ambassador delivered his manifesto; and it is certain upon a moderate computation, there will not be less than 700 sail of privateers and Letters of Marque fitted out at the different ports in this Kingdom."

The *Richard*, Captain Lee (a Letter of Marque), of 150 tons, 16 guns, and 70 men, owned by Messrs. Rawlinson & Co., brought in with her a large schooner, laden with sugar, cotton, coffee, molasses, &c., bound from Guadaloupe to America, which she had captured in her passage from Tortola. The *Juliana*, Captain Robinson, on her passage to Antigua, took two prizes, bound from France to the West Indies, and the *Ranger*, Captain Adams, took a Dutch ship bound from Marseilles to St. Valery, and sent her to Liverpool. On the 26th of August, a French snow called the *Chamont*, bound from Beaufort, North Carolina to Nantz, with tobacco, naval stores, and indigo, arrived in Liverpool, having been captured by the slave ship *Blossom*, Captain Doyle, on her passage to Africa. The slave ships were often lucky in taking prizes. The *Nancy*, Captain Hammond, a slaver of 250 tons, 20 guns, and 59 men, belonging to Messrs. Fowden & Berry, arrived at Jamaica with 430 slaves; and the *Nancy*, Captain Nelson, belonging to Messrs. Pringle & Co., a vessel of 150 tons, 16 guns, and 50 men, with 420 slaves. The latter captured and carried in with him a valuable prize, a Guineaman with 200 slaves. The *Gregson*, Captain Jolly, belonging to Messrs. Boats & Gregson, also arrived at Jamaica with a valuable prize.

Early in July, 1779, the *Amazon* privateer, of 14 nine-pounders, and 95 men, Captain Charles Lowe Whytell, returned from a cruise, with a Portuguese brig, which she had taken, bound from Lisbon to Havre. A few weeks later a much finer stroke of good fortune befell the *Amazon*. A letter, dated Cork, September 9th, 1779, gives the following particulars:—

"Yesterday was brought into Cove, the *Sancte Incas*, Don

Fr. Renosso, a Spanish man-of-war, from Manilla to Cadiz, deeply laden with gold, silver, coffee, china, cochineal, and indigo, 800 tons burthen, mounting 18 twelve-pounders, and some small guns, but pierced for forty guns, and had 130 men. She was taken the 23rd ult., off the Western Isles, by the *Amazon* privateer, of Liverpool, and the *Ranger,* of Bristol, of 16 guns each, after an engagement of two hours. A cask of gunpowder taking fire on board the prize, forty of her hands were blown up, which threw them into such confusion as to give the brave English tars an opportunity of boarding her, with the loss only of one man. She is deemed the most valuable prize taken since the rich Acapulco ship by the late Lord Anson. In her after hold, the King of Spain's cargo is stowed, which is supposed to be gold and silver, but not yet opened. The captain and crew were not permitted to see it when shipped, as she was laden by porters, which is the usual custom at Manilla; nor is it supposed it will be examined until the owners of the privateers from England arrive here."

Captain Whytell's * own account, written at sea, is as follows:—

On Tuesday, the 24th August, we saw a ship, which proved to be a Spaniard; and at five minutes after twelve o'clock p.m., began to engage her. She looked exceeding large, and shewed fifteen guns on a side, but we could not tell whether they were metal or not until we tried; so run up and received her fire, and found she had only fourteen metal guns, but they were heavy ones. We gave her two broadsides for one, and continued the engagement for three glasses very briskly, and then lost sight of her for ten minutes in a cloud of smoke, and feared she had sunk. When it cleared up we perceived her endeavouring to make her escape, and gave chase to her again; came up with, received her broadside, and returned only a few guns before she struck her colours. A Bristol privateer (which

* Among the obituary notices of 1795, is the following: "On the 13th of June, 1795, in Hamoaze, on board his Majesty's ship *Standard*, Capt. Ellison, Mr. Charles Lowe Whytell, lately tide surveyor at Hoylake."

we afterwards found to be the *Ranger)* came up during the engagement, but kept aloof, and never fired one gun. We having received much damage in our rigging and sails, and our yard teacles shot away, the Bristol privateer took the advantage and boarded her first, and received the Captain's sword and papers, which they did not deserve.

"She proved to be the *St. Agnes*, a Spanish frigate, commanded by Fernando de Reynosa, from Manilla bound to Cadiz; larger than any of our thirty-six gun frigates, and pierced for forty guns. She had two eighteen and twelve nine-pounders mounted, and upwards of 150 men, of whom forty-seven were killed and wounded in the action, and in an explosion of gunpowder; (thirty-three of the forty-seven are dead).

"We only lost one brave fellow (the master's mate) who had his arm shot off by an eighteen-pounder, close to his shoulder, and he died in about an hour. My officers and ship's company all behaved like men of true courage during the whole engagement. I believe the prize is very rich; but know not yet what she is loaden with, therefore cannot ascertain her value."

"The prize," says the editor, "is since arrived safe in Cork, and the *Amazon* is come into Hoyle-lake. Letters from Ireland say the above prize is worth one million."

Amongst other curiosities exhibited in Liverpool in the year 1780, was a zebra, shipped on board the *St. Inez*—for that was the real name of the *Amazon's* prize—at the Cape, as a present for the King of Spain. An elaborate description of this zebra, as an animal exceedingly rare and curious, appeared in the columns of the local papers.

The paper of August 6th, 1779, stated that the *Charming Kitty* privateer, Captain Williams, had captured and sent into the Mersey, a Spanish brig laden with dollars and provisions, and was left in chase of a large ship. A week later, we read that the *Charming Kitty*, "cruising in lat. 41° 10′, long. 10° 30′ took a Spanish brig laden with 125 barrels of beef, 762 quintals of rice, 744 quintals of calavences,

904 quintals beans, 2505 bottles of oil, two trunks and one bale merchandize, and 130 dollars."

A letter from Fougères Castle, Ille-et-Vilaine, dated August 16th, 1779, says:—

"I have the happiness of conveying to you a letter by a prisoner who is to be exchanged in room of the Americans, and take the opportunity to let you and my friends know in what manner we have been used. In the first place we have a pound and half of bread, such as is the cause of all the sickness, beef is but just good enough for dogs, sometimes it amounts to half a pound a day, but more often to six ounces, sometimes we have peas, and those so bad that one half of them are as hard when they come out of the furnace as when first put in. The worst of usage in England for the prisoners is absolutely too good. The great havock it made in Dinan last winter is astonishing! Thirty died of a day, in the whole about 1600. They were put into a cart, a pit dug, and were thrown in like dogs. We have nothing to lie on but straw full of vermin, which deprives us of rest. The beds we had at first are taken away, and we are now treated as if we were horses. We dread the thought of another winter, and expect nothing but to fall victims to death."

In September, 1779, the following advertisement appeared in the Liverpool papers:—

"Port of Milford. To all Commanders of his Majesty's ships of war, cruisers, Letters of Marque, etc. For the more speedy condemnation of American, French, and Spanish prizes, captured by the said ships, the High Court of Admiralty hath issued out commissions appointing commissioners, etc., for taking depositions, etc., for condemnation of such prizes in the said port of Milford. W. WRIGHT, Actuary."

On Saturday, the 18th of September, 1779, the *Molly*, late Captain Seddon, arrived in Liverpool from Tortola. Her people had a sad story to tell. On the 7th September, she had encountered an American privateer, of 22 guns on one deck, besides quarter-deck and forecastle guns, a force greatly superior to that of the *Molly*. After a hot engagement which lasted upwards of an hour, the privateer sheered off, having received considerable damage, and, it was supposed, with many of her people killed and wounded. On board the *Molly*, Captain Seddon and five of his brave crew were killed, and seven wounded. "True courage," says the *Liverpool Advertiser*, "was never more conspicuous than in Captain Seddon's conduct during the engagement, nor was ever young man more deservedly respected, not only in the capacity of a commander, but in private life, by all who knew him."

With so many local instances of the heroism of British seamen coming under his observation, well might Liverpool's blind poet champion the cause of

> "A race renown'd in story:
> A race whose wrongs are Britain's stain,
> Whose deeds are Britain's glory.
> By them when Courts have banish'd peace,
> Your seagirt land's protected,
> But when war's horrid thunderings cease,
> These bulwarks are neglected."

The *Defiance*, Captain Thomson, took the *Francisco de Paula,* laden with wool, hides, and dollars, but the prize, *minus* the dollars, was re-captured by the notorious *Dunkirk* (alias *Black Prince)* privateer, of Dunkirk, which also took the *Three Friends*,* Captain Maine, on her passage from

* The following notice appeared in the papers:—

LIVERPOOL, *September 24th, 1779.*

"Whereas, Samuel Maine, master of the *Three Friends*, Letter of Marque, of Liverpool, bound to New York, was captured the 19th inst., in Lalliman's Bay, in the Island of Jura, by the *Dunkirk* privateer, of Dunkirk, commanded by J. B. Royer; who afterwards disposed of to the beforementioned Samuel Maine, a sloop laden with kelp, which the said privateer had likewise taken, after the crew had totally deserted her. This is inserted with a wish to apprize the owners of the said sloop, that unless application is made for the sloop and her cargo, and all demands and expenses incurred upon her discharged. before the expiration of twenty-one days, from this date, the sloop's cargo will then be absolutely sold by public auction, to defray such charges as may have accrued from the time of her capture, to release the hostage given by Captain Maine. Every proper enquiry will be fully answered and attended to by Crosdale, Barrow & Co."

Liverpool to New York. The *Defiance* was taken on her passage to the West Indies by a French privateer.

In September, 1779, a Special Council was held, to take into consideration the most effectual method of putting the town into a state of defence against a possible descent by Paul Jones, or any other invader.* Orders were passed "to remove the gunpowder from the magazines in Cheshire to the New Fort, and St. George's Battery; to apply to the Government for a thousand stand of arms for the use of such gentlemen and privates who may offer themselves to serve as volunteers in case of the enemy landing on the coast; that steps be immediately taken to receive names for volunteer service; that application be made for the removal of the French and Spanish prisoners now confined in the gaol at Mount Pleasant to the Castles of Chester, Carlisle, etc., and for the removal of the prisoners now on parole at Ormskirk and Wigan to some more inland situation; that a pilot-boat be sent out to cruise off Point Lynas, to give intelligence upon the appearance of an enemy, and that boats be stationed at the different buoys along the coast to sink them in case of imminent danger."

On the 29th of October, 1779, as the *Stag* privateer, Captain Wilson, ready for a cruise, was sailing up and down the river, to the admiration of a crowd on shore, as was the custom of privateers about to go forth against "the

* "Though the present Administration cannot be called felicitating, or that we now sit quietly under our vine and fig tree," says the paper of September 10th, 1779, "yet the public need not be apprehensive that an invasion of consequence can take place till our fleet be first destroyed. We are to consider that to effect the Revolution, the Prince of Orange (afterwards the Glorious King William), had with him 52 men-of-war, and 25 frigates, with 400 large Dutch transport ships, for the bringing over of 3,660 horse, and 10,692 foot. From this it would seem, that 800 transports should be necessary (besides men-of-war) to land in these kingdoms 30,000 men. This number of ships the French and Spaniards have not ready, nor will they venture so hazardous an enterprise till Sir Charles Hardy's fleet shall be discomfited. This, we trust in the great Disposer of Events, will not be the case. From the preparations, spirit, and unanimity that now appear, none can seriously believe that even 30,000 troops could conquer Britain or Ireland. Away then with false, unmanly fears. Let magnanimity and fortitude, vigilance, activity, and the love of our country animate us to the noblest actions."

enemies of Great Britain," she got aground near the Codling Gag and bulged. Most of the ship's stores were saved. On the 18th of February, 1780, the *Stag* captured a French ship, bound from Bordeaux to Martinico, laden with wine, provisions, bale goods, etc. Early in 1781, she took two prizes valued at £14,000 currency, and carried them into St. Kitts. The *Stag*, "upon a cruise in the West Indies," on the 14th of March of the same year, took a ship of 18 guns and 65 men, bound from Martinico for America, loaded with dry goods and some produce, valued at about £12,000.

The *Vengeance,* Captain Graham, took the *St. Maria* from Campeachy for Valencia, laden with logwood, etc. On the 21st of October, 1779, the *Who's Afraid,* Captain Moore, in company with the *Benson*, Captain Ball, 360 tons, 20 guns, and 79 men, belonging to Messrs. Rawlinson & Co., took *La Jeanne Lucy*, from Martinico to Marseilles, laden with sugar, coffee, and cocoa. A month later the *Diligence*, a prize to the *Who's Afraid*, arrived in the Mersey. In March, 1780, we read of the *Who's Afraid*, with two more prizes, being at Jamaica.

On the 24th of November, 1779, the frigates *Telemachus*, Captain Ash, and the *Ulysses*, Captain Briggs, both from Liverpool, on a cruise, took a Spanish frigate of about 600 tons burthen, called the *Soladad*, pierced for 26 guns upon her main deck, and carrying 170 men. She was bound from the South Seas for Europe, and had been three years out. The prize narrowly escaped an American privateer, of 30 guns, off Mizen Head, as she was making for Crookhaven.

"The first entry of licensed goods from England," says the paper of October 15th, 1779, "made in the Isle of Man after it was annexed to the Crown, was made by Paul Jones, he having imported the first rum there. His name stands first in the Custom House books at Douglas."

During this critical part of the war, when France and

Spain united with America, and presented a most formidable coalition in arms against Great Britain, the Royal Liverpool Blues—named in memory of the battalion raised in 1745—formed part of the garrison of the beautiful island and rich colony of Jamaica, which was in the utmost danger, until Rodney's great victory gave the English forces complete ascendancy in the West Indies. The Liverpool Blues, raised in 1778, principally at the expense of the Corporation, was a regiment of the line, commanded by Lieutenant-General Calcraft, as Colonel; Major Pole, as Lieutenant-Colonel; the Honourable Thomas Stanley, Major; Banastre Tarleton (afterwards General Tarleton), William Greaves, Bryan Blundell, Thomas Dunbar, Richard Cribb, Lieutenant Pigot, and Lieutenant Andrew Despart, as Captains; Mr. Buckley, as Captain-Lieutenant; George Headlam, as Lieutenant; and Christopher Graves, George Leigh, Thomas Leigh, and James Smith, as ensigns. These subalterns were principally Liverpool gentlemen. There was a comic side to the achievements of the valiant Blues of '45—the march to Warrington, and the nocturnal charge, that would have fired the soul of Don Quixote, and satisfied the stomach of Sancho—but the brief story of this line regiment is one grim tragedy. On the 25th of May, 1778, the Liverpool Blues mustered 1100 strong on the sands near Bank Hall, where they were reviewed, and presented with their colours. On the 4th of June, the birthday of George III., they were reviewed in front of the Goree warehouses. On the 17th of the same month, they marched from Liverpool to Warley Common, Essex, being ultimately sent to Jamaica, where nearly the whole regiment succumbed to the climate. On the 9th of February, 1784, the poor remnant, reduced to 84 in number, returned to Liverpool in the ship *James*, belonging to Messrs. James France & Nephew, and deposited their colours in the Exchange. Thus Liverpool men, by land as well as by sea, freely gave themselves in defence of their King and Country. On the departure of the Blues, the first division of the Leicestershire militia, commanded by the Marquis of Granby, was stationed in Liverpool, hence, no doubt, the connection of the future Duke of Rutland with privateering. A little later, a regiment of Yorkshire militia, commanded by Sir George Saville, did duty in the town. At the close of 1779, Sir George gave £50 to the infirmary, £50 to the dispensary, and £50 for the relief of the French and American prisoners. An appeal for subscriptions in aid of the latter object concluded with these modest words: "and as the town of Liverpool is already the terror of our foes, they will, by this means (at the time that they acknowledge our spirit and bravery) be obliged to reverence our virtue and humanity."

On the 22nd of January, 1780, the *Lively*, Captain Watts, sailed from London for Liverpool, and, two days after leaving the Downs, they fell in with the *Black Prince* (called an Irish pirate vessel), to whom they were forced to strike. The sea at that time ran so high that the enemy could not board the *Lively,* but ordered them to follow, which they did, till night coming on, and the gale continuing, they got away from her. Two days after separating, the *Lively* had the misfortune to fall in with the *Monsieur*, a frigate of 44 guns, who made a capture of her, took the captain and all the people overboard, except three boys, and put on board a French officer and twelve seamen. Some time after they parted company with the frigate, the *Lively* grew very leaky, so that it was with difficulty she could be kept up. On the 4th of February, when all but three of the Frenchmen, greatly fatigued with working and pumping the ship, were asleep, the three boys seized on two cutlasses, the only arms on board, and recaptured the ship, "and" says the paper "preserved the power they had taken with amazing resolution." The day following they arrived off Kinsale, and making a signal of distress, were conducted into port by two pilot boats, where Captain M'Arthur, in the *Hercules* (Letter of Marque), took possession of her, "after beating off the savages of our own realm, who came in shoals to plunder. More than 50 of those unprincipled villains, taking advantage of the signal of distress, had actually got on board, and had already begun the shameful business which so repeatedly practised fixes an eternal stigma on the coast which shelters such abandoned miscreants."

On the 16th of January, 1780, the *Antigallican* privateer, Captain Butler, of Liverpool, in company with the *Alert*, of London, took the snow *Diana,* from Philadelphia to Bordeaux, laden with tobacco, logwood, staves, etc., valued at £12,000. Both privateers then chased a large ship of 30 guns, laden with 600 hogsheads of tobacco, etc., which had parted company with the snow only five hours before she was taken. The *Antigallican* mounted 20 eighteen-pounders and 2 long sixes, and had 120 men. The *Alert* carried 12 eighteen-pounders, 4 long nines, and 70 men. It is not clear whether this was the *Antigallican* privateer, which sailed from Shields on the 13th of March, 1779, "on a six months' cruize against the enemies of Great Britain, completely fitted and manned," and was "universally allowed by every competent judge to be the finest ship for that purpose yet fitted out from England."

In February, 1780, the *Sparling*, Captain Jackson, on her passage to New York was taken by the *Thorn* sloop-of-war, 22 guns and 150 men, after an engagement of nearly an hour, in which Captain Jackson, the first and second lieutenants, and seven men were wounded, and three killed. The *Sparling* was carried into Boston.

Early in 1780, the *Hero*, Captain Wilcox, bound for Guinea, was taken by the French, and retaken within a league of Cherbourg by his Majesty's ship *Champion*. On the 1st of May, the *Hero* was again taken, 16 leagues south of Cork, by a French privateer, and again retaken from the enemy. Once more the *Hero* essayed to reach Africa, only to fall a prey to the combined fleets of France and Spain, who sent her to Cadiz.

The *Bridget*, Captain Gilbody, on the voyage from St. Kitts to Liverpool, re-captured the *Brothers*, Captain Hasseldine, which was proceeding to France in charge of a prize crew. In the summer of 1781, Captain Hasseldine was again taken within one day's sail of New York, and carried into Providence.

On the 8th of February, 1780, the *Pallas*, 16 six-pounders, Captain Townsend, took the *La Anna*, from Bordeaux to St. Domingo, laden with 642 barrels of flour, 180 barrels of beef, 216 barrels of pork, two pipes and 30 casks of oil, 141 barrels and cases of wine, 55 cases of brandy, besides large quantities of butter, salt, pease, prunes, soap, hoops, medicines, and women's shoes. On the 10th of March, the *Pallas* brought into the Mersey, the ship *La Victoire*, of 16 six and nine-pounders, and 100 men, laden with naval stores, captured off Cape Finisterre, as she was making for Corunna. This prize had previously been taken by one of Admiral Digby's fleet, but the French prisoners had put the prize crew in irons and retaken the ship, about thirty hours before Captain Townsend fell in with her. The ship *La Vulture*, from L'Orient for Maryland, another prize taken by the *Pallas*, was totally lost on the coast of Ireland, and several of the crew drowned. In May, the *Pallas* sent into Liverpool a Spanish schooner, laden with iron, oil, brass and steel ware, etc., captured off Bilbao. A few weeks later, the *Pallas* was taken by the *L'Aimable*, frigate, and carried into Rochefort. "Tuesday," says the paper of May 18th, "was married Captain Holland, in the African trade, to Miss Townsend, sister to the brave Captain Townsend,* of the *Pallas*."

* Commenting on the superiority of the new prepared cartridges over the old fashioned or common sort, the paper of December 17th, 1779, relates the following

At the Annual Board of the Infirmary, on March 6th, 1780, thanks were voted to the late President, Nicholas Blundell, Esq., for a benefaction of forty guineas, and to the owners of the *Enterprise, Terrible*, *St. George*, and *Dragon* privateers, who had sent benefactions towards the support of the charity.

The *William*, Captain Wignall (a Letter of Marque), cruising on her passage from New York for Liverpool, captured a schooner laden with lumber, which she sent to Liverpool. On the 8th of February, 1780, the *William* fell in with a French cutter, of 18 nine-pounders, 20 swivels, and full of men, from Havre-de-Grace, on a cruise, with which she had a stout engagement of an hour, but by the gallant behaviour of the officers and men of the *William*, the cutter was obliged to sheer off with much damage. Captain Wignall had one man killed, seven wounded, and two blown up by a cartridge taking fire. The loss of the cutter was believed to be greater.

On the 4th of March, 1780, there entered the river Mersey, under peculiar circumstances, a vessel appropriately named the *Happy Return*. She was a cartel ship, commanded by Captain Webb, bound from L'Orient to Plymouth, with 300 prisoners of war, to be exchanged. These men had taken possession of the vessel during the passage, and brought her into Liverpool, in hope of escaping the press. Among them were the crew of the *Bess*, Captain Walker, which had been taken by the *Monsieur*, three days after she sailed from Liverpool for Tobago.

In March, 1780, two additional frigates and two cutters were stationed in the Irish Channel, in consequence of a

incident:—"In a smart engagement with an American privateer, in which Capt. Townsend, of this place, lost his leg, in the heat of action to save time, a brave, high-spirited boy dared to endeavour to load one of the guns on the outside of the vessel, though the privateer was alongside, but when putting the cartridge down with his arm into the gun just discharged, the burning paper left behind, set fire to the new one, and blew the bold fellow into the sea, where he was drowned."

petition of the Liverpool Merchants, in which they stated that the force previously on the station was insufficient for the protection of trade.

On Friday, March 17th, 1780, between ten and eleven o'clock at night, the press-gang assembled before the house of James Richards, in Hackins-hey, where a number of sailors had resorted to protect themselves from impressment; and upon Richards refusing to open the door a general firing ensued, which continued about half an hour. In the affray, Richards received two dangerous wounds in the face. A soldier, who happened to be in the house, was shot through the body and died next morning.

The *Modeste*, Captain Bewsher, in her passage from Liverpool to New York, took a prize with tobacco, lumber, etc., and sent her for the Bermudas; and on her passage from New York to Jamaica she captured a sloop and a schooner.

Occasionally we discover an element of comedy amongst the sanguinary records of the sea at this period, as in the following experience of the dignified Captain Gurley and the irreverent rover, as reported in the paper:—

> "On Friday morning, April 14th, the *Hussar* wherry, Capt. Gurley, a revenue cruizer under the inspection of Charles Lutwidge, Esq., of Whitehaven, sailed on a cruize, but returned ingloriously to port in the evening, having met with a large buccaneering cutter, mounting 18 carronades, twelve and eighteen-pounders, off the Abbey Head, about four miles distant from Kirkcudbright. The cutter fired several guns into the *Hussar*, shot away her colours and the main haulyards; made several holes in her mainsail and foresail, and lodged several shot in one of her masts and in the hull. Captain Gurley had a part of his hat and wig taken off by a ball, and one of the men was also in the same perilous situation, but happily neither of them received much bodily injury. The lawless rover was very near them, and had the insolence to call out

and order them to go home, which they were under the necessity of complying with, being much inferior in strength. The cutter had an English ensign flying."

In May, the *Tonyn*, Captain Wade, took a prize which sold for £1,300; and the *Ceres*, Captain Cook, on her passage from Liverpool to Archangel, took an American ship called the *Governor Johnson*, from Bergen to Baltimore, with salt and sail canvas.

Instances of cowardice in face of the enemy were happily rarer in the armed mercantile marine of Liverpool than in the Royal Navy itself. In the following interesting letter from Captain William Garnett, of the *Vengeance* privateer, to the owners, Messrs. Jonas Bold & Co., dated Port L'Orient, June 19th, 1780, we find a serious and unusual charge of poltroonery brought against two of his officers:—

"GENTLEMEN: I take this opportunity to inform you that on the 13th inst., fell in with two large ships, which we took to be loaded, and chaced them under Belleisle. As one of them got under the battery, we bore down on the other, which we soon brought to action, in doing which, we began to suspect we had got into a very disagreeable situation, but were too far advanced to retreat, therefore kept down upon her, and received three broadsides without any great damage, reserving our fire until we found it absolutely necessary to engage, as we were to prevent the other ship if possible from raking us, she having bore away for that purpose. Thus after two hours contention with the *Magician*, of 36 guns and 300 men, and the *Etourdie*, of 24 guns and 180 men (both King's frigates), was obliged to surrender, on a false alarm of our magazine being on fire, our ship having from four to five feet water in the hold, her fore-topsail-yard hanging in two pieces, her foremast wounded, and in short her hull, rigging and sails much shattered; and was very much surprised from the situation we were in when we struck, that we had but one man killed and fourteen wounded, all but two or three of which were slightly so. It is impossible for me to do justice to our people's behaviour;

some of whom, when we had struck, lay down by their guns, and shed tears for downright vexation. We have lost all but what we had on us, otherwise we were tolerably well treated while on board, which was till this morning, when they landed us at this place, from whence we are to set out on the 20th to travel to St. Malo, in order to be exchanged, and have for company the *Alert's* crew, and the crew of a Jersey privateer, all taken this week. We had but two persons whose behaviour during the action, I am sorry to say, was exceptionable."

On the 21st of June, the Captain wrote again, from Portevin :—

"I wrote you the 19th, advising you in a particular manner of our being taken on the 13th inst., after an action of two hours with a frigate of 36 guns and 300 men, and one of 24 guns and 180 men, in which we had one man killed and fourteen wounded, and our vessel shattered to pieces. Our people behaved in a very brave manner, and had several encomiums paid them by the Captain of the large frigate, whom we struck to, particularly Mr. Given and Mr. Walker, two better or braver officers never went on board a ship ; the latter of whom, after being knocked down twice with splinters, and his right arm broke, still kept encouraging the men. The conduct of two of our officers during the action was highly exceptionable ; there is no knowing a man until he is tried. I have not spoke to either of them since, which conduct I shall observe. I have all the other officers along with me ; and as they are brave fellows, think they have a right to every assistance in my power to afford them. We are now forty miles on our way to Dinan, and hope we shall not be long detained. I write you from hence in preference to any other place, having met with a Monsieur ———, who, to offering me any money I should want, takes the trouble of having this conveyed to you in the most expeditious manner. We were very well treated on board the frigate, our wounded in particular with the greatest care and humanity. We have left six wounded in the hospital at L'Orient, where they are taken the greatest care of, are all in high spirits, and have every good symptom. The rest are along with me, Mr. Given and Mr. Walker in particular, who reckon nothing of walking twenty miles per day."

The *Albion*, Captain Hutchinson, from Liverpool to Archangel, was taken on the 8th of June, 1780, by three American Letters of Marque, bound to Amsterdam, with tobacco; the *General Washington*, of 18 six-pounders and 73 men, with 160 hogsheads; the brig *Alexander*, of 12 four-pounders and 50 men, with 110 hogsheads; and the brig *Maryland*, of 10 four-pounders and 50 men, with 120 hogsheads. They also took the *Speedwell*, from Peterhead to Norway, with oatmeal, and gave her to Captain Hutchinson to carry the prisoners to Inverness, where they landed on June 15th. Captain Hutchinson reported that the ship *Ashton* and three brigs, which sailed with her from Liverpool, had been taken by American privateers. The *Albion* was sent to Boston. Some of the privateers were very unlucky. On the 10th of June, 1780, the *Alert* privateer, Captain Chapman, was taken by the *Venus*, French frigate of 40 guns, and carried into L'Orient. The *Alert* had sailed from Whitehaven in March, 1779, and returned to Liverpool in the following July, without taking anything. Captain Chapman died about September, 1780, imprisonment and disappointment probably hastening his end.

In February, 1780, the *Sally*, Captain Denny, on her passage to Barbadoes, took a Spanish brig from the Grand Canaries, loaded with sugar, barley, and household furniture.

The *Watt* (Letter of Marque), Captain Coulthard, on her passage from Liverpool to New York, took two prizes; one called the *Nancy*, from Virginia to Nantz, laden with 103 hogsheads of James River tobacco; and the other, the brig *Le Pegase*, of 16 guns, bound from Bordeaux to St. Domingo, laden with provisions. The best contested battle fought by any of the British privateers during this war was undoubtedly that fought by the *Watt*, and the American ship

Trumbull, Captain Nicholson. The following account of the engagement appeared in the New York paper of June 16th, 1780:—

"Monday arrived the Letter of Marque ship *Watt*, Captain Coulthard, in twelve weeks from Liverpool. On the 1st instant, in lat. 35. 54 long. 66. she fell in with and engaged a rebel frigate of 36 guns, 12-pounders, for upwards of seven glasses. The rebel ship was crowded with men, and fought 19 guns on a side. The *Watt* mounts 32 twelve and six-pounders, some of them carronades, and had only 164 men on board, eleven of whom were killed, and several wounded. The action was obstinate and bloody, and the carnage on board the rebel frigate amazing, as the vessels were a considerable time yardarm and yardarm, and the *Watt*, by the superior skill of her officers, and the alertness of her crew, had the opportunity of twice raking her antagonist fore and aft, which made her a perfect slaughter house. Her stern was drove in almost down to the water, many of her guns dismounted, hundreds of shot through her sides, her foreyard and topmast shot away, and all her sails and rigging greatly damaged. She at last put before the wind, and run from the *Watt*, which chased her eight hours; but having a cargo on board, and her masts so damaged that she could not venture to carry a great press of sail, she lost sight of the chace on the 2nd inst. The *Watt* has a great number of shot holes through her sides and sails, four of them through her powder magazine. She has certainly fought a more glorious battle than any private ship of war since the commencement of hostilities. The most exalted encomiums are inadequate to the merit of the brave Captain Coulthard. The determined courage he exhibited during the action, and the cool, deliberate manner in which he issued his orders, does him the highest honour; nor ought the approved behaviour of his gallant officers and crew remain unnoticed; they richly merit, and will certainly receive applause from every man who has the glory of his country at heart." A later issue supplied the following particulars:—

"By a flag of truce arrived last night from the Eastward, we are informed that the Lieutenant of the *Trumbull* Rebel frigate, had with much difficulty got the ship into New London, after being torn down to a mere wreck in an engagement on the first instant, with the Letter of Marque ship *Watt*, of Liverpool, commanded by the truly invincible Capt. Coulthard. We have as yet only been able to learn that Capt. James Nicholson, the *Trumbull's* commander, was killed at the first broadside received from the *Watt*, and that there were fifty-seven men killed on board the *Trumbull*; the number of wounded has not yet been declared. Our last *Gazette* gave the particulars of the glorious behaviour of Capt. Coulthard and his Crew of HEROES."

In connection with this sanguinary drawn battle, after which both vessels were nearly sinking when they got back into port, the following appeal appeared in *Williamson's Advertiser*, of August 3rd :—

"To the Humane Inhabitants of Liverpool, and others.

"Phebe Rigby, widow of Nicholas Rigby, late a mariner, on board the *Watt*, commanded by the truly magnanimous Captain Coulthard, claims your attention to her present distressed condition, having lost her husband in that memorable engagement with an American frigate of 36 guns, wherein the *Watt* had thirteen men killed and seventy-five wounded, and the loss of the American was considerably more (an engagement which does very great honour to the intrepid Captain Coulthard, and casts an additional splendor on the British flag). This destitute widow thus deprived of her husband (who was the chief support of her and two helpless children, and an expert and courageous sailor) is really a deserving object of charity, and claims that kind and liberal attention of the humane and generous, which ever distinguishes Britons from other nations, and which extends its munificent hand to all ranks of deserving objects, but more especially to those who have sustained the irreparable loss of an industrious husband and affectionate parent, and that, too, in fighting for his country.

"By applying to Mr. Tate, hair-dresser, in Church Street, the public may be assured of the reality of this case; and contributions for the widow may be left with him, or at Mrs. Williamson's and Mr. Gore's, printers, in Liverpool.

"Descend, sweet Charity, celestial maid,
And to the widow lend thy fav'ring aid,
Whose valiant husband, under Coulthard's sway
(Coulthard, the dauntless hero of the sea)
Asserting Britain's glory, lost his life,
And left two helpless children and a wife.
Celestial Charity, thy hand extend,
Be now the widow and poor orphans' friend."

In the Liverpool paper of July 6th, 1780, appeared the following paragraph :—

"The *Ellen,* Borrowdale, arrived at Antigua in May with a Spanish sloop of war called the *St. Ann Gracia,* Don John Morallas, commander, mounting 16 guns, and full of small arms, bound from the island of St. Thomas to Cadiz, which he took on the 26th of April, in lat. 30. 30 N. long. 37. 38 W. after an engagement of three hours. The Spaniard had eleven men killed and two wounded."

Such is the bare record of a notable engagement which has been selected by Professor Laughton as worthy of a place in his "Naval Studies," and, which we venture to quote here :—

"The *Ellen,* which mounted 18 light six-pounders, and had on board 64 men, all told, of whom many, including a Captain Blundell of the 79th regiment, were passengers, was making a passage to the West Indies, under orders of urgent haste. Her small complement shows that she had no aggressive intentions; but, when overhauled by the Spaniard, she prepared to defend herself. She shortened sail, and, to prevent the enemy opening fire at long range, and thus getting the advantage of a presumably heavier armament, hoisted American colours. At the same time, her guns were double-loaded with round shot and grape; and Borrowdale, encouraging his men, 'recommended to them a cool and determined courage, entreated them to fire quick, to take good aim, and to fight the ship to the last extremity.' We seem almost to have before us the old sea-dog described by Captain Marryat:—

"The Captain stood on the carronade; 'First Lieutenant,' says he,
'Send all my merry men aft here, for they must list to me;
I haven't the gift of the gab, my sons,—because I'm bred to the sea;
That ship there is a Spaniard, who means to fight with we;
That ship there is a Spaniard, and if we dont take she,
'Tis a thousand bullets to one, that she will capture we.'

"And so as the Spaniard ranged up alongside to windward, he hauled down the American colours, hoisted the English, and poured into her his whole broadside, with a volley of musketry. The astonished and entirely disabled Spaniard fell to leeward, and received the *Ellen's* other broadside, in the same fashion, after which she put before the wind and endeavoured to make off. But the privateer held on to her advantage, and after a running fight of an hour and a half the *Santa Anna*, a commissioned sloop of sixteen guns—heavy 6-pounders—exclusive of swivels, and 104 men, hauled down her colours, and accompanied the *Ellen* to Jamaica."

In regard to the quotation from Marryat, Professor Laughton informs us that carronades were not used in the time of William and Mary. They were first ordered for use in the navy in 1779. The *Ellen* had not any, but, in 1780, she might have had if her owners had chosen. The "lawless rover," who attacked the *Hussar* wherry, had 18 carronades on board. On the 26th of June, 1782, the *Isabella*, from the Isle of Bourbon for France, laden with cotton, coffee, pepper, cloves, etc., arrived in the Mersey, having been captured by the *Ellen*, on her passage from Liverpool to Jamaica. In October of the same year, the *Ellen* arrived at Hoylake from Jamaica, having on board forty-two of the ship's company of the *Ramilies*, flagship of Rear-Admiral

Graves, which had foundered. The Mayor received a letter from the Admiral requesting him to convey to the captains of the merchant ships belonging to Liverpool, who were the preservers of the lives of the Admiral, officers, and company of his Majesty's ship *Ramilies*, the approbation of the Lord's Commissioners of the Admiralty, of their humane conduct.

On the 10th of August, 1780, the *Snapper* privateer, Captain Taylor, returned to Liverpool from a successful cruise. On the 24th of July, she met with a fleet of seven ships, off Bordeaux, under convoy of a 20 gun ship, which chased the *Snapper* till night came on, when the *Snapper* altered her course, and the next morning fell in with the seven sail, four of which she captured, and run the three others on shore. The four prizes consisted of a snow from Bayonne, laden with bombs, mortars, and oak plank; a sloop from St. Sebastian, with iron and iron hoops; a brig from Bayonne to Rochefort, with anchors, hemp, and canvas; and another brig from the same place to St. Malo, with pitch, tar, and oak plank. In December, the *Ann*, with fish and oil, and *La Santa Louisa*, both prizes to the *Snapper*, arrived in Liverpool.

Early in the year, the *Tartar* cutter, Captain Whytell, took a French snow bound from Guadaloupe to America, with a cargo of sugar, and carried her into St. Kitts. On the 22nd of August, the *Tartar* arrived in Liverpool from a cruise, and brought in with her a prize called the *St. George*, laden with flax, iron, etc. In September, the *Tartar* had a smart engagement with a French cutter privateer of 16 guns, which resulted in the Frenchman accompanying the *Tartar* into Penzance. After taking another prize, a Dutch vessel, from Ostend to Bordeaux, with 420 hogsheads of tobacco, the *Tartar* had the ill-luck of being herself captured by two French frigates, one of which was commanded by "Monsieur Le Viscount Mortimer."

In the summer of 1780, the *Black Princess*, and other French privateers, were very active in the Channel, picking up, at a small risk, a large number of vessels, many of which they ransomed for sums ranging from 100 guineas to £6,000. One Sunday in July, the *John*, of Newcastle, Captain Rawson, fell in with the *Black Princess*, off the Mull of Galloway, and was obliged to strike, after having one man killed and the captain and second mate wounded. He was ransomed for £1,000, "to which," says the paper of July 20th, "he was compelled, at the hazard of his life and the lives of his crew. The inhuman villain who commands the *Black Princess* would not permit his surgeon to dress the wounded, and on Captain Rawson hesitating to ransom for so large a sum, was preparing to burn the ship, and, horrid to relate, the people also." The "inhuman villain" in this case happened to be an Irishman named Edward Macartney, who had lived twelve years in France. In 1781, we find him, together with his second captain, and first and second lieutenants, enjoying the hospitality of the British Government in Mill Prison, Plymouth.

The first cigars introduced into Liverpool are said to have been brought in some French prizes, from the Island of St. Domingo, taken during this war.

In July, 1780, the *Porcupine,* a private ship of war, of Liverpool, John Walker, commander, in company with the *Tartar* cutter, of Folkestone, took the ship *Elizabeth*, from Bordeaux to Bilbao, with a cargo of sugar, chocolate, indigo, wine, etc., and sent her to Falmouth. The *Eagle*, Captain Ashton, on a cruise, took two prizes; and the *Peggy,* Captain Leigh, captured three prizes in the West Indies.

"Jenny" was a favourite name for Liverpool vessels, there being at this period about half-a-dozen "Jennys" sailing from the port. The *Jenny*, Captain Gill, and the *Jenny*, Captain Walker, had a smart engagement for upwards of five hours with an American frigate of 28 guns, off the Banks of Newfoundland. They shot away the frigate's main-mast, and otherwise damaged her so much, that she made the best of her way from them. The *Jenny,* Walker, had four men killed, and one wounded ; the *Jenny,* Gill, had two men killed. The armament of these two vessels is described in the following advertisement, which appeared in the paper of December 28th, 1780:—*

"The ship, *Jenny*, a Letter of Marque, Thomas Walker, commander, is now fitting out to cruize for four months against the combined enemies of Great Britain, and will proceed to her station as soon as possible in order to intercept some valuable Dutchmen, that are soon expected to arrive from the East and West Indies.

The *Jenny* carries 14 guns, six-pounders, swivels, and small arms, is copper bottomed, and has every convenience for the comfort and accommodation of her crew, being about 5 feet 6 inches between decks.

Captain Walker invites all brave seamen and landmen that are willing to try their fortune in the *Jenny*, to apply to him immediately, at his house, No. 13, in Paradise-St., or to Daniel Backhouse, who wants a few good seamen and landmen for the brig *Jenny*, Capt. Wm. Gill, now in the river, and will sail on Saturday or Sunday next for St. Kitts, and from thence upon a cruize. She is copper bottomed, sails like the wind, and carries 16 guns, six-pounders."

The *Jenny*, Gill, took the *F. Coleux*, of Boston, with wine, flour, etc., which arrived in the Mersey in January, 1782.

In 1780, the *Mars*, Captain John Forsyth, a slave ship belonging to Messrs. Wm. Earle & Sons, on her passage to Africa, took a Dutch snow laden with French brandy, wine, and corkwood. In January, 1782, we read that the *Mars* was herself taken on her passage from St. Kitts to

* In the same issue of the paper, an advertisement appears, offering a reward of £20, in addition to the King's reward of £40, for the apprehension of the highwaymen who infested the roads in and near the town.

Liverpool, and carried into Boston. She was retaken and carried to Jamaica.

The *Emperor,* Captain Wm. Wilson, owned by Mr. John Galley, took the brig *Jupiter,* from Newberry, with tobacco, staves, etc.; also the brig privateer *L'Impromptu,* of 14 guns. In 1781, the *Emperor,* and the *Telemachus,* Captain Sherwood, on their passage to Jamaica took a brig from Salem, laden with lumber, etc.

"By an Act passed this Session," says the paper of May 25th, 1780, "merchant ships are allowed to have three-fourths of their crew foreigners; and all foreigners who shall have formerly served, or shall hereafter serve, two years on board any of his Majesty's ships, or any privateer, or merchant ship, being British property, shall be deemed a natural born subject of Great Britain, and enjoy all privileges and immunities thereto belonging."

On the 5th of June, 1780, the *Vengeance, Hypocrite,* and *Surprise,* three Liverpool privateers, captured off Belleisle, the *Dauphine* snow, from L'Orient to the Isle of France, with wine, brandy, flour, cordage, etc., on the French king's account. A month later, the *Hypocrite,* Captain Beynon, returned from a cruise, and brought into the Mersey a Genoese snow, from St. Andero to Cadiz, with 250 tons of wheat. In August, another prize laden with wheat, taken by the *Hypocrite,* arrived in Liverpool; and early in the following year, Captain Beynon, in his passage to St. Kitts, took a valuable schooner, bound from St. Eustatia to Marigalante. While cruising in the West Indies, the *Hyprocrite* was taken by a French privateer, after a severe engagement, in which Captain Beynon was killed.

In November, 1780, the *Hawke,* Captain Smale, took *La Jeune Emilie,* from Rochefort to Martinico, laden with wine, brandy, etc.

The paper of October 26th, 1780, stated that in the action at Camden, Lieut.-Colonel Tarleton had killed nine

Americans with his own hand. Though this may have been an exaggeration, the Liverpool hero certainly covered himself with glory during the war, and the fame so won, together with the dexterous display of a maimed hand, was of immense value to him in later years for electioneering purposes.

An idea of the unsightly, narrow, and mean appearance of Liverpool streets and alleys at this period may be gathered from the following account, extracted from the Journal and Letters of Samuel Curwen, Judge of Admiralty, etc., an American refugee in England, who visited the town on the 12th of June, 1780:—

> "Entered the City of Liverpool so celebrated for its commercial character; houses by a great majority in middling and lower style, few rising above that mark; streets long, narrow, crooked, and dirty in an eminent degree. During our short abode here, we scarcely saw a well-dressed person, nor half a dozen gentlemen's carriages; few of the shops appear so well as in other great towns; dress and looks more like the inhabitants of Wapping, Shadwell, and Rotherhithe, than in the neighbourhood of the Exchange, or any part of London above the Tower. The whole complexion nautical, and so infinitely below all our expectations, that naught but the thoughts of the few hours we had to pass here rendered it tolerable. The docks, however, are stupendously grand, the inner one, called Town Dock (*a*) lying in the centre of it, and filled with vessels exhibiting a forest of masts; besides this, are three very large ones (*b*) lying in front of the city, communicating with each other by flood gates, intermixed with dry ones for repairing (*c*); the lower or new one (*d*) has a fine, wide quay on its outer side; an agreeable walk, being lined with trees on either hand; (*e*) below this, on the river, is now building, nearly finished, a circular battery, (*f*) with embrassures for thirty cannon. Parade and barracks are in hand,

• (*a*) The Old Dock. (*b*) George's Dock, the Dry Dock (now part of the Canning Dock), and the Salthouse Dock. (*c*) The Graving Docks. (*d*) George's Dock. (*e*) The North Ladies' Walk, (*f*) The Old Fort.

and when completed, will afford a charming walk and prospect, if allowed to the inhabitants."

On the 7th of April, 1781—a notable month for the arrival of prizes this year—the *Adylet*, from Curacoa, a prize taken by the *Lookout* and *Prince of Orange* privateers, came in, and, about the same time, the brig *Venus*, Captain Quayle Fargher, from Bordeaux to America, laden with cordage, etc., a prize to the *Terror* privateer. On the 10th, arrived the *Success*, laden with 3000 bushels of salt, 115 boxes of lemons, 14 boxes of hats, 300 pairs women's shoes, and about "4000 weight" of cordage, a prize to the *James and Mary*, Captain Preston. She had been taken before by a Dartmouth privateer, who left on board nine men and a boy, and eleven Frenchmen; the latter had overpowered the privateer's people and got possession of her again.

The *Betty*, Captain Wilson, on her passage to Greenland, took and sent into Lough Swilly, the *Johannes*, from St. Eustatia for Amsterdam, laden with 292 hogsheads of sugar, 100 hogsheads of tobacco, 158 bags of coffee, 103 bags of cocoa, and 9 casks of indigo.

On the 14th of April, Captain Butler, of the *Tartar*, then a prisoner in Bayonne Castle, wrote to his owners in Liverpool that he was captured on the 15th of March, by the *Eagle*, French ship-of-war, of 28 guns, twentyfour-pounders, and 430 men, after a chase of eight hours and an engagement of one hour and a quarter. The *Eagle* had captured nine prizes in three weeks, amongst which were the *Stately*, Captain Fisher, and the *Fly*, Captain Byrne, both of Liverpool. Captain Fisher was getting better of a long sickness. Men who laughed at the perils of the deep and faced death without flinching in the stress of battle, soon succumbed to chagrin, prison fare, and close confinement in a foreign land. In this respect the more vivacious Frenchmen suffered less during their temporary sojourn in the land of "perfidious Albion." The self-reliant spirit of the Scot in adversity, is happily illustrated in the following letter, written to his wife, by a Scotch sailor, who was evidently a humourist and a philosopher :—

> "Dear Jenny : This is to let you know that I am well in a dungeon at Dunkirk, God be blessed for it, hoping to hear from you and all friends. Tell Mrs. Ross I bought her stuffing ; but it is gane. Let Jean know that I bought her a gown, and it is gane too. I bought an anker of brandy and gin to ourselves ; but Jenny, they are gane too, and a's gane : for the French dogs unrigged me in an instant, and left me nought but a greasy jacket of their ain ; but Jenny, I have saxpence a day from the King of England, God bless him ; and I have bread and water from the French King, God curse him. Out of the saxpence a day, I have saved as much as bought me a knife, a fork, and a wee Coggie. Jenny, keep a good heart, for I'll get out of this yet, and win meikle Siller, and get a bottom of my ain too ; and then have at the French dogs. I am, &c."

The vivacity of the French prisoners in Liverpool is mentioned by the Rev. Gilbert Wakefield, who resided in the town at this period, and who did not think it beneath his dignity to write an anonymous letter to the Mayor on their behalf. He says :—

> "The American and French war had now been raging for some months, and several hundred prisoners of the latter nation had been brought into Liverpool by privateers. I frequently visited them in their confinement, and was much mortified and ashamed at their uniform complaints of hard usage, and a scanty allowance of unwholesome provision. What I occasionally observed in my visits gave me but too much reason to believe the representations of this pleasing people, who maintained their national sprightliness and good humour undamped even by captivity. I kept my suspicions secret ; but wrote an anonymous letter to the Mayor, stating my observations and sentiments on the subject.

"I was happy to learn very soon, from the prisoners themselves, the good effects of my interference; and the Commissary, the author of their wrongs, was presently superseded; whether in consequence of my detection of his iniquities, I could never learn; but when I met him in the street there was fire in his eye, and fury in his face.

"Towards the conclusion of one of my sermons, preacht at Liverpool, I was led by the proximity of the subject to condemn, in terms of the utmost asperity, and somewhat hyper-tragical, the horrid practice of aggravating the calamities of war by the rapine and injustice of private hostility. This, in the grand mart of privateering during that war, and of the African slave trade, excited, of course, no small degree of resentment against the author of such outrageous doctrine. I was acquainted at that time with no other effect of my interference besides malignity against myself; but learnt some years afterwards, that the nerves of one lady were so agitated by the thunder of my lecture, as to allow herself and husband no rest till he had sold his share in a privateer."

In April, 1781, the *Balgrove*, Captain Thompson, was taken by a French privateer, and recaptured by the mate and four of the crew, who overpowered sixteen Frenchmen, and carried the ship into Cove.

By a singular coincidence, the *Alert* privateer, of Alderney, took the *Reine Jeanne*, from St. Domingo to Nantz, which proved to be the former *Alert* privateer, of Liverpool. A much more remarkable circumstance happened before the close of the year, when a Captain M'Bride discovered in the father and son, who commanded two Dutch privateers taken by him, the very men he had captured, under similar circumstances, twenty-one years before.

In May, 1781, the *Ferret* privateer, of 10 guns, Captain Archer, was taken by the French, and retaken by the *Vulture* privateer, of Jersey; and on the 31st of the same month, the *Patsey*, Captain Dooling, was taken, off the Western Islands, by the *Fripon*, French frigate of 44 guns and 400 men, after an engagement of one hour and-a-half in which Captain Dooling, the sailing master, and six of the crew of the *Patsey* were killed, and several wounded.

The slave ship *Essex*, Captain Potter, on her voyage to Guinea, took two Dutchmen, from St. Eustatia to Amsterdam and Rotterdam, laden with 400 hogsheads of sugar, 119 hogsheads of tobacco, and 800 bags of coffee. *La Fortune,* one of the prizes, was totally lost near Wexford, and all the crew perished; the other prize, called the *Golden Tea*, arrived safe in Liverpool. The *Fly* privateer took a Dutch brig, from St. Eustatia, and carried her into Kinsale. The *Stormont*, Captain Dawson, took the *Henry and Maria*, of Amsterdam, from Salonica, with 462 bales of cotton. Another Dutchman, called the *Vleyt*, from Curacoa to Amsterdam, was taken by the *Lookout,* Captain Wright, and sent into Scilly. The cargo consisted of 794 bags of coffee, 77 casks of indigo, 110 hogsheads of sugar, 43 bags of cocoa, 140 bags of tobacco, hides, sarsaparilla, and 800 pieces of wood. The *Minerva*, Captain Ryder, took another valuable Dutch ship, called the *Good Friends*, from St. Eustatia to Amsterdam, with 504 hogsheads and tierces of sugar, 524 bags of coffee, 137 hogsheads and 244 bags of tobacco, 16 bales of cotton, besides elephants' teeth, etc.

The *Industry*, Captain Moore, on her passage to New York, had an engagement of 75 minutes with a privateer of 16 guns, which, having much shattered, she beat off. In March, the *Woolton*, Captain Backhouse, took and carried into the Shannon, a ship from St. Domingo, called *La Sartine*, of 350 tons burthen, 16 guns, and 58 men. She engaged the *Woolton* three hours and-a-half, and had eight men wounded, three of whom afterwards died of their wounds. The *Woolton* had only one man wounded. The prize, which entered the river Mersey on April 10th, was laden with coffee, sugar, etc., valued at £15,000. The *Barbara*, Captain Perry, on her passage from St. Eustatia, met with a French privateer, of 24 guns, which, after an engagement of three glasses and-a-half, she beat off. The *Barbara* captured a brig, from Curacoa to Rotterdam, which arrived in Liverpool on April 10th, 1781.

The *Townside*, Captain Bonsall, cruising in company with the *Rodney* and *Union*, of Barbadoes, captured three Dutch prizes from Demerara, one of which, a large ship with 190 bales of cotton, 456 hogsheads of sugar, 1447 bags and 111 casks of coffee, they sent to Tobago; another, with a cargo of 206 bales of cotton, 241 casks and 3015 bags of coffee, they ordered to Barbadoes; and a schooner, with rum etc., into St. Lucia. The *Townside* and the *Rodney*, in company with two sloops of war, were concerned in the further capture of four ships in the harbour of Demerara; and in Essequibo of several other vessels. A little later on, the *Townside* had a narrow escape from capture when the French fleet appeared before St. Lucia, but she cut her cables, slipped out, and arrived safe at Tortola.

In May, Captain Fayrer, in the *Harlequin* privateer, cruising in sight of the Azores, took a Swedish brig, and, by stratagem, discovered that "she was sent out to give advice to the East Indiamen." He afterwards took and detained another from Ostend, upon the same errand. The *Harlequin* and the *Cæsar*, of Bristol, in company, took a ship from Curacoa, and sent her to Bristol. In August, the *Harlequin* arrived in Liverpool with two prizes, the *Swallow* and a French snow. In the summer of 1783, the *Harlequin* arrived at St. Lucia, from Africa, after a severe engagement with a French privateer of 20 guns, in which Captàin Fayrer behaved with great courage.

"Tuesday se'nnight," says *Williamson's Advertiser*, of the 6th of February, 1783, "was determined in the Court of Admiralty, a cause long depending between the owners of the ships *Patsey* and *Harlequin* of Liverpool, and the *Cæsar* of Bristol, respecting the right which the former claimed as joint captors of the ship *Eendroght*, bound from Curacoa to Amsterdam, the most valuable West Indiaman taken during the course of the war, upwards of £40,000 of her proceeds being lodged in the Court of Admiralty. It was adjudged in favour of the Liverpool ships, which are the property of Henry Rawlinson, Esq., member for this town, and Messrs. Earle & Co., merchants."

The *Rawlinson* and the *Molly* arrived in Liverpool at the beginning of June, 1781, having parted from the fleet, which left Jamaica on the 17th of March, under convoy of 4 line of battle ships, a 50 and a 44. In consequence of information received at Jamaica, the fleet sailed through the Windward passage, to avoid 20 French and Spanish ships of the line said to be at the Havannah waiting for them. The convoy fell in with a French 64, called the *Marquis de la Fayette*, bound from France to America, with 80 pieces of brass cannon, clothing for ten regiments, stores for two ships, and about 2,000,000 livres in specie, which they captured. They also retook the *James and Rebecca*, from Liverpool to New York, which had been captured by an American privateer, of 18 guns, and was proceeding with her for America. The privateer was chased for eighteen hours but outsailed her pursuers. The portion of the fleet destined for Liverpool consisted mainly, of course, of slave ships, bringing sugar, rum, and other commodities, purchased with the proceeds of the human cargoes which they had carried from Africa to the West Indies.

In the first week of July, the *Prosperity*, a ship of 300 tons, laden with lumber, etc., for Teneriffe, arrived in Liverpool, having been captured by the *Lydia*, Captain Fell; also the *Resolution*, laden with brandy, Geneva, etc., a prize to the *Lurcher*, Captain Doyle. The *Seacombe*, Captain Pagan, arrived on the coast of Africa, from Liverpool, "with five spermaceti whales, and a large Dutch ship, her prize." The *Kitty*, Captain Clough, on her passage from Liverpool to Jamaica, had an engagement of seven hours with a privateer, which she beat off. On the homeward voyage, the *Kitty* captured a prize, which entered the Mersey with her in September; and in the same month there came in a vessel called the *Johannes*, laden with tobacco, coffee, etc., prize to the *Betty*.

About this time, threats of a French invasion helped to intensify the horrors and miseries of war. In the beginning of September, the following alarming dispatch was received by the commanding officer in Liverpool:—

"DUBLIN, in Homoaze, *August 30th*, 1781.

"SIR,—I think it necessary to acquaint you, by express, that on the evening of the 28th inst., the combined fleets of the enemy (French, Spanish, and Dutch), consisting of thirty-four or thirty-five sail of the line, were seen five or six leagues to the east of Scilly, and that there is great reason to apprehend that they are now in the Channel; in order that you may make the same known to the captains of any of his Majesty's ships that may be within your reach, as well as the merchants of Liverpool, to prevent any of their trade from falling into their hands. Vice-Admiral Darby, with his Majesty's squadron under his command, is now in Torbay. I am, Sir, yours,

SHULDAM."

Captain Campbell, of the *Dick*, of Liverpool, writing to his owners from Staten Island, on the 29th of July, 1781, gives the following account of an engagement between the *Dick* and an American vessel:—

"On the 17th of June, at nine in the morning, lat. 39. 40; long. 54. 30; we fell in with an American ship of 20 six-pounders, which engaged us from nine till a quarter past 11, when he made sail from us. We immediately gave chace, but could not come up with him, our ship sailed so heavy. When we got upon the coast of America, saw two or three sail every day, sometimes five privateers in a day, sloops and schooners. His Majesty's frigate *Orpheus* fell in with us off the

Hook, and pressed three of our best men, nine more were pressed at the Hook. Our officers and men behaved very gallantly all the time we were engaged. We fired twenty-seven broadsides; only one man wounded, and one boy his arm broke.

"Amount of shot which took place from the privateer: In the jibb, 13, great and small; fore topmast staysail, 2; fore-sail, 14; fore topsail, 8; main topmast staysail, 31; main top-sail, 30; main sail, 42; main staysail, 12; mizen, 11; main topsail, 13; in the hull, 36; main mast, 2; mizen mast, 1; fore shrouds, 3; main shrouds, 2; main stay, 1; mizen stay, 1; fore and main topgallant rigging all cut away; fore top-mast shrouds, 3; main top ditto, 2; back stays, 3; a number more, not ascertained, one shot went through the side and through a butt of water."

In September, 1781, the *Lightning* privateer, Captain Walker, took a large Swedish ship of about 500 tons, from Bordeaux to St. Domingo, laden with bale goods, wine, flour, etc., value as per invoice, 330,118 livres. In March, 1782, the *Lightning* captured a Spanish packet, from the Havannah to Cadiz, with 12,000 dollars on board, and sent her into Lisbon. On the 30th of May, 1782, the *Maria*, from L'Orient, with wine, salt, etc., another prize to the *Lightning*, arrived in Liverpool. The *St. George* schooner, from Rochelle for Martinique, with wine, flour, oil, and bale goods, also captured by the *Lightning*, narrowly escaped being recaptured by a large cutter privateer, which chased her into Kinsale. In September, the *Lightning* took a vessel with 77 hogsheads of tobacco, and on December 21st, off the island of St. Michael, she captured a French East India packet, from the Cape of Good Hope, for St. Malo, with passengers and despatches for France. The mails were thrown overboard, and narrowly escaped being taken. The vessel was formerly the English privateer, *Resolution*, re-christened *Le Mars* by her French captors.

The paper of October 4th, 1781, stated that the *Quaker*, Captain Evans, had arrived at Newfoundland from Liverpool, with a rebel privateer of 13 guns, which he had captured. Early in 1782, the *Quaker* took three prizes, and carried them into Antigua, where they sold for £21,000. On his passage to Newfoundland, in the autumn of the same year, this very pugnacious *Quaker* fell in with a French 44-gun ship, exchanged a broadside with her, and got clear by dint of sailing, after an exciting chase of twelve hours. The *Quaker* had one boy killed, and another wounded, but received no other damage. In the paper of February 6th, 1783, we read that the *Quaker* had captured in the West Indies, a brig with a Letter of Marque, from Martinico to France, laden with sugar, coffee, and cocoa, valued at £10,000, and sent her to Tortola.

England was now at war with Holland, as well as with the United States, France and Spain. The English were by this time disgusted with the folly of their rulers, and weary of the unnatural strife with their own kinsmen beyond the seas. The enemies' privateers were doing excellent business on our coasts. When the *Count de Guichen*, French privateer, was taken by the English frigate *Aurora*, Captain Collins, she had on board the following ransom bills, or promises to pay ransom, given by British merchant ships to the French commander. The *Peace*, of Whitehaven, 2000 guineas; the *Spooner*, of Glasgow, 1800 guineas; the *Fortitude*, of Greenock, 1500 guineas; the *Six Sisters*, Isle-of-Man, 1500 guineas; the *William*, of Bristol, 1500 guineas; the *Sally*, of Strangford, 500 guineas; the *Lark*, of Workington, 300 guineas; the *Glory*, of Workington, 150 guineas; and the *Elizabeth*, 110 guineas; a total of 9360 guineas during one cruise. This probably fell short of the mischief actually done to British commerce by this single ship, as it was the habit of privateers to plunder, burn, or sink vessels which were not ransomed, or which were too insignificant to send home in charge of a prize crew to be condemned. So great was the boldness of the enemies' privateers, that the Dublin linen ships, said to be worth £150,000, were convoyed from Dublin to Chester fair, by the *Boston* frigate and two armed cutters, lest the linen should be diverted to a French market.

In September, 1781, the *Heart of Oak* privateer, Captain Ash, recaptured the *Alexander* privateer, of Liverpool, which had been taken by an American frigate. In March, 1782, we read of the *Heart of Oak* taking a Dutch privateer, which was cruising off the Humber, and carrying her into Hull. The Dutchman had taken two colliers and a corn vessel, the ransom of the latter being 1200 guineas.

The *Tom* privateer, of Liverpool, captured the *Countess de Maurepas*, French privateer, 16 guns and 120 men, which had been cruising in the Channel, and taken the *Blessing*, of Workington, which she ransomed for 450 guineas.

In the summer of 1781, the notorious Pat Dowling was doing a "roaring trade" in the Channel, when he took the *Olive Branch*, from Liverpool to Charleston, which he ransomed for 7700 guineas. It was said he had 17 ransomers on board, and had taken on the Irish coast upwards of twenty vessels, five of which he had sunk, as the people would not ransom on his terms. He took a vessel from Maryport, and ransomed her for 750 guineas; the *William*, from Bristol to Liverpool, ransomed for 900 guineas; the *Elizabeth*, from Liverpool to Cork, ransomed for 800 guineas; and the *Sally*, for Guernsey, which he released for 700 guineas. It is difficult to understand what the King's cruisers were doing while this enterprising Irishman, and others of his countrymen, were serving France so effectually in the Channel. Well might the paper of October 4th, observe that "the safe arrival of the Leeward Islands fleet is a circumstance which must diffuse a general joy through this country, and ought to excite its gratitude, when it is considered from what an host of foes it has escaped."

"In the history of England," says the paper of January 24th, 1782, "many are the intervals where she was surrounded with imminent dangers; yet did her native spirit prove ultimately superior to and surmount every peril. Let her admirals and generals rouse into a true spirit of action, her people be united and lay by at the present alarming crisis all party animosities, and act with one heart and with one arm, and there is no doubt but the ensign of Albion will again wave to victory, to fame, and to honour." The country as a whole, however, had by this time become sick of the war. Petitions and addresses against the further prosecution of it began to pour in. In January, 1782, the Corporation of Bristol, not content with merely voting a petition to the House of Commons against the continuance of the struggle with America, went a step further and requested the House to advise the King to a total change of the unhappy system which had involved the nation in such complicated misfortunes. In March, 1782, a resolution was moved in Parliament, and passed without a division, declaring that the House would consider as enemies to his Majesty and the country all who should advise the prosecution of offensive war in North America.

On Sunday evening, March 3rd, 1782, as two of the press-gang were conveying a man, whom they had just impressed, to the press-room, he suddenly turned upon them, drew out a loaded pistol, shot one of them dead, and escaped. The coroner's jury brought in a verdict of manslaughter.

In April, 1782, the *Venus*, Captain Brown, arrived at Hoylake, from St. Lucia, having on the voyage taken a valuable prize, with 87 hogsheads of tobacco, naval stores, etc., on board.

At six o'clock on Monday morning, the 19th of May, 1782,

an express from the Admiralty arrived in Liverpool, with dispatches for the Mayor, announcing Admiral Rodney's victory over the French fleet, in the West Indies. The "great and glorious news" was received with delight in the town, which was deeply interested in Jamaica and the West India Islands, both on the ground of property, and because the Liverpool Blues were on military duty in the former island. Joy bells were set ringing for the rest of the day, and flags were displayed on all the public buildings. At noon, a royal salute of twenty-one guns was fired, from George's Battery. At one o'clock, the Westminster Militia were drawn up in front of the Exchange, and fired three volleys, amidst the acclamations of thousands of spectators. After this great victory, which restored confidence to the public mind after so many disasters, all the West India ships sailed for England, and arrived in safety. Amongst them were eleven rich Jamaica ships for Liverpool, which arrived in July.

In May, the *Kitty*, Captain Clough, on the voyage from Liverpool to Jamaica, took a prize from Guadaloupe for Cadiz, which she convoyed to Londonderry. The cargo consisted of 130 hogsheads of sugar, 7 hogsheads, 25 tierces, and 150 bags of coffee, 55 bags of ginger, 39 bales of cotton, 151 bags of cocoa, 2 barrels of copper, etc. On the 25th of the same month, the *Bridget,* Captain Gilbody, from Liverpool for the Leeward Islands, took and sent into Londonderry, the American brig *Dove*, from St. Domingo to Cadiz, laden with 35,500 lbs. of cocoa, 4000 lbs. turtle shell, 4020 lbs. indigo, besides other articles.

The *Jenny*, Captain Collison, and the *Tom*, Captain Briggs, on their passage from St. Lucia to Liverpool, had an engagement with an American privateer, and beat her off, with a loss to the *Tom* of two passengers—an officer and his servant—who were killed. A little later they took an American vessel called the *Fox*.

The ship *Rumbold,* Captain Molyneux, on the middle passage from Africa to Jamaica, with slaves, beat off a French privateer, of 24 guns, after a severe action of about an hour, in which three men were killed and eleven wounded on board the *Rumbold*, which had only 16 guns and 45 men.

The *Quest*, Captain Ogden, in company with a Jersey privateer, took a vessel from Havannah to Cadiz, laden with sugar, etc. The *Quest* afterwards took the *Good Design*, an American brig, laden with fish, molasses, etc., and carried her into St. Lucia, to be sold. On the 3rd of November, 1782, the *Quest*, in company with the *Iris*, of Tortola, captured the American brig *Thoroughgood*, with a cargo of rum, salt, and dry goods. On the 16th, they engaged four American vessels, one of which carried 18 nine-pounders. The *Iris* in this affair kept aloof. After a gallant action, the *Quest* had to sheer off, much shattered, having four men killed, and four wounded, amongst the latter being the brave Captain Ogden, who afterwards died of his wounds.

The *Liverpool*, Captain Webster, on her passage to Africa, took a French ship, of 400 tons, 14 twelve-pounders, and 63 men, bound from Bordeaux to the Isle of France, with cordage, wine, brandy, etc., but the prize was recaptured by two French frigates. The *Mossley Hill,* Captain Hewan, captured off Cape Mount, on the coast of Africa, an outward bound East Indiaman from Toulon, and despatched her to *Tortola*, where the *Mossley Hill* arrived in due course, with a cargo of 723 prime negroes. The *Spy*, Captain Burrows, while proceeding from Africa to the West Indies, with 250 slaves and about 6 tons of ivory, was taken by two French frigates, and carried to Dominica. The *Stag*, Captain Butler, was more fortunate, having on her voyage to the coast, taken a ship, bound from Barcelona to Buenos Ayres, valued at £8000, which, added to the profit on a cargo of 700 slaves carried to the West Indies, no doubt satisfied the owners.

The *Molly*, Captain Jordan, from Jamaica for Liverpool, was captured off the Tuskar, by the *Terror of England* privateer, of 22 guns, commanded by an Irishman named Kelly, after an engagement of three hours, in which Captain Jordan and four of his people were killed. Afterwards, a gale of wind came on, when the prize crew, whom Kelly had put on board the *Molly*, not knowing what to do with her, delivered her up to her own crew, who had been left on board, and by them she was carried into Greenock. A few days later, the *Terror of England* was captured by the *Stag* frigate, Captain Cooper, and carried into Dublin, where Kelly does not appear to have been received with the amount of fraternal love which an "enemy of Great Britain" naturally expected on Irish soil, even prior to the Union.

A Dublin newspaper referred to the captain in the following unsympathetic terms :—

> "Captain Kelly," says the Journalist, "seems to be not in the least affected with his present situation, and considers this change of fortune as a mere bagatelle, beneath making any impression upon a gentleman of his spirit and humanity. The Captain imagines that by the assistance of Le Roy de France, whose signature he displays to a scrap of parchment, he will be able to baffle the utmost efforts of the King's lawyers, and once more be liberated to plunder the property of the subjects of his native land with impunity. There does not seem to be a doubt entertained but Kelly will add one more to the numerous throng that occasionally make their exit from that tree which so often has promoted the good of the community by ridding the world of villains disgraceful to human nature. The horrid treatment of Lieutenant Vickers, of the *Hope* cutter, with his brave crew, is recent in every memory. This renegade refused quarter to these tars, after fighting him nobly with an inferior force, and continued, when they had pulled down their colours, pouring in his broadsides. Some of his infernal crew, after

they boarded the cutter, cut and abused, in a shocking manner, several of the men. He can scarce escape the reward due for his inhumanity and piracy, as the most positive evidence can be produced of his being born in this country."

In August, 1782, there arrived in Liverpool, from Jamaica, a self-made man, one of whose first acts on landing in his native land entitled him to be called one of Nature's noblemen. Richard Watt, a poor boy, like Richard Whittington, and with some of that hero's grit in him, came to Liverpool from Standish, near Wigan, probably about 1740, and, according to Smithers, was hired by Mr. Geoffry Walley,* to look after his horse and chaise, the only carriage then kept in the town, except the coach of "Madame" Clayton. His master sent him to an evening school, and, finding him tractable and industrious, advanced him to the counting-house, and employed him as supercargo to Jamaica, where he settled and acquired a large fortune. Time had not effaced the memory of his old master's kindness, and on his return to Liverpool, after an absence of about forty years, his first enquiries were after the survivors of his former employer's family. He found two maiden sisters—one account states two widows—in poor circumstances, upon each of whom he settled £100 a year for life. He was head of the firm of Messrs. Watt and Walker, and built the mansion of Oakhill, Old Swan. His nephews, Richard Watt and Richard Walker, to whom he left upwards of half-a-million, resided in Duke Street. Mr. Watt died in 1796, aged 72, and was interred at Standish, having been born at Shevington in that parish. Mr. Richard Watt bought the manor of Speke from Charles George, son of Topham Beauclerk, the friend of Dr. Johnson, and Miss Adelaide Watt, the representative of Richard Watt, is the present lady of the manor. Her residence, Speke Hall, is

* Brooke states that his employer was James Dimmock, or Dimoke, horse and chaise hirer, Fenwick Street.

a fine specimen of an ancient Lancashire manor house, and was for ages the home of the Norris family.

The *Molly,* Captain M'Kown, on her passage to Jamaica, took two prizes, but the *Molly*, Captain Lloyd, bound for St. Lucia, had the ill-fortune to fall in with the *Holker*, American privateer, which, after a smart engagement, carried her to Martinico. Captain Lloyd had four men killed and 13 wounded.

Two items of intelligence in the newspapers of October 17th, 1782, spread universal joy throughout the nation. The first was, that the grand attack of the Spaniards and French on the fortress of Gibraltar had been totally defeated by sea and land by General Elliot; the second was, that Messrs. Fitzherbert & Oswald, on the part of Great Britain, had exchanged credentials with Messrs. Franklin & Jay, the ambassadors of the United States, preparatory to the arrangement of the terms of peace between the two countries.

The year 1782, at the end of which we have now arrived, was memorable for the great losses and defeats sustained by both the French and Spaniards. A Liverpool poet, Edward Rushton—like Roscoe, the friend of liberty, irrespective of colour—published the following stanzas:—

Britain! thy fame in eighty-two
    Outswells the boast of fifty-nine,
Gallia was vanquished then, 'tis true,
    But now a host of foes combine;
A host combine to pull thee down,
And strip thee of thy nautic crown;
Whilst proud rebellion towers on high,
And millions from their duty fly:
Never, oh, Britain! did the warring storm
Howl round thy rocky coast in such a threatening form.

To keep such numerous foes at bay,
Is one continued victory;
But now old ocean owns thy sway,
And vanquished foes confounded fly.
As skims the flying finny brood,
When by the Albicore pursued,
So, in this great, this wondrous year,
When 'bove themselves thy sons appear,
Proud Gallia's navy fled in dire dismay,
And had Cordova dared, so had he winged his way.
Britain! 'tis done, and grim despair
Has fastened on each vengeful foe;
The Rock's* relieved; and through the air,
Hark! how the sounds of triumph flow.
And now, ye unassisting powers,
What think ye! is the trident ours?
Ye baffled foes, what arts, ah! say,
Can wipe the foul disgrace away?
For wondering Europe ey'd the important deed,
And, spite of every boast, beheld your foes succeed.

The strong contrast afforded between the kindly feelings cherished in England towards the Americans, and the rooted animosity entertained for their allies, "the hereditary enemies of Great Britain," is reflected in the following song "On the prospect of peace with America," to be sung to the tune of "Hearts of Oak, etc.," printed in the Liverpool paper of the 9th of January, 1783:—

I.

Hark! the lion is roused, and the cannons they roar,
Like the thunder of Jove, from the main to the shore.
Tell the false sons of France, and their neighbours of Spain
We'll teach them to dance to the old tune again.
France and Spain then shall know,
That their topsails shall bow;
If we meet them, we'll hail 'em,
Like Britons assail 'em;
We'll fight or we'll die, still Lords of the main.

*Gibraltar.

II.

Let the Cherubs of bliss now diffuse it around,
And the Seraphs of concord exult in their sound,
That America comes, with mild peace in her train,
While the olive re-blooms, and we're friends once again.

France and Spain, etc.

III.

The Maid of the Colonies puts forth her hand,
Bids commerce to flourish once more in this land;
Britannia she bends, and with joy in each eye,
Cries, let us be friends, and the world we defy.

France and Spain, etc.

The *Mermaid*, Captain Reynolds, on her passage from Antigua, took a brig, laden with lumber, etc. The *Antigallican*, Captain Corran, on her passage to Tortola, captured off Porto Rico two prizes; one an American brig, loaded with salt, the other a Spanish vessel, of 182 tons, bound to Cape François with flour, wine, oil, soap, canvas, cordage, bale goods, etc. The *Rover*, Captain Latham, from Africa for the West Indies, with 209 slaves, was captured by an American privateer, and carried into Martinico.

The *Bella*, Captain Burgess, was taken by her crew, a day or two after she sailed from Jamaica for Liverpool. The mutineers killed the chief mate, and landed the Captain and two of his men upon a rocky island, called the Jordans. The vessel was retaken by the *Harlequin*, bound from Jamaica for Liverpool, but as she had five feet of water in her hold, and was in a bad condition, they sunk her, the pirates being placed in irons on board the *Harlequin* and the *Neptune*, and carried to Liverpool.

On the 24th of January, 1783, Mr. Secretary Townshend announced, in the House of Commons, that preliminaries of peace had been signed with France, Spain, and the United States of America. The definite treaty of peace was published in the Liverpool papers on the 30th of the same month. "The Mercantile World," says *Williamson's Advertiser* of that day, "is in a hurry and bustle unknown at any former time. The merchants are endeavouring to outstrip each other in the race of traffic. European goods, and particularly the produce of England, being greatly wanted in the ports of America, the destination of many of the vessels now in the river is altered from the West India Islands to the American ports, where it is expected the cargoes will sell at an immense profit."

Thus ended, at last, to the joy of the English speaking peoples throughout the world, and for the future blessing of mankind, a most disastrous, disgraceful, unnatural, unnecessary, and expensive war, that might have been averted had a single grain of common sense been admitted into the councils of the obstinate old King.

"Many of the American privateers," says a London paper of March 20th, 1783, "have been cruising all the war without commissions, and others have been frequently suspected of having had forged ones; there is therefore every reason to apprehend that piracies will continue for some time, notwithstanding that hostilities are over. This is the general opinion among the captains at Lloyd's." News travelled slowly in those days, and no doubt captures were made in good faith on both sides for some time after the conclusion of the war, especially in distant waters. The slave brig *Fancy*, Captain Greaves, of Liverpool, was taken at Cape Mount, on March 22nd, 1783, (the very morning she sailed for the West Indies, with a cargo of 390 negroes, two tons of ivory, and a quantity of rice), by a French 50 gun ship, and carried to Cape François.

On the 4th of March, 1783, the *Count Belgioso*, Captain Pierce, a fine new ship, lately launched, sailed from Liverpool, for the East Indies, with a fair wind and fine weather; but a violent storm of wind, and a great fall of snow coming on, she was lost upon the Kish Bank three days after leaving port, and all on board, comprising 147 persons, perished. She was said to be one of the richest ships that ever sailed from Liverpool, being valued at £130,000. She had 100,000 dollars on board, besides a great value in ginsang, bale goods, and 300 tons of lead.

The general effect of the American war of independence on the position of Liverpool, was to put an entire stop to the commercial progress of the port, during seven long and disastrous years. The foreign trade of the port, which had doubled itself between the accession of George the Third, in 1760, and the commencement of hostilities, in 1775, declined in all its branches, from the beginning of the struggle, to its close in 1783. The customs revenue of the port, which amounted to £274,655 at the commencement of the war, had fallen to £188,830 in 1780, the sixth year of the contest. The tonnage declined from 84,792 tons to 79,450, of which a large part consisted of privateers. The population decreased from 35,600 to 34,107; and the condition of the inhabitants was deteriorated so greatly in the latter years of the war, that, at its close, not less than 10,000 of the poorer class, were supported either by the parish, or by charitable donations. "The seven years of the first American war," as Baines truly observes, "were the seven lean years of Liverpool, and the only seven years of the eighteenth century during which the port did not increase in population and wealth." While the war lasted town improvements were mostly suspended. Beyond the occasional bustle of numerous sales by auction of the cargoes of prizes taken from the enemy, there was little business transacted in the port. "The manners of the common people at this period," says Troughton, "made a retrogression towards barbarism, rather than a progress in refinement or virtuous habits. This was the natural consequence of that spirit of enterprise cherished by the proprietors of privateers; for successful adventurers, upon their return to port, spent in excess what they had obtained with danger. As for the public amusements, the theatre was opened every summer, and the people were also sometimes gratified by the occasional visits of Breslau, and other jugglers, whose dexterity disencumbered them of their superfluous cash." One remarkable but natural effect of the war was to destroy for a time the popularity of Liverpool as a bathing place, but on the return of peace, there was a great influx of visitors from the interior, whom fear of the press-gangs had deterred from visiting the town during the war. "For the last week," says the *Liverpool Advertiser*, "the town has been uncommonly crowded with country people from the vicinity of Rochdale, Blackburn, Manchester, etc. It is computed that there were upwards of 3000 of them. They came to bathe and drink salt water. During the war, very few of them durst come down, on account of the warmth of the impress; and it is therefore supposed that this is the most crowded bathing season ever known here."

CHAPTER V.

Liverpool Privateers and Letter of Marque Ships during the Wars of the French Revolution.

The ten years of peace which succeeded the war with America, were years of extraordinary activity and prosperity, during which the trade of Liverpool increased more rapidly than it had done during any former decade in its history. The population in 1793 was estimated at about 60,000. The marvellous progress of the port is seen from the fact that, in 1716, the whole tonnage which entered and cleared from English ports, was 456,309 tons, of which 18,371 tons cleared from Liverpool; in 1792, the whole tonnage cleared from English ports was 1,565,154, of which 260,380 cleared from Liverpool. Thus, in a period of 77 years, Liverpool's share had increased from the twenty-fourth to the sixth part of the whole. But the peace and plenty which England enjoyed from 1783 to the end of 1793, were doomed to be followed by years of war and scarcity. On the 21st of January, 1793, Louis XVI., deposed King of France, was guillotined at Paris. In common with other European Courts more concerned about the safety of Kings than the rights of the people, England, under the administration of Pitt—then Liverpool's favourite statesman—alarmed by the progress of the French Revolution, declared war with France. This war continued until 1815, when it culminated in the Battle of Waterloo. It was the cause of untold misery, the destruction of an appalling number of human lives, and of an incalculable amount of property on sea and land, and cost upwards of £831,000,000. The two main results of this war were to deliver France to despotism again, and to hinder our own march of progress at least half a century. During its continuance the commerce of Liverpool was exposed to all the dangers and chances of war, with only one short interval of rest during the peace of Amiens.

When the news of the execution of Louis XVI. arrived in Liverpool, it produced a general feeling of pity, horror, and despondency. The colours at the Exchange and Custom-house were hoisted half-mast high, and the shipping in all the docks exhibited the same signs of mourning. But sorrow soon gave way to anger, and to resolute preparations for war with "republican and regicide" France. At the invitation of Mr. Pitt, a deputation of merchants proceeded to London, to consult with the Government on the "protection necessary to be afforded to the shipping of the port." For the purpose of depriving the enemy of naval and military supplies, and of arming the British fleet more rapidly, an embargo was laid on vessels taking out naval and military supplies. Greenland ships proceeding to the northern fisheries were ordered to be well armed; and Letters of Marque were issued against French ships and commerce. The old fighting instinct of Liverpool was revived in full force, but it does not appear that privateering was carried on to so great an extent, comparatively, from the port, as during the American Revolutionary War. Many acts of bravery were, however, performed, and valuable prizes taken by the officers and crews of Liverpool privateers and armed merchant ships during this long war.

On Wednesday afternoon, the 20th of March, 1793, a most distressing sight was witnessed by a number of people from the Pierhead and its vicinity. The *Pelican* privateer, of 20 guns and 100 men, having that day been launched, full rigged, with all her guns and stores on board, was cruising to and fro in the river, with a moderate breeze, according to custom in such cases, with about two hundred persons on board, including the shareholders and their friends, and women and children. While they were making merry, and enjoying themselves to their hearts' content, to the strains of music, the ship, on being put about opposite Seacombe, suddenly capsized, filled with water through the lee-ports, and sunk in ten minutes with all on board. Seventy or eighty persons were drowned—accounts vary as to the number—and the rest either swam ashore, or were rescued by boats. Twenty-five persons were saved by Mr. John Starkey, excise surveyor, who went from the Pierhead in his boat, and subsequently received the Humane Society's medal for his activity. Amongst the saved was James Creasey, the pilot of the ship, who was tried at the Lancaster Assizes for manslaughter, as it was said that the accident was the result of his negligence. He was, however, acquitted. The ship was never raised, and the top of her masts stood above the water for years after the fatal event. "A young man, who was saved at the sinking of the *Pelican* privateer," says the *Naval Chronicle* (which however gives the date as 1783, instead of 1793), "had the singular affliction of losing his mother, sister, wife, and two children, who had come on board to take a long, a last farewell. The grief of a son, a brother, a husband, and a parent, on being thus suddenly deprived of all his dearest relatives, may more easily be conceived than described."

On the 5th of April, 1793, the first French prize taken by any vessel belonging to the port of Liverpool, since the commencement of hostilities, was brought in by the ship *Harriet* (Letter of Marque), Captain Caitcheon, belonging to Mr. Thomas Barton. She was a fine Bermuda-built brig, raised upon a cedar frame, and copper bottomed, about 200 tons burthen, called *L'Agréable*, laden with coffee, sugar, indigo, and cotton, and was taken on her voyage from Port-au-Prince to Bordeaux. The value of ship and cargo was variously estimated at from £6,000 to £10,000. The French manifests seldom expressed the quantity of goods contained in the vessel. The people on board the prize stated that they sailed from Port-au-Prince, with 31 more French vessels;—"good information this for our brave tars," remarks the *Advertiser,* "as we hope, and are led to believe, that few of them will reach their destined port." Both the ship and cargo were sold by auction at Messrs. Ewart & Rutson's office, in Exchange Alley. On her next voyage, the *Harriet* chased a French Guineaman into Martinique, but being fired upon by the fort, was obliged to desist. "The ship *Harriet*, belonging to Thomas Barton, Esq.," says the paper of January 8th, 1798, "has made 33 voyages from hence to Barbadoes and back in the last ten years and three months; has taken and retaken some vessels, and rescued others, and has been lengthened in the time—an instance of commercial expedition, we believe, scarcely to be paralleled."

On the 28th of March, 1793, the *Ann*, Captain Worthington, belonging to Messrs. Boates and Seaman, captured the brig *La Porkin*, a privateer of 10 guns and 79 men, out three days from Nantz, and the property of Messrs. Margerin, Reneau, & Co., of that port.

On the 6th of April, 1793, the *Thomas* privateer, Captain Huston, took the French ship *La Expeditif*, from Charleston to Havre, with rice, indigo, deerskins, etc., valued at about £10,000. On the 15th of the same month, the *Princess Elizabeth* privateer, Captain Beasley, took the French ship *Les Bons Freres*, about 400 tons, from Port-au-Prince for Bordeaux, laden with coffee, indigo, and sugar; and towards the close of the year, she brought into the Mersey the *Amsterdam Packet*, from New York to Havre, with a cargo of tobacco, coffee, sugar, cotton, and pearl-ashes.

On the 20th of May, 1793, the brig *Victoire* from Guadaloupe for Havre, laden with 125 hogsheads of sugar,

80 casks of coffee, and 18 bags of cotton, was brought in a prize by the *Earl of Derby* privateer, Captain Perrin.

The *Prince of Wales* privateer, Captain Thompson, after being out three weeks, arrived at Hoylake on the 14th of April, 1793, bringing in with her the French ship *Le Federatif*,* from St. Domingo for Bordeaux, which she had captured on April 5th, in latitude 46°. Her cargo was valued at about £32,000.

On the 8th of October, the *Prince of Wales* captured the brig *Maryland*, from Baltimore to Bordeaux, with 135 hogsheads, 78 tierces, 39 barrels, and 604 bags of coffee, 9 hogsheads of sugar, and 5,000 hogsheads' staves, and sent her for Montserrat. In December, she recaptured, and brought into the Mersey, the *Best*, from Lancaster for the West Indies, which had been taken by a French man-of-war. Early in 1794, the *Prince of Wales* captured and brought in the *Flugan,* of Malmö, from Bordeaux for St. Domingo, laden with wine, brandy, and bale goods.

The *Gipsey*, Captain Tobin, captured at Loango, on the 30th of May, 1793, a French ship, *Le Hirondelle*, having on board 122 slaves and 8 guns, which prize was sent to Mayomba, where she captured a French schooner, *Le Pourvoyeur*, with 51 slaves, 500lbs. of ivory, and a cargo of 70 slaves more. This prize was given up to the prisoners after taking out the cargo. On the 3rd of June, in company with the *Isabella*, of Bristol, and the *Lord Charlemont*, Captain Pinder, of Liverpool, the *Gipsey* captured, at Malimba, a French ship, *Le Emilie*, with 241 slaves, and sent her to Grenada. *Le Hirondelle* was retaken by the French and carried to St. Lucia. Early in 1794, on the

* A French officer, M. Thiballier, was coming home a prisoner in the *Le Federatif* when she was captured. He was Lieut.-Col. of the 4th Regiment of Provence, and had been made Colonel in St. Domingo, and Governor of the fourth part of the Island. Being an intimate friend of M. Blanchelande, the Governor, he was suspected of supplying the rebellious negroes with arms, etc., which so enraged the Democrats that they caused him to be seized and sent to France for trial.

passage from Jamaica to Liverpool, the *Gipsey* took an American vessel with a cargo of provisions for Martinico, and sent her to Jamaica.

On the 13th of April, 1793, the cutter *Dudgeon* privateer, Captain Gullin, took the French brig *St. Roman*, from Charleston for St. Valery, with 730 barrels of rice, 2 hogsheads of tobacco, and 105 cow-skins, valued at about £5,000; and on the 17th of the same month, she captured a French snow, from Cayenne for Havre, laden with sugar, coffee, cotton, and indigo, valued at £15,000. On May 4th, the *Dudgeon* recaptured, and sent into Milford, the brig *Argyle*, of Greenock; and on the 9th of June, she brought into the Mersey a Spanish brig, from Caracas for Spain, laden with cocoa, indigo, and hides. In company with the *Jenny* privateer, the *Dudgeon* made a prize of *L'Esperance*, a French vessel, from Lisbon for France, which was sent North about, and arrived in Liverpool on June 19th. On the 3rd of February, 1794, we read that the *Dudgeon*, Captain Egerton, returned from a cruise, after throwing her guns overboard and receiving other damage at anchor in a gale at the N.W. Buoy. In March, the *Dudgeon* recaptured the Danish galiot *Unge Simon*, from Lisbon to St. Petersburgh, with sugar, oranges, figs, and almonds. A few weeks later, the *Dudgeon* and the *Ann and Jane*, a ship of 500 tons, from Liverpool, were carried into Brest. The *Dudgeon* was afterwards fitted out by the French as a National vessel, and sent to sea to prey on British commerce, but in September, 1794, she was captured and carried into Falmouth.

In June, 1793, the *Ann* privateer, Captain Flanagan, recaptured the ship *Harriet*, from Honduras for London, with wood, etc., and took an outward bound vessel, from Old France to the West Indies, laden with bale goods, wine, etc. The *George W. Lutwidge*, from Baltimore for Havre, with flour, another prize to the *Ann*, was retaken by two French privateers, and again captured by the *Mary*, of Liverpool, Captain Pince, arriving safe in the Mersey on July 1st. The Frenchmen on board the prize informed Captain Pince that the same privateers had retaken another prize of the *Ann*. On the 7th of August, a sloop, from Bayonne, bound to Brest, with resin, arrived; and on the 15th of October, the *L'Augustine*, from Guadaloupe to Havre, with a cargo of sugar, coffee, and cotton, both prizes to the *Mary*, Captain Pince.

The *Mary*, Captain Thompson, took the *Le National Pavillon*, about 500 tons burthen, from Guadaloupe to France, and recaptured the brig *Diligent*, from Jersey for Quebec, with cordage, etc., and the *Franklin*, loaded with provisions, from Dublin for Cadiz. The latter had been twice recaptured before the *Mary* fell in with her. Early in 1795, the *Mary*, Captain Thompson, was captured on her passage from Liverpool to Leghorn, and carried into Brest.

The *Mary,* Captain Mollineux, took the French lugger privateer *La Carnagmolle,* and recaptured an East India brig, which was lost near Baltimore in Ireland.

The *Favourite*, Captain Bradley, recaptured from the French, a very fine brig belonging to Leith, and bound to Cadiz, with glass bottles and iron hoops; and, in company with the *Bess* privateer, of Bristol, recaptured a Swedish brig, from Barcelona for Ostend, laden with 283 pipes of brandy. On the 19th of January, 1794, the *Favourite* arrived at St. Eustatia, having taken ten prizes; three she sent to Montserrat, and seven to St. Christophers. In 1798, on her passage from Demerara to Liverpool, the *Favourite* was taken by the *Bougainville* French privateer, and carried into L'Orient after an action of three hours, in which one man was killed, and the captain and several men wounded.

The *Loyal Ann* chased a French West Indiaman on shore near Bordeaux, and took a fine new sloop, from

New York for Havre, laden with sugar, cotton, and potashes.

The *Brothers*, of Liverpool, recaptured the ship *Community*, with wine, oil, and cocoa, from Cadiz to Ferrol, which arrived in the Mersey on June 30th, 1793; as did also the ship *Three Brothers*, of Dartmouth, laden with stock fish, from North Bergen for Venice, recaptured by the *Dispatch* privateer.

The *Brothers*, Captain Fleming (Letter of Marque), captured the *Hebe*, from the Southern Fishery to France, with 130 tons of oil, which arrived in Liverpool in October, 1793.

On the 4th of May, 1793, the *Pilgrim* (Letter of Marque), Captain Hutchinson, fell in with the *La Liberté*, French East Indiaman, from Bombay for L'Orient, a very fine ship, Danish built, about 800 tons burthen, carrying 12 six-pounders, and 60 men. An obstinate engagement took place, on the second evening of which the French captain was killed, and on the following morning his ship struck to the *Pilgrim*, and was carried to Barbadoes, where she arrived on the 29th of May. The ship having been three years in the country, her cargo turned out a most valuable one, realizing £190,000.* We have already seen two Liverpool estates—the "St. Domingo," and "Carnatic-hall"—named in grateful commemoration of fortune's favours granted to privateers, and the present capture must be added to the list. A certain shoemaker, who flourished when a comparatively small portion of Everton had been brought under cultivation, enclosed a considerable tract of

*The cargo, as enumerated in the Liverpool paper, consisted of the following:—138,557 pieces yellow and white Nankeens, about 150 hogsheads of sugar, 71 chests of china ware, 18 chests mother of pearl, 139 chests cinnamon, 183 bales of Surat goods, 2 chests Nankeen silks, 1 chest cotton woollen stuff, 4 bales niccanees, 17 bales casileys, 1 bale tapsel, 1 bale muslin, 500 cardels of pepper, 500 chests tea, 20 cases images, 2 bales coral, 2 chests silk manufactory, 1 case Nankeen calico, 1 chest painters' paper, 108 sacks Malabar pepper, 3 bales white linen, 90 bales cotton, 13 bales Bejuta pants of Surat, 1 bale Bengal goods, 1 bale embroidered waistcoats, 1 parcel medical roots, 6 parcels sugar-candy, 1 parcel Fontanagu lacca.

land in the neighbourhood of Sleeper's Hill, and modestly called his estate "Cobbler's Close." This property was bought and re-named "Pilgrim," by Mr. Barton, who, in conjunction with Mr. Thomas Birch, had an interest in the fortunate "Letter of Marque" of that name. The Pilgrim property was afterwards bought by Mr. Atherton, who sold it to Mr. Woodhouse, the agent for Lord Nelson's Bronté estate in Sicily, and that gentleman re-named the property Bronté, a name it retained until it was invaded by enterprising builders. The Bronté and the St. Domingo properties adjoined each other, and it is a rather curious circumstance that the two valuable estates should have been the products of two rich privateering adventures.

Many vessels were fitted out of the American ports, under French colours, manned chiefly with Americans, and they captured many prizes. Captain Morgan, of the *Jean*, from Jamaica for Liverpool, put into Philadelphia to refit, having been chased a whole day by the *L'Ambuscade* frigate, which cruised off the Hook, speaking most vessels that passed in or out. Captain James, of the *Halifax Packet*, applied to the Governor and Council to prevent the frigate sailing immediately after him, but they would not comply with his request. He then, through a friend, applied to the French Consul, who politely gave him four days' start of the *L'Ambuscade*.

A gentleman in Philadelphia, writing to a merchant in Liverpool, on May 13th, 1793, says:—

> "What can all your frigates, of which we are told you have such an immense number in commission, be about, to permit the French frigate, *L'Ambuscade*, Citizen Bompard, commander, to insult your flag, take your merchantmen, and "ride triumphant o'er the western waves." She is now abreast of our city, and has taken five or six prizes since her departure from France, two of which are at present alongside of her, the *Little Sarah*, of Kingston, Capt. Laury, built in Liverpool, taken ten leagues

at sea, and the *Grange*, Hutchinson, of your port also. This vessel it is expected will be delivered up, as she was taken at anchor with the pilot on board, ten or twelve miles up the Capes, and we were two days ago informed that the President, minister of state, of war, and of the treasury, and the attorney-general, have given it as their unanimous opinion that she was illegally taken, and therefore no prize, she being within the jurisdiction of the United States, and of course, under the protection of a neutral country. The business, however, will not be determined until the arrival of the French Minister, M. or rather *Citizen* Genet, who is daily expected from Charleston, where he was landed by the above frigate. The ship *William*, of Glasgow, Capt. Nageto, is just sent up as a prize to a little privateer of six guns."

The *Grange* was eventually given up, and the captain and seamen all liberated.

The French privateers from Martinico, Guadaloupe, and St. Lucia, captured, in a short time, 70 sail of British, Dutch, and Spanish merchant vessels, and carried them into those islands.

The ship *Swift*, Captain Roper, was taken on the 26th of May, by a French privateer, who took out of her thirty-three male slaves, and 224 elephant teeth. They returned to Captain Roper his ship, and the remainder of his cargo, for a ransom of £1000 sterling, and took the second mate as a hostage for the same.

Captain Heavysides arrived from Philadelphia, and reported that Congress were exerting themselves to hinder vessels, fitted out as privateers, sailing from any of the American ports.

The *King Grey*, Captain Cash, arrived at Jamaica, from Africa, having been captured on the passage by a French privateer, and retaken by his Majesty's frigate *Hyæna*, who ran the privateer on shore at Hispaniola, where Captain Cash and his people released themselves from their irons, and, getting possession of the Frenchman's boat, got on board the frigate.

Captain Raphel, of the *Polly*, arrived at Jamaica, from Liverpool, reported that about fifty leagues to the westward of Madeira, he met with a French brig privateer, without her main mast. On boarding her, he was informed she had been dismasted, six days before, by the *Christopher*, Captain Molyneux, of Liverpool, who, after she had struck her colours, sent on board and dismantled her of her guns, powder, shot, and all kinds of arms, took out what stores (except provisions) they stood in need of, and proceeded on his voyage. The privateer had 75 men on board, and was only six weeks off the stocks. The *Christopher* afterwards captured a valuable prize at Angola.

The *Robust*, Captain Forrest, recaptured the *Little Joe*, Captain Jones, and the *Echo,* Captain Kelly (the latter with 120 negroes on board), two Liverpool slavers, which had been taken on the windward coast of Africa by the *Liberty*, of Bordeaux, which also took the *Union,* Captain Farrington, the *Mercury*, Captain Hewitt, the *Hazard*, Captain Rigby, and the *Prosperity*, Captain Kelsall, all engaged in the man traffic. The *Mercury* was retaken by the *Seaflower* cutter, and sold to Captain Hewitt, who, by the way, lived in Murray-street, Williamson-square, "adjoining the rope walk," when he was not prosecuting his humane mission in Africa, etc. The *Prosperity* was also retaken by the *Andromache* frigate, and carried into Barbadoes. The *Robust* had the good fortune to capture a French ship, with about 200 slaves, at Cape Mount, and in November, 1793, we read that she took, on the coast, a large French ship, called *Le Patriote Soldat*, with 260 slaves and a cargo of goods, and carried the prize to Dominica.

The slave ship *Minerva*, late Captain Moore, arrived at Jamaica from Africa, with a cargo of prime negroes, "without burying either black or white, the master excepted"—a feat by no means of common occurrence. The *Minerva* privateer, Captain Williams, recaptured a Swedish vessel, and took the *Ann and Margaret*, from Riga, with deals.

The *Lord Stanley*, Captain Farquhar, captured the *Julie Chere*, from Guadaloupe for Bordeaux, with sugar, coffee, cotton, etc.

A letter from Dominica, to a merchant in Liverpool, dated May 6th, 1793, mentioned that the sloop *Amity*, Captain Spellin, of 6 guns and 34 men, had brought in there the ship *Bon Menage*, pierced for 22 guns (only six mounted), and 38 men, belonging to St. Malo. She had been thirty-three months on a voyage to the East Indies, and afterwards called at Malimba for slaves, 674 of whom she had on board. Her hold was full of trunks and bales of India goods, brandy, gold dust, etc., supposed to be worth £100,000.

On the 9th of June, 1793, Captain S. Bower, of the *Active* privateer, captured by the French, wrote as follows, from Morlaix, to his owners in Liverpool :—

"It is with concern I inform you of our being captured on the 21st ult., by the French frigate *Semillante*, of 44 guns and 300 men, who took me on board. She also captured, the next day, a brig privateer (the *Betsey*, of Guernsey), of 10 guns and 55 men. On the 27th, she fell in with an English frigate, whom she engaged two hours, had twenty men killed and forty wounded: Amongst the former was the captain, first lieutenant, and a petty officer, when he bore away, having five feet water in her hold, and was chased by the English frigate, whose main-top giving way, the Frenchman (I am sorry to say it) escaped, for could the frigate have come up with her again, she would have struck immediately. She proceeded directly for Brest, where we arrived the 2nd of June, and where I have been in prison until yesterday, when I was marched for Dinant, with 112 more English prisoners, and this day arrived at Morlaix on our road thither. We have been just now joined by two men belonging to the *Allanson*, Capt. Byrne, private ship of war, taken in a prize captured by that vessel."

The *Active* was retaken, and carried into Guernsey.

The *Golden Age*, Captain Fayrer, from Jamaica to Liverpool, was taken by the French frigate *Citoyen*, of 36 guns, which two days later captured the *Courier*, Captain Rigby, also from Jamaica to Liverpool, and, after plundering her, ransomed her to the captain for £300. The lieutenant of the *Citoyen* was an American, and 27 of her crew were English, but called themselves Americans. The *Citoyen* had lost her captain and 63 men in an engagement with an English frigate, all of whom had been killed outright or died of their wounds.

The *William*, of Liverpool, Captain Ward, on her passage from Virginia, in company with the *Hector*, of London, the *Fanny*, of Greenock, and the *Joseph*, of Appledore, fell in with a schooner privateer, fitted out in America, under French colours. A desperate engagement ensued, which lasted three hours, resulting in the capture of the *Joseph;* the rest escaped. The privateer was twice beaten off, but in the third charge, one of the guns of the *Joseph* exploded, by which unfortunate accident Captain Prance lost both his hands, and was obliged to strike his colours. He also received a wound in the thigh, and had one of his eyes much injured; his recovery was despaired of. The mate also was wounded. Only three persons were wounded on board the privateer. The *William* arrived in Liverpool on the 3rd of July, 1793.

By the 1st of July, 1793, no less than sixty-seven Liverpool privateers were armed and manned, and were either at sea or preparing to sail. Great numbers of privateers were fitted out afterwards, and an extraordinary number of prizes was taken. The French were too much distracted by internal dissentions, and attacks from abroad, to carry on this mode of warfare with any success. In three or four years their commerce was swept from the ocean; whilst, from the commencement of the war, the English commerce was carried on in tolerable safety, under the protection of ships of war. The number of British vessels employed in commerce at this period was said to be 23,600.

The *Savannah*, Captain Wrigglesworth, recaptured a Dutch ship, from Cadiz for Amsterdam, laden with 250 lasts of bay salt.

On July 10th, 1793, the *Tarleton*, Captain Gilbody, from Liverpool, in company with the *Eliza*, Captain Cannell, for Africa, took the *Le Guerrier*, a brig privateer, of 8 guns and 72 men, belonging to Bayonne.

On the 18th of June, 1793, the *Fancy*, Captain Robinson, recaptured the brig *Margaret*, belonging to Leith, bound from Alicant to Dublin, laden with 456 bales of barilla, 47 pipes and 4 casks of wine, about a ton of saffron, and 7,000 reeds. Some months later, on her passage to Jamaica, the *Fancy* took, and sent into Kingston, a brig laden with 700 barrels of flour. A few days after, the *Fancy* was attacked by a French privateer, of 16 guns and upwards of 100 men. The engagement lasted five hours and-a-half, the Frenchman sheering off, leaving the *Fancy* much shattered in her sails, rigging, etc.

The *Union* privateer, Captain Nicholson, took a Swedish vessel, laden with brandy, from Barcelona for Calais.

The *Colonel Gascoyne* and *Margaret*, privateers, captured *The Sisters*, from New York for Havre, laden with coffee, indigo, etc.

The *Duke of Leeds*, Captain Purvis, took a Danish vessel, from Guadaloupe, laden with sugar, coffee, and cotton, and carried her into St. Kitts. The *Duke* was captured early in 1794, on the passage home from the West Indies.

The *Philip Stephens* privateer brought into the Mersey, in September, 1793, the ship *Sarah*, one of the Jamaica fleet, which she had retaken from the French.

The *Oporto*, Captain Hamilton, captured an American ship called the *Birmingham*, laden with sugar, coffee, and cocoa, from Baltimore for Amsterdam, which arrived in the Mersey on September 29th, 1793. In March, 1794, the *Oporto* recaptured a ship of 400 tons, with salt, from Alicant, which had been taken by the *Tribune* French frigate.

The *Alert*, Captain Hollywood, captured on the coast of Africa, a ship of about 500 tons burthen, from Bordeaux, with 50 slaves, and a cargo of goods on board; also a sloop with 50 slaves, and a schooner with 40 slaves.

The *Mercury*, Captain Mellanby, on his passage to the West Indies and Virginia, was attacked by two small privateers, who came alongside and fired into him, which he returned with a broadside that caused them to sheer off, and make the best of their way from him. The *Mercury*, on her passage home from Virginia to Liverpool, was wrecked near the Orme's Head, and the captain, with fifteen of the crew, perished ; ten hands were saved.

The *Hope* privateer, Captain Hall, recaptured *La Mahon*, a Spanish brig, from La Guira for Cadiz, laden with 500 bales of tobacco, about 700 quintals of cocoa, coffee, indigo, and hides. The *Hope*, in company with the *Thought*, of London, also recaptured the *Neptune*, from Dominica, and took an American ship, bound to France, with a cargo of sugar, coffee, indigo, and cotton, and a number of French passengers on board. On the 5th of September, 1793, the *Hope* was taken, in latitude 48°, by two luggers, one of them, the *Hook*, of 20 guns and 125 men, formerly an English revenue vessel. The *Hook* came up with the *Hope* (which only carried 12 guns and 44 men) about seven o'clock in the morning, and fired one gun, which Captain Hall answered with a broadside. The lugger then fired two guns, and bore away. The *Hope* gave her a second broadside, when the other lugger came up, and the *Hope* engaged her for an hour; but, after having the carpenter, boatswain's mate, and a seaman killed, the first and fourth lieutenants with six men wounded, Captain Hall was forced to strike.

The *Nereus* (Letter of Marque), of 16 guns, Captain M'Iver, on her passage from Liverpool to New York, re-captured a Spanish brig, from the Havannah, laden with 850 boxes of fine sugar, beeswax, honey, etc. This vessel, when retaken, was in tow of a French privateer, of 16 guns, which the *Nereus* beat off. The privateer had unfortunately taken 50,000 dollars out of the prize before the *Nereus* fell in with her. On the 12th of July, 1793, another prize taken by the *Nereus*, arrived in the Mersey—the brig *Two Friends*, with oil, from the fishery at Falkland Islands, for Dunkirk; and on July 17th, the *Joseph* came in (probably the Spanish brig before mentioned).

In August, 1793, Mr. Asburner, agent at Barbadoes for Mr. James Kenyon, of Liverpool, was captured on the passage to Dominica, by the *Sans Culotte* privateer, and carried into Martinique. The Frenchmen treated him "exceeding ill, taking everything from him." He was confined with the common sailors in the prison ship, under Fort Republique; and what must have intensified his humiliation, although it savoured of poetic justice, was the fact that the commander-in-chief at that place was "a mulatto man."

On the 24th of September, 1793, the *Olive*, of Liverpool, Captain Pennant, was taken in lat. 45°., long. 28°., by the *La Felicité*, French frigate, of 40 guns.

> "The French captain," says one of the officers of the *Olive*, writing from Brest, "behaved in a most villainous manner, sheering up alongside and pouring nine of his heavy guns right into us before he hailed, which killed one man and wounded another. We found it impossible to get away from her, she sailed so much faster than us. They boarded and stripped us of every article but the clothes on our backs, and in that state we were landed at Brest, the 5th inst., and marched to a

hospital five miles in the country. In two days we are to march to Dinan Castle, 80 leagues from this place. The *Lydia*, Captain Crow, was taken on the 26th ult. by the same frigate. Captain Crow is sick in the hospital, having his arm dislocated."

One Friday evening, in October, 1793, an affray took place at the bottom of Redcross Street, between Mr. Felix M'Ilroy, master of the sloop *Ann*, of Newry, and some of the press-gang belonging to a tender lying in the river, one of whom drew a pistol and shot the captain dead upon the spot. The murderer got on board the tender, but was arrested next morning. The coroner's jury brought in a verdict of "murder," but, at the Lancaster assizes, the man got off with one month's imprisonment. On the Saturday evening following the crime, a large body of sailors assembled, and, out of revenge, it was supposed, for the death of Captain M'Ilroy, attacked the "rendezvous" of the press-gang, in Strand Street, and, soon after, the one on New Quay, which they completely gutted, cutting open the beds, and throwing the feathers, bedding, and household furniture of every description into the street. They tore down the wainscoting, mouldings, cornices and doors, which, as well as the windows, shutters, etc., they utterly demolished, leaving little more than the walls, floors, and roof undestroyed. At the solicitation of the Mayor and Ex-Mayor, who appeared upon the scene, the mob desisted from further outrage and dispersed. That bright and genial gossip, "An Old Stager," has left so racy a picture of the press-gangs in his native town of Liverpool, and their doings, of which he was an eye-witness, that we cannot resist the temptation of quoting his own words:—

"We had a venerable guardship in the river, the *Princess*, which, we believe, had originally been a Dutch man-of-war, and if built to swim, was certainly never intended to sail. There she used to lie at her moorings, opposite the old George's Dock pier, lazily swinging backwards and forwards with the ebbing and flowing of the tide, and looking as if she had been built expressly for that very purpose and no other. Her very shadow seemed to grow into that part of the river on which she lay. But, besides her, we had generally some old-fashioned vessel of war which had come round from Portsmouth or Plymouth to receive volunteers or impressed men. Those who live in these "piping times of peace" have no idea of the means which were employed in the days of which we are speaking to man our vessels of war. The sailors in our merchant service had to run the gauntlet, as it were, for their liberty, from one end of the world to the other. A ship of war, falling in with a merchant vessel in any part of the globe, would unceremoniously take from her the best seamen, leaving her just hands enough to bring her home. As they approached the English shore, our cruisers, hovering in all directions, would take their pick of the remainder. But the great terror of the sailor was the press-gang. Such was the dread in which this force was held by the blue-jackets, that they would often take to their boats on the other side of the Black Rock, that they might conceal themselves in Cheshire; and many a vessel had to be brought into port by a lot of riggers and carpenters, sent round by the owner for that purpose. And, truly, according to our reminiscences, the press-gang was, even to look at, something calculated to strike fear into a stout man's heart. They had what they called a "Rendezvous," in different parts of the town. There was one we recollect in Old Strand-street. From the upper window there was always a flag flying, to notify to volunteers what sort of business was transacted there. But look at the door, and at the people who are issuing from it. They are the press-gang. At their head there was generally a rakish, dissipated, but determined-looking officer, in a very seedy uniform and shabby hat. And what followers! Fierce, savage, stern, villainous-looking fellows were they, as ready to cut a throat as eat their breakfast. What an uproar their appearance always made in the street! The men scowled at them as they passed; the women openly scoffed at them; the children screamed, and hid themselves behind doors or fled round the corners. And how rapidly the word was passed from mouth to mouth, that there were 'hawks abroad,' so as to give time to any poor sailor who had incautiously ventured from his place of concealment to return to it. But woe unto him if there were no warning voice to tell him of the coming danger; he was seized upon as if he were a common felon, deprived of his liberty, torn from his home, his friends, his parents, wife or children, hurried to the rendezvous-house, examined, passed, and sent on board the tender, like a negro to a slave-ship. And so it went on, until the floating prison was filled with captives, when the living cargo was sent round to one of the outports, and the prisoners were divided among the vessels of war, which, were in want of men. Persons of the present generation, have certainly heard of the press-gang, but they never attempt to realise the horrors by which it was accompanied. Nay, the generality seem to us to hardly believe in its existence, but rather to classify it with "Gulliver's Travels," "Don Quixote," "Robinson Crusoe," or the "Heathen Mythology." But we can recollect its working. We have seen the strong man bent to tears, and reduced to woman's weakness by it. We have seen parents made, as it were, childless, through its operation; the wife widowed, with a husband yet alive; children orphaned by the forcible abduction of their fathers. And yet, there were many in those days, not only naval men, but statesmen and legislators, who venerated the press-gang as one of the pillars and institutions of the country. In those days, indeed! We much fear that, if even now we could look into the heart of hearts of many a veteran admiral and captain, we should find that they have, in the event of war, no other plan in their heads for manning the navy, but a return to this dreadful and oppressive system. We would, however, recommend those in whose department it lies to be devising some other scheme, as we are strongly impressed with the conviction that public opinion will not in these days tolerate, under any plea or excuse of necessity, such an infringement upon the liberty of the subject. But we are not writing a political article, but only describing our old-world fashions.

Pretty rows and riots, you may suppose, now and then occurred between the press-gang and the fighting part of the public; and not a few do we remember to have witnessed in our younger days. On more than one occasion, we have seen a rendezvous-house gutted and levelled to the ground.

"Sometimes the sailors and their friends would show fight, and as the mob always joined them, the press-gang invariably got the worst of it in such battles. Sometimes, too, the press-gangers would 'get into the wrong box,' and 'take the wrong sow by the ear,' by seizing an American sailor or a carpenter, and then there was sure to be a squall. The bells from the shipbuilding yards would boom out their warning call in the latter case, and thousands would muster to set their companion at liberty. A press-gangman was occasionally tarred and feathered in those days when caught alone. We remember, as if it were only yesterday, walking down South Castle Street (it was Pool Lane then), with the Old Dock, where the Custom House now stands, before us. It was, for some reason or other, tolerably clear of ships at the time. We well remember, however, that there was one large vessel or hulk somewhere about the middle. Before we tell what happened, we must observe that, attached to the Strand Street press-gang, there was one most extra piratical-looking scoundrel, named Jack Something-or-other. Perhaps, as is often the case, 'they gave the devil more than his due'; but if one half of the things said against this Jack was true, he deserved to be far and away Prince and Potentate and Prime Minister in Madame Tussaud's Chamber of Horrors. Well, as aforesaid, the Old Dock was in front of us, when all at once we heard a noise behind us, which told us that the game was up, and the hounds well laid on and in full cry.

"At the same moment, Jack shot past us, like an arrow from a bow, while hundreds of men, women, and children were howling, shouting, screaming, yelling, threatening close behind him. Every street sent forth its crowd to intercept him. There was no turning until he reached the dock quay, but there the carters and porters rushed forward to stop him. What was to be done? How was he to escape? The dock, as we said before, was in front, and there was the vessel in the middle. Without a moment's hesitation, the terrified wretch took the water, dived, like Rob Roy, to baffle his pursuers, and soon gained the deck of the hulk. Some talked of boarding her, and dragging him from his concealment; but the majority of the mob decided that justice was better than vengeance, and, satisfied with Jack's fright and ducking, concluded that although he was a bad one, he was game, and would make them more sport another time, and so dispersed."

Desperate press-gang encounters took place in Pool-lane, now South Castle Street, so that it was hardly safe to pass along it at times, certainly not in the night season. It was much frequented in war-time by the privateersmen, who spent their prize money freely amongst the dingy denizens of the locality, in the public-houses, and slop-sellers' shops. When Jack came in from a cruise, with his cutlass at his side, and his pockets full of plunder, Mistress Quickly received him like a mother, and Doll Tearsheet wept crocodile tears of joy at his hairbreadth escapes. He was her Othello for the nonce, but when his money was spent, or stolen, they could always betray him to the press-gang, or the crimps, for a Guinea voyage. At Seacombe, in those days, there were but few houses, and only a farm or two between it and the Magazines, near which, and on the site of the villa erected and now occupied by Mr. Joseph Kitchingman, at the bottom of Caithness Drive, on the Promenade, was a little public-house, kept by "Mother Red Cap," where the sailors fled from impressment, and where the privateersmen lodged their gains with the landlady, who earned her title by the red cap she always wore. Many curious stories are extant about this old woman and her inn. There is a tradition that the caves at the Red Noses, in addition to penetrating as far as Chester, communicated in some way, and somewhere, with

Mother Red Cap's house, and it was in their recesses that the hunted sailors, in war-time, used to be put into safe hiding. The caves undoubtedly penetrated to a considerable distance in the direction of the Magazines, as there is now living at Wallasey, an old man who explored them in his youth. In preparing for the foundations of Mr. Kitchingman's villa there was a good deal of excavation, and perhaps a reasonable expectation of finding "treasure-trove," but it produced nothing of greater interest than a nine-hole stone, on which the old lady's customers used to amuse themselves.

The striking contrast between Mother Red Cap's humble hostelry, filled with a wild crew of blood-stained men of the sea, drinking, smoking, singing, quarrelling, and fighting, as the humour took them, and the present beautiful home of the artist, is typical of the moral transformation which the people on both sides of the "Silent Highway" have undergone since the last private ship of war sailed out of the Mersey.

Stonehouse, speaking of Woodside Ferry, also bears testimony to the character of both the privateersmen and the press-gang:—

> "The last century was a lawless time in its history, for it swarmed with fierce privateersmen, inhuman slavers, reckless merchantmen, and violent men-of-war's men, who all conspired to make the sailor element of the town 'thick and slab.' In these days of peace we have no conception of the uproar, the violence, the turbulence, as well as the merriment that prevailed when men came home from some short voyage with large sums to receive, the results of their rapacity upon, and robbery of their neighbour, that war gave countenance to and justified. These men's hearts were hardened against the cry of humanity. After some great engagement, when men were scarce and the strength of the navy was enervated, the press-gangs stalked through the town, seizing anyone to whom they took a fancy; and though such an one might have been able to show himself to be a simple landsman, or, if a sailor as having

protection, if the service was hard pushed but small consideration was used in any case. A man was a man, and away he went on board the tender. It was no uncommon circumstance in those days for persons to be unaccountably missing, men in really respectable positions in life, who would after a year or two suddenly turn up, having been impressed and sent to a foreign station. The atrocities of the press-gangs we read about, but can scarcely credit."

A rare instance of the press-gang lion and the sailor lamb lying down, or rather sitting together, happened in July, 1794, when a benefit performance took place at the Theatre Royal, for the relief of the widows and orphans of the brave fellows who fell on the "glorious first of June." Prior to the performance a letter appeared in the *Advertiser*, from a "J.B.," of Tarleton Street, in which occurred the following passage:—

"The only thing to be lamented on the occasion is that a set of men are precluded from attendance, whose principal characteristic is a most unbounded generosity. I mean the sailors, who dare not appear, in order to show their liberality, being deterred by the apprehension of being impressed; this very circumstance I should presume will materially affect the receipts of the gallery, unless the gentlemen in power would step forward and generously guarantee the personal safety of these hardy heroes for twenty-four hours, to commence this day at noon and to continue till the noon of the following day; a circumstance this, I should presume, that would in no respect injure the general purpose of government."

Immediately below it appeared the following re-assuring letter, probably inspired from the "rendezvous" of the press-gang:—

"LIVERPOOL, July 21st, 1794. Mr. Tim Mainstay—Being well acquainted with the disposition of the Regulating Captain, I will answer for his not suffering any man to be impressed at the time of his going to, or returning from the Theatre, this evening; therefore all jolly tars may subscribe their mite to the widows and orphans, and at the same time enjoy the evening's amusement, without any apprehension of being pressed. The officers, press masters, &c., will partake of the amusement, without the least intention of interrupting the performers or the audience.—Yours, TOM BOWLINE."

The receipts at the Theatre on the occasion in question amounted to £208 18s. 6d.

The press for seamen for the navy was very hot this year, but Tim Mainstay, Tom Bowline, and the rest of the tarry fraternity were not the only people who had cause of complaint. The press-gang, not content to scour Jack's usual haunts and hiding places on both sides of the river, went inland, invading the merriment of wakes and fairs, and carrying off every eligible landsman they could lay their hands on. This practice carried indescribable grief and misery into many a home.

Even a stage coach, dashing along in charge of an armed guard, and presided over by a masterful and dignified Jehu, the admiration and awe of all pedestrians, was not sacred in the eyes of the hardened ruffians, who hunted for sailors with the tenacity of sleuth hounds.

In the newspaper of May 12th, 1794, we read as follows:—

"On Thursday evening, about six o'clock, a press-gang stopped Sherwood's boat coach from Warrington, at Low Hill, near this town, and cut the bridles of the horses, to prevent the carriage proceeding, when they examined the outside passengers in expectation of some of them proving seamen. The York mail coming up at the time, they also stopped it, the guard threatened to fire at them, and in getting his pistols, one of them by accident went off, and the shot passed through the back of the coach, wherein were two gentlemen, but hurt neither of them: the report of it frightened the horses of the other carriage so, that they set off with it at full speed, without the driver. After gaining a distance of about 400 yards, a gentleman of the law of this town, threw himself out of the coach and has been confined to his room ever since, being dreadfully hurt on his head and one of his legs. Two ladies were also inside, but stuck by the carriage and were not hurt. The offending parties were carried before a county magistrate on Saturday, and after examination were bound over to the Sessions to answer for the outrage."

During the year 1794, the navy of France received a blow from which it never recovered, and her principal West Indian Colonies fell into the hands of England, opening up for Liverpool, with its West India and African trade, prospects of great extension of commerce. On the 28th of April, the Mayor of Liverpool, Mr. Henry Blundell, whose brother, Lieutenant-Colonel Blundell, had distinguished himself at the capture of the island of Martinique, gave a grand banquet in honour of that event. On Friday, the 21st of May, the bells of the churches rang all day in honour of the conquest of Guadaloupe; and early in July, news arrived of the capture of Port-au-Prince, in the island of St. Domingo. But the crowning event of the year was the great victory gained by Lord Howe over the Brest fleet, on the 1st of June, a victory which greatly diminished the apprehensions of invasion, but did not altogether remove them.* "On Friday morning, when the news of Lord Howe having defeated the French fleet arrived in Liverpool," says *Williamson's Advertiser* of June 16th, "it gave rise to the most unbounded joy. The bells of the different churches rang incessantly; flags were displayed from the ships and steeples; the ships in the different docks were gaily decorated; pendants and ensigns were hung out from the various dwellings throughout the town; and where those could not be obtained, quilts, handkerchiefs, curtains, etc. At one o'clock, a royal salute was fired from the great guns of the fort. On Saturday, the flags were again displayed."

* The Corporation headed a county subscription, to raise volunteer regiments, with a contribution of £1000.

Whether rejoicing, fighting, or slave trading, the old Dicky Sams did nothing by halves.

On the 18th of July, 1794, the brig *Three Brothers*, Captain Hanna, on her passage from Liverpool to New York, was taken and burnt by two French frigates, which, a month later, also captured the brig *Hawk*, of Pool. Captain Hanna, with one of the passengers he had in the *Three Brothers*, a Dutchman, and two small boys, with several Frenchmen and boys, were put on board the *Hawk*. On the 28th of August, Captain Hanna, with the assistance of his companions, got the Frenchmen made fast in the forecastle, and carried the brig into Dor Sound, in the Orkneys. She was richly laden with bale goods, tanned leather, butter and cheese, teas, etc. The two frigates captured twenty-five prizes while Captain Hanna was on board, burning and sinking all but two, which they sent to Ireland with 130 men. They had previously taken the *Hound*, sloop of war, and sent her to Brest, and burnt a Spanish vessel.

The ship *Fame*, of Boston, Captain Davies, on her passage from Virginia to Liverpool, was taken, off Cape Clear, by the French frigate *L'Agricole,* of 44 guns and 550 men. She was set for France, under the care of a prize-master and six men, and her crew were taken on board the frigate, except Captain Davies, two men, and a black boy. Three days after parting from the frigate, Captain Davies and his men rose upon the Frenchmen, secured them, and carried the ship safe into Cork and thence to Liverpool.

The *Polly*, Captain Jones, arrived at Jamaica, from Liverpool, after a severe engagement, off Martinico, with a French privateer, of 14 guns and 120 men, whom she beat off.

The French privateer *Sans Culottes*, of 20 guns (supposed to have been formerly the *William*, of Liverpool), was captured and carried into St. Kitts by the *La Blanche* frigate.

The *Mary*, Captain Bonsall, and the *Agnes*, Captain Parker, captured eight vessels from Martinico and Guadaloupe, laden with sugar, coffee, and other commodities, and carried them into Monserrat. The *Agnes* also had a claim upon another prize, being in sight when it was taken. Soon after, the *Mary* captured a French privateer, of 12 guns and 55 men, and carried her into St. Christophers.

The ship *Christopher*, Captain Smith, captured three vessels laden with West India produce, and destroyed a French privateer, of 12 guns and 44 men, belonging to Martinico.

The *Elizabeth*, Captain Fletcher, arrived at Jamaica, from Liverpool, after beating off two French privateers, of 16 and 14 guns, who attacked her near Isle de Vache.

The *James*, Captain Brailsford, on her passage to Africa, was taken, off Cape Clear, by two French 40-gun frigates, and a sloop of war. The French commander put eleven of the frigate's crew on board, leaving only the mate and two boys belonging to the *James* in her, and ordered the prize-master to take her to Brest. He being incompetent for the task, requested the English mate to steer for that port. The mate, however, very wisely altered her course for Bristol, at which place he would probably have arrived, had not the *Castor* and *Peggy* tenders relieved him of his charge, and carried her into Plymouth.

At the beginning of August, 1794, there arrived in Liverpool, James Scallon, carpenter of the *Ellen*, Captain Raphel, which had been taken, in March, by the French frigate *La Proserpine*. On the 1st of July, he, with six others, had escaped from the prison at Quimper, in Brittany, from whence, after a march of seventeen nights (being obliged to conceal themselves by day), they reached the sea coast, where they fortunately found a small boat, in which they embarked, in the hope of reaching England. Having made a mast with a strong pole taken from a neighbouring wood, they converted their shirts into a sail, and with a piece of board for a rudder, without water or food, they committed themselves to the ocean, and in three days, to their inexpressible joy, landed at Sidmouth Beach, in Devonshire. Scallon stated that he left 2,700 English in the prison at Quimper. In a small house near the prison, was confined, under a guard, the Lady Anne Fitzroy, who had been taken some time before in the packet returning from Lisbon, about whose fate there had been great uncertainty. Among the prisoners was Lieutenant Robinson, of the *Thames* frigate, whose leg had been shot off in the action when the *Thames* was captured, but who was sufficiently recovered to move on crutches. Some of the crew of the *Thames* had been shot in attempting to escape from prison. Since the defeat of the French fleet, the English prisoners had been treated with much more severity than before. One of Scallon's fellow prisoners, who was sent to Brest, in order to be examined for the condemnation of a ship, saw, whilst he was there, the Bishop of Quimper, and 25 other persons guillotined in the space of eleven minutes. It was stated that in the action of the 1st of June, the French Admiral's ship *Montague* had 1,500 men on board, 500 of whom were killed.

The *Gregson*, Captain Gibson, was taken on her passage from Liverpool to Africa, by the *La Robuste*, of 22 guns and 160 men, and carried into L'Orient. On the 2nd of July, 1794, Mr. Pince, second mate, and Mr. Jones, surgeon's mate, of the *Gregson*, were ordered, with other prisoners, about 45 miles up the country. During the preparations in weighing their baggage, etc.,—each having previously provided himself with a National cockade—they passed through the crowd unobserved, and got on board a Danish vessel, where they put on a disguise. They then went into an American vessel, but the captain being unwilling to keep them, they returned to the Danish ship, only, however, to board the American a second time, unknown to the skipper. They were concealed in the hold eight days, and, when discovered, were treated by him with the greatest civility. A few days later, when off Portland Point, they boarded a brig, bound for Bermuda, which was standing in, with the intention of getting on shore, but the Captain refused to assist them, and they were forced to return to the American vessel. Soon after, they met with a pilot boat, which landed them at Portsmouth Point. Captain Tristram Barnard, the kindly commander of the American ship, paid their passage, and gave them four guineas to enable them to reach Liverpool. They reported that the French were re-building several ships of the line and frigates, a number of the latter being specially intended to cruise against the African vessels belonging to Liverpool.

During the year 1794, the following recaptures were effected by Liverpool ships :—

The *Enterprize*, Captain Young, recaptured the *Virginie*, 400 tons burthen, from River Plate for Cadiz. The *Sarah*, Captain M'Ghie, recaptured the *Mary*, Captain Taylor. The *Swan*, Captain Hall, recaptured the brig *Nancy*, of Belfast. The *Fortune* recaptured the *Two Brothers*, of Yarmouth. The *Mary* recaptured the *Active*, from Lisbon for London. The *Hawke*, on her passage to Barbadoes, recaptured the *Penelope*, of Greenock. The *Old Dick*, Captain Bird, recaptured the brig *Martin*, of Whitehaven, and the brig *Ilfracombe*, with 198 pipes of wine, 30 bags of shumac, and 40 quintals of cork. The *Old Dick*, the *Betsey*, Captain Corran, and the *Jenny*, Captain Smith, had narrow escapes of being taken by French frigates, but escaped by dint of sailing. The *Barton*, Captain Hall, recaptured the brig *Mentor*, of Aberdeen, with fruit from Lisbon. The *Othello*, Captain Christian, recaptured the ship *Minerva*, from Martinico for St. Domingo. The *Edgar*, Captain Kendall, took possession of a fine brig, off Cape Finisterre, laden with salt, wine, and fruit, but without a soul on board.

The *Commerce*, Captain Bosworth, cruising off Desseado, took four prizes, and the *Allanson*, Captain Byrne, recaptured the brig *Minerva*, of Belfast; and, on her passage to Barbadoes, took the *Cleopatra*, an American vessel, from Mauritius for Boston, with many French passengers on board. In August, 1794, the *Allanson* was herself captured, homeward bound, by a French frigate.

The *Molly*, Captain Ford, on her passage to Virginia, was attacked by a French privateer, of 12 guns, and full of men, whom she beat off.

The *Kitty*, Captain Taylor, arrived in Liverpool from Guernsey, having been retaken from the French by the *Hero*, of that island.

Seven slave ships belonging to Liverpool, together with others from different ports, were taken and destroyed by the French squadron on the Gold Coast, which also devastated the settlement of Sierra Leone, leaving the settlers, 1300 in number, without provisions or necessaries, but such as could be obtained from the natives. The store houses at Isles de Los were also burnt.

The *Kitty*, Captain Mount, on her voyage to the coast of Africa, made the island of Madeira, and observed three ships close in with it. Captain Mount ran down on the weathermost of them, and found her to be a French privateer, mounting 24 guns on one deck, and seemingly full of men. He made sail and stood away, after exchanging a few shots. Two of the vessels followed him, and continued the chase two days and nights, pulling a number of sweeps, and had nearly come up with him, when a breeze sprung up, and the *Kitty* sailed away from them with ease. The *Kitty*, and the *Sally*, Captain Woods, were in company with the *Clemison*, Captain Jones, when she was taken by a French frigate, of 44 guns, and saved themselves by bearing down on the Frenchman, causing him to make off with his prize.

On the 10th of January, 1795, three ships belonging to the French squadron, which had committed considerable ravages on the coast of Africa, sailed up to Duke Town, Old Calabar, and one of them, carrying 20 twelve-pounders, attacked the slave ship *Kitty*, Captain Walker, of Liverpool, which received and returned three broadsides. Finding that only one of their number could engage the *Kitty* at one time, owing to her position, the enemy sheered off, and made no further attempts. On receipt of the news in Liverpool, a subscription was started for the master, officers, and crew of the *Kitty*, for their gallant behaviour. The *Kitty* was an "amazingly fast" sailer, and on her voyage to Africa, in April, 1796, on two occasions escaped from different squadrons of frigates which chased her.

The *Mary Ann*, Captain Bushell, left Jamaica on the 14th of January, 1795, with a crew of 19 men, and two passengers. On the 29th, a French schooner privateer, of 6 guns and 70 men, which had been hovering in sight for two days, twice attempted to board them. The Frenchman ran on board the *Mary Ann's* starboard quarter, where he lay for half-an-hour, but was repulsed. The contest was very severe on both sides, and the slaughter on board the schooner was dreadful. The *Mary Ann* had two men killed and four wounded. The two passengers, Lieutenant Wall, of the *Belliqueux*, and Lieutenant Ford, of the *Penelope*, were complimented by Captain Bushell for their bravery and gallant conduct, having been foremost in danger throughout the action, and unremitting in their exertions to animate his little crew against "their numerous and inveterate foe."

Nine of the crew of the *Cochrane*, Captain Wiseman, of Liverpool, fearful of being impressed in the West Indies, went on shore in the boat at St. Vincent. They unfortunately landed in a small bay near where the French and Charaibs were encamped, who, on seeing the boat, rushed to the beach and made them all prisoners. After remaining in durance two days, the unlucky mariners were relieved from their perilous situation by the British troops storming the French and Charaibs' camp.

On the 10th of April, 1795, off Abaco, one of the Bahamas, the *Crescent*, Captain M'Gauley, fell in with a Republican privateer, of 10 guns and about 70 men, who maintained an obstinate action for two hours, and then sheered off.

The *Eolus* was attacked off Nevis by a French privateer, of 10 guns, and full of men, with which she kept up a running fight of three hours, and at length beat her off, after inflicting considerable damage and receiving none.

Some curious expedients were adopted to baffle captors searching for treasure. Amongst the packages on board the French ship *Hermione*, captured by his Majesty's ship *Argonaut*, were some marked "verdigris," which, on being opened, were found to contain 46 ingots of silver, which sold for upwards of £12,000 sterling.

The ship *James*, Captain Warren, arrived at Montego Bay, on June 22nd, 1795, with a cargo of slaves. On the passage from Barbadoes, he was attacked, off St. Lucia, by a French schooner privateer, which he beat off after an engagement of near two hours. Neither the ship nor crew received any damage, but seven negroes died afterwards from fright, taken during the conflict.

On the 31st of August, 1795, at five a.m., the ship *Mary Ellen*, Captain Grierson, bound from Liverpool to Barbadoes, discovered a sail in chase, astern, upon which they made all the sail they could to get clear of her, but to no purpose. She came up with them at seven a.m., hauled down English colours, hoisted French National colours, and proved to be an armed French brig. She fired her bow guns into the *Mary Ellen*, then came alongside and poured a broadside of nine guns into her, which was returned "as quick as possible." Then a continual fire was kept up on both sides for six hours, within pistol shot, when the stays, topsail sheets, and steering sail booms of the brig being shot away, she dropped astern for some time; but the *Mary Ellen* being at the same time almost a wreck—the hull, masts, yards, rigging, and sails much shattered, and the rudder shot away, so that they could not work her as they wished,—the Frenchman took advantage of their situation, run alongside the ship again, and poured a whole broadside into her, which being immediately returned, he dropped astern, and hauled his wind to the northward, leaving the *Mary Ellen* in a most shattered condition. They then returned to the repairs of the ship, put her in the best condition they could, made sail, and arrived at Barbadoes on the 3rd of September. This gallant action gained Captain Grierson and his crew universal approbation. On his return to Liverpool, the captain was married at St. Thomas' Church, to "Miss Stringfellow in Park Lane."

On the 30th of October, 1795, Captain Farquhar, of the ship *Lord Stanley*, wrote from Havannah as follows :—

"On my way from St. Kitts, down the north side of St. Domingo, I fell in with a French schooner privateer, of 12 guns, and full of men. He engaged us, but in about forty-five minutes got so severely handled, as to haul off; his sails tore, main-topmast and the tricolour flag hanging heels up; his mainsail down, and the mast so wounded, that he made no sail on it while in sight. An intelligent lad in our main-top, saw them heave seven dead bodies overboard. Thank God, we have received no damage, but some shot through the sails. A pair more of them came to look for us off the Matanzas, but the ship's appearance when the sloop of 8 guns—the largest—was within hail, prevented any firing; and I had the satisfaction of protecting a Danish ship, which had been before plundered and much illtreated, by a privateer from Port-au-Paix."

On his arrival in Liverpool, Captain Farquhar brought the intelligence that the *San Lorenzo*, of 80 guns, had arrived at the Havannah from St. Domingo, bearing "the coffin, bones, and fetters of Christopher Columbus, which the Government were preparing to receive on shore for re-interment, with the highest military honours."

Captain M'Quay, of the *Stag*, arrived from the Havannah, and brought the news that five Spanish sailors on board the *Hibernian*, of Dublin, Captain Wilson, had cut the throats of the captain, second mate, and carpenter, and thrown their bodies overboard. The chief mate had three cuts across his throat, but, with two boys, he brought the ship to the Havannah. The five Spaniards took the boat, and landed at the Moro. The Governor immediately despatched a party off, who took them, and in a short time they were all executed in the presence of Captain M'Quay, their bodies quartered, and their heads hung in cages. The Governor took all the property, intending to dispose of it, and remit the amount of sales to the owners in Dublin.

Captain Hart, of the *Bolton*, of 20 guns and 30 men, left Jamaica for Liverpool, on September 22nd, 1795, in company with the *Union*, a London ship, of 20 guns and 40 men. Off Cape Corrientes, on the 27th, in a dead calm, a French ship-rigged privateer, of 18 nine-pounders and 140 men, commanded by an Irishman, named O'Brien, with the aid of 24 sweeps laid him alongside of the *Bolton*, and engaged her for an hour and-a-half. Unfortunately the *Bolton* was between the privateer and the *Union*, so that the latter vessel was unable to fire a gun, for fear of damaging her consort. A breeze springing up, the privateer out with his studding sails, and by the help of his sweeps, made off. The *Bolton* arrived in the Mersey on November 8th.

On September 25th, 1795, the *Jamaica,* a Liverpool vessel in the Government transport service, commanded by Captain James Farmer, sailed from Gibraltar with the homeward bound fleet, and on the 7th of October, was taken with the rest of the convoy. The same afternoon, six of his seamen deserted from him in a boat, but with the remainder of his crew, the captain retook the ship, made the French prize crew prisoners, and sailed for Portsmouth, where he arrived in safety.

In November, 1795, the *Wilding*, Captain Pemberton, engaged a French privateer, of 18 guns, and full of men, for two hours, when the privateer blew up, and all on board perished. Captain Pemberton died of his wounds received in the action. As a just tribute paid to his memory, a tablet, bearing the following inscription, was erected in St. James' Church, Toxteth Park, by Mr. Moses Benson, owner of the *Wilding* :—

"To the Memory of Captain George Pemberton, Commander of the ship *Wilding*, of Liverpool, who died on the 20th day of November, 1795, of the wounds He received in a most Gallant Action with a French Privateer, of superior Force, when bound on a voyage to Jamaica, In which Captain Pemberton did Honor to the Character of a British Sailor. This Monument is erected by the order of Moses Benson, In testimony of the High respect he entertained for Captain Pemberton, during many years' faithful services."

The *Mersey*, Captain Jones, sailed from St. Thomas's on the 2nd of February, 1796, in company with the *Aurora*, *Ceres*, *Diana*, and *Atlantic*, all for Lancaster. On the 5th, they fell in with a French privateer, from Charlestown, mounting 16 four-pounders, and 95 men, which boarded the *Diana*, Captain Fox. After a sanguinary fight on the deck of the *Diana*, the privateer was obliged to make off in a shattered state, having received a heavy fire from the other ships, leaving thirteen of her people on board very badly wounded, and some dead on the deck. Three of the *Diana's* brave crew were killed, and thirteen wounded. The privateer had, in fact, caught a Tartar, through believing the statement of a Danish skipper that the English ships were badly manned, having only twelve men each, which encouraged the Frenchman to board the *Diana*. The Danish ship had been lying alongside the others at St. Thomas's, and, after giving the Frenchman the above information, she bore down to see the fight.

In January, 1796, within five hours' sail of Jamaica, the *Elizabeth*, Captain Jacob Fletcher, of Liverpool, was attacked by a French privateer of superior force, which she beat off after a smart action, in which the Frenchman had about 30 men killed and wounded. In recognition of his gallant and seamanlike conduct, the underwriters of London presented Captain Fletcher with a piece of plate, of the value of 100 guineas.

The merchants and shipowners of Liverpool petitioned the Lords of the Admiralty, to send three or four frigates, or fast-sailing sloops of war, to cruise, from March to December, on the coast of Norway, from the Naze along that coast to the eastward, and across the Slieve to the coast of Jutland. Likewise, for two or three frigates, or fast-sailing sloops of war, to cruise from Duncan's Bay Head, on the north-east coast of Scotland, to the Shetland Islands, for the protection of English vessels trading to and from those parts. The merchants and shipowners of Hull also petitioned to the same effect.

The *Diana*, Captain Pince, from Africa and Demerara for Liverpool, captured the *Rosanna*, from Surinam to Amsterdam, and sent her to Falmouth.

The *Ranger*, Captain Wilson, from Africa to the West Indies, was taken to windward of Barbadoes by a French privateer, of 16 guns, after an action of two hours, and carried into Curacoa, where Captain Wilson died soon after.

Two of his men were killed in the action. The *Ranger* was recaptured and carried to Barbadoes.

"On the 16th inst.," says a letter from Madras, dated January 23rd, 1796, "arrived the schooner *Fame*, Captain Robertson, from Calcutta. On the 7th, Captain Robertson, being in the latitude of 18 North, fell in with the *Modeste* privateer, when she immediately gave chase to the *Fame*, but the latter, being a swift sailer, fortunately escaped. Captain Robertson, a few hours after, saw a large ship standing towards him, which brought him to by firing a shot athwart his forefoot, and a boat, with an officer, being sent on board the *Fame*, the strange vessel was found to be the *Sally*, Captain Brown, one of the Company's extra ships, a beautiful vessel of 600 tons, and mounting 24 guns, 9 twelve-pounders, five months from Liverpool, and had not touched at any place, or seen a single sail during the whole voyage. Captain Robertson having, to the inquiry of 'what news'? answered that the *Modeste*, a French privateer of 20 guns, was within a few hours sail of them, the crew of the *Sally* instantly gave three cheers, loudly exclaiming: 'Captain Brown; you have forty-two old privateersmen on board: only run us alongside this Frenchman, and we will shew him what can be done for the honour of Liverpool.' Captain Brown instantly complied, the guns were run out, the ship cleared for action, and in five minutes the *Sally* proceeded, under all the canvas she could crowd, in chase of the enemy; and should she be so fortunate as to fall in with the *Modeste*, we have no doubt but that, 'for the honour of Liverpool,' the *Modeste* will accompany the good ship *Sally* to her moorings off Calcutta."

On the 30th of March, 1796, the *Ferina*, with a cargo of salt, from Liverpool to Riga, a new ship belonging to "Mr. Gladstones,"* and commanded by Captain James

* Probably Mr. (afterwards Sir) John Gladstone, father of the Right Hon. W. E. Gladstone.

Handyside, was captured off the coast of Norway, by a small French privateer, of 8 guns and 28 men, and carried into Farsund.

To be "a free and accepted Mason" was not without its advantages, even in dealing with French privateers, at this period, as we find in the case of Captain May, of the *Susannah*. About half-past six in the morning of the 6th of March, 1796, he discovered two sail ahead, bearing W., distant two miles, lying to under bare poles. Perceiving they were Frenchmen, he made his ship ready for action. They proved to be two privateer schooners, one of 16 guns (nine and twelve-pounders) and 100 men; the other of 6 four-pounders and 50 men. At seven, one of them came up and attempted to board him. He poured a broadside into her, and she dropped under his stern. The other vessel bore down on his starboard quarter, and both privateers hoisted the bloody flag on the foretop-gallant mast. In this situation, with one enemy raking him fore and aft, and the other laying on his quarter, Captain May found it impossible to sustain the engagement any longer, or to make his escape, and struck, after an action of three quarters of-an-hour. Finding him to be a Freemason, the French commanders allowed Captain May to depart in an American vessel, which was present during the action. The *Susannah* was recaptured by the *Favourite* sloop, and carried into Barbadoes.

All owners and masters of vessels taking on board guns for their defence, were required to procure a license for that purpose, from the Lords of the Admiralty. The Act passed in the 24th of George III., only allowed them to have two carriage guns, not more than four-pounders, and two muskets, for every ten men. Vessels found with more on board, without a license, were subject to seizure and forfeiture. A caution to this effect was published in April, 1796.

Captain Wright, of the *Ann*, writing from Barbadoes, on the 21st of March, 1796, to his owners in Liverpool, says:—

"Yesterday, to windward of this island, I was attacked by a long schooner privateer, of 16 guns, with a great number of swivels and small arms, and full of men, which, after an engagement of two hours and a half, we beat off. He seemed much disabled, as he shewed no other sail after he left us, whilst in sight, but a piece of foresail. The disabled state of my own vessel totally prevented me following him, could I for a moment have had such an intention, the fore-mast being shot one-third through, the fore top-gallant-mast shot away, fore and main topmast stays and mizen stay gone, main yard shot through in the slings, the braces, staysail haulyards, chief part of the running rigging and sails cut to pieces, and the boats stove; though fortunately none of my small crew (only 25 in number) was killed, and but two slightly wounded. Every encomium is due to my officers and ship's crew; and too much cannot be said in praise of their bravery and good conduct on the occasion."

The *Brothers*, Captain Cudd, was captured by the *Morgan Rattler*, French privateer, of 14 guns, commanded by John Coffin Whitney, of L'Orient. The privateer, and six prizes which he had taken, were all captured by the *Suffisante* British sloop of war, 14 guns, and it was a fine sight to see the little vessel sailing into Plymouth on the 30th of June, 1796, with her seven prizes. The *Morgan Rattler* was originally a Liverpool privateer.

The *Nereus*, Captain Williams, arrived at Port-au-Prince, from Liverpool, after beating off two French privateers. In November, 1797, on her passage to St. Domingo, the *Nereus* had an engagement of two hours with a French privateer, of 16 guns, which she beat off with the loss of one man killed, and one wounded. The *Recovery*, Captain Needham, had a narrow escape from a privateer, which kept up a running fire for two hours, but found the *Recovery's* two stern chasers too heavy, and dropped the pursuit.

The *Fame*, Captain Bennett, recaptured the brig *Bernard*, laden with coffee and cotton for Messrs. Neilson & Heathcote.

The schooner *Thomas*, Captain Bosworth, of Martinique, belonging to Messrs. Thomas Gudgeon and Co., sent to cruise off Surinam against the Dutch, fell in with a Dutch fleet, from Surinam bound for Holland, without convoy. Captain Bosworth gave chase, and brought twelve of them to, but he could only man five, which he carried into St. Pierre. They were described as "amazing large ships, and five of the richest prizes taken this war." Their united cargoes consisted of 1,240,682 lbs. of coffee, 671 hogsheads of sugar, 244 bales of cotton, and a quantity of cocoa. The *Thomas* had previously captured a valuable ship, laden with coffee, from the same place. The remaining part of the fleet was taken the same day by two British frigates. Out of 72 sail, 69 were captured.

At ten p.m., on the 12th of August, 1796, on her voyage to the coast of Africa, the slave schooner *Harlequin*, Captain Topping, belonging to Messrs. T. & W. Earle, fell in with a French privateer, who came up within gunshot, and fired his bow chasers at them, which was returned with their two stern chasers, whereupon the enemy shortened sail, watching them all that night. They altered their course several times in hope of escaping, but the Frenchman kept so close that, with the advantage of night glasses, he prevented them getting clear. Captain Topping then determined to try to beat the enemy off, and got everything ready for action. At three a.m., he made her out to be a long, black brig, pierced for 16 guns, and she then made sail to run alongside of him, with the intention of boarding, but was prevented by the play of the *Harlequin's* stern chasers. A general action immediately ensued, which was kept up with equal spirit on each side for thirteen hours. During this engagement, the Frenchman attempted to board several times, but Captain Topping himself taking the helm, by a watchful steerage frustrated his design, either of boarding or doing any material damage to the *Harlequin*, which, with all sails up, made the best of her way, the enemy in hot pursuit. When a lucky shot from the slaver damaged his rigging, the Frenchman dropped astern to repair, and again followed them, and this happened repeatedly during the action. Captain Topping being informed by the crew that all their shot was expended, gave orders to fire with copper dross, which was accordingly done for some time. Finding, at length, that it was in vain to resist without proper ammunition, and hard to be shot at without a return, Captain Topping, at the request of all the crew, who, till then had behaved with the greatest bravery, ordered the colours to be struck, and the privateer took possession of the *Harlequin*. She proved to be the *L'Aventure*, of Bordeaux, of 14 guns and 90 men, commanded by Pierre Lautorine, who kept Captain Topping and his men nine days, and then put them on board a Swedish dogger, which landed them at Figueira. The *Harlequin* was recaptured by the *Sugar-Cane*, of London, and carried to Cape Coast. She afterwards traded on the Windward Coast, under the command of Captain Higgin, but, early in 1797, we hear of her again, under the command of Captain Topping, recapturing from the French a Swedish ship, which she sent to Lisbon; and in June, 1797, taking, after a running fight of an hour and-a-half, off Cape Finisterre, the Spanish brig privateer *Signora del Carma*, of 9 guns (nine-pounders), a number of brass swivels, and 70 men. On her passage to Angola, in February, 1798, the *Harlequin* beat off a French privateer, of 14 guns, and full of men, after an engagement of three hours. On the 20th of December, 1798, the *Harlequin*, bound to Africa, was taken by the *La Mouche* French privateer, of 18 guns and 200 men, of Bordeaux, which, on the 17th, had taken the *Union*, of Lancaster, Captain Thompson, after a severe action of three hours and-a-half. The Frenchman gave the *Harlequin* to Captain Thompson, with 89 English prisoners, on condition that he should proceed with them to England, and get them exchanged for the same number of French.

Captain Smerdon, of the *Bud*, writing from Jamaica to his owners in Liverpool, in October, 1796, says:—

> "We were chased on our passage from the Windward Islands by a French privateer, and as I found he sailed faster than we did, after we had got everything prepared to receive him, I hove to. On his coming up, he fired several broadsides at us, before I returned him a shot, as from the length of his guns his shot went over us, when ours would not touch him. At last, the fellow in going about missed stays, and was obliged to wear, which brought him close to us, and I immediately gave him the contents of our starboard guns, then wore round and gave him the larboard ones, which were well loaded with round and grape. We did him considerable damage, as he immediately made sail from us, and as the wind was very light, he was able to get away. I chased him about an hour, and then bore away. Some of his shot went through our sides, just above the bends, but he did us no other damage."

On the 8th of October, 1796, the ship *Backhouse*, Captain James Flanagan, on her voyage from Liverpool to Martinique, was chased by a French cutter brig, of 16 guns, full of men, from eight in the morning till nine at night, when she came up, and began to fire at them. Being dubious of her all day, Captain Flanagan had made every preparation to give her a warm reception, and when he found really what she was, he illuminated the ship with his side lanthorns to every gun; in with all his small sails, backed his main-topsail, and fired a shot, reserving his broadside till the Frenchman came alongside. Contrary to expectation, the enemy sheered off, but followed them till daylight next morning, compelling them to keep their quarters all night, for fear of attack. The next morning, the Frenchman renewed his visit, under English colours, which he hauled down three times, and fired a shot each time, as much as to say, "Strike!" Captain Flanagan, writing to his owners, says:—

"But we never minded ; we kept everything clear, and the guns pointed at him, waiting his coming close to us, so that we were sure our shot would tell, as we had none to waste, when he down English colours altogether and up French; sheered under our quarter, and gave us a broadside, which we returned directly, which staggered him very much, and I believe wounded his mainmast, as we afterwards saw him repairing. But his men, four or five, came tumbling out of their main-top in a terrible hurry; nevertheless he continued his fire about one hour successively at us, till our last two cartridges were handing out from the cabin by one of the boys, who said, 'Sir, here is the two last cartridges,' which struck me, but not with fear, when I exclaimed, 'Never mind, there is luck in those two, I hope.' I had not well spoke, when he made sail from us, on which we made sail after him, and continued our chace till he got clear; when we resumed our course, but he came down on us again on Monday, after repairing his mainmast, but we again met him; and he, seeing our intention as he supposed, he sheered off again, and we after him till he was out of sight. Commodore Blanket was kind enough to spare me seven casks of powder, which I paid him for with a stock of potatoes, etc. Our engagement, lat. 28.30 N. long 24 W."

In July, 1797, on the passage from St. Vincent to Liverpool, the *Backhouse* had an engagement with a French schooner privateer, of 16 six-pounders, and full of men. Captain Flanagan's crew consisted only of 15 men, including the officers, by whose steady and brave conduct, aided by the gallant intrepidity of three gentlemen who were passengers, he fortunately beat her off, after a warm action of two hours and-a-half, without a single man on board being hurt, but with his running rigging much cut by the enemy's shot, and his ammunition nearly all expended.

Captain John Mills, of the slave ship *Sally*, of 8 four-pounders and 23 men, writing from River Rionoones, on October 10th, 1796, tells how he baffled Monsieur Renaud:—

"I arrived at Isle de Los the 25th of August. On the 5th of September, being thick, hazy weather, the ship *Mentor*, M. Renue, from Goree, carrying 20 guns, nine and six-pounders, was observed standing in to the harbour, under English colours, and as we expected the *Manchester*, and *Falmouth*, to arrive daily from Liverpool, we took her to be one of those ships, but when she came close alongside of us, she hauled down the English and hoisted French colours, and gave us a broadside.

"The people being all in good health and spirits, we determined not to give the ship up, but immediately cut the cables, and set the sails to the best advantage, although the shot came very fast upon us. As soon as that was done, we fired a broadside, and hauled our wind to beat out of the harbour, and get clear. The second tack of the privateer, away went his main-top-gallant-yard in the slings, and then the *Sally* gained on him fast, till we got clear of the islands, when the flood tide making, he could not get out after us. We stood out to sea for six days, and lay to four more, in order to give them time to get away. I then bore away for the River Riopongos; but on making the land on the 17th, the privateer and her prize, the *Manchester*, hove in sight, almost within gunshot, about three o'clock in the afternoon.

"How to get clear of them then, I did not know, but hauled my wind to the southward, till dark, and then wore right round to the northward, in order to get in shore of her, which I luckily effected, and got into the River Rionoones safe in the morning, where I now lie 300 miles up the river. I lost three boats, two anchors, and cables, but have got another anchor since my arrival here. I have been told by Mr. Jackson, of Isle de Los, that Renue declared, if he took us, he would put us all on shore on a desolate island, with a biscuit round each of our necks, for daring to engage him. He did not destroy any property on shore, but cut the buoys from our anchors, lest we should recover them again; and had heard of our going to Riopongos, which induced him to cruize off there for us. I expect to be off the coast in all January next."

Monsieur Renaud's squadron, having taken and destroyed many British ships on the coast of Africa, two Liverpool vessels were sent out to punish him. On the 25th of December, 1797, the *Ellis*, Captain Souter, and the *St. Anne*, Captain Jones (both belonging to Liverpool); the *Dedalus* frigate, Captain Ball, and the *Hornet*, sloop-of-war, Captain Mash, arrived at Isle de Los, after sinking the *Bell*, and doing some damage to the town and fort of Goree. The *Ellis* and the *Hornet* cruised off that place, and took the *Ocean* and the *Prosperity*, two of Renaud's cruisers, and recaptured the *Quaker*, with 388 slaves on board, and also an American ship called the *President*. Early in 1798, it was stated that the two Liverpool ships had totally destroyed Renaud's squadron, with the exception only of his own ship, which managed to escape. On their passage to Africa, in the same year, the *Ellis* and the *St. Anne* recaptured the *Hannah*, from Mogadore for London. Captain Souter, writing to his owners from Barbadoes, on the 3rd of July, 1798, gives the following account of an affair with a French frigate, on the coast of Africa:—

"On the 30th of May, lying at Cape Mount, saw a large ship coming from the southward; made the signal to the *St. Anne* to get under way immediately. The *Pilgrim*, having a copy of our signals, got under way also. As soon as the *St. Anne* came up, I took my station astern of her, finding it was impossible for her to escape if I left her, thinking better to risque an action, than bear the name of a runaway; the *Pilgrim* being a long way astern shortened sail for her to come up. The French frigate (as I was afterwards informed by Captain

Mentor, who was unfortunately captured by her), was called the *Convention*, Captain Roscow, two months from Dunkirk, carries 32 guns, and 200 men, had taken four ships. She bore down on the *St. Anne* and *Ellis*, and commenced a brisk fire, which we returned with all the strength we were masters of. After a little, finding us rather stronger than she expected, she filled, I supposed with an intention to rake the *St. Anne.* I immediately filled, and shot ahead of the *St. Anne* to leeward and met her, being little more than a good musquet shot apart, and received her broadside, being well prepared to pay her for her trouble. She, finding our shot heavier than she expected, made sail in a greater hurry than she took it in. My running rigging being very much cut, she got out of reach of my guns before I could set my top gallant sails; chased him till dark, then wore round, and joined the *St. Anne*, who was a long way astern."

The *Ellis,* in 1800, recaptured the ship *La Fraternite,* Captain Rockliffe, which had been taken on her passage to Africa, by a French privateer of 22 guns and 200 men.

On the 1st of November, 1796, in N. lat. 42.30. W. long. 16, a French privateer, of 18 guns (12 nine-pounders, and 6 six-pounders), with 2 swivels, and a crew of 90 men, ranged alongside of the slave ship *Ann,* Captain Catterall, hailed her and then sheered off, but came up again the next morning under their starboard quarter, when the action began on both sides, and continued with great spirit for about an hour. The *Ann*, having sustained several broadsides from the enemy, was greatly disabled in her sails and rigging, had her boats stove, and received very considerable damage. Her crew, seeing the great superiority of the enemy, fled from their quarters, and ran below, leaving the captain and his officers alone to defend the vessel, which, as it was then impossible, obliged them to strike their colours. "We are sorry to state," says the Liverpool paper, "what may at first appear repugnant to the character of British seamen, which, we trust, will not be tarnished, when we inform our readers that the crew was in great part composed of Americans and foreigners, not interested in the preservation of that exalted name." After the lapse of a century, England is more than ever dependent on foreigners to man her merchant navy. Will they, in the next great naval war, think as much of preserving her exalted name, as of quietly running her ships into an enemy's harbour?

The *Cornwallis*, Captain Tate, from Liverpool, arrived at Jamaica in 34 days, after beating off two French privateers, full of men, both of which attempted to board her.

The *Swan,* Captain John Walls, one of the London cheese ships, gave a French privateer such a warm reception, on mere suspicion of his intentions, that he bore away before the wind, without attacking the *Swan,* or the *Apollo*, which was in company with her.

The following letter, dated November 30th, 1796, was written off Barbadoes, by Captain Ratcliff Shimmins, of the slave ship *Tarleton*, belonging to Messrs. Tarleton & Rigg, of Liverpool :—

> "On the 28th instant, about forty leagues to the eastward of Barbadoes, at daylight in the morning, we fell in with a large French schooner, of 12 guns; after giving him a broadside, he bore away. Same day at meridian, rather hazy, saw a ship to the S.W. standing to the northward, about six miles distant. As we got nearer, perceived her to be a ship of force. Did not like her appearance, but found it impossible to avoid her, and to induce him to shew colours, hauled our wind, hoisted an ensign, and fired a gun to windward. On which, he hauled up his courses, down stay sails, and fired two guns to windward, then hoisted the bloody flag at the fore-top-gallant masthead. We then saw what he was; kept our wind, which he perceiving, made after us. Finding my people all healthy and well disposed (particularly my officers), and with the assistance of the best of our slaves, prepared for action, and about two o'clock he got alongside

of us, hoisted his French ensign, and before there was any time for hailing, gave us a broadside, which we returned warmer than he wished. The action continued without ever ceasing, till five o'clock, when he sheered off, and stood to the northward. The only damage we received was in our sails and rigging; not a man hurt. She was as handsome a frigate-built ship as I have seen, mounted 20 guns, nine-pounders, on her main deck, and eight guns on her quarter deck; had much the appearance of the *Princess Royal*, formerly of Liverpool.

"My people were in high spirits, and if we could have got alongside of him again, we would, I am certain, have saved them the trouble of taking down their bloody flag, but our rigging and sails being a good deal cut, partly prevented us. He was much more shattered than us, and his hull pretty well moth eaten, his quarter was at one time so well cleared, with our eighteen-pounders, that we suppose a number of them slept under their arms. Nothing but his superior sailing saved him at last. We expended five barrels of gun-powder, and the next afternoon, about five o'clock, made the Island of Barbadoes."

"Captain Peter M'Quie, who commanded the ship *Thomas*, of Liverpool," says Brooke, "was as brave and respectable a man as ever commanded a vessel sailing out of Liverpool; and he several times signalised himself in engagements with vessels of the enemy, of superior force. The *Thomas* carried 16 guns, of heavy calibre, and sailed from Liverpool, under his command, with a crew of 78 men, and besides being adapted for the regular trade* in which she was employed, she was completely equipped as a privateer. On the 2nd of January, 1797, she encountered a French National corvette, mounting 18 guns, twelve-pounders, and four carronades of very heavy metal, and

* The slave trade; in dealing with which, we shall have more to say of this vessel.

between 200 and 300 men, and a severe engagement took place, the vessels being so close together that the enemy's bowsprit was entangled in the foreshrouds of the *Thomas*, and so remained forty-seven minutes, whilst the enemy threw on board hand grenades, stink pots, and other missiles, besides keeping up an incessant fire from the tops upon her deck. After making some attempts to board, and after sustaining considerable injury, and much loss of life amongst her crew, the French vessel was beaten off."

In a letter to his owner, Mr. Thomas Clarke, dated at sea, in lat. 38° 56′, January 5th, 1797, Captain M'Quie himself supplies full particulars of the engagement, which are as follow :—

"On the 2nd instant, in lat. 37. 40 scudding under easy sail, the man whom I had stationed at the masthead, gave the signal of a sail ahead, and bearing right down for me; I, however, judged it most prudent to keep the course I was then steering. On the vessel approaching nigher, I discovered her to be an armed vessel. Of course, I made the necessary arrangement to act on the defensive, for the preservation of the *Thomas* and cargo. The vessel having come within gun-shot of the *Thomas,* I fired a gun, and hoisted my colours, to learn who or what she was, when I found her to be a French National corvette, mounting 18 guns, twelve pounders, with four carronades, of very heavy metal, with from 200 to 300 men. The shot of one of the carronades made a hole in the side of the *Thomas's* cabin, of ten inches diameter, but no material injury accrued therefrom.

"But to commence with a detail of the whole action. The corvette steering right down upon me, I hoisted my colours, giving her a shot, which for some time was not answered. I, however, took every necessary precaution. The corvette being now abreast of me, I gave her a full broadside, which was answered by several guns, miserably conducted, and from which I received no damage. The corvette kept her course for some time, and I expected had no further intention of engaging, which was wished for on my part, being agreeable to your instructions. I therefore continued my course.

"In a few hours the Frenchman about ship, hoisted his bloody pendant for boarding, made sail, and in a short time (he sailing comparatively speaking, two feet for my one), came off my larboard quarter, and in a very peremptory manner ordered me to haul down my colours, otherwise he would grant me no quarter whatever. I hailed him through my Linguist that if he would come alongside, I would treat upon more amicable terms, but to no effect. He then, like a man, laid his ship alongside of me, with his bowsprit entangled in my fore-shrouds, when the action became general, and for forty-seven minutes remained in this position, with a determined resolution to board me on his part, and a determination on mine to resist him to the last. His bowsprit being thus entangled I with my own hands, lashed my shrouds to his main-top-mast back-stay, which, if the lashing had not been cut, I am convinced you would have had a good account of her. The men were all armed with tomahawks, etc. Her tops were all crowded with men, and from so well continued and kept up fire of small arms, I am surprised the injury was not greater. The enemy threw on board hand granadoes, stink-pots (five and twenty or thirty stink-pots and hand granadoes I have now on board), marling-spikes, boarding-pikes, and even the arm of his ship's head.

"My first broadside, I am assured, injured her masts very materially, his foretopmast and jib-boom being both shot away. In the general part of the action, my quarter guns tore him to pieces, the carnage was dreadful, sweeping every thing before them, being both well loaded with grape, ball, and canister shot, and well conducted. After the smartness of the action was over, the fellow gained on me much, and shot ahead of me like an arrow (in plain truth, I never saw a vessel sail so remarkably fast in all my life), and soon about ship, and went astern of me, I suppose to repair the injury sustained from my guns. The same evening he came several times down, I believe with an idea of finding me unprepared, and to board me, but I was ever ready to receive him, my men always resting on their guns

"The following day, the 3rd of January, the fellow hove down upon me, as if to engage, but the cowardly scoundrel never came so near as that one of my shot could tell. I therefore kept them in reserve. The whole of that evening, till four o'clock in the morning of the 4th, the fellow kept pestering me by turns. What must my feelings be when I inform you that my surgeon, Mr. James Beatty, was shot through the head, and died instantaneously at my feet, on the quarter deck, after having fired several muskets at the enemy. I had also one seaman shot through the head (John Stile); my ship's steward, Thomas Bevington received a shot through his leg, but is in fair way of recovering. My gunner's mate (James Hogat), received a shot through the arm, but will soon be of service to me again. Several others of my hearty crew received small wounds, but of no material consequence. I should be wanting in feeling was I not to observe with what firm resolution the whole of my small ship's company, consisting of forty-seven, behaved. I am particularly indebted to Mr. Gullin for the grand manner in which he worked the stern chasers and quarter guns, which much injured the enemy. Mr. Douglas, who commanded the main-deck guns, his conduct was such that will ever reflect honour upon him, as well as Mr. Crabbe.

"My boatswain behaved in a grand manner, going through the most imminent danger. I recommend him to your notice; in fact, the whole of my small crew behaved in the most gallant and heroic manner. At two o'clock a.m., observed a fleet to the S.E. From the number of vessels, I judged them to be an English fleet; four o'clock came within hail of a small sloop, who gave me to understand that the fleet was from England, under convoy of the *Sheerness,* James Cornwallis, Esq., commander, to whom I am particularly obligated, he having sent his surgeon, after finding my situation, to examine the wounds of my people, with a promise of every assistance."

This well-fought action was soon followed by another, which occurred off Monte Video, in April, in the same year, when the *Thomas* fought a Spanish vessel of war, full of troops, and mounting between 30 and 40 guns. The action commenced at eight in the morning, and lasted until half-past twelve, with scarcely any intermission, at no greater distance than musket or pistol shot. The *Thomas* suffered considerably in her hull and rigging, and in the loss of several of her brave crew; and her quarter-deck at one time took fire, in consequence of an explosion of gun-powder; yet Captain M'Quie succeeded in preserving his vessel, and beating off the enemy's ship. In the same year he captured a ship, from Buenos Ayres, laden with hides, tallow, etc.

On the 3rd of January, 1797, at 7 p.m., the ship *King Pepple*, Captain James Brown, in her passage to Barbadoes, fell in with a French brig, of 18 guns and full of men. At 7.30 they commenced a smart action, and kept up a hot fire until 10 o'clock, when the privateer ceased firing and hauled her wind about two miles from them. At 6.0 in the morning she bore down upon them again, and both ships maintained a warm fire for about four hours, when the brig hauled away, seemingly in very great confusion, her sails and rigging much shattered, and with great slaughter amongst her crew. The three last broadsides from the *King Pepple*, with double charges of grape and langrage, went home with great effect. "I could plainly see the people either drop or dodge from the fire," writes Captain Brown. "She having much the advantage in sailing, I thought it useless to follow. We expended nine barrels of gunpowder. I cannot say enough in behalf of my officers and people, no men could behave with more spirit and good conduct; fortunately had nobody hurt."

"On Tuesday last," says the *Advertiser*, of February 20th, 1797, "was launched from the building yard of Mr. Edward Grayson, a remarkable fine three-decked ship, called the *Watt*, pierced for 22 guns on her gun deck, built for Richard Walker, Esq., and intended for the Jamaica trade. The tide was very high, the launch very fine, and having a large band of military music on board, playing martial tunes, the whole proved highly gratifying to a vast concourse of spectators which had assembled on the occasion." Thus were war and commerce harmoniously blended together.

The *Fair Penitent* privateer, Captain Dunlop, captured and sent into Liverpool, the *Clara A. Norbeg*, from Lisbon for Bilbao, with salt, cocoa, sugar, etc.; also a prize laden with anchors, cables, and naval stores, from Altona; and the brig *Seahorse*, from Havre to Cadiz, laden with linens and other merchandise. The *Forbes*, of Liverpool, captured and sent into Martinico, the *Neptune*, from Surinam for Amsterdam.

In March, 1797, the *Barton*, Captain Richard Hall, having parted company with the *Agreeable*, Captain McCallum, on the passage to Barbadoes, was attacked by a heavy Spanish privateer, of 16 guns and 120 men, which was repulsed after a smart action of twenty minutes. The Spaniard kept about half a mile astern of the *Barton* all night, but on the *Agreeable* appearing in sight at daylight next morning, the privateer bore down on both ships, when a warm action was fought for an hour and forty minutes, resulting in the privateer sheering off, much damaged in her sails and rigging. The *Agreeable* had two men wounded during the engagement.

The armed brig *Swallow*, Captain John MacIver, of Liverpool, whilst cruising off Leogane, to prevent supplies being carried in there, sent into Port-au-Prince a large brig and schooner, laden in America, with French property on board. He took several other vessels, and saved the *Fame*, of Liverpool, from being captured when parted from the fleet.

The owners of the *Swallow*, were Thomas Twemlow, Peter MacIver, Samuel McDowall, Iver MacIver, of Liverpool, merchants, and the commander, John MacIver. In the Letter of Marque* granted to Captain MacIver on the 12th of July, 1796, to cruise against the French, the *Swallow* is described as of about 256 tons burthen, British built, square stern, scroll head, and two masts, mounted with 18 carriage guns, carrying shot of six pounds weight, and no swivel guns, and navigated by thirty-five men, of whom one-third were landsmen. In the commission granted to cruise against the Spaniards in January, 1797, she is said to have twenty carriage guns, carrying shot of six and twelve pounds weight, cohorns, and swivel guns, and to be navigated with 80 officers and men. The *Swallow* was not an ordinary privateer, or Letter of Marque ship, but an armed vessel, specially hired by Government, as will be seen by the following letter, addressed to Captain MacIver by Mr. Huskisson :—

"PARLIAMENT STREET, *December 7th*, 1797.

"SIR,—I am directed by Mr. Dundas to desire that, on the receipt of this letter, you will put yourself under the orders of Captain Lane, of His Majesty's ship *Acasto*, and obey such directions as you may receive from him, until the period of your arrival at St. Domingo, which you will immediately report to the Officer commanding His Majesty's Troops there, and obey such further orders as you may receive from him. You will, previously to your sailing from Portsmouth, receive on board, Colonel de Cambefort, with his lady and family, and such other officers as may be furnished with letters from me for that purpose.

"I am, Sir, Your most obedient, humble servant,

W. HUSKISSON.

"To the Officer commanding the *Swallow*, hired armed vessel."

As special interest attaches to this vessel, owing to her principal owners and commander being members of that

* By the courtesy of Messrs. D. & C. MacIver, we are enabled to give, in another part of this volume, the full text of the Letter of Marque.

famous Clan Iver, which has given to Liverpool several merchant princes, whose foresight and enterprise have contributed greatly to the prosperity of the port, we append the following account of the Clan, condensed from an elegant and interesting *brochure*, printed for private circulation :—

The MacIvers were of Scandinavian origin, but of Iver, the progenitor of the race which bears his name, nothing is known. Perhaps he landed upon the shores of Scotland from his own private ship of war, and found some dear 'Highland Mary,' who lured him from his wild sea life. Anyhow, his grandson or great-grandson, Donald MacIver, lived in the reign of Alexander II., A.D. 1219, and was the father of Iver Crom, the conqueror of Cowal. The ancestors of the race were among the chieftains, who, in 1221, fought under Alexander II. against Somerled the Younger, and were rewarded with Baronies in Argyll formed out of the lands which they had conquered. The Ordinance of King John Baliol, dated at Scone, 10th February, 1292, shews the decendants of Iver to have been settled there as an independent family, holding their lands of the Crown in the thirteenth century ; thus assigning to them as high an antiquity in that district as can, on any certain historical ground, be claimed for the name of Campbell. The MacIvers always maintained in Argyll the character of a brave and energetic Clan, and constituted a formidable division of the forces of the House of Argyll. The Chieftains of the Clan were hereditary keepers and captains of the Castle of Inverary. The Clan Iver formed part of the vanguard of the Scottish host on the fatal field of Flodden, when Archibald, Earl of Argyll, with his cousin, Sir Duncan Campbell, and all the flower of Argyll, fell valiantly fighting in front of their King. The main body of the Clan Iver exchanged their ancient patronymic for that of Campbell, and the greater number of the Ross-shire MacIvers migrated to Lewis in the seventeenth century ; from these are descended the MacIvers of Uig, and of the MacIvers traceable to Uig, the most important are the MacIvers of Liverpool. A member of the Clan settled in Uig had two sons, Iver and John; from Iver the late Rev. Wm. MacIver, of Lymm, Cheshire, was descended. John, the son of Iver, had three sons, named Iver, Peter, and William. Iver and Peter settled in Liverpool, and became prosperous merchants and shipowners, having at one time almost a monopoly of the trade between Liverpool and Glasgow. They were joined by their brother William, who, after the death of both without issue, became the head of the house. He married Anne Clark, by whom he had (besides a daughter) an only son, the Rev. Wm. MacIver, who died in 1863, leaving six sons and five daughers. Charles MacIver, the progenitor of the MacIvers of Liverpool, was the son of Captain John MacIver, brother of the great-grandfather of the Rev. Wm. MacIver. This Charles MacIver, the grandfather of the late Charles MacIver, of Calderstone, also commanded a ship. He had seven or eight sons, of whom only three grew up. The eldest of these, John, earned a very high reputation by his skill and gallantry in command of the *Swallow*, a ship of 18 guns, and in other armed vessels in the Government service. He died without issue, as also did a younger brother who served under him with the same credit, and afterwards commanded a ship. These two brothers were uncles of the late Charles MacIver, of Calderstone. One of them is referred to in the following paragraph from the *Edinburgh Advertiser*, of March 23rd, 1795:—'The armed ship in his Majesty's service, *King Grey*, commanded by the gallant Captain MacIver, was sunk by a bombshell, and part of her crew drowned.'

The only son who left issue was David, who, like the other members of this family, was an intrepid and skilful mariner, and who perished in command of a ship in the Bay of Biscay, in 1812. He married Jane, daughter of John Boyd, of Port Glasgow, who, when in command of a merchant ship, volunteered his services on board of a man-of-war of the convoy, on the occasion of an attack by a French squadron. The attack was successfully repelled, but Captain Boyd was killed in the action. The before-mentioned Captain David MacIver was the father of Messrs. David and Charles MacIver, the founders of the firm, so well known in connection with the Liverpool and Glasgow steam trade. In conjunction with Sir Samuel Cunard, Bart. (then Mr. Cunard), and Messrs. James and George Burns (now Sir George Burns, Bart.), of Glasgow, Messrs. David and Charles MacIver established the Trans-atlantic Royal Mail Steam Service, which is now known as the Cunard Line. The firm of D. and C. MacIver, of which the two brothers were the original partners, managed this Trans-atlantic service at Liverpool, from the year 1840, until their retirement from the management in 1883. David MacIver died unmarried in 1845, at the age of 38 years. Charles, the head of the house of the MacIvers of Calderstone, and representative of the family in Liverpool, died in 1885. He has left numerous descendants to hand on the honourable traditions of the race.

The *Elizabeth*, Captain Johnston, on the passage from Barbadoes, beat off a French privateer, of 14 guns and full of men, after an engagement of two hours, in which the Frenchman's fore-top-gallant mast was shot away.

The *Lord Rodney*, Captain Joseph Campbell, took, and carried to Montego Bay, a valuable Spanish prize, bound to Cadiz with cotton, coffee, cocoa, hides, etc.

The *Eliza Jane*, Captain Hayward, on her passage from Africa to St. Kitts, had an engagement of four hours with a French privateer, whom she beat off.

The *Dart*, Captain Clare, on her passage from Liverpool to Africa, took a French privateer, of 6 guns, dismantled her, and gave her to the crew; and afterwards had an engagement with another privateer, of 12 guns and 90 men, which she beat off.

The *Lucy*, Captain James, from Liverpool for Demerara, and the *Cornbrook*, beat off a French privateer, and a Spanish cutter of 14 guns.

The *Posthumous*, Captain Leigh, of Liverpool, recaptured the *Plumper*, from Jamaica for London, which had been taken by the French.

The *Hinde*, Captain Mullion, on her passage to Africa, was chased five different times by French cruisers, but escaped by superiority of sailing.

The *Molly*, Captain Tobin, from Liverpool to Africa, captured a Spanish ship of 300 tons, bound from Cadiz to the River Plate; and the *Gudgeon*, Captain Boardman, on the passage from Africa to Demerara, had an engagement with a French privateer, which she beat off. The *Ocean*, Captain Harrison, on her voyage from Liverpool to St. Domingo, had the good fortune to take the *La Victoria*, a fine Spanish brig, from Buenos Ayres to Old Spain, laden with hides, oil, and copper, valued at £10,000. The *Eagle*, Captain Wright, homeward bound from St. Croix, was sunk in an engagement with a French privateer. Several of the crew were wounded, and Captain Wright was carried prisoner to Nantz.

On the 25th of August, 1797, the *Ranger*, Captain Bell, on her passage from Liverpool and Providence for the Caicos, was taken by a French privateer, after an engagement of two hours, in which Captain Bell was killed, and the *Ranger* carried into Cape François.

The ship *Susannah*, Captain Gladstone, on her passage from Riga, in company with the *Jane*, of Workington, fell in with a French privateer, of 14 guns and full of men, which he engaged for an hour and-a-half, although he had only 8 guns and 14 men. The privateer sheered off with the loss of her mizen mast, and otherwise much disabled, and appeared to have lost a number of men. Captain Watson, in the *Jane,* with 2 guns, rendered every assistance in his power. There appeared in sight during the engagement twelve sail of merchant ships, which made their escape.

The *Isabella*, Captain Rogers, from Liverpool for Africa, was taken on November 23rd, 1797, by the *Ferret* privateer, of 16 guns and 190 men, from Bordeaux, after an action of one hour, in which the mate was killed and the captain badly wounded. The *Isabella* was carried into Bordeaux, where Captain Rogers died of his wounds.

On the 9th of October, 1797, the *Backhouse*, Captain James Hunter, from Liverpool for Africa, in lat. 45°, long. 11°, fell in with a French cutter, mounting 16 guns (twelve-pounders). Captain Hunter, writing at sea, gives the following account of the engagement :—

> "At ten in the morning commenced action within a cable's length, and continued so until one p.m. A heavy fire was kept up on both sides. About meridian, unfortunately, had our fore topmast shot over the bows, but notwithstanding our dismantled state, we kept so well directed a fire that a little after one he thought proper to sheer off, being compleatly beaten ; which plainly appeared by his not being able to take an advantage of our crippled state, having laid by us until four p.m. when he made sail and came up again, attempted to rake, and do us all the damage he could, but did not prevail as we kept firing random shot as well as him, from our after guns. But the truth is he was so much disabled that he would not risk a close action again, therefore at six o'clock he hauled away to the N. W. and left us. We have suffered greatly in our rigging and sails, not a mast or yard in the ship that has escaped his shot. It has taken until this time to repair our damages, and hope by tomorrow we shall have every thing in order again. I have great reason to be thankful we suffered no more in the ship's company, having only two killed and three wounded, one of the latter I fear will prove mortal. I was slightly wounded early in the action, but it proved no detriment to maintaining the engagement. My officers behaved with truly becoming courage, and are deserving of every notice ; indeed my ship's company all, to a very few, behaved gallantly, and would have supported me to the last in defending the ship. In the sails are 170 shot holes, besides a much greater quantity in the rigging and hull."

On November 26th, 1797, the *Elizabeth*, Captain Graham,

belonging to Messrs. Henderson and Sellar, on her passage to Africa, had a close engagement for an hour and twenty-five minutes, with a ship of 16 or 18 guns, which sailed away much damaged, and in evident confusion, to join a brig—her consort. The *Blanchard*, Captain M'Gauley, on her passage from Africa to Barbadoes, took a Spanish brig, laden with oil, skins, etc.

In December, 1797, the *Eliza*, Captain Bird, on the voyage from Africa to the West Indies, was blown up, after an engagement with a French privateer, which had struck to her. Only seven of the *Eliza's* people were saved, being picked up by the privateer. As the *Eliza* was apparently a slave ship, the catastrophe must have been a heart-rending one.

On the 30th of December, 1797, the *Lovely Lass*, Captain William Lace, belonging to Mr. Thomas Parr, and the *Agreeable*, Captain Hird, on their passage from Liverpool to Africa, had an engagement for upwards of two hours with two privateers, one a blacksided ship of 22 guns, and the other, a yellowsided ship of 18 guns, which they beat off. Captain William Lace, though engaged in the slave trade, enjoyed the friendship of Roscoe, one of the most zealous enemies of the traffic.

The ship *James*, Captain Miller, on her passage to Africa, fell in with a French privateer, and engaged her from seven o'clock in the morning till half-past eleven, when she sheered off. Four days later, the *James* was attacked by a French National brig, of 14 guns and 100 men, and taken after an action of three hours and-a-half at close quarters, in which Captain Miller and the boatswain were killed, and five men badly wounded. The *James* was nearly a wreck, the Frenchman having fought her on both sides, and raked her fore and aft. She was shortly afterwards recaptured by the *Magnanime* frigate, and carried into Cork.

On the 17th of February, 1798, the *Barbara*, Captain Dickson, belonging to Messrs. Edmund Chamley & Co., was taken, by boarding, after a hard fought engagement of sixteen hours, by the *Zemly* corvette cutter, of 14 guns and 170 men. Two men were killed, and ten wounded on board the *Barbara*, amongst the latter being the Captain, wounded in seven places. The *Barbara* was so much disabled, that it was with the greatest difficulty she was got into Guadaloupe, the action having been fought within twelve leagues of Martinique. The *Zemly* had two long 18-pounders on her forecastle. In the same month, the *Young Dick*, Captain Smith, was captured, full slaved, by a Spanish privateer, of 16 guns and 120 men, at Cape Mount.

Captain Williams, of the *Abigail*, writing at sea, on St. David's Day, 1798, says:—

> "At two p.m., saw a sail to the northward, standing towards us with all sail set; at three p.m., took in steering sails, and hauled our wind to meet him; at four p.m., got within gunshot, when he fired a gun, and hoisted National colours. We manned our guns, and gave him three cheers. She proved to be a large schooner of 14 guns, and upwards of 180 men, as I am informed by the bearer of this letter. We fought him within pistol shot, for seven hours, and kept a steady and well directed fire with grape, doubleheaded and langridge shot. He attempted boarding us three different times, but we repulsed him with small arms, and three hearty cheers. During the action we carried away the privateer's maintopmast, shot her foresail to rags, and killed and wounded a great number of her people. My officers and men behaved as Englishmen, steady and collected. In the middle of the action, they all came aft, and declared they would stick to their guns, and be true to me, for which I thanked them—they instantly returned to their quarters, and behaved like heroes. Am sorry to inform you my poor carpenter was wounded, but not dangerous; he received a shot through the leg, went below

to the doctor, and as soon as the wound was dressed, came up again and behaved like a man. Our hull and sails are much shattered, boltsprit and sprit-sail-yard severely wounded, but we will soon put all to rights again. We received 23 grape and musket shot through our fore and aft main-sail.

P.S. Depend upon it I will not give the ship away."

The *Abigail*, on her passage from Africa to Jamaica, recaptured an American vessel.

The *Governor Williamson*, Captain Kelsick, and the *Eliza*, Captain Bird, recaptured a large Portuguese brig, laden with tobacco and rum. The former vessel was subsequently lost going into New Calabar, the crew and part of the cargo being saved. The *Brothers*, Captain Thompson, was lost in Old Calabar river, and the crew saved.

In March, 1798, Mr. Gladstone (little dreaming that from his loins should spring the greatest man of the great Victorian age) presided at "a very elegant entertainment," given at Bates's Hotel, by the merchants and shipowners trading to Hamburg and Bremen, to Captain Paget, commander of his Majesty's ship the *Dart,* and his officers, in consequence of her being appointed by the Lords of the Admiralty to convoy a number of valuable ships from Liverpool to the Elbe and Weser. The Mayor, Bailiffs, "several naval characters," and many of the leading merchants, were of the party. In the following April, Mr. Gladstone became a widower.

Captain I. H. Morgan, of the brig *Betsey and Susan*, writing to his owners, from Port-au-Prince, on April 5th, 1798, says :—

"In lat. 41. long. 18 30. fell in with a large French privateer brig, which shewed 18, but mounted 16 nine and six-pounders. I made a running fight for about one hour, but finding she would come alongside me, I prepared every thing for close action, which lasted above two hours within pistol shot, when she sheered off. I was in a most shattered state, main and gib-boom shot overboard, main and forestays gone; and almost every rope in the vessel cut to pieces, several dangerous shot in my hull, and my masts and yards much wounded. She was full of men, not less, I suppose, than 100. I had three men slightly wounded, but am in hopes they will soon get the better of it. During the action, my officers, and men behaved as becomes Britons on all such like occasions—"Remarkably well."

"We have much satisfaction in stating," says the paper of May 7th, 1798, "that the mode so generally recommended on the sea coasts, is likely to form a very essential part of our voluntary armament, as we hear that Messrs. Thomas and William Earle are completely fitting up at their own expence a very formidable gunboat of 60 tons burthen, carrying 24 pounders on her bows, for the public service; which we hope will be followed by many others. This, in addition to the naval force to be stationed at the entrance of the port, will be a very important and effectual additional protection to this town and neighbourhood."

On the 9th of May, 1798, the *Hind*, Captain Mackenzie, in her passage to the West Indies, took a sloop privateer, of 4 carriage and 4 swivel guns, and 41 men.

On the 30th of May, 1798, the ship *Henry*, Captain Samuel Every, saw a sail, which tacked and stood towards them, hoisting a French ensign. All hands were called to quarters, and the privateer, which proved to be the *Caroline*, of Nantz, 14 guns and 120 men, came up and fired a broadside into the *Henry*, which was immediately answered, and the engagement continued for two hours. The *Henry* was then obliged to strike, having had one man killed, the mate and four men wounded, and her hull, sails, and rigging considerably damaged. "We were all that evening on board the privateer, and with great reluctance I came out of the old *Henry*," says Captain Every. Next day, a British frigate stood towards them, and on the Frenchman asking

Captain Every what he thought she was, he replied, "An American." The privateer then stood towards the stranger to see for himself, but finding his mistake, took to his heels. After a chase of three hours, the frigate came nearly up to the privateer, upon which the English prisoners rose on deck, and Captain Every had the satisfaction of hauling down the French colours. The *Henry* being then just in sight from the masthead, the frigate gave chase and recaptured her.

On the 14th of June, 1798, the *Maria*, Captain Martin, the *Mersey*, Captain Molyneux, and the *Africa*, Captain Smerdon, three vessels bound to Africa, captured the Spanish xebeck *Soliadad*, from Cadiz, laden with wine, brandy, iron, etc.

"Lloyd's Lists of last week," says *Billinge's Advertiser*, "announce the arrival of 192 ships from the West Indies, exclusive of those at Liverpool, Lancaster, and Whitehaven. There is not a missing ship of either fleet—a circumstance unparalleled in any former war. What a delightful view of the vigour of our navy, and of the prosperity of this country, to see our fleets of merchantmen arrived safe in the midst of war."

The *Agreeable*, Captain M'Callum, belonging to Mr. Barton, was captured on the 20th of September, 1798, by a schooner privateer, of 14 guns and 120 men, and taken into Guadaloupe. She was carried by boarding. The privateer ranging up, put upwards of 60 men into her, over her quarters, and through the cabin windows. Thirteen of the *Agreeable's* people were killed in the action, three of them passengers, and a great number of her crew were wounded. The French put 18 twelve-pounders and 210 men on board the *Agreeable*, and sent her to cruise off Barbadoes. As she was a match for any merchantman, and sailed very fast, it was feared she would do much mischief. The *Concorde* frigate and the *Amphitrite* were sent after her, the latter with orders not to return without her. She was ultimately retaken by a sloop-of-war, and carried into Tortola.

On the 27th of September, 1798, the *Bud*, of 10 guns and 30 men, Captain Robert Tyrer, bound from Liverpool to the coast of Guinea, was taken in latitude 37°, longtitude 18° N., after a very severe action of half-an-hour, by the *President Parker* privateer, of L'Orient, of 8 brass guns (thirty-six-pounders), 1 long nine-pounder, and 65 men. The *Bud* had two men killed, and two wounded. She was retaken, on the 4th of October, with the privateer, by his Majesty's ships *Flora* and *Caroline*, and sent to Lisbon.

The *Forbes*, Captain Pince, and the *Charlotte*, Captain Crow, recaptured the *Portland*, from Virginia for London.

On the 7th of October, 1798, a shot fired by one of the homeward-bound ships saluting the town, carried off the arm of Robert M'Combe, an old cooper, standing near the Old Dock Gates; tore open the breast of William Treasure, a fine young man, mate of the *William*; and killed Dennis Burns, an apprentice, standing near the bridge of the Old Dock. Treasure died in a quarter of an hour after the accident. After this, vessels were forbidden to fire in the river nearer the town, on the north side, than lineable with the North Battery, nor on the south side, than Birkenhead Point. Since that accident no vessel can salute the town under a penalty of £10 a gun, as was found by the Captain of the *Hannah,* who was fined in June, 1799.

The *George*, Captain Hackney, from Liverpool to Africa, was taken on the Coast, by the *Republican* French privateer, of 32 guns.

The *Swallow*, Captain White, escaped the same privateer, in a squall, after a running fight of an hour. The *Swallow*, having, on her passage from Liverpool to the West Indies, captured a privateer, from the Isle of France, was herself taken by the prisoners, and sent to Cayenne.

The *King Pepple*, Captain Phillips, recaptured the *Prince of Wales*, from Bristol to Africa.

The *Brooks,* Captain Williams, on her passage from Jamaica, recaptured the *Clermont*, from North Carolina, laden with tar, turpentine, etc.; and the *Mary*, Captain Erskine, on the voyage to Africa, recaptured the *Maria*, with fruit and wine, from Malaga. The *Two Brothers*, Captain Cummins, recaptured the *Astrea*, Captain Tinkman, from Liverpool for Boston.

The slave ship *King William*, of Liverpool, Theophilus Bent, master, having on board only 15 effective hands able to stand to their quarters, was, on the morning of the 11th of October, 1798, at the distance of 180 miles from Barbadoes, chased by a French privateer of 16 guns, six and four-pounders, and 170 men. Captain Bent, finding that he could not avoid fighting, brought the enemy to close action, which lasted two hours and-a-half, when the privateer, having sustained considerable damage, and an immense loss of men, sheered off, leaving the *King William* almost a wreck, having received 602 shots, and her rigging cut to pieces. She had one of the crew killed, and four wounded, besides eight male slaves below, two of them mortally.

The *Otter*, Captain Grierson, and the *Beaver*, Captain Murray, on their passage to Africa, took a brig bound to Bilbao, with naval stores.

On the 21st of October, 1798, Cape Clear, bearing E.N.E., distance 235 leagues, at one a.m., Captain Brelsford, of the ship *Mary*, 12 guns and 29 men, saw a brig to the northward, which followed close astern till daylight, when she brought the *Mary* to action, and, after a contest of one hour and twenty minutes, sheered off, with her foretop-sails a good deal dismantled. The *Mary's* principal damage was in the mainsail, with some of the running rigging cut away. In consequence of her good quarters, her crew escaped scathless. The privateer was pierced for 18 guns, and fought 12, with from 90 to 100 men; they were so numerous as to fire muskets from the bowsprit.

Captain Alexander Speers, of the slave ship, *Amelia and Eleanor*, writing from Barbadoes, on the 26th of October, 1798, to his owners, Messrs. W. Brettargh & Co., Liverpool, says:—

"On the 1st inst., I fell in with a French privateer, of 18 guns, six and nine-pounders, in lat. 3½ S. long. 22 W. He hailed from London, bound to Angola. At eleven a.m., the action commenced, and continued till half-past two p.m. Early in the action, I lost my bowsprit and foremast, close by the rigging. When he found I was disabled, he renewed the action with double vigour, and hoisted the bloody flag at his main-top-gallant-mast head, steered alongside within pistol shot, and hailed me, "*Strike, you* ——! *strike!*" which I answered with a broadside, which laid him on a creen. He then stood away to the northward, to plug up his shot holes, as I could see several men over the side. In about twenty minutes, he came alongside again, and gave me a broadside as he passed. He then stood to the southward, and got about a mile to windward, gave me a lee gun, and hauled down his bloody flag, which I answered with three to windward. I have received a deal of damage to my hull; on my starboard bow, two ports in one; several shot between wind and water. I had not one shroud left forward, but what was cut to pieces, stays, etc. I lost all head sails, and my after sails much damaged. I lost one slave, and four wounded; four of the people wounded; two are since dead of their wounds. I shall not be able to proceed from hence till January, as my hull is like a riddle."

In a letter from Barbadoes, dated December 1st, 1798, we have the following spirited description of an engagement between the ship *Barton*, Captain Cutler, which had arrived there in 51 days from Liverpool, and a French privateer:—

"In the afternoon of Monday, about 20 leagues to

windward of the Island, she discovered a sail standing to the southward, which in the close of the evening stood for her, and coming within gunshot, kept in the wake of the *Barton* most part of the night, receiving her constant fire of stern chasers, without returning a shot. At daybreak, the enemy, (which proved to be a French privateer schooner, of 18 guns, nine and six-pounders), spoke an American brig astern, and at sunrise bore down with a press of sail upon the *Barton*, who again opened her fire as soon as she came within shot, and soon after a close action commenced, which lasted two hours and an half, the schooner repeatedly attempting to board; but by the heavy and well directed fire from the ship, was prevented from getting near enough to effect their purpose, and was at last so dismantled in her rigging, that she sheered off; but having refitted, commenced a second attack at noon, with a most sanguinary design of boarding, and notwithstanding the incessant cannonading from the ship, ran plump on board, and endeavoured to throw her men into her. But well prepared to receive the enemy, the whole of the *Barton's* crew being assembled on the quarter deck, and headed by their gallant commander, who was spiritedly seconded by his passengers, an attack, sword in hand, commenced, and the enemy were driven back with considerable loss, many of them being spiked from the netting and shrouds of the ship, while by a well directed fire from the cabin guns, numbers were swept from their own deck; and great part of her rigging being cut away, she dropped astern and gave over the contest, amidst the victorious huzzas of the British tars, whose bold commander, calling from his quarter deck, defied the vanquished Republicans to return to the attack. Captain Cutler's conduct on this occasion cannot be too highly spoken of, and such was the enthusiasm of all on board the ship, that his passengers bear a proportionate share of honour, while his mates have a just claim to the approbation and applause of their merchants, whose well-known liberality is ever ready to reward the merit of every man in their employ. The second mate, and three seamen were wounded on board the *Barton.*"

The following intelligence was communicated by a gentleman, who went out passenger in the ship *Benson*, Captain Croasdale, for Jamaica:—

"At daylight in the morning of Thursday, the 6th December, 1798, St. Kitts, N.N.E. about 18 leagues, we descried two vessels on the starboard bow, which at eight we could plainly discover to be a ship and a brig, under a press of sail, standing towards us. At half after ten the latter passed us about a mile astern, under American colours, standing to the southward; and the ship, which we could by this time observe to be a vessel of force, upon our weather quarter, coming up with us fast, under English colours. At a quarter before eleven she fired a shot at us, and showed the tricoloured flag, when we in studding sails, and laying to for her coming up, prepared to give her a warm reception. At eleven the action commenced, within pistol shot of each other, and continued without intermission till about thirty minutes past twelve, when the firing ceased, and both vessels, which had been ungovernable, lay to for the purpose of refitting.

"At twenty minutes past one, the action again commenced and continued till about a quarter past two, when our opponent hauled his wind to the southward, and left us in such a crippled state in our rigging, masts, sails, as to be unable to follow. Fortunately no lives were lost in the contest, from the excellent quarters our wood hoops afforded, and the enemy chiefly aiming to disable us aloft. A neutral vessel we spoke the same evening, informed us the ship we had engaged was a National Corvette, lately from France, and that she mounted 20 nine-pounders, and was manned with 170 men. This was afterwards corroborated by a gentleman, a prisoner at that time on board, who got down to Jamaica shortly afterwards, and says that they had twelve killed, and ten wounded."

On the day following the action, the *Benson* fell in with a large schooner privateer, of 12 guns, and full of men, which she drove amongst Cape Roxen shoals, the west end of Porto Rico. On the 11th of December, she chased a

French cutter of 17 guns, which had an American ship, her prize, in company. The privateer liberated the prize, on seeing the *Benson* gaining upon her, but the wind dying away in the evening, the cutter out sweeps and escaped.

On a former passage to Jamaica, the *Benson* took a Spanish prize, valued at about £7000.

The slave sloop *Henry*, of 6 three-pounders, 2 two-pounders, and 14 men, Captain Cusack, on her passage from Africa, was chased by the Spanish packet, *St. Roselia*, of 10 eighteen and twelve-pounders and 75 men, which dropped astern after an engagement of forty minutes. Coming up again shortly after, a close engagement took place for about three-quarters of an hour, when the Spaniard sailed away. At 1 p.m. on the following day, he again came alongside, and gave the *Henry* a broadside, which was returned, and an engagement within pistol shot followed for three hours, resulting in the capture of the *Henry*, which was heavily damaged and ungovernable. After taking possession of the sloop, the Spaniards ran her on shore, about seven leagues to leeward of Cape Maize, where all the prize crew and slaves perished, except 27 negroes, who swam on shore. Captain Cusack and his crew were well treated by the Captain and officers of the *St. Roselia*, but in prison, at Havannah, the Captain was only allowed three-sixteenths of a dollar per day to live on. The Nassau paper, of February 22nd, 1799, contains the following curious intelligence:—

"On board of the sloop *Henry*, Cusack, from Africa for this port, captured by the Spanish schooner, *St. Roselia*, Captain Monasc, were two African youths of about twelve years each, one named John, the son of King George, and the other, Tom, son of King John Qua Ben, both having extensive domains on the river Gaboon. These youths their fathers had committed to the charge of Captain Cusack, to be carried to Liverpool, to be there educated. They were both taken from

Captain Cusack, to be sold as slaves, in spite of all his remonstrances, and at Havannah, he was told by a respectable Spanish merchant, that they would not be delivered up. The owner of the Spanish vessel is Francisco Maria Cuesto, who must consign his name to eternal obloquy, should he persist in refusing these unfortunate youths their freedom. A representation on this business, we have reason to expect, will be made to the government of Cuba."

On the 17th of April, 1799, Captain John Ainsworth, of the slave ship *Polly*, wrote to his owners in Liverpool, from Jamaica, as follows:—

"In lat. 3. 46 S., long. 22, W., I fell in with a large Spanish brig, and after a running engagement of four and-a-half hours, captured her, called the *St. Antonia*, from Teneriffe to Buenos Ayres. We expended 160 cannon cartridges, and upwards of 400 musquet and musquetoon cartridges. Our sails and rigging were much cut, and several of our slaves slightly wounded by a shot that went through our side under the main-chains, and broke two stanchions of the bulk head of the women's room. On the 12th of March went into Barbadoes to land the prisoners, being 22.

"I left Barbadoes on the 16th March. In the morning of the 17th, fell in with a French schooner privateer, who chaced us till 2 p.m. I then hove to for him, on which he shortened sail, and seemed consulting with his officers. Soon after he made sail, and came up under our quarter, when I gave him what guns I could get to bear. We had a number of our men slaves with small arms, which they fought very well, and killed and wounded several of the privateer's people. She then attempted to board us on the quarter, and carried away our main-sheet. At this time only small arms were fired, and if our people had been at the cabin guns we must have sunk her. In their attempt to get up the side, I took a boarding pike, and threw it at them, which went through the side of one man, into the thigh of another and they both fell. He then sheered off. I can safely say he had 20 men, or upwards, killed and wounded, his decks being full of blood. We gave them three cheers, and chaced him in our turn, but could not come up with her. She was full of men, but cannot say what force. I had one man wounded, our hull full of musquet shot, and our sails and rigging very much cut and shattered."

The *Townley*, of Liverpool, was, on the 4th of July, 1799, captured by a French privateer, of 14 guns, which took out her crew, except Mr. W. Atkinson, the chief mate, and John Overton, and put six men on board her. On the 7th, Mr. Atkinson, assisted by Overton, took an opportunity to fasten three of the Frenchmen below, and attacked the rest. The prize-master fired his pistols without effect, and fell in the conflict, when his men submitted; and on the 14th, the two Englishmen took their ship safe into Viola Sound, in Shetland.

On the examination of the French prize-master of the *Polly*, Captain Thompson, for Liverpool from Lisbon, recaptured by the *Sylph*, 18 guns, Captain Dashwood, it came out in evidence that the convoy of the fleet, a Portuguese frigate, of 44 guns and 300 men, suffered the *La Bellone*, French privateer, of 22 guns and 130 men, to come into the middle of the convoy, capture and man five sail, worth £10,000 each, and carry them off, without making any effort to retake them.

At two p.m., on the 10th of July, 1799, the ship *Planter*, of 12 nine-pounders, 6 six-pounders, and 43 men, Captain John Watts, on her passage from Virginia to Liverpool, espied a lofty ship to the southward in chase of them. Captain Watts, in a letter dated off Dover, July 15th, gives the following account of the subsequent proceedings :—

"By her appearance we were fully convinced she was an enemy, and being likewise certain we could not outsail her, at four p.m. had all ready for action, down all small sails, up courses, spread boarding nettings, etc. At half-past five p.m., we backed our main top-sail, and laid by for her, all hands giving her three cheers. She then bore down under our starboard quarter, fired one gun into us, and showed National colours. We found her to be a privateer of 22 guns, twelves, nines, and sixes, with small arms in the tops, and full of men. We immediately rounded to, and gave her a broadside, which commenced the action on both sides. The first broadside we received cut away all our halyards, top-sheets, and braces, and killed three men on the quarter-deck. We kept up a constant fire for two glasses and-a-half, when she sheered off to repair damages; and in about one glass returned to board us, with his Bloody Flag hoisted. We were all in readiness to receive him, got our broadsides to bear upon him, and poured in our langrage and grape shot with great success. A heavy fire kept up on both sides for three glasses this second time. In all, the engagement continued firing for five glasses. At last he found we would not give out, and night coming on, sheered off and stood to the south-west. His loss, no doubt, was considerable, as the last two glasses we were so nigh each other that our fire must have done great execution. My ship's company acted with a degree of courage which does credit to the Flag. I cannot help mentioning the good conduct of my passengers during the action: Mr. M'Kennon and Mr. Hodgson, with small arms, stood to their quarters with a degree of noble spirit; my two lady passengers, Mrs. Macdowall and Miss Mary Harley, kept conveying the cartridges from the magazine to the deck, and were very attentive to the wounded, both during and after the action, in dressing their wounds and administering every comfort the ship could afford; in which we were not deficient for a merchant ship.

"When he sheered off, saw him heaving dead bodies overboard in abundance. Our ship is damaged in the hull; one twelve pound shot under the starboard cat head splintered the sides much; one double-headed shot through the long boat; sails, rigging, spars, prodigiously injured. We had four killed, and eight wounded."

A letter from Whitehaven supplies the following additional particulars:—

"Mrs. Macdowall and Miss Mary Harley, who lately distinguished themselves so much in the gallant defence of the ship *Planter*, of Liverpool, against an enemy of very superior force, off Dover, are now at Whitehaven. These ladies were remarkable, not only for their solicitude and tenderness for the wounded, but also for their contempt of personal danger, serving the seamen with ammunition, and encouraging them by their presence. The merchants of that town have accordingly acknowledged their services in the handsomest manner, and have also instituted an enquiry for the parents of one William Aickin, a native of that town, who was killed in the action, after signalising himself in a most exemplary manner. Early in the conflict he received two wounds, one of which almost separated his hand from the arm, notwithstanding which, without any other assistance than the application of some styptic, and a bandage by Mrs. Macdowall and her companion, he returned to his station and continued his exertions in defence of the ship, till he fell in a manner covered with wounds, from a broadside too successfuly directed by the adversary. He was then carried below, where he expired in a few minutes after requesting Mrs. Macdowall to convey his duty to his parents, and to let them know that 'he died in a good cause.'"

The *Dick*, private ship-of-war, Isaac Duck, commander, on her passage from Liverpool to Gibraltar, beat off eight gunboats, after an action of three hours and-a-half. Later in the year, she arrived at Barbadoes with three prizes. On the 13th of October, 1799, on her passage from St. Bartholomew's to Liverpool, the *Dick* fell in with a National corvette of 22 guns, with which she came to close action, the enemy keeping up a smart fire of musketry from his tops and quarter deck for two hours, when the *Dick's* langrage and grape shot cleared her tops. Finding they had received some shot between wind and water, and having four feet water in the hold, they bore down and came within half pistol shot abaft the corvette's beam, and kept up a regular and well supported fire at her for an hour, when she made all sail possible and run ahead. They then brought their ship by the lee to plug the shot holes, and found, although they had the weather gage, a twelve-pound shot had gone through the lower part of their bends. An hour and-a-half later, they made the pumps suck, and at five p.m. (eight hours after the commencement of the action) had their rigging stopped and sails set, and all ready for engaging. The enemy laying to ahead, seeing them coming up ready for action, made sail, and run to the S.W. The behaviour of the ship's crew, many of whom were wounded, was extremely steady and valiant. Mr. Hugh Morris, the first mate, specially distinguished himself in the engagement. Nearly five years later—in the paper of August 20th, 1804—we read that the underwriters had presented Captain Duck with 200 guineas, in recognition of his good conduct and bravery, in beating off a corvette of 22 guns and 200 men, after a close action of three hours—possibly the same affair.

At the close of the year 1800, the Emperor Paul, of Russia, declared war against England, and suddenly seized on all the English vessels in Russian ports. Russian vessels in English ports were promptly seized and confiscated by way of reprisals. The *Angola*, the only Russian vessel then in the port of Liverpool, was seized, and the crew sent to prison, by Captain Hue, commander of his Majesty's ship *Actæon*, who thereby made £800 prize money. The Admiralty ordered the release of the crew. In January, 1801, Captain Hue took possession of eight Danish and Swedish vessels in the port. During the eighteenth century, a great number of girls and women entered the army and navy as soldiers, sailors, and marines, doing duty, and fighting side by side with the sterner sex, without being suspected, until some unlucky accident, or severe wound, revealed the jealously guarded secret. Sometimes the fair aspirant for military or naval honours or a violent death, was detected on the threshold of her anticipated glory. A romantic affair of this sort happened on board the *Actæon.*

"Some few weeks since," says Billinge's *Liverpool Advertiser,* of May 12th, 1800, "a young person who had the appearance of a boy, solicited to be brought on board his Majesty's ship, the *Actæon,* and continued in the ship upwards of seven weeks, performing the duty of his station the same as other boys, when by means of a letter sent to some friends it was discovered that this pretended boy was a fine girl, about 18 years of age. The loss of a mother, and neglectful father, was the only reason she would ever acknowledge for such a step. During the time she was on board the *Actæon,* she conducted herself with the greatest propriety, that no one had the least suspicion of her sex. She was sent on shore again, dressed in proper clothes, with a handsome collection made for her by the officers and ship's company."

On the 13th of January, 1800, a French brig privateer, of 14 guns, entered Torbay with the Gibraltar fleet, and remained six days. She was several times boarded and questioned what she was, but her hands, to the number of 50, being concealed, the few on deck (who spoke good English) said she had been a French privateer, but was bought by some Liverpool merchants. On the sailing of the convoy, she also got under weigh, and in the night would doubtless have captured the most valuable; but a signal was made, which she being unable to answer, of course, was suspected, boarded, and taken possession of by the *Namur.*

The underwriters presented Captain James Sturrock and the crew of the ship *Pursuit,* five per cent. on the value of the ship and cargo, for their gallant defence against a French privateer, of considerable force, on the 5th of January, 1800.

On the 6th of October, 1800, the *Dick*, Guineaman, mounting 20 guns (four and six-pounders), and 42 men and boys, Captain W. Grahme, sailed from Liverpool for the Coast of Africa. On the 15th of October, she had the misfortune to fall in with the *La Grande Decide*, a famous French privateer corvette, mounting 22 guns (nine and twelve-pounders) on one deck, and 176 men. After as desperate an action as ever was fought, lasting about seven hours and-a-half, the *Dick*, reduced to a mere wreck, was forced to strike to superior force. The brave Captain Grahme and ten of his crew were severely wounded. The captain died six days afterwards on board the privateer. The first lieutenant of the corvette was killed, and 39 of the crew killed and wounded in the action. Soon after, the *Clyde*, of 44 guns, Captain Cunningham, hove in sight, took possession of the *Dick*, and carried her into Plymouth, while the *Fisgard*, of 48 guns, chased the privateer. Captain Cunningham took every care of the wounded men, entering them as supernumeraries, and by that means, procuring their admission into the royal naval hospital. A letter received in Liverpool, probably from one of the officers of the *Dick*, gives the following account of the engagement:—

"An action commenced a few minutes past one o'clock at noon, which was most gallantly defended on both sides within pistol shot. About five o'clock the *Dick's* standing and running rigging, bracings, and bowlines were cut to pieces; sails all in rags, topmasts gone, lower masts crippled, and several shots betwixt wind and water. It was about this time that an unfortunate canister shot struck poor Grahme and took away all the upper part of his skull; in this situation he was carried below. To revenge his death, which his brave crew anticipated, and for the honour of the British ensign, one of the brave tars nailed the *Dick's* colours to the stump of the mizen mast, and they one and all were determined to fight the vessel as long as she could swim; and without dread or fear the chief mate and crew fought on till near eight o'clock, having at that time their noble captain and ten men wounded, their ammunition expended, every gun dismounted, spars and rigging shot away, 3 feet 10 inches water in the pump well, both pumps going, vessel expected to go down, and the enemy upon their quarter in the act of boarding, when Captain Grahme advised them, to prevent every man from being put to the sword, to strike their colours. He delivered up the vessel in the most courageous manner; and even had the presence of mind to desire the third mate to fling his rifle-piece, pistols, sword, &c., overboard, saying no other man should ever use them. He manfully walked overboard his own vessel into the enemy's boat, refusing aid or assistance, saying to his men, 'My brave fellows, you have done your duty like Britons,' adding (meaning his own vessel) 'Poor *Dick* thou hast done thy duty likewise, but obliged to strike to superior force—I only wish thy guns had been heavier metal.'

"The French first lieutenant was killed; the enemy had also 27 killed and wounded, and several of her crew died after the action. She was much hurt in her masts and hull, and several holes in the side, which they were obliged to plug up with lead. On Grahme's arrival on board the *La Grande Decide*, he was allowed a cot in the Captain's cabin, who behaved to him like a brother. The French doctor attended him night and day, his own chief mate was always with him, and his crew allowed frequently to see him. He was insensible after the first twenty-four hours, and on the 21st of October, about three o'clock in the afternoon, he departed this life, universally respected by all who knew him. He fell like a hero and a British sailor, fighting under the influence and for the honor of his country's proud ensign! God rest his soul in peace and happiness. He was launched into the deep same evening, sewed up in his cot, in as decent a manner, as the situation would admit of."

Captain Samuel A. Whitney, of the ship *Hiram*, writing to his owners in Liverpool, from Fort Royal, Martinique, on the 22nd of October, 1800, gives the following account of events in real life, that have a foretaste of Stevensonian romance about them :—

"I have a very unpleasant account to give you of the *Hiram*, which, after being twice taken and retaken, arrived here the 13th inst. after being one hundred and two days at sea; the circumstances are these: On the 13th September, being in long. 55. and lat. 30. I was overtaken by a French sloop of war brig, called the *Curieuse*, Captain Ratlett, from Cayenne, on a cruise of two months, and then to France, who after an examination of my papers, pronounced the greater part of my property to be English. They then took out all my people, (except my brother, one green hand, and a boy of 12 years of age), .and put on board two officers and eight men, and ordered us for Cayenne, and after keeping us company for two days, and robbing us of a lower yard, a cask of water, a ship glass, and sundry small matters, they left us. I, on first discovering her to be French, went below, loaded my pistols, and hid them away in a crate of ware, which if I had not done I should have lost them, for no less than three different times was my trunk searched, my brother's chest and the cabin all over, and were as cautious as though they read my determination in my face. The officers would not allow the men to go off deck at any time, and they eat, drank and slept on deck themselves, never suffering but one at a time to go off deck; therefore, I found I had no other chance but to engage them openly by daylight. I directed my brother to have a couple of handspikes in readiness, and when he saw me begin, to come to my assistance. Therefore, at four o'clock on the afternoon of the fourth day after being taken, I secured my pistols in my waistbands, went on deck, and found the Prize-master asleep on the weather hen-coop, his mate at the wheel, and their people on different parts of the main deck, my brother and man on the lee side of the windlass. Under the circumstances I made the attempt, by first knocking down the mate at the wheel. The prize-master jumped up so quick that I could get but a very slight stroke at him. He then drew his dirk upon me, but I closed in with him, sallied him out to the quarter rail and hove him overboard, but he caught by the main sheet, which prevented his going into the water. By this time I had the remaining eight upon me, two of whom I knocked backwards off the quarter deck; by this time my people got aft with handspikes, and played their parts so well that I was soon at liberty again. I then drew a pistol and shot a black fellow in the head, who was coming to me with a broad axe uplifted, the ball cut him into the skull bone and then glanced, but it stunned him and amazed all the rest, who had no suspicion of my having pistols. By this time the mate whom I first knocked down, had recovered and got a loaded pistol out of his trunk, and, apparently, fired it directly in my man's face, but the ball missed him. The prize-master got on board again and stabbed my brother in the side, but not so bad as to oblige him to give out until we had got the day. In this situation we had it pell mell for about a quarter of an hour, when at last we got them a running, and followed them so close, knocking down the hindermost as we came up with them, until part made their escape below. The rest then began to cry for mercy, which we granted on their delivering up their arms, which consisted of a discharged horseman's pistol, a midshipman's dirk, a broad axe, a handsaw, and two empty junck bottles. We then marched them all aft into the cabin and brought them up one at a time, and after examining for knives, etc., we confined them down forward. By this time it was quite dark, and my brother was obliged to give out, and lay in extreme pain for forty-eight hours, expecting every moment to be his last, but he afterwards recovered astonishingly, and was soon able to keep his watch. My man got so drunk that I could not keep him awake at night, so that there was only my little boy and I to work the ship, watch the French, and attend my brother. I kept a French lad upon deck, the only one that was not wounded, and kept him at the wheel all night. The weather was extremely fine and the Frenchmen quite peaceable, so that I met with little difficulty. Thus we kept possession of her for ten days, when we had reached within two or three days sail of Savannah, being in the long. of 75. On the 27th September, was again overtaken by a French privateer, from Guadaloupe, who, without any ceremony of examining papers (only to find out the contents of my packages) came immediately on board, broke open the hatches, and filled the deck with bales, trunks, cases, etc., and after examining for the most valuable goods, sent them on board the privateer. As her cruise was nearly at an end, having sent off their men, they hove overboard all their empty water casks and lumber of all kinds, and filled themselves as full as an egg out of us, not leaving room for their people to sleep below. They were two days at work upon us. They then took out my brother, man and boy, (leaving me on board) and all the former French crew, except four men, and put on board eleven more of their own men, and after plundering me of part of my cloaths, brass hanging compass, carpenter's tools, spare cordage, deep-sea line, and many other like stores, they left us, ordering us for Guadaloupe; and after being forty-six days longer in their hands, we were taken by his Majesty's ship *Unite*, and sent into Martinique."

Billinge's *Liverpool Advertiser*, of February 9th, 1801, records the death of Captain William Hutchinson, in the following terms:—

"On Saturday, universally lamented, Mr. William Hutchinson, aged 85. Of him, it may be truly said, that he steered through the voyage of life, under the direction of the great Captain of our Salvation, without ever deviating a point from moral rectitude; he was a friend to the fatherless, and made the widow's heart sing for joy; to his indefatigable exertions, we are indebted in a great measure, for the superior advantages we enjoy as a commercial port, and the instituting of the laudable society for the relief of the widows and families of Masters of vessels, will ever make his remembrance be held dear, by that useful body of people."

On the 5th of March, 1801, the *Bolton* (Letter of Marque),

280 tons, 20 guns, and 70 men, Captain J. Watson, on her passage from Demerara for Liverpool, engaged for an hour most gallantly, a large French privateer, of 26 guns and 260 men, called *La Gironde*, of Bordeaux, which ran her on board, and she was obliged to strike to a superior force. The Frenchman had a great advantage in the action, owing to the number of men he was able to keep at the musketry, to the great annoyance of the *Bolton's* people. Captain Watson and five of his crew were wounded, and two passengers were killed. Both ships were considerably damaged, but the French had none killed or wounded. In addition to a valuable cargo of sugar, coffee, cotton, elephant teeth, etc., which was plundered by the privateer, the *Bolton* had a very fine tiger on board, and a large collection of birds, monkeys, etc. She was retaken on the passage to Bordeaux, by the *Leda*, of 38 guns, and sent to Plymouth.

The *General Keppel* privateer, Captain James Finlayson, recaptured an American ship, and took another from Cadiz. On the 14th of June, 1801, he had an action with the *La Mouche* privateer, of 22 guns and 250 men, with her prize, the *Hiram*, of Liverpool, in company. The latter, manned with 60 Frenchmen, soon sheered off, as did the privateer, after a warm engagement, in which she had her second captain killed, several men wounded, and her hull, masts, sails, and rigging considerably damaged. The *General Keppel*, was captured in the Rio de la Plata, on November 20th, 1801, by a Spanish frigate of 44 guns, after a severe engagement of three hours.

On the 14th of August, 1803, the ship *Juno*, of 18 six-pounders, and 44 men and boys, Captain Affleck, was taken 70 leagues from Wilmington, after an action of two hours, by the French frigate *Poursuivant*, mounting 22 French twenty-four-pounders, 12 nines, and 350 men. The *Juno* had two men killed, the mate wounded, and her hull, masts, sails, and rigging, very much shattered. When she struck, the Frenchmen gave her three cheers, and Captain Affleck, when he stepped on board the frigate, was very kindly received by the French commander, who returned him his sword, and let him have part of his own cabin, expressing surprise that he had fought so long against such a superior force. "But," says Captain Affleck, in a letter to his owners, "knowing I had a set of the bravest fellows that ever swam salt water, I was determined to defend the ship to the last extremity." The French captain, finding the *Juno* too much damaged to proceed to France, made for Charleston, but the American government refusing to allow the frigate and her prize to enter the port, the Frenchman took the cargo out of the *Juno* and burnt her. "This brilliant action," observes Billinge's *Advertiser*, "reflects immortal honour on Captain Affleck and his brave crew, and will no doubt meet that admiration and applause, we conceive they are so well entitled to."

The Underwriters, of London, presented Captain Affleck with a valuable bowl bearing the following inscription :—

> "The ship *Juno*, of Liverpool, commanded by Captain Lutwig Affleck, of 18 guns, six-pounders, and 44 men, being captured off the coast of North America, on the 14th August, 1803, by the French frigate *Poursuivant*, of 22 twenty-four-pounders, 12 nine-pounders, and 350 men, after a well fought battle, the Underwriters of London present Capt. Affleck with this token of their estimation of his skill and bravery, in maintaining a long and gallant action, with a ship of such superior force."

The following communication was sent to Captain Affleck, by Mr. (afterwards Sir John) Gladstone, on behalf of the Liverpool Underwriters :—

> "UNDERWRITERS' ROOMS, LIVERPOOL, *15th August*, 1805,
>
> "Sir,—By the direction of the Underwriters of Liverpool, and with particular satisfaction to myself, I beg leave to

enclose a Bill on London for the sum of £120. It is their desire this money may be employed in the purchase of a suitable piece of plate, of which they request your acceptance, as a mark and testimony of the high sense they entertain of the high skill and gallant conduct displayed by you, when Commanding the ship *Juno*, of this port, armed with 18 six-pounders, and defended by a crew consisting of 44 men and boys, in the action which you maintained for two hours, off the Coast of America, against the French National frigate *La Poursuevante*, mounting 22 long twenty-four-pounders, 12 long nine-pounders, and 350 men, tho' at last compelled to submit to superior force. I have the honour to be, Sir, your most obedient, J. GLADSTONE, Chairman of the Underwriters' Committee.

"Ten Guineas of this sum was subscribed by Messrs. Davies, Dale and Co.

"To Captain Lutwidge Affleck, late of the ship *Juno*, at Greenock."

Messrs. Davies, Dale & Co., were, no doubt, the owners of the *Juno*. Captain Affleck acknowledged the honour in the following terms:—

"GREENOCK, *22nd August*, 1805.

"DEAR SIR,—Your esteemed favour of the 15th curt., enclosing a Bill on London per £120, I received by last post, and beg you will assure the Committee of Underwriters, at Liverpool, that I cannot find words to express the gratitude I feel, for so great a mark of their regard.

"I have ever considered it my duty to defend the property of others, entrusted to my care against the enemy, as long as there was any prospect of advantage to be gained by resistance. Yet, I cannot help feeling much gratified by the high opinion which so respectable a body of men, have been pleased to express of my conduct in the defence of the *Juno*. I have the honor to be, Sir, Your obliged humble servant,

LUTWIDGE AFFLECK."

On the 20th of October, 1805, the ship *Harmony*, of Greenock, Captain Affleck, fell in with a brig, supposed to be Spanish, but showing no colours, off the island of Teneriffe. She mounted 16 guns, and appeared to have 170 men on board. Captain Affleck was determined to attack and take her if possible, but it being light winds, could not come up with her. He, however, manned three boats, with himself and 24 men in one, and 22 men in the other two. They pulled off, and soon got alongside the brig, when a heavy fire of musketry took place on both sides, but after a severe conflict, the boats returned without success, Captain Affleck, with eight men, having been killed, and 18 wounded in the contest. The report of this affair leaves us in doubt whether the commander was Captain Affleck, formerly of the *Juno*, or a relation—possibly a brother.

In October, 1803, the *Ainsley*, Captain Every, brought into the Mersey, a prize called the *L'Aimable Lucile*, a large French Indiaman, from the Isle of France to Bordeaux, valued at £80,000.

The *Margaret and Eliza*, Captain Barry, outward bound Guineaman, captured, on the 5th of September, 1803, the ship, *Maria Alletta*, from Batavia for Amsterdam, valued at £45,000.

The *Sarah*, Captain Sellers, and the *Ann Parr*, Captain Baldwin, took the French ship *City of Lyons*, 400 tons, from the Isle of France for Bordeaux, laden with coffee, pepper, indigo, etc., valued at about £26,000.

The peace of Amiens, which had caused great rejoicings in Liverpool, proved to be nothing more than a truce, or short breathing-time between two desperate conflicts. A series of military victories, culminating in the triumph of Marengo, had placed the continent of Europe at the feet of France, or rather under the heel of Bonaparte. The naval conquests of the 1st of June, of St. Vincent, Camperdown, and the Nile, with innumerable smaller victories, had made

Great Britain mistress of the ocean, had placed the colonies of France at her mercy, and inflicted upon the military and commercial navies of France, Holland, and Spain, in the first ten years of the war, the loss of 81 line of battle ships, 187 frigates, 248 smaller vessels of war, 934 privateers, and 5,453 merchant vessels. Thus the commerce of Europe was lost to Havre, Bordeaux, Cadiz, Rotterdam, and Amsterdam, and ultimately to Hamburg and Bremen, and concentrated in London, Liverpool, Bristol, Hull, the Clyde, and the other ports of the British empire. The war, which had ruined the allies of both, had left the principals in possession of immense strength, unbroken courage, and with additional causes of irritation and jealousy. Hence grounds of difference sprang up almost immediately, and after a stormy scene between Bonaparte and the English ambassador at Paris, both parties began to prepare for war. Bonaparte collected an army at Boulogne for the invasion of England. The threat was received with shouts of defiance by the people of Great Britain. Letters of Marque and Reprisals were issued on the 16th of May, 1803, and the King's Declaration was dated May 18th. The armed vessels of England scoured the channels, sinking every gunboat that ventured to leave Boulogne, and even attacking them under the batteries; whilst hundreds of thousands of volunteers rushed forward to defend their country. Liverpool, true to its fighting instincts and its renown on the sea, did not yield to any town in the empire in the energy and efficiency of its patriotic preparations against the invader. As this is not a military history of Liverpool, it would take too long to relate in detail how nobly the merchants and people of "the good old town" did their duty at this great crisis in the world's history—when the liberty of the nations hung on the attitude and pluck of Britons. It is, however, due to their patriotism to state briefly the result of their efforts. Mr. John Bolton, of Duke

Street, one of the wealthiest merchants, raised and clothed, at his own expense, a regiment of volunteer infantry, of which he became Colonel. All the boatmen of the river Mersey, who were secured from impressment, came forward and offered to assist in working the great guns of the forts, and were formed into a regiment of artillery, under the command of Peter Whitfield Brancker, Esq. Two regiments of infantry were formed, one commanded by Lieut.-Colonel Williams, the other by Lieut.-Colonel Wm. Earle. There was also a Rifle Corps and a Custom House Corps, and the Liverpool Light Horse. Lieut.-Colonel Hollinshead raised and clothed a company of pioneers at his collieries to serve with Lieut.-Colonel Williams's regiment, of which he was second in command, and Mr. Ford North presented the same regiment with two brass guns, completely equipped for service. The Corporation subscribed £2,000 from their own funds, and £1,000 from the funds of the Docks. The drilling of the regiments was incessant, the enthusiasm great, the people being of one mind—to save the country or nobly fall in its defence. At a review of the Liverpool Volunteers, on the sixty-seventh birthday of George the Third, the number of officers and men who appeared in the field was as follows: 1 Colonel, 6 Lieut.-Colonels, 8 Majors, 54 Captains, 111 Subalterns, 221 Sergeants, 152 Musicians, and 3,313 rank and file. From this crude outline of the defensive attitude of the old "Dicky Sams," it would appear that a propensity for privateering and slave trading in a community is not incompatible with self-sacrifice and an exalted patriotism—or, at any rate, was not in old Liverpool.

In 1803, Messrs. J. & H. Parry, merchants, presented a piece of plate, with the following inscription, to Captain Thos. Nicholson:—

"Presented by John Parry and Henry Parry, of Liverpool, Merchants, owners of the *Anna and Ellen*, private ship of war, to her commander, Thomas Nicholson, in grateful testimony of his unwearied exertions for their interest, in his able and active conduct as an officer, in capturing two valuable French merchantmen, and of his judicious management in bringing them safe into port."

The paper of January 24th, 1804, warns "persons liable to the impress service and all others," against two practical jokers, or "extraordinary informers," as they are called, one a tobacco manufacturer, and the other a clerk in a salt warehouse, who sported with the feelings of their acquaintance, by causing them to be seized and carried to the rendezvous of the press-gang, where, with considerable property upon them, they were detained several hours among a company with which few would associate by choice.

Captain Richard Sherrat, of the ship *Caldicot Castle*, captured by the French, gives the following account of the affair, in a letter dated Barbadoes, 18th April, 1804:—

"I sailed from Demerara on the 27th February, and on the 8th March, being then about 200 miles to the eastward of Guadaloupe, I fell in with two French privateers, a ship and a schooner, who came alongside about eight in the evening, and opened a very heavy fire upon us, which we returned, and in about fifteen minutes disabled the schooner, when she sheered off. We continued the action with the ship until about twenty minutes past nine, when she sheered off also, but continued in sight during the remainder of the night, in which time we were employed repairing damages and getting the ship in a proper state of defence. About half-past six next morning the ship came within pistol shot, and opened a tremendous fire of great guns and small arms, which we returned, and continued in action for about fifteen minutes, when finding our sails and rigging cut to pieces, the ship very much hulled, several shot having gone through her, our wheel shot to pieces, two others and myself wounded, one (my second mate) mortally, we were obliged to strike. The ship proved to be *Le Grand Decide*, of Guadaloupe, mounting 20 nine-pounders and 2 brass twelves, with 160 men. I have been here these two days, and will go home either in the *Venerable* or *Barbadoes*, with the protest regularly done. I am nearly well of my wound; it was a musket ball which entered my right hip, and came out near my backbone. I have nothing more to inform you of, but hope by the above account there will not be any blame attached to either my men or me, as they all to a man behaved in a very gallant manner."

The following letter, relating to the sale of East India prize goods brought into Liverpool, was received by the Mayor :—

"LONDON, *6th June*, 1804.

"I have this day had a final hearing of the Lords of Trade, on the subject of the Petition of the Mayor and Corporation and Merchants of Liverpool for leave to sell at that port, the cargoes of those East India Ships which have been taken in there, and I have their Lordships' authority to acquaint you, that the Petition *has been granted*, and that there is no objection to the Owners proceeding to advertise and dispose of the said Cargoes. I beg the favour of your making this generally known.

"I have the honour to remain, Dear Sir, Very truly yours,

S. COCK.

"To his Worshipful the Mayor of Liverpool."

The paper of July 2nd contained an advertisement stating that in consequence of an unexpected opposition to the Bill then pending in Parliament, for permitting the East India prize cargoes to be sold in Liverpool, the sale was unavoidably postponed.

On the 4th of August, 1804, the ship *William Heathcote*, of Liverpool, Captain Thomas Phillips, about 600 tons burthen, carrying 20 guns and 30 men, from Demerara bound to Liverpool, with a valuable cargo, consisting of 1,400 bales of cotton, and 125 casks of sugar, said to be worth £80,000, had the misfortune to encounter in the Irish Channel, the French dogger privateer, *General Augereau*, of 12 guns and 192 men. After a very severe action of half-an-hour, the Frenchmen made use of their only superiority, which consisted in their number, when by running their ship alongside, they carried the *William Heathcote* by boarding with nearly their whole force. Captain Phillips was killed after the Frenchmen got on board; they rushed upon him and stabbed him in many places, and he died, encouraging the mate to fight the ship as long as possible, but the mate was soon after mortally wounded. The captain's son, a lad about twelve years old, behaved nobly when the French were boarding. He was, however, mortally wounded, and thrown overboard before he expired. A passenger and a seaman were also killed, and another passenger and seven seamen, besides the mate, were badly wounded. According to one account, the owner's son was one of the killed. The ship's sails and rigging were much damaged, especially in the after part of the vessel. The privateer suffered considerably in her hull, and had several men killed, and the captain and five men wounded. After the exchange of prisoners, the Frenchman bore away for a Spanish port, and had arrived near St. Andero, when the *Nautilus* sloop, of 18 guns, Captain Aldham, recaptured her, and carried her into Plymouth. The *General Augereau* was taken on the 13th of February, 1805, by H.M.S. *Topaze*, Captain Lake. On the 2nd of October, 1804, the *Cockatrice*, of 18 guns, escorted the *William Heathcote* to Liverpool.* The latter's

* The following letter was written by Mr. Bamber, one of the officers of his Majesty's ship *Nautilus*, Captain Aldham, and prize-master of the *William Heathcote*, to Captain Moses Joynson, of Liverpool:—

"HIS MAJESTY'S SHIP *Nautilus*, 24th *August*, 1804.

"DEAR SIR,—I have to congratulate you upon the recapture of the *William Heathcote*, by his Majesty's ship the *Nautilus* on the 9th inst., which ship I am now master of. Knowing your great partiality to the employ of Messrs. Neilson & Heathcote, and your prepossession in favour of the ship, determined me to inform you of the recapture of her, as I know you would be very happy to be the first to congratulate Messrs. Neilson & Co. on this subject. I am likewise proud in saying, average was settled for the recapture, by the agents for the *Nautilus* and the underwriters, at £36,000 for the cargo, and £8,000 for the hull, stores, guns, and tackle.

Captain Leavy, of the ship *Britannia*, writing to his owners, Messrs. France, Fletcher & Co., from Jamaica on September 1st, 1804, gives the following description of a well-fought battle between the *Britannia* and the *General Erneuf*, French privateer:—

"On 3rd July, we fell in with a French corvette of 22 guns, in lat. 41, long. 13, who ran from us; on the 5th, fell in with the same corvette, who at first seemed determined to attack us, but desisted on our chacing, and again run away. After this, nothing particular occurred until Sunday the 5th of August, at 8 a.m., in lat. 17, Antigua, W. 200 miles, saw a strange sail which we soon perceived to be a cruizer, by making all sail after us, which we took no notice of, not wishing to lose a good breeze which we had not been favoured with for several days. This encouraged the robber to make boldly for us, our guns being then in and our ports down, he thought we should be a good prize for him. At half-past 3 p.m. found him coming up fast, took in our steering sails, prepared for action, and hauled our wind towards him. At four, he hoisted the Tri-coloured Flag, and gave us a salute with a 24 lb. shot, in ten

that she was by no means given away, as they gallantly defended her till the last, against a superior in number, in which Captain Phillips, Mr. Shepley, and two men were killed. Mr. Fraser, a passenger, Mr. Kewley, the mate, and several men were badly wounded. Mr. F. was fortunate enough in being left on board the *William Heathcote*, with three of the wounded men, and the major part of the ship's company; and I am happy to say they are all in a fair way of recovery. The French officers taken on board the *William Heathcote* were loud in their plaudits of the bravery of the Captain, Mr. Shepley, Mr. Fraser, Mr. Kewley, and the ship's company. Mr. Kewley, they say, killed three men with his own sword. He is on board the privateer, which I am very sorry for, as his brother was a most particular friend of mine. Mr. F. received two musquet balls, which was nearly affecting his life; one our surgeon extracted since he has been on board this ship. I am happy to say he is nearly well of his wounds. The privateer was in sight at the time we recaptured the *William Heathcote*, and did not make sail until she saw her haul down her colours, and we could not go in chase of her. I hope I shall have the pleasure of seeing you at Plymouth to take the *William Heathcote* to Liverpool. She is very much shattered in her hull, and her deck and ropes are steeped with blood, much to the honour of them that fell in defence of her, and those poor fellows who are wounded, and in fact, her whole crew."

minutes he grappled our quarter, when a brisk and well directed fire commenced on both sides, with great guns and small arms. He continued fast to the *Britannia* 40 minutes, during which time he twice attempted to board, and was beat back with great slaughter. All this time we could only get our stern and quarter guns to bear, which cleared his rigging, and shot away his boarding booms and grapplings.

"After he found he could not succeed in boarding, he attempted to haul off, and get on board his tacks, which enabled the *Britannia* to get her side to bear, and in ten minutes the enemy was a complete wreck, his main-mast shot away close to the cap, his fore-top sail sheets and fore-top gallant ditto all cut, his sails in tatters, his side drove in, and his fire compleatly silenced, his tops and decks that were before full of men, scarcely one to be seen. With the remains of his shattered foresail, mainsail, and mizen, he kept the wind. Perceiving night coming on, and the *Britannia's* fore, and fore-topsail braces, mizen-boom, mizen-stay, and mizen-topsail all shot away, it was some time before the ship could be brought upon a wind. Immediately after getting braces reeved, we gave chase and passed him to leeward, having to make a tack for want of after-sail, the ship was long in stays, and before we could come up with him, it being dark and squally, we lost sight of him. We then hove to, in full expectation of falling in with him at daylight, expecting from his crippled state he must run before the wind, but was much disappointed in not seeing him, and I much regret, that after fortune had so far favoured us, we had not daylight to take possession of him, who intended to make a prey of us. My people were in high spirits, and fought like English seamen. I am confident the proudest of Frenchmen with equal numbers must have humbled to them. I met with a great loss from the enemy's first fire, having my boatswain, carpenter, and two of my best men stationed with me to work the ship, wounded; four of my people run from me at Madeira, and two sick, which made some of the guns to be weakly manned. My passengers volunteered their services, and am truly sorry to say, one of them lost his life, but on the whole our loss was not so great as might be expected.

"The *Britannia* has suffered much in her rigging and sails, most of our lower rigging dreadfully cut, as well as almost every running rope in the ship. The masts are full of small shot. We were obliged to bend an entire fresh set of sails, but am happy to say the ship is not much injured in her hull. The day after our engagement, we were spoke by Commodore Hood in his Majesty's ship *Centaur*, who very politely sent his surgeon on board to examine the wounded, and also supplied us with medicines we were in want of.

"On the 9th, we fell in with a schooner under Danish colours, who informed us that the privateer we had engaged was the *General Erneuf*, carrying 4 long brass twenty-four-pounders, and 12 eighteen-pound carronades, and had on board, when she sailed from Basseterre, 170 to 190 men. From the description given, she must be the same vessel, but, to all appearance, must have had more men on board, they being as throng as they could well stand.

"List of killed and wounded:—R. Rishton,* passenger, died by a shot in his side. Captain Leavy, D. M'Call, J. Newman, John Grey, and Edward Audley, wounded, but fast recovering."

In February, 1806, while Captain Leavy was on shore, the *Britannia* blew up in Cork harbour. A lady passenger and others on board perished.

On the 9th of October, 1804, the *Barbadoes*, Captain Lewis, on her passage from Barbadoes to Liverpool, beat off a French privateer of 14 guns and full of men, after a smart action of two hours, in which the *Barbadoes* had two men severely wounded, one of whom afterwards died of his wounds. During the engagement, the privateer hoisted the bloody flag and attempted to board.

* "Richard Rishton, aged 21, son of the widow Rishton, at the Waggon-and-Horses public-house, in Blackburn."

The sloop *Dick*, of Chester, laden with slate, from Carnarvon to Portsmouth, was captured, near the Land's End, by a French privateer, and retaken by the mate alone, "a fine daring Welshman," who was left on board the sloop with four of the captors. The Frenchmen, being frightened at a gale of wind, the mate, who evidently had some of the polite and persuasive qualities of Davy Llewelyn, told them he was well acquainted with a port under their lee, and unless they would give up the helm to him, every soul of them would perish. They consented, and he bore away for England. He then, with the blandness of the heathen Chinee, enticed them to go below, make a good fire, and take tea, and when they were down, kept them there, having previously secured a musket and hanger. He carried the vessel safe into Torbay.

On the 26th of December, 1804, the ship *Lord Nelson*, Captain Maginnis, the *Harmony*, Captain Reed, of 20 guns, and the *Nymph*, Captain Heinsen, of 10 guns, sailed in company for mutual protection, from St. Thomas's for Liverpool. A few hours after they sailed, they fell in with a large schooner privateer of 10 guns (two of them long 12 pounders) and 100 men, all of whom, as well as the captain, were blacks. Captain Maginnis, seeing that the privateer was making a stretch to cut off the *Nymph*, directly hove to, to give her time to come up with him. In this, however, he was disappointed, as the privateer succeeded in boarding and carrying her. The *Lord Nelson* then continued her course, the privateer in chase, which she kept up the whole of the night, and at three p.m. the following day the privateer came up and the *Lord Nelson* prepared for action, which soon commenced by the enemy attempting to board. This manœuvre Captain Maginnis evaded by heaving his ship in the wind, and giving the privateer his broadside of star and grape shot. The action was then continued within pistol shot, with great warmth on both sides, for upwards of an hour and-a-half, the blacks making several attempts to board. At length the privateer crowded what sail she could, and bore off in the most shattered condition, her rigging being very much cut, her main boom shot away, and all her bulwarks entirely gone. As she sheered off, the officers and crew of the *Lord Nelson* gave her three cheers. The *Harmony*, Captain Reed, bore away before the action commenced, when the privateer was endeavouring to cut off the *Nymph*.

The schooner *Lancaster*, Captain John Pettigrew, having captured the *Die Vigilante, L'Union*, *Les Deux Anges*, *Der Guteman*, and *Vrow Esther*, a dividend was paid in full to the owners on January 16th, 1805, and also to the crew on the 23rd, at the office of Messrs. Gabriel James & Co., 59, Parr Street. This captain was probably the same who, on the 9th of July, 1801, wrote the following letter, from Barbadoes, to Evan Nepean, Esq., Secretary to the Board of Admiralty :—

"SIR,—I have the honour to acquaint you, for the information of the Lords Commissioners of the Admiralty, that on the 22nd day of June, in N. lat. 10 deg. 25 min., W. long. per accompts. 40 deg. 18 min., on board the ship *Intrepid*, of Liverpool, bearing Letters of Marque, under my command, having in company the ships *Dominica Packet* and *Alfred*, I had the good fortune to capture, after a running engagement of nearly two hours, the Spanish frigate-built ship *La Galga,* commanded by Francisco de Pascadello, and mounting twenty-four heavy sixes, and seventy-eight men, bound to Cadiz or any port in Spain, loaded with hides, cocoa, indigo, and copper in bars, the quantity not yet known. I am happy to say we sustained no other loss than that of one of my brave men, and our sails and rigging a good deal cut. The other ships have not sustained any damage, except the prize, which has suffered considerably in both hull and masts, and rigging. I arrived here on the 4th of July, with the prize and above-mentioned ships, I have the honour to be, etc.,

JOHN PETTIGREW.

"P.S.—The *Galga* has been at different ports, but was last from Rio de la Plata."

In March, 1805, there arrived in the Mersey, the Spanish ship *St. Ana*, alias *Nostra Hermanos*, from Vera Cruz and the Havannah for Cadiz and Malaga, laden with 60,000 dollars, 242 chests sugar, 1,800 pieces logwood, 368 cwt. cocoa, 69 bags wool, etc., captured on the 14th of February, 1805, off St. Mary's by the *Lady Frances*, private ship of war, Captain Hawkins, of Liverpool. The *Westmoreland*, Captain Goodall, had the good fortune to capture a Spanish ship from Vera Cruz, laden with sugars, dollars, etc., and valued at about £25,000. She also recaptured the *Eliza*, of Waterford, which had been taken by a Spanish privateer, off Cape Clear.

In April, 1805, the *Westmoreland*, Captain Reed, an outward bound Letter of Marque, of Liverpool, was taken, after a desperate action of two hours, by the Spanish ship privateer *Napoleon*, of St. Sebastian, pierced for 20 guns, and mounting 10 nine-pounders, and 4 eighteen-pound carronades, with 180 men on board. Captain Reed died of his wounds soon after his vessel struck. Six of his crew were killed. The *Napoleon* was captured by H.M.S. *Topaze*, Captain Lake.

The Underwriters of London presented to Captain Lewtas a valuable silver cup, with the following inscription engraven thereon:—

"This cup is presented by the Underwriters of Lloyd's Coffee-house, to Captain William Lewtas, of the ship *Venerable*, of Liverpool, as their Testimony to the Bravery of his conduct in twice repulsing, with great slaughter, a French privateer, carrying 16 guns, and 104 men, on his voyage from Liverpool to Barbadoes, in March, 1805."

In May, 1805, an order for a general embargo on shipping was issued from the Admiralty. It was followed by a very hot press for seamen and even landsmen. Protections were altogether disregarded, and ships were stripped of their hands, except such as were absolutely necessary to preserve them. The paper of May 13th, thus referred to the subject:—

"The immediate augmentation of our naval force is thought a matter of such pressing necessity, that all considerations of individual suffering must, for the present, give way. The order for an embargo at this port was announced from the Custom-House on Thursday; and, during the whole week, the press gang had been indefatigable in their exertions. Persons of all professions, as well as seamen, have been occasionally taken; though many have been released, on proper application being made. In the early part of the week about forty Irishmen, just landed from a Dublin packet, and who were proceeding up the country in search of employment, were pressed, and immediately taken on board the tender; but most of them are since liberated. The embargo extends to all vessels bound to foreign parts, including Ireland and the Isle of Man, with the exception of ships belonging to foreign powers, provided they have no British seamen on board. It extends, likewise, to coasting vessels of every description, except such as are laden with coals and grain."

The cause of this extraordinary press for seamen was, that the French and Spanish fleets were at sea, prepared to strike a great blow either at the Colonies, at Ireland, or at England itself, and it was thought urgently necessary to be prepared at every point.

On the 11th of August, 1805, Mr. Joseph Whidbey, late chief mate of the *West Indian*, Captain Dunn, wrote the following letter to his friends in Liverpool, from Oporto:—

"It is with much concern I inform you of our being captured by the combined fleets, on the 8th of June last, they being 20 sail of the line, 7 frigates, and 2 brigs, the day after we left Antigua, under convoy of the *Netley* schooner, who escaped the enemy by superior sailing. The French finding that the prizes could not beat up to Guadaloupe, and fearful of Nelson overtaking them, they the next morning dispatched 5 frigates (having troops on board) to destroy them and afterwards land the troops at Guadaloupe, which they effected, putting my captain and others on shore with them. It was a distressing sight to us to see our ships and cargoes burnt and sunk, when two English frigates were bearing down on them, but too late, the 5 French frigates returning at the time to join the fleet. We were stripped of every thing but the cloaths we had on. On the 22nd of July, to the southward of Vigo, we fell in with the British fleet, consisting of 13 sail of the line and 1 frigate. We were crammed below at six in the afternoon, when the British Tars gave us three cheers, which was returned by the cowardly Frenchmen, and a heavy cannonade commenced on both sides. I was on board the *Bucentaur*, the French Admiral's ship, of 90 guns; she, with one of 80 guns, engaged the English Admiral's ship, which unluckily got dismasted. A brave 74 going to engage the ship in our line ahead of us, sheered alongside and poured such a broadside into us that occasioned not a few to be brought to the doctor (where I was), without arms or legs, and caused numbers to fall on the decks, headless, and no doubt our poor fellows suffered greatly also. Two Spanish 74's got dismasted and were taken; the French suffered much in their rigging and people. At half-past nine at night, being very thick weather, the firing ceased, and the English fleet,* the next morning, lay to leeward of the French line, but the supper the Frenchmen got that night made them afraid of getting a similar breakfast, and indeed I, myself, was fearful of a renewal of the engagement, the English not being a match for such a superior force. The French bravadoed to me, and said they would bear down on board the English. I told them to go, they were ready to receive them, although the three-decker

*Commanded by Admiral Sir Robert Calder.

of the English was dismasted, but the thick weather coming on, the French made the best of their way to Vigo Bay, where we arrived three days after, and landed the prisoners, which were marched into the Portuguese dominions, where I now am, sufficiently distressed."

The *Mersey* privateer, Captain Baldwin, captured the *La Asia* from Lima, bound to Cadiz, laden with 282,151 dollars, 46 marks, 3 oz. of plate silver, 1497 chests of cascarilla, 3068 cargas of cocoa, 583 bars of copper, 792 bars of pewter, 19 bags of beaver, 18 bales of Spanish wool, and 1 bale of carpets. The paper of September 2nd, 1805, stated that the log of the *Mersey* was then on board Lord Nelson's ship, the *Victory*, having been taken from on board an American vessel on the 15th of July, the day before he made Cape St. Vincent. The American Captain reported that when he left the *Mersey*, she was water-logged and on fire, and had evident marks of having been employed in towing a large vessel which was, no doubt, the Spanish prize afterwards recaptured by the combined fleets.

The merchants of Liverpool have ever been prompt and liberal in recognising the gallantry of their captains. That they rewarded pluck and faithfulness, apart from success, is proved by the following correspondence which appears in *Billinge's Advertiser*, of September 30th, 1805:—

"CAPTAIN WILLIAM DEAN, Dear Sir,—We have the pleasure to inform you, the owners of the *Bellona* privateer commanded by you, on a six months' cruize, have desired us to present you with One Hundred Pounds, as a token of the high opinion they entertain (notwithstanding you have been unsuccessful) of your good conduct, and zeal for the concern, during the cruize. We are, Sir, your most obedient servants, LAKE & BROWN, Liverpool Packet Office, September, 1805."

"Messrs. LAKE & BROWN, and Owners of the *Bellona*, Gentlemen,—I have to acknowledge your letter of this day's date, and must say, the handsome and liberal manner in which you, with the other owners of the *Bellona*, have been pleased to testify your approbation of my conduct, leaves me quite at a loss how to express myself on the occasion, but however inadequate I may be to such an undertaking, I shall ever feel the most mortifying regret that my exertions were not crowned with the success due to such liberal minds, and I am proud in having the honour to subscribe myself, your very obliged and very faithful humble servant, WM. DEAN."

In February, 1806, the ship *Shipley*, belonging to Messrs. Shipley, Williams & Co., of Liverpool, and commanded by Captain Wilson, on her passage to the West Indies, was attacked by a French three-masted schooner privateer (late his Majesty's schooner *Demerara*), mounting 14 guns and full of men. Waiting until the breeze was dying away, she attempted to board the ship, but was repulsed with the loss of several of her men. She then, by means of her sweeps, dropped under the *Shipley's* larboard quarter, and commenced a very hot fire of great guns and musketry. From the position in which the ship lay, it was impossible to get any of her guns to bear, and the calm rendered the vessel unmanageable. In that situation she engaged the privateer with small arms for an hour and-a-quarter, until four of her men were killed, Captain Wilson and Mr. Holden, the first mate, besides the steward severely wounded (the former shot through the shoulder and his hand much shattered, the mate having his thigh broken), when the men, after a most gallant defence, and having no officers to command them, were obliged to strike. The French had their second captain and five men killed and many wounded. The *Shipley*, after having been plundered of much valuable cargo, which was carried on board the privateer, was recaptured by H.M.S. *Galatea*, and sent to Barbadoes.

The private ship of war *Mars*, John White, commander,

took the Dutch brig *Jong Vrow Maria*, the prize money of which was distributed on February 25th, 1806, at the counting house of Messrs. M'Iver, M'Viccar, and M'Corquodale, in Pownall Street. The following letter from the ill-starred French Admiral Villeneuve to his chief, explains the fate of the *Mars* :—

"On board the *Bucentaure*, off the Azores,
on the 4th Messidor.

"My Lord,—I have the honour to inform your Excellency, that yesterday morning the advanced frigates discerned two sail, to which they gave chase and came up with. One was an English privateer, the *Mars*, of Liverpool, of 14 guns and 50 men ; the other was a Spanish ship, the *Minerva,* which had been captured by the privateer, and which he was escorting. The ship was coming from Lima, having been at sea nearly five months, with a very rich cargo. Independent of 420,000 piastres, her cargo consisted of bark, cocoa, etc. ; the whole estimation at from five to six millions (French). The privateer being much damaged from boarding, Captain Lameillerie, of the *Hortense*, set it on fire, after taking the crew on board. The *Didon* manned the other, and I have her under my protection.

"I entreat your Excellency to accept my respects.

VILLENEUVE."

The brig *Hope*, Captain Higgins, of Liverpool, bound from Oporto to Dublin, laden with wine, was captured shortly after leaving port by a French privateer, who took out all her hands, except the captain and one man, and, leaving the prize in charge of six Frenchmen, the privateer bore away. The captain gave them plenty of wine to drink, with which they became so intoxicated as to render them quite helpless. When in this state, the captain and his assistant secured the arms, and confined four of the sailors in the hold. The remaining two they left on the quarter deck to become sober, and then compelled them to work the vessel till they arrived in an English port, when they were sent to prison. Having completed his complement of hands, the captain proceeded on his voyage, but only to meet with worse disaster. On the night of the 6th of August, 1806, the weather being remarkably hazy, the good brig *Hope* struck on a reef of rocks off the Point of Greenore and immediately foundered. The crew were saved.

A remarkable and interesting example of courage and perseverance is afforded in the escape from a French prison, of Mr. M'Dougall, lieutenant of the *Laurel* privateer, of Liverpool, captured on the 14th of June, 1803, and Mr. Samuel Mottley, a midshipman in the Navy. Mottley was taken in the *Minerve*, Captain Brenton, on the 3rd of July, 1803, off Cherbourg, where she had grounded while in chase of some vessels. He and other officers of the ship were marched to Verdun, after having been hurried from one prison to another, where they had the liberty of the town. Mottley got into some scrape with the townspeople, and was sent a close prisoner to the fort of Bitche, in Lorraine, and confined in a "*souterrain*," many feet below the floors of the prison. Here he remained, treated with excessive severity, from the 24th of May till the 22nd of August, 1806, on which day he, and three of his fellow prisoners, got leave to go to the town to settle some affairs. They were conducted under a guard, and therefore lay under no obligation not to escape, and it struck the midshipman that the thing might be accomplished. He communicated his thoughts to M'Dougall, who appeared to him the most enterprising of the other three, and therefore the fittest to share in the daring of the undertaking. Their minds were soon made up, and a lucky opportunity offered itself. The party asked the guard permission to bathe, which was granted, and Messrs. Mottley and M'Dougall left the river before the other two. Dressing themselves, they told the guard they would go on to the hotel, and provide dinner, the guard remaining to attend the others to the house. In a word—they ran off, and got away without hurt. When they had marched about six miles, they heard alarm guns firing, and they pushed on about an hour longer, and then concealed themselves in a wood, where they remained till half-past nine o'clock in the evening. They then made the best of their way till towards dawn, and then again lay down in the woods. This method was steadily pursued till the evening of the 27th of August, when they crossed the Rhine in a boat they seized on the bank. Their sufferings were extremely great during the six days we have been speaking of. They avoided the high road and habitations, and tasted no food whatever but fruit, which they stole occasionally. During the first night and the last they waded up to the middle in swampy ground, and suffered much from lying wet each of the days, and not daring to take exercise. From the Rhine to Stuttgart they proceeded on their former plan, only they ventured to obtain food, and one night, from excessive fatigue, they slept in a bed in a village. At Stuttgart, an English gentleman advised them to make the best of their way to Cassel, Hanover, and Hamburg, which they did. The route to the town of Hanover was performed on foot, on the same system as before, only a little relaxed, and they slept oftener under cover. They walked about 600 miles without shoes or stockings. At Hanover, they took a carriage to Hamburg, where they saw the English Consul, who furnished them with money and a letter to the English agent at Husum. They sailed from Husum in the *Lark* packet, and landed at Harwich on the evening of the 1st of October, 1806. They slept there and proceeded next day to London, where these companions in a hardy enterprise separated for their respective homes.

The great event of the year 1806 was the battle of Jena, which crushed the Prussians, on the 14th of October. On November 20th, Bonaparte issued his famous Berlin Decrees, forbidding France, and all her allies, to trade with Great Britain, declaring all British ports to be in a state of blockade, all British subjects wherever found prisoners of war, all British goods lawful prizes. All the Continental ports under French influence were thus closed against British ships; all neutral vessels which had touched at a British port were excluded. Bonaparte, deprived of his navy by the glorious victory of Trafalgar, hoped, by means of the Decrees, to strike a heavy blow at British trade—the secret of British strength, as he well knew. Great Britain retaliated by an Order in Council, dated the 7th of January, 1807, which declared all the ports in the French Empire in a state of blockade, and prohibited all neutrals from trading with the enemy;—that no vessel should trade from one enemy's port to another of a French or French allies' coast closed against British vessels. In November, 1807, another Order in Council enacted that no vessel whatever should enter a French port unless she had previously touched at a British one; and claimed the right of searching neutral bottoms for the purpose of carrying out this regulation. This was out-heroding Herod, and Great Britain, being mistress of the seas, was able to effectively blockade the Mediterranean, the Baltic, and the French ports, and practically to sweep from the ocean the commerce of France and her allies. On the 17th of December, Bonaparte issued his Milan Decree, which declared all merchant vessels of whatsoever nation, which should submit to the British Orders in Council, to be lawful prizes to the French. Consequent on this, a number of American ships were seized and confiscated in the ports of France and Italy. It was a war of commercial extermination. These high-handed proceedings found favour at first with the commercial community of this country; but ere long the pressure had to be mitigated by the grant of licenses exempting particular ships from the operation of the law, and this opened the door to forgery and fraud. Unfortunately this was not all the mischief caused by the Orders in Council. Their enforcement made us enemies of neutral states who wished to trade with France. The Americans naturally resisted the assumption of the right of search, and passed Acts in retaliation. The united result of the Berlin Decree, of the Orders in Council, and of the American embargo, was to suspend and, for a time, destroy the commerce of the United States. "Our commerce at this moment" said the *Boston Centinel*, "is like a poor flying fish, pursued from below by a couple of dolphins, and from above by a couple of hawks. While the French blockading decree, and the English retaliatory Order in Council, pursue it on one side, the non-importation act and the general embargo assail it on the other." The evil effects of this policy were soon felt in Liverpool. In one year its commerce declined by the amount of 146,000 tons, or nearly one-fourth of the entire trade. In spite of this drawback, the Liverpool docks were soon after found to be insufficient for the accommodation of the commerce of the port, in consequence of the opening of the trade with Spain and Portugal, and with their colonies in America. So it has ever been, and so may it ever be, with Liverpool trade—the closing of one door has been but the prelude to the opening of another; and the merchant on shore, as well as the sailor on the sea, has been distinguished by courage, resourcefulness, and endurance, in every crisis in the history of the port.

On the 17th of May, 1807, Captain Frears, of the ship *Fortitude*, wrote to his owners in Liverpool, from Port Royal, Jamaica, as follows:—

"Nothing material occurred after our leaving St. Thomas', until the morning of the 14th inst., at daylight, the Port of Jaquemel, N.N.W. distant 11 miles, saw two schooners close in with the land. At half-past five o'clock, perceived them to be armed vessels, pulling a considerable quantity of sweeps. As there was not a breath of wind, came up very fast, just gave me time to get in readiness to receive them. At seven, hoisted French colours, and continued sweeping towards us, and firing their great guns at intervals. At a little before eight, commenced our fire with what guns I could get to bear, which made them retreat out of gunshot. At half-past eight, swept up again on either quarter to board. In this situation remained until ten, keeping constant fire at them with what guns I could get to bear. At a quarter-past ten, got their boats out, and grappling up to the square-sail yard-arms. At eleven, clapt me alongside, one on each quarter. As there was no wind, I could not work the ship to get our guns to bear as I could wish. The fire, believe me, Gentlemen, was tremendous. My two after-most guns, with the carronades, were all the guns of service to me, with my small arms—but alas! what was my musketry to contend with 185 men, some on one side, and some on another. At meridian, boarded me, cutting up every person who could not get out of their way. I am sorry to say that my loss is so great, 4 men killed, 8 dangerously and 4 slightly wounded. Every praise is due to my officers and men; they behaved like Englishmen to the last moment. I am sorry to see so many suffer, although I suffer most myself, having received at boarding a ball through the thigh, and a dangerous cut on the cheek and ear. At six p.m., of the same, the ship was recaptured by his Majesty's ship *Heureux*, and sent for this port, where we arrived this day.

"The following are the names of the killed and wounded on board the *Fortitude*:—

"*Killed* — Mr. Charles M'Adam, Junr., supercargo, James Harrington, William Williams, Francis Frederay, seamen.

"*Wounded dangerously*—Robert Frears, captain; Hugh Rogers, boatswain; Thomas Williams, William Catton, Jacob Peterson, James Hamilton, Donald Mark, Nathaniel Hunt, seamen.

"*Wounded slightly*—Joseph Dunn, Joseph Edwards, John Jones, John Tyrer, seamen."

The paper of October 5th, in recording the death of the brave Captain Frears, at the early age of 30, observes, "the severe wounds he received in his gallant but fruitless attempt to preserve his ship from the grasp of the enemy (two French privateers of superior force), off St. Domingo, on the 14th of May last, brought on a fatal illness, which has at length terminated his existence, and left his family and friends to lament his loss."

At sunset, on the 1st of August, 1807, the brig *Pope*, of Liverpool, Captain Masheter, carrying 12 six-pounders and 25 men, on her passage from Liverpool to Barbadoes, when within a day's sail of her destination, and soon after speaking one of the King's cruisers, fell in with the French privateer schooner *Le Jeune Richard*, mounting 8 twelve-pounders and 120 men. Taking advantage of the night, while the privateer kept aloof to reconnoitre, Captain Masheter battened down his hatches, made the best arrangements for the attack, and nailed his colours to the mast. At sunrise, the action commenced, and was kept up with unabating spirit on both sides for an hour and-a-half, during which Captain Masheter lost his right leg, and his left arm, and some of his men were desperately wounded. The enemy then boarded, and carried the brig, which lost, in the stubborn and sanguinary contest, besides the gallant Captain, who was now completely cut down, the first and second mates and three seamen killed, and four seamen wounded. The privateer and her prize arrived at Point-a-Petre (Guadaloupe) on the 4th of August. The loss of the privateer was said to be only three or four men, which is probable enough from the great disparity in numbers of those opposed to her. Two months later the privateer encountered Nemesis in the shape of the *Windsor Castle* Packet, Captain Rogers. A passenger on board the packet, writing from Barbadoes on the 3rd of October, 1807, gives the following account of the affair :—

"We are just landed here after an unpleasant passage of 37 days, and experiencing one of the most desperate actions which has been fought this war, though, thank God, we have been victorious, and have cleared those seas of one of the fastest sailing privateers out of Guadaloupe, which had in the last six weeks taken no less than six fine running ships, viz.—the *America* and *Clio* in company, the *Margaret*, the *Pope*, the *Portsea,* and another. When we met her she was six days on a fresh cruize, with 86 men, and 6 long sixes and 1 long thirty-two-pounder gun. Our force consisted of 6 guns, short sixes, and 30 men, including 3 passengers. We lost 3 men killed, and 7 wounded, the first broadside ; but I am happy to say that with the remainder, in an hour and forty minutes, such was their gallantry, that they carried the privateer, after killing 26, wounding 30, and making prisoners of 30 not wounded, in all 60 prisoners, almost treble the number we had left for duty. We have therefore, as you may suppose, had little comfort for the last three days, not having had our clothes off, and being obliged to sleep upon deck in order to secure the prisoners. But I have so little time for the *Barbadoes*, and am so nervous, that I cannot enter more into detail by this opportunity, and can only say that if any man has deserved a token of merit from your Underwriters, Captain Rogers deserves it in the highest degree. He is a young man, his first voyage as Acting Captain (the Captain being left at home), and has therefore nothing but his merit to depend upon. He was left with only 10 men about him for the last half-hour, rallying them to their duty, with a determination to carry the prize, which repeatedly endeavoured to clear from the packet, but was too fast lashed by her bowsprit to escape, and he boarded her at the head of four men, and charged her decks with a gallantry never excelled and seldom equalled. The officers of the man-of-war here are astonished when they look at the two vessels and their crews, and, instantly in the handsomest manner relinquished all claim to the prize."

"His Majesty's Post-Masters General," says *Billinge's Liverpool Advertiser*, of February 1st, 1808, "have appointed Mr. Rogers, the Acting-Captain of the *Windsor Castle* Packet, to a command. Few instances can be found of more determined bravery than that shown by the whole crew, which consisted of only 28 men and boys. The muster roll of the French privateer had 109 men, of whom there appeared, on the arrival of the packet at Barbadoes, 61 killed and wounded, forty of whom were mowed down by the last fire." The same paper published the following paragraph on the 4th July, 1808 :—

"A few days since, an elegant silver cup, value sixty guineas, was presented to Captain William Rogers, of his Majesty's Packet *Windsor Castle*, with the following inscription engraven thereon :—

"Presented by the Underwriters of Liverpool to Mr. William Rogers, Acting-Captain of his Majesty's packet *Windsor Castle*, as a testimony of their high sense of his distinguished gallantry in defending that vessel with a force of 28 men and boys against the French privateer *Le Jeune Richard*, with a crew of 92 men, which he bravely boarded at the head of five followers, and captured on the 1st of October, 1807, after an action of four hours, in which he had 13 men killed and wounded; the enemy, 54 killed and wounded. Thus in the hour of battle displaying to his countrymen an example inspired by the soul of the immortal Nelson, that England expects every man will do his duty."

"In addition to the above was added £130 from the merchants and Underwriters of Liverpool, to be distributed amongst the officers and crew of the *Windsor Castle*, as a testimony of their high approbation of the great bravery displayed by them on the above memorable occasion."*

* Captain Rogers acknowledged the presentation in the following letter dated Falmouth, 20th July, 1808 :—

On the 28th of August, 1807, the *Diana*, Captain Lewis, a Liverpool Letter of Marque, bound to Port-au-Prince, was attacked by the *La Vengeance* French schooner privateer, Captain Bligh, from Guadaloupe, mounting 12 nine-pounder carronades, and 100 men. The privateer attempted to board, but was repulsed, four of the boarders being taken prisoners, two of them badly wounded, and the privateer's foremast and bowsprit shot away. The *Diana* received considerable damage, but managed to rejoin the *Hannah*, which had been unable to take part in the engagement. Captain Lewis was shot through the thigh, and wounded in the face, while several of his men were also wounded.

In a letter from Captain James, of the ship *Glenmore*, written to his owners from Madeira on the 12th of November, 1807, we have an account of another gallant and successful defence made by a Liverpool Letter of Marque:—

"On Wednesday, 21st October, a suspicious sail appeared on our wake, about seven p.m., when we beat to quarters. At eight she fired into us, which we returned with 2 nine-pounders. I immediately hailed him, but the answer he returned was not satisfactory; however, he sheered off and kept without the reach of our guns all night. At five a.m. saw him bear down towards us, at seven he was in our wake, and observing no stern guns, no doubt was determined to keep us end on, so as to drive us from our quarters, but he was mistaken, for as soon as he came within gunshot, we lowered down the jolly-boat and fired 2 heavy

"To the Merchants and Underwriters of Liverpool.

"GENTLEMEN,—I beg leave to return you my most grateful thanks for the distinguished honour you have so generously conferred on me, by presenting me with a piece of plate, for the service I performed on board the *Windsor Castle* packet, in capturing the French privateer *Le Jeune Richard*, and be assured, whenever an opportunity offers, I shall not be found wanting in the duty I owe to my country, to support the high opinion you have entertained of my conduct.

"I remain, with great respect, Gentlemen,

"Your most obedient and most humble servant,

"WILLIAM ROGERS."

long nines into him, which in the course of the night we had placed there. But notwithstanding a well directed fire from our stern, he still persevered in his attempts to board. At half past eight he attempted our starboard quarter, but we rounded to and gave him a broadside. He immediately wore round, expecting to get on our larboard side, but he found we were ready to receive him there; he was then within pistol-shot of us. A continual fire on both sides continued until half-past nine, when we drove them from their quarters, and not a man was to be seen, nor a single shot from him. I then hailed him and enquired why he should attempt a second time such a ship as ours? His answer was, 'I wished to try what you was,' and immediately upwards of 100 men made their appearance. From the small number I had on board, was afraid to attempt boarding him. As he had the superiority in sailing, he immediately made sail and stood to the west. She was a beautiful vessel, pierced for 16 guns, and mounted 14 brass six-pounders, as the wad found on our decks was covered with verdigrease, and about the size of our sixes; she was apparently a new lugger. For the gallant behaviour of my sailors I have promised them five guineas reward to drink, which I certainly think they deserve, as in the very heat of the action they gave three cheers, and sung out '*Conquer or Die.*' I am happy to say not a man was hurt on board our ship, though many shot have gone through our bulwarks and cloths, some of our running rigging is cut, and several shot through our sails. We must have done him much damage, as we fired 75 shot, 18 of which were from our stern guns."

In July, 1807, an important lawsuit arising from Liverpool privateering was tried in the Court of King's Bench, before Lord Ellenborough and a special jury. Messrs. Hobsons and others, the owners of the *Eliza* privateer, Captain Keene, of Liverpool, claimed £2888 10s. 6d. from the Hon. Captain Blackwood, being a loss sustained by the plaintiffs in consequence of the act of the defendant, who, being in command of H.M.S. the

Euryalus, in 1805, sent a lieutenant on board the *Eliza* (which the plaintiffs had sent to sea three weeks before, fitted out for a cruise, with Letters of Marque, etc., manned with 41 men and boys), impressed four of her men, and carried them off, although the captain of the *Eliza* produced the usual protection from the Lords Commisioners of the Admiralty. In a few days after, that is, on the 4th of April, 1805, the *Eliza*, in company with the *Greyhound* privateer, of Guernsey, fell in with and captured, off the Azores, after an action of one hour and forty minutes, a rich Spanish ship, called *La dos Amigas,* 24 guns, bound from Lima to Cadiz. She was about 700 tons burthen, and laden with 179,935 dollars, 473 marks of worked silver, 561 chests of cascarilla, 54 bags 3 serons of wool, 40 serons of sea-wolf skins, 9 serons of indigo, 1 chest of drugs, 17,507 cargas of Guayaquil cocoa, 1,745 bars of copper, 3,398 bars of pewter. The prize-money which this rich haul enabled the privateers afterwards to divide, amounted to upwards of £151,000. In the distribution of the prize-money, in captures of this description, the owners of the privateers shared three-fourths, and the crew the remaining one-fourth ; and where there was a joint capture, each ship, upon such distribution, was entitled to her portion according to the number of hands on board at the time of the capture. In consequence, therefore, of Captain Blackwood's high-handed proceeding in impressing four of her hands, the *Eliza's* share of prize-money became proportionally less by upwards of £3,000, to recover their dividend of which, the plaintiffs brought the action. It was contended for the defence that the certificate and protection produced by the captain of the *Eliza* were frauds upon the Admiralty, and justified the taking of the men in question. It appeared upon the evidence of Lieutenant Methuen, of the *Euryalus*, that Captain Blackwood, having orders to impress able seamen for his Majesty's service, sent the witness on board the *Eliza*, to examine her hands, and impress any of them that might be liable. He accordingly had the crew mustered on the deck, and upon comparing them by name, age, and description, with the license from the Admiralty, he found that some of them did not in any way answer the description given, either as to age, name, or appearance, and by the account given of their ages, by the four men whom he did so impress, he found they varied three, four, and five years from the description in such license. It also appeared, that the Letter of Marque was originally granted for a complement of 50 men; but from the certificate granted at the Liverpool Custom-house, the parties had sworn only to a complement of 35 men and boys. It appeared, however, that from the difficulty of procuring men for such service in the port of Liverpool, the parties were not limited to 35 men, but had the power to engage more if they could be procured. Lord Ellenborough, in summing up the evidence to the jury, observed that the first question was whether the defendant was duly authorised to impress men? Of this there could be no doubt, as the good of the service required that a certain description of persons should be liable to be impressed, and in almost every case captains of his Majesty's ships had such power vested in them. The next question was whether these men so impressed, answered the description (in point of age and appearance) given in the license. Captain Blackwood, his Lordship continued, could have no sinister purposes to answer in taking these men. It was a part of his duty, and from the well-known fame and character of that gallant officer—whose life must always make a prominent feature in the naval annals of this country,—there could be no doubt, if he had acted improperly, or rather illegally in impressing these men, he could have done it with no other motives but with a view to the public service. His Lordship did not mean this as any compliment to that honourable officer, whose services were too well known to require any eulogium. At the same time, if any indulgence was to be made for an error of this nature, certainly no man was more entitled to have such indulgence than the defendant. But it was their duty to decide wholly upon the facts before them in evidence; and they would consider whether he was justified in this proceeding, and if not, what compensation in damages the plaintiffs were entitled to. The jury brought in a verdict for the plaintiffs—damages, £2,888 10s. 6d.

Captain Phillips, writing from Guadaloupe, on March 20th, 1808, gives the following account of the capture of the *Robert*:—

"We sailed from Africa, 5th February. Nothing particular occurred until the 6th of March. On that day we saw a sail, and immediately knew her to be a cruizer, upon which we made every preparation to engage her, determined to resist being taken, or sell our vessel dearly. About 7 p.m. she was right astern, and commenced firing from a long artillery eighteen-pounder (whilst the *Robert's* stern chasers would not reach him). He kept in that position for an hour, then run close up under the stern and quarter, pouring in his small arms with an intention of boarding. Fortunately a shot from the *Robert*, at this moment (as I afterwards found) killed two men, and wounded three others on board the privateer, and materially damaged his foremast. He then sheered off, but kept in sight all that night, during which we were every moment expecting him to renew the engagement. However, at 7 a.m., being daylight, he came up under the larboard quarter, and kept up a constant fire from his gun and small arms, for upwards of an hour, when the fire from the privateer caught the arm-chest on the poop, which exploded, and made sad work amongst such as were near it, who were dreadfully scorched by the cartouch boxes exploding about them. In this situation, the privateer boarded, and after a quarter of an hour's fruitless resistance, they succeeded in gaining possession; they then hauled down our colours. Mr. Youd, the chief mate, was dangerously wounded by a musket ball near the temples, but the ball is since extracted, and I hope he will recover. Stephen Baker, William Gray, Philip Crawley, and John Post, seamen, were wounded, and now in the hospital, where great attention is paid them. From being immediately hurried on board the privateer, I cannot exactly inform you of the fate of those who remained on board the *Robert;* but by a French gentleman who arrived here from Martinique, I am informed the *Robert* arrived there the 12th inst.; that the Doctor was, it is feared, mortally wounded, and that several of our crew are since dead of their wounds."

The following is an extract from a letter written by Mr. J. L. Forrester, on board the armed ship *Active*, of 12 guns, Captain Teed, of Liverpool, dated harbour of Chaquaramas, Trinidad, July 18th, 1808:—

"On the 16th inst. we made the river Demerara, off which we fell in with a vessel, which we supposed to be a Demerara dogger, or pilot boat; but on hailing, she proved to be a Spanish privateer, and a running fight commenced, which ended in her sheering off into shoal water. This lasted about forty minutes, and in bearing up for our port, we found ourselves about 12 or 15 miles to leeward of Demerara; and knowing the impossibility of beating to windward with a ship so deeply laden, our bends being actually under water, we resolved to bear up for this island, to which we were welcomed at six o'clock this morning by a French privateer of considerable force, who seemed certain of her prize. On our firing a gun for her to shew her colours, she had the impudence to run alongside, sent a hand to the mast-head, who either lashed or nailed her colours, and then returned us two guns, and a volley of musquetry. The latter seemed a continual shower during the whole of the engagement, which lasted an hour and a quarter. We had no one hurt on board, many shot-holes through our sails, and some trifling injury to our rigging."

Captain Bibby, of the *Juliana*, writing to his brother, in Liverpool, on the 7th of October, 1808, from sea, the Lizard bearing E. by N. distant 15 leagues, says:—

"This is to inform you of the death of Captain Bosworth, which took place on the 4th ult., after an illness of eight days, after which nothing material occurred until the 2nd inst., when at three p.m. we fell in with a French privateer, full of men, he having an English Jack hoisted at the main. We took him for an English gun-brig; we, however, cleared for action, and he sheered up under our quarter, till within half pistol shot, when he hoisted French colours, and without hailing, fired into us with his great guns and small arms, his deck crowded with men, which we returned with a broadside, our guns being loaded with round and canister shot, when he hauled his wind, firing his musketry, we firing our great guns as long as they could reach him. At five p.m., she having left us, we stood on our course, not being in a condition to follow him, being damaged in our hull, rigging, and sails, and the ship making a considerable quantity of water. At six p.m., lost sight of the privateer. At half-past eight a.m., in lat. 49. 44. long. 12. 6. saw a brig ahead, cleared for action; at half-past eleven a.m., hoisted a French ensign, when she hove to, and shewed Hambro colours. We then fired a gun, pulled down the French, and hoisted English colours, and went on board to overhaul her, when finding the Captain had different papers, one not agreeing with the other, we took possession, and ordered her for Liverpool. At half-past one p.m., discovered a strange sail; at seven p.m., she came up with us, and proved to be H.M. sloop of war brig *Mutine*, Capt. Hugh Steward, who carried me and the captain of the brig, with the papers of both vessels, on board his ship, saying he would take the brig from us. He then sent his boat and took Wm. Gourley and the men we had left on board the brig, out of her, putting his own people on board, not suffering mine to take a single article of their cloathing except what they had on; nor can we get to speak to the Captain of the man-of-war to get the people's cloaths overboard. Fortunately, not a man of our crew was hurt. N.B. The name of our prize is *Johanna*.

The following letter was written at sea, in lat. 45, long. 13. 18 W., by Mr. William Hymers, commander of the snow *Shaw*, to her owners, Messrs. John D. Case & Co., of Liverpool:—

"GENTLEMEN,—I am sorry to inform you that on Wednesday, November 3rd, 1808, I saw a roguish-looking brig to windward, cleared ship for action, at half-past three p.m., he bore down into our wake, under a press of sail, with an English Jack flying. I shewed Spanish colours. At five p.m., he down English colours, and up French ensign, and wore round and gave us his larboard broadside. We commenced firing our stern chasers, continued firing two glasses; he dropped astern, having been a little disabled. At daylight the privateer appeared off our starboard quarter; at half-past eight a.m., commenced close action, and continued without intermission until half-past eleven a.m. He then out sweeps and sweeped from us. We then gave him three cheers, and when I came to look over the shot I had left, I must say that I was heartily glad that he had sheered off, as I had only six rounds, 13 cannister, and 15 langridge shot, and no cartridges. I have cut up all my stockings, and the ship's company followed my example; I then tied up all the carpenter's nails and tools that would go into a gun, and the cabouse lead. As the privateer was only laying out at gunshot, I perceived that he had a mind for another tack as soon as he was ready. I saw him get his stink-pots on his main yard-arm, and his grapplings on his fore yard-arm. I then got a spare main-yard athwart abaft to prevent him getting on our quarter. At one p.m., he crowded all sail and sweeped up in our wake. As soon as our stern guns would reach him, we slapped away, and shot away his gaft and hauled (hulled?) him several times. At three p.m., he gave us his whole broadside and sheered off. She is a brig, pierced for 16 guns, but only 14 mounted; she was full of men, she has two narrow yellow streaks, and all the rest black. I cannot say too much for Mr. Jackson (chief mate) for his manly support, and to do the crew every justice, they fought like Englishmen. Having no shot, I thought it my duty to call at Madeira for more. I am happy to say that none of my people are hurt, only sails and rigging suffered materially; the grape shot played like hail."

The *Lord Cranstoun*, Captain Gibson, for St. Croix, with 50 men and 22 guns, and the *Lydia*, Captain Lewis, for St. Kitts, with 22 guns and 45 men, sailed in company from Liverpool, on the 9th of November, 1808, and on the 16th, were parted in a heavy gale of wind, thunder and lightning. On the 29th, in lat. 26. 30, long. 31. 26., the *Lydia* was chased by a large frigate under Spanish colours, which, when within half-pistol shot, gave her a broadside, which was returned by the *Lydia*, and an action commenced which lasted about 25 minutes, when Captain Lewis thought it prudent to strike, having 1 man killed and 4 wounded, his masts so crippled that it was impossible to carry sail, and his rigging and sails completely cut to pieces. The enemy proved to be the French frigate *L'Amphitrite*, of 44 guns (28 eighteen-pounders, 12 thirty-six-pounders, and 4 long nines) and 450 men, 200 of whom were soldiers bound for Martinique. On the 3rd December, in lat. 23. 35. N. long. 37. 30. W., they fell in with the *Lord Cranstoun*, with which the frigate exchanged two broadsides. The *Lydia* being in company, and on his starboard side, the frigate on his larboard, Captain Gibson had the mortification to receive two broadsides from his old consort. The *Lord Cranstoun's* masts, sails, and rigging, being completely shattered, she was obliged to yield to such superior force. The Frenchmen threw overboard from both prizes the least valuable articles, transferred part of the *Lydia's* cargo to the *Lord Cranstoun*, and scuttled the former. Having captured an American brig they gave her up to the prisoners (96 Englishmen and 12 Portuguese) as a cartel, with a small proportion of provisions. Fearing a long passage to a British port under such conditions, the prisoners steered for the Isle of Flores, where they arrived on the 24th of December. Having victualled they sailed for Liverpool, where they arrived on the 16th of January, 1809.

In consequence of the number of captures made by the enemy's privateers in the Channel, the Government gave directions for the adoption of a system of alarm gun signals, intended to serve as an intimation to the men-of-war that a privateer was on the coast, and to point the very place where it might be found.

The following account of a gallant and successful stand against fearful odds is extracted from a letter dictated by Captain Spence, of the *Lascelles*, at Palermo, 21st of August, 1809, and received by the owners in Liverpool:—

"It will now be proper to inform you, that after seeing many privateers in the Mediterranean, we at last had one to engage, close under the island of Galitor. It was on the 7th inst., about 12 o'clock at noon, we perceived a large sail in the offing bearing down upon us from W.N.W., apparently an enemy, but we still continued our course. At two p.m., coming up with us very fast, we immediately beat to quarters and cleared away the decks for action. He still coming up, with a long pendant at his main, and an Algerine flag on his mizen, we shortened sail ready to engage. We gave him a gun and hoisted our colours, which he immediately returned with a broadside, his French colours hoisted. Then we came to a general action within pistol-shot, with our great guns and small arms. He fought very hard for about an hour and-a-half with his great guns, but we suppose that by our driving them from their quarters, they betook themselves to small arms, which they continued to do until the end of the action, having all their sweeps out on both sides, endeavouring to get away as fast as possible, we still continuing to keep up our fire upon them.

"She was a very large vessel or ship, and much longer than the *Lascelles*, shewed 16 ports, and mounted 14 guns, and we cannot conceive that she carried less than 150 men.

"We then have the pleasure to say, that we succeeded in

beating her off entirely; and by a peculiar Providence not one soul of us lost our lives, but five of us were most shockingly burnt (particularly myself and one of the sailors,) who have suffered in a most excruciating manner, and are far from being recovered. All that the ship could afford in point of relief was administered to the injured; and when arrived here, Mr. Gibbs immediately sent a surgeon on board to attend us all; and now we have great hopes of our recovery.

"My running rigging and sails are very much shot away; standing rigging and the hull of the ship likewise have received many a shot. In the course of a few days, I hope to be so much better as to be able to enter into a protest on this account, which I shall send you in my next in course."

The foregoing letter was written by Mr. H. Le Resche, passenger on board the *Lascelles*, and dictated to him by Captain Spence, who was disabled by the action he fought from writing himself, as the following letter from Mr. Le Resche to the owners of the *Lascelles*, shows:—

"DEAR SIRS,—I wrote you a few lines by a ship at sea, on the 17th July, viz., *La Rose Duncannon*, belonging to Messrs. Rogers & Bownas. Now I have the pleasure to drop you a few lines more from hence. You see what an awful encounter we have had, and the effects of it are such that Captain Spence cannot use his hands, therefore he has begged of me to write you as above. It's a great blessing that we all escaped with our lives. All the passengers were equally engaged in the action. I and Mr. Cougan were employed in working a six-pounder all the time.

"I have now to inform you, that you have chosen a very good Captain; he has your interest very much at heart; he thinks of nothing else. I hope you will keep him long in your employ; and that you will give him every encouragement, as he well deserves it."

The *Alexander Lindo*, Captain Pince, on her passage from Rio Janeiro to Liverpool, having thrown 10 guns overboard in a heavy gale of wind, afterwards encountered a French privateer, of 14 guns. Captain Pince having only 4 guns left, called his men together, and addressed them on the danger of their situation without their individual exertion, offering 20 guineas to the first man who would disable the enemy's vessel. The first gun that was fired by Mr. Patterson, the chief mate, shot away the privateer's main-yard, upon which she hove to and clued up her sails. Two days later the *Alexander Lindo* was attacked by a French schooner privateer, full of men, who bore down upon her, but after having received her fire made sail and bore away.

On the 22nd of April, 1810, as a boat from the ship *Earl of Chester*, just arrived from Madeira, was putting off with some of her crew for the shore, it was pursued by a boat belonging to one of the King's ships lying in the river, for the purpose of impressing the seamen. The man-of-war's men wantonly fired several shots at the boat, which was running in the direction of the Parade Walk, then crowded with pedestrians. One of the shots took off part of the finger and lodged in the thigh of an elderly woman then on the walk. This occurrence increased the hatred of the people towards the impress service.

In July, 1810, about 200 American sailors assembled at the Queen's Dock, and having armed themselves with staves, proceeded to the rendezvous houses of the press-gang, in Cooper's Row and Strand Street, where they broke the windows and furniture, and liberated some seamen who had been impressed. Two of the ring-leaders were apprehended, and committed for trial at the Lancaster Assizes.

On the 9th of November, 1810, several hundred people on the heights of Dover, had the privilege of seeing a Liverpool Letter of Marque engaging six French lugger privateers, full of men. The *Mary*, Captain Barry, was on her passage from Pernambuco to Liverpool when chased by the luggers, four of which were within half-pistol shot of her stern, and the other two on her lee quarter, though not within range of musket shot. Owing to the fatigue of the crew, and the vessel being under close reefed fore and main-top sails and reefed foresail, caused by the severe gales of wind, they could make sail but slowly, and the privateers had nearly got alongside before they could get her sails trimmed. That done, Captain Barry got the *Mary's* guns well supplied with round and grape shot, and by two well directed broadsides caused the two headmost luggers to drop astern, until they were again supported by their consorts. At last, having drawn close in to the land, where they saw the English gun brigs making sail, the privateers made off, but were so daring that they chased the *Mary* almost within gunshot of the men-of-war brigs, the commander of one of which complimented Captain Barry on his perseverance and consequent escape.

An atrocious and deliberate outrage, far exceeding any wild, practical joking ever indulged in by Joe Daltera and his Committee of Taste, was perpetrated in Liverpool, in November, 1810. Half-a-dozen fellows, assuming the character and authority of a press-gang, seized a very respectable gentleman of the town, who never was at sea in his life, and took him to a public-house, where they shut him up in a room, and confined him as an impressed man. In this miserable apartment he was forcibly detained two days and nights, without food or refreshment of any kind, and was not released till his captors had extorted from him a sum of money as the price of his liberty. The gentleman immediately complained to the Regulating Captain of the port, of this unexampled outrage, and was assured by that officer that every exertion would be used to discover the authors of it, but that the offence had certainly not been committed by any of the press-gangs under his command, nor had any such person been brought to any of the houses of rendezvous under his direction. It does not appear that the daring ruffians were ever brought to justice.

The ship *Brothers*, of Liverpool, Capt. Geo. Powditch, on her voyage from Bahia to London, was captured on the 13th of March, 1811, by the French privateer *Diligente*, of 14 eighteen-pounders and 150 men. Monsieur Garceau, the commander of the privateer, put on board 15 men, leaving in the *Brothers* only the steward, John Murdock, who selected such of the privateer's people as he found inclined to his purpose, and recaptured the ship. The prize-master was ordered to carry the *Brothers* into a port in Norway, and the people who aided Murdock in the recapture were chiefly Norwegians, pressed into the French service. The *Brothers* arrived in Liverpool on the 20th of April, 1811. The threatened disappearance of the British seaman from the mercantile marine of Great Britain is suggestive of a bad time for our shipowners and the country in the next naval war.

When the French privateer *La Cupidon* was taken on the 24th of March, 1811, by the *Amazon* frigate, four Englishmen were discovered amongst the crew. They represented themselves to be Americans; but some suspicions arose, and they were taken into custody as traitors, and tried at the Old Bailey Admiralty Sessions, in February, 1812. In their defence, the prisoners stated that they had suffered much in a French prison, and their only motive in getting on board the French privateer was to seek an opportunity of returning to their native country. Far from wishing to aid the enemy, they had actually engaged with other Englishmen and Americans to overpower the crew of the privateer, and lodge her in a British port. This was corroborated, but it unfortunately turned out that two of the prisoners had also served in the *Napoleon* French privateer. They were found guilty, and sentenced to death, the Judge observing that the distress of the prisoners was no excuse for serving the enemy. In March, 1812, two seamen were executed for the same offence—high treason; five more, who had been sentenced to death at the same time, received the royal pardon. Long confinement and hard usage in French prisons undoubtedly drove many British seamen to take service on board French privateers, while others entered the enemy's ranks for baser reasons.

During this long struggle, Liverpool became a depôt for prisoners of war. The gaol in Great Howard Street, which had been erected in 1786, but not occupied, was used for this purpose. In January, 1799, there were 4009 French prisoners in Liverpool, out of a total of 30,265 in Great Britain. The mortality amongst them was very considerable, and the hearse was constantly in requisition to convey from the gaol the corpse of some poor Frenchman to the portion of St. John's Churchyard then used as a public cemetery. Among the 1100 French prisoners liberated after the Peace of Amiens, was one who had made 300 guineas during his confinement, by his skill and industry in manufacturing toys.* With their usual ingenuity, the French manufactured a variety of trinkets, rings, snuff boxes, slippers, crucifixes, baskets, little carved boxes, and toys, which were exhibited on a stand in the entrance of the gaol, and sold for their benefit. Though ill-clad, dispirited and miserable, they were not always sad. Occasionally they performed plays in a small theatre within the walls, to which the public were admitted, the admission money raised in one night being, in some instances, as much as £50. Once an unrehearsed tragedy took place ; one of the Frenchmen, while dancing and singing on a Sunday evening, in July, 1793, dropped down and expired immediately. A prisoner named Domery, a Pole by birth, possessed a marvellous and

* This was used as one of the arguments in favour of Mr. Gregson's plan for the encouragement of mechanical drawing and design in the Blue Coat Hospital, and all the public schools of Liverpool. Referring to the advantages to the boys from such training, the paper of April 12, 1813, says : "Should they incline for sea, and be taken into a French prison, their ingenuity there may enable them to sustain their confinement with more comforts than usually fall to the lot of a British tar."

insatiable appetite. In one day he consumed 14 lbs. of raw beef, 2 lbs. of candles, and drank twelve bottles of porter, and felt fit for more. The capacity of the deposed King of Babylon for eating grass has not been recorded, but from a medical report published on Domery's case, we know that he could eat grass weighing 4 lbs. or 5 lbs. at a time. Cats, dogs, and rats, were mere tit-bits for him, and his sufferings from the want of what is vulgarly called a really "square meal" must have been terrible. Felix Durand, one of the Frenchmen confined in the Tower, in Water Street, about the middle of the eighteenth century, had some romantic experiences. He worked for a tradesman in Dale Street, the go-between being a young lady, who became sufficiently interested in the prisoner to herself make a survey of the rooms adjoining his place of confinement, and in consequence of the information so gained, Durand and several of his compatriots made their escape. After wandering about the country for some time, pretending to be deaf and dumb, and surprising the country-people by the clever workmanship he turned out in return for their hospitality, he one day, being in hiding, overheard a young lady expressing her admiration of the scenery in the French language. Unable to suppress his emotion, he rushed forward and poured forth his sorrows in his native tongue, and, as he thought, into a sympathetic ear. Unfortunately, he was recognised by the lady's companions and attendants as the deaf and dumb man who had sought employment a a few days previously. In spite of the lady's pleading, a gentleman of the party arrested the poor Frenchman, and carried him before a very gruff old justice at Ormskirk, who sent the prisoner back to Liverpool. One true heart in that town was not sorry to see him once more, and Monsieur Felix Durand, having discovered that fact, was in due time united in holy matrimony to Miss P——, of Dale

Street, the young lady who had facilitated his escape. His compatriots had been retaken before him.

The French privateer *L'Amelie*, described as schooner-rigged, with a yellow streak, and white bottom, showing no guns, but carrying 14, and 100 men, and commanded by Captain Lacroix, sailed from St. Malo for a cruise, and very judiciously chose a station commanding the entrance into three channels—St. George's, Bristol, and the English. She had been three days out, when, on the 25th of November, 1811, she encountered the ship *Sally*, of Liverpool, George Knubley, master. Captain Knubley took every precaution, and made every disposition which human foresight could suggest, for the preservation of his vessel, and during the action kept the quarter-deck amidst a shower of musket-balls, endeavouring to encourage his men to an effectual resistance. But, after a sharp action of about twenty minutes, the *Sally* was carried by boarding, and the crew, with the exception of three, taken on board the privateer, where, to the honour of the commander and his officers, they experienced every possible kindness compatible with their unfortunate situation, being allowed to preserve the whole of their private property, and indulged in all the comforts and luxuries which the privateer afforded. During the action, the first lieutenant of the *L'Amelie* was killed, and several of the crew wounded. The *Sally* had five wounded, one dangerously. She was ordered for France, and parted company with her captor next day. Captain Lacroix promised to Captain Knubley that he should have his liberty, and the first ship of little value which the *L'Amelie* should take, upon condition of his giving his parole for the exchange of an equal number of French prisoners in England, to be sent to France as soon as possible after his arrival at an English port. This promise Captain Lacroix had an opportunity of fulfilling on the 28th of November, when he captured the brig *Noah*, of Dundee, Captain Bowman. After taking possession of the brig, and offering Captain Bowman to ransom her, which was refused, the Frenchman agreed with Captain Knubley about the exchange of prisoners, and having filled up the necessary papers, and given him the sole command of the *Noah*, set him, and the crews and the passengers of both vessels, at liberty, declaring at the same time, that if the exchange of prisoners was honourably made on Captain Knubley's part, he would set every Englishman, whom the fortune of war should throw into his power, free the first opportunity.

During this war, commerce, like politics, continued in a state of extraordinary excitement, being too often a mere lottery, prices depending on the course and result of events which no sagacity could foresee. A victory or a defeat made one man, who was rich in the morning, poor at night, or suddenly raised another from poverty to riches. May Great Britain never again experience the horrors of such a prolonged struggle ; but if her own liberties and those of mankind call for a similar magnificent effort of courage and endurance, may the sons and daughters of the most powerful empire the world has ever seen, do their duty as valiantly and as successfully as their forefathers, who held the bridge of liberty against the Arch-tyrant in "the brave days of old." Passing over an innumerable series of minor engagements, captures and recaptures, which would only weary the reader, though representing great bloodshed and immense gains and losses to the combatants, we proceed, in the next Chapter to chronicle the leading incidents and the fading glories of privateering during the second war—by the grace of the Prince of Peace may it be the last war—with the United States of America.

CHAPTER VI.

Liverpool Privateers during the Second War with America.

The relations of Great Britain with the United States had been of the most unsatisfactory character ever since the first issue of the Orders in Council, in 1807, which compelled all vessels on their way to blockaded ports to touch at British harbours, and asserted a right of seizing British sailors found in American vessels. The United States, highly exasperated, met this step by the announcement that all intercourse with Great Britain and her dependencies was at an end. Although the embargo was withdrawn in 1809, and the trade with this country for a time resumed, the friction still continued. In spite of the remonstrances of the American Government, of the American merchants of Liverpool and elsewhere, and of many of the ablest men in this country, the British Ministry persisted in enforcing the orders until June, 1812. President Madison, in his address to the American people, stated that upwards of a thousand American vessels were seized under these orders on the high seas, carried into English ports, where many of them were condemned, and all subjected to heavy losses. At the beginning of the year 1812, as the commercial and manufacturing distresses became greater, it appeared that a perseverance in the unwise policy would produce a war with America. The great attainments and powerful eloquence of Mr. Brougham were for four years ranged against the orders, and on the 16th of June, 1812, his efforts were crowned with success, Lord Castlereagh announcing to Parliament that the ministry had decided to suspend the orders. The concession came too late. Two days after Lord Castlereagh's announcement and three weeks before the news could reach America, President Madison had issued a declaration of war against Great Britain. The war lasted about two years and-a-half, inflicting enormous losses on both belligerents, whilst their successes were so nearly balanced that both nations were heartily glad to accept the mediation of the Emperor of Russia to put an end to the unnatural conflict. Perhaps there never was a contest where the amount of political and commercial benefit received on either side was so ridiculously disproportionate to the frightful material and moral damage inflicted by the belligerents upon one another.

From a return made to the House of Lords, it appears that from the 1st of October, 1812, to the 1st of May, 1813, 382 British ships were captured by the Americans, of which 66 were retaken and 20 restored, leaving a loss of nearly 300 British ships in seven months—a most unsatisfactory result of a naval war for the mercantile classes. It is difficult to arrive at a correct estimate of the losses on each side. "In the course of the conflict," says Baines, "from eight hundred to a thousand English merchant ships were taken by the American privateers and ships of war, and at least an equal number of American merchantmen were taken by the British cruizers." The American privateer commander, Captain Coggleshall, however, puts down the number of American vessels taken and destroyed by the British, at not more than five hundred sail. He points out that most of the American losses occurred during the first six months of the war. After that period the United States had very few vessels afloat, except privateers and Letters of Marque. A large portion of their merchant-ships, he says, returned home within the first two or three months after the commencement of the war, and were laid up out of reach of the enemy. Some of them were taken up the navigable rivers, and others dismantled in secure places. The same authority claims that the little navy of the United States, with the aid of privateers and Letters of Marque, captured, burnt, sunk, or destroyed about two thousand sail of British shipping, including men-of-war and merchantmen. This statement does not include captures made on the great lakes, which would swell the number to a much larger figure. It has been roughly estimated that of these two thousand vessels, two-thirds, or say thirteen hundred and thirty sail, were taken by American privateers and private armed vessels, and the remainder by United States Government ships. The British, according to Captain Coggleshall, entered the contest with a navy of 1060 men-of-war, 800 of which were in commission, and were effective cruising vessels. To oppose this immense force, the United States had but seven effective frigates, with some twelve or fifteen sloops of war. Of the latter, the greater part were lying in the dockyards repairing.

In the latter part of the war, the risk of capture was so great that the freight on cotton, from Savannah to France, rose to 10d. a pound. The insurance on coasting voyages in America, rose to the rates of from 6 to 25 per cent.* according to circumstances. At the close of the contest, upwards of 200,000 bales of cotton (then more than a year's supply) was piled up in the warehouses of America, whilst

* *The Liverpool Mercury*, of May 7th, 1813, quoting from an American insurance list, says, "the following is a statement of the premiums of insurance on the coasting trade from Boston, on the 3rd ult.—To Eastport, 7 to 10 per cent.; other eastern ports, 2 to 5; to New York, £6 to £7 10/-; to Philadelphia, 10; to the Chesapeake, 12 to 15; to North Carolina, 17 to 18; to South Carolina, 21 to 28; to Savannah, 22 to 25. With regard to foreign trade, it is emphatically stated in the insurance list that there is none remaining, except to France, and the premium upon voyages to that quarter is 30 to 50 per cent.!" Another account stated that at Halifax insurance had been absolutely refused on most vessels; on others 33 per cent. had been added to the former premiums.

England was suffering distress for want of it. The highest quotations of the war, for cotton, were those of March 19th, 1814, as follows:—New Orleans, 3/- to 3/2½ per lb.; Sea Island (April 9th), 3/11 to 4/1; Pernambuco, 2/11½ to 3/1½; Surat, 1/9 to 2/-.

American privateers swept the Atlantic, and even penetrated within a few leagues of the mouth of the Mersey. The merchants and shipowners of Liverpool, instead of fitting out private armed vessels with the energy which had characterised them in former days, put their trust in the Lords Commissioners of the Admiralty, and found, too late, that the King's cruisers, like the modern policeman, were too often absent from the spot where their services were most required. The depredations of the American privateers on the coasts of Ireland and Scotland at length produced so strong a sensation at Lloyd's, that it was difficult to get policies underwritten, except at enormous rates of premiums. It is said that thirteen guineas for £100 was paid to insure vessels across the Irish Channel.

Liverpool suffered greatly in 1812, the diminution in the number of ships entering the port (compared with 1810) being 2130, representing a fall in tonnage of 287,603 tons, and in dues of £21,379.

The following is a copy of a letter from Captain Affleck, of the ship *May*, to his owners in Liverpool, dated St. Lucia, 8th of August, 1812:—

"I am happy to inform you of the arrival of the ship *May* here on the 5th. Nothing materially occurred on the voyage until the 3rd inst., at 2 p.m., when a vessel was seen from the masthead, bearing W.N.W. standing to the S.E. the wind at the time E.N.E. a light breeze, our course west, being at the time in the latitude of this island, and about 160 miles to windward. At 4 p.m., had neared this strange sail so as to see his hull distinctly, and perceived him to be a large schooner, and apparently a vessel of war. Ordered all hands to quarters, and had everything clear for action. At five, he tacked to the northward, and at half-past he tacked again, and came into our wake, when he immediately bore up after us under all sail, with English colours hoisted, and not wishing to let him come too near, fired the stern guns at him, which were immediately returned by his broadside of 4 guns, and was answered by the *May* in the same manner. At 7 p.m., he hoisted a light, and hailed—"Where is that ship from?" Answered, "Falmouth," and demanded to know what schooner that was. He replied—"A British man-of-war," and ordered me on board with my papers immediately. I told him if he attempted to come a yard nearer, he should receive my broadside, but at that distance I would send my boat on board, which I did, with my chief mate and two men. His boat immediately returned with six men and an officer, all armed, none of whom were allowed to come on board, except the latter; one who attempted had a pistol put to his breast, and he immediately sat down in the boat. The officer, on coming on board, told me he was a British privateer, belonging to Bermuda, and insisted on my going on board his boat with my papers. I told him I was a British Letter of Marque, and would not quit my vessel, unless to go on board one belonging to his Majesty, and ordered him out of the ship, at the same time, desiring him to send my mate and people on board. His boat soon after returned with the following note: "Captain Taylor presents his regards to the master of the ship, and insists on his coming on board with his papers, otherwise he may abide by the consequences." My answer was as before, and sent his boat off. He then hailed, declared he would sink me, should I refuse to comply with his request. My answer was, "Fire away!" which was put into execution as soon as his boat reached him, by his broadsides, and showers of musquetry, and was as quickly returned by the *May*. I had no longer a doubt of his being an American privateer, and on the dawn of day my suspicions were confirmed by his colours. From this time, half-past 7 p.m., till 9 a.m., a fire, with very little intermission, was kept up by both vessels; and it appeared during this long action to be his intention to board the *May*, which was always frustrated by rounds of grape, until at last he was obliged to haul off in the greatest confusion with his sails, rigging, and hull dreadfully cut up, and indeed we are in the same situation, having six of our lower shrouds shot away, forestay, main-top mast back-stay, three shots between wind and water, the main topmast wounded, and the sails and running rigging cut to pieces, one man killed, and two wounded. And it affords me the greatest pleasure to say, that nothing could exceed the coolness and bravery of the few people I had the honour to command.

"I am, Gentlemen, your most obedient servant,

"WILLIAM AFFLECK.

"P.S. The above is a confused account of the action with the American privateer, as I had only a few minutes' notice of this opportunity to write you by way of Martinique. I have, however, only to add, that had the *May* been armed with any other guns than those on Col. Congreve's plan, she must inevitably have been captured, from the small crew I had on board—one man having been killed, and one wounded, by the first discharge from the privateer, after the return of his boat. The privateer mounted eight guns and full of men.

"Killed—Joseph Rummona, seaman. Wounded, J. B. Hanna, second mate; Wm Walker, apprentice; both slightly, and they are doing well.

"Prisoners on board the privateer—Samuel Hazelhurst, chief mate; John Erick and James Antonia, seamen."

* In this case we have the advantage of presenting the enemy's version of the affair. The following is an extract from the log of the American privateer schooner *Shadow*, of Philadelphia, Captain Taylor, which evidently met, without catching, a Tartar in the *May*:—

"On the 4th of August, at half-past twelve (meridian), saw a sail to the eastward standing westward; made all necessary sail in chase. At half-past five p.m., carried away the square-sail boom; cut the wreck adrift; rigged out the lower studding-sail boom, and set her square-sail again, coming up with the chase. At six p.m., being within gunshot, she commenced firing from her stern guns. At seven p.m., came up with her, and commenced an action; at half-past seven, the ship hoisted a light in her mizzen rigging, which was answered by a light from us; at the same time hailed her. She hailed from Liverpool, when Captain Taylor ordered her to send her boat on board with her papers, which she in part complied with, by sending her boat with an officer and two men, whom we detained, and gave directions to man the boat with our crew, board the ship, and demand her papers. These orders were delivered by Mr. Thomas Yorke, who received for answer, that such a demand would not be complied with, at the same time handing

The *Shadow* arrived at Philadelphia, was refitted, and soon sailed on another cruise.

As a mark of their sense of his gallantry, the Underwriters of Liverpool presented Captain Affleck with an elegant silver cup of the value of forty pounds, and at a general meeting passed a resolution allowing him free access to their rooms. Thus did the merchants, shipowners, and underwriters of Liverpool at all times foster the spirit of gallantry and fidelity in the merchant navy of the port, and helped to render it formidable in war and unrivalled in peace.

On the 13th of December, 1812, Captain G. Howard, of the private ship of war, *John Tobin*, writes from Bahia to his owners, Messrs. Hughes and Tobins, of Liverpool, describing an action between that ship and the American privateer *Alfred*, as follows :—

"GENTLEMEN,—I am happy to inform you of my safe arrival at Bahia, after a pleasant passage of 46 days. Nothing material occurred until the 21st Nov., being in the latitude 8. 10. S., and longitude 33. 30. W., being a degree or two to the eastward of Cape St. Augustine, at one o'clock in the morning, being moonlight, a vessel was seen under a very heavy press of sail in our wake, coming up fast. I continued

him a note addressed to Captain Taylor, purporting that his ship was a British Letter of Marque, called the *May*, from Liverpool, bound to St. Lucia, commanded by Captain Affleck, mounting fourteen guns and fifty men. He also stated that the Orders in Council had been rescinded, and a change of Ministry taken place in England. The note was handed to Captain Taylor. The boat was again sent on board, with a note from our captain, demanding his papers, which were refused. At half-past eight o'clock a brisk fire commenced on both sides, during which time William Craft, sailmaker, was wounded. At ten p.m., dropped astern, with the intention of lying by all night within gunshot; at intervals kept up a brisk fire; weather, squally and dark.

"At daylight, ranged up under her stern and commenced a severe action, when we received a shot in our starboard bow which shattered the wooden ends, started the plank shear, and broke several timbers. At half-past seven a.m., received another in our larboard bow; struck the larboard after-gun-carriage, killed six men and wounded three. At half-past eight a.m., our commander received a ball in his left temple, which instantly terminated his existence, to the inexpressible regret of all hands. About the same time a shot struck under the larboard fore-chains, between wind and water, which caused the vessel to leak badly; found three feet water in the hold on sounding the pumps."

on our course until four o'clock, when I called all hands to quarters, took in steering sails, stay sails, hauled up the courses, and prepared for action, she being then on my weather quarter, and took in all her small sails and prepared for action also. At five o'clock, I hauled up for her, being daylight, and hoisted Spanish colours, with a gun. At a quarter past five, she being within gunshot, hoisted American ensign and pendant, and gave us a shot. I thought it prudent to keep up the Spanish colours until he came a little nearer, as the *John Tobin's* guns are short, and I did not wish to let him know we were English until my guns would tell. He soon gave us another shot between the fore and main masts. I then down with Spanish colours and displayed the British, which he no sooner saw than he began to fire away with round and grape shot as fast as he could well load and discharge, and we returned it as quick as he sent it, from a quarter before six until a quarter after nine o'clock, when we both desisted in order to repair damages, having the chief part of my running rigging and sails shot away, also two guns disabled during the action. At seven o'clock, he shot away my ensign halliards, and our colours came tumbling down. It was not long, however, before they were up again, and a second time shot away, on which, one of my people volunteered to go to the mizen topmasthead and nail them up, which was done, although the shot was flying in all directions. I then hoisted the red flag forward, and gave them three cheers. At a quarter before ten, we commenced firing again, and shot away his boom. He then thought it best to make off, and making all sail, kept close by the wind. At a quarter after ten, she tacked and stood to the northward. The *John Tobin* tacked also after her, firing as long as our guns would have any effect ; but he soon got out of their reach, owing to his superior sailing. At eleven o'clock, I wore ship and stood on my course, not being able to come up with him.

"Gentlemen, I cannot speak too highly of Mr. Cannon, my chief mate, likewise the rest of my officers and ship's company, not forgetting Mr. Toole, passenger, for his gallantry during the whole of the action. I am happy to say that the *John Tobin* has not suffered much in her hull. The privateer mounts nine guns on a side, with two stern chasers, and full of men. Her guns are nine-pounders, having some of her shot on board, which we have taken out of the *John Tobin's* bends. She was a long, low ship, with a billet head, yellow sides, and three royal yards rigged aloft, exactly the appearance of a small sloop of war. You will be astonished, Gentlemen, to hear that I had not a man or boy hurt on board, although the shot were flying about us like hail. Since my arrival here, I am told that there are three American privateers on this coast, two of them brigs, and the other (the one we engaged) a ship. I shall be under the necessity of purchasing some more powder before I sail. I suppose we shall have another dust coming home, with one of the brigs; as for the ship, she will not come near us again."

The Underwriters of Liverpool presented Captain Howard with a silver cup "in testimony of his gallant and seaman-like conduct in defending his ship against the American privateer *Alfred*, on the 21st of November, 1812, off the coast of Brazil."

The following details of an action between the *Bridget*, of Liverpool, Captain Archibald Kennan, and an unknown armed schooner, which fought under English colours, off Surinam river, were supplied by a person on board the *Bridget*, writing from Demerara :—

"On the 6th December, 1812, at 4 p.m., saw a sail to the northward, standing to the southward; at 5 p.m., made it out to be a large hermaphrodite brig or schooner. At half-past five, when he was on our weather quarter, he bore up before the wind and stood towards us. We then cleared ship for action, supposing him to be an American privateer. At six, we took in all steering sails, and hauled to the northward to see what he was before dark. On this he took in his steering sails also, and came down with English colours flying. He hailed and asked what brig it was? Captain Kennan asked what schooner? He answered, H.M. schooner——, and desired us to send our boat on board, which the captain refused. He then said, if our captain did not comply, he would take him to the gangway and flog him. The captain answered, and said it was more than he dared to do. He hailed the fourth time, and said if we did not send our boat on board, he would fire into us. Captain answered, if he did we should return it. He instantly fired a broadside of round and grape, with musketry, which we returned. The action commenced at a quarter-past six p.m., and lasted until five minutes past eight. He fought under English colours during the whole of the action. We had two men killed and six wounded. After the first broadside, we ran the brig on board of him, between his main-mast and fore-mast. He mounted 18 guns, and a very large one amidships, which did us a great deal of damage in our bows, several shot going through our bows, carrying away timbers and breast-hook, and going through our upper-deck beams and deck, our jib-booms both carried away; masts, sails, and rigging very much injured, cut-water shattered to pieces, and our figure-head very much damaged. From the situation of the privateer or pirate in fighting, he must have received considerable damage from our vessel's bows falling on him. When he got clear of us, he stood off immediately, and as we could not get the brig to wear, for the purpose of following him (by reason of our anchors being shot from the bows, and hanging by the cables), and the night being dark, we soon lost sight of him. During the action, he attempted to set fire to us with a bag of combustibles, with a view to board us at the same time, but he was received different to what he expected. The crew are extremely thankful to the passengers for their heroic exertions during the whole engagement.

"List of Killed and Wounded.

"John Burns, seaman, and Alexander M'Keller, killed; James Sanders and Samuel Turner, severely wounded; Daniel Dunn, passenger, had a musket ball through his ankle; Daniel Ross, passenger, Thomas Capper, and Captain Kennan, slightly wounded."

The Underwriters of Liverpool presented Captain Kennan with a silver cup in recognition of his gallantry.

The Underwriters of Liverpool also presented a silver cup to Captain John Irlam, of the ship *Maxwell*, "in testimony of his gallant and seamanlike conduct in defending his ship against the *General Armstrong* American privateer of 18 eighteen-pounders, and 1 forty-four-pounder, and 130 men, on the 29th of November, 1812, off Demerara, in which action he was severely wounded." The Committee also voted a free admission to Lloyd's Rooms, in Liverpool, to the three gallant commanders, Captains Irlam, Howard, and Kennan. The *General Armstrong*, which belonged to New York, captured one of the most valuable prizes made during the war—the ship *Queen*, 16 guns and 40 men, from Liverpool, with a cargo worth from £70,000 to £100,000 sterling. She was bound to Surinam, and was bravely defended, the captain, his first officer, and nine of his crew being killed before she was surrendered. The prize was wrecked off Nantucket. The *General Armstrong* took the brig *Lucy and Alida*, with a full cargo of dry goods, which was retaken by the Liverpool Letter of Marque ship *Brenton*, and again recaptured by the *Revenge*, of Norfolk, and sent into that port.

The brig *Henry*, 6 guns, 200 tons, from Liverpool for Buenos Ayres, laden with 300 packages of dry goods and other valuable articles, invoiced at £40,000 sterling, was taken and sent into New York by the *Governor Tompkins* of that port. The bounty (or reduction of duties) allowed by the United States on this prize, amounted to about 35,000 dollars. The *Governor Tompkins* was a very formidable vessel and made many prizes. On her first cruise, she was commanded by Joseph Skinner, of New London, and subsequently by Captain Shaler. She suffered severely from the shot of a British frigate, but finally made her escape.

The brig *Nancy*, from Liverpool for Halifax, laden with dry goods, was captured by the *Portsmouth* privateer, of Portsmouth, N. H., divested of 318 bales and packages of goods invoiced at £27,000 sterling, and ordered into port. The *Portsmouth*, commanded by John Sinclair, was a conspicuous cruising vessel, and made a great many valuable prizes. The *Fox* privateer, belonging to the same port, captured and burnt the schooner *Brother and Sister*, and the brig *Dove*, from Liverpool ; and sent to Norway the sloop *Fox* and the brig *Chance*, both from Liverpool. The *Stork* sloop-of-war, and the *Fortune* frigate, cruised between Achill Head and Cape Clear, and off Tory Island, in vain quest of the *Fox*. Some of the *Fox's* people had the audacity to go ashore at Sligo and Newport, in uniform, and, personating British officers, procured supplies of fresh provisions, etc., at both places, and gave the requisite drafts for the payment of the amount. The name of the Captain of the *Fox* was said to be Stewart. He was formerly master of one of the regular traders between Londonderry and Liverpool. The *Fox* mounted 20 guns, and had a crew of 150 picked seamen. The *Thomas* privateer, also of Portsmouth, N. H., captured the ship *Dromo*, of 12 guns, from Liverpool for Halifax, with a cargo invoiced at $70,000 sterling. The *Macedonian*, of Portsmouth, captured and burnt the brig *Britannia*, from St. John's, N. B., for Liverpool, laden with 195 tons of ship's timber and other articles.

The *True Blooded Yankee* privateer, of 18 guns and 160 men, cruising in St. George's Channel, captured the *Margaret*, of Hull, the *Fame*, of Belfast, with linen for London, the *George*, from Kinsale, a Liverpool Letter of Marque (name unknown), of 14 guns, bound for Spain, and three other vessels. The *Margaret* was recaptured and carried into Plymouth. The *True Blooded Yankee*, was formerly the *Challenger* gun-brig, and her crew were said to be chiefly British. The captures were made between Holyhead and the Skerries. Captain Coggleshall, himself a distinguished privateer commander, thus refers to the vessel in his History:—

"The famous brig privateer *True Blooded Yankee*, carrying 18 guns and 160 men, was owned by an American gentleman, residing in Paris, by the name of Preble. She had an American commission, and sailed under the American flag, but always fitted and sailed out of French ports, viz., Brest, l'Orient, and Morlaix. This vessel was very successful. She cruised the greatest part of the war in the British and Irish Channels, and made a large number of rich prizes. These she generally sent into French ports; sometimes, however, she sent a few to the United States. During one cruise of thirty-seven days, she captured twenty-seven vessels, and made two hundred and seventy prisoners. While on this cruise she took an island on the coast of Ireland, and held it six days; she also took a town in Scotland, and burned seven vessels in the harbour. She was soon after fitted out to make another cruise, in company with the *Bunker Hill*, of 14 eighteen-pounders and 140 men. When the *True Blooded Yankee* arrived in France, she was laden with the following spoils:—18 bales of Turkey carpets, 43 bales of raw silk, weighing twelve thousand pounds; 20 boxes of gums, 46 packs of the best skins, 24 packets of beaver skins, 160 dozen of swan skins, 190 hides, copper, etc."

An account of a gallant action with the American privateer *Snap Dragon*, is given in the following extract of a letter from William Hill, master of the ship *Liverpool*, to Messrs. Hughes & Tobins, the owners, dated Demerara, 25th of March, 1814:—

"GENTLEMEN,—On the 3rd inst. the entrance of Juramac [Saramac?] River bearing S.S.W. six or seven miles, whilst standing in shore, endeavouring to get to windward, saw a schooner on our larboard bow, standing the same way under her foresail, but immediately made all sail. At half-past six a.m. she bore up for us; a quarter before seven she fired a shot, and hoisted American colours. At half-past seven, finding her shot going over us, opened our fire on him. At eight, the enemy nearing us, and making every attempt to get under our stern. At 9 she opened her broadside, still keeping up a hot fire from his long gun, whilst we annoyed him much from our quarter guns. At ten the enemy hoisted a red flag at the fore, gave us a volley of musketry, three cheers, and again bore up for us, which was returned with a broadside and musketry. Finding from her superiority in sailing I could no longer keep her on either quarter, I bore up before the wind, and set topgallant sails, and got the two aftermost guns through the stern ports. At 11, he dropped astern, frequently cheering; at half-past he made sail. At 12, came up with a drum and fife playing, and keeping up a hot fire of grape and musketry, but firing high, which we returned with grape and cannister. A quarter before one, the enemy closing fast, I ordered the helm a starboard, to bring the larboard guns on him, when he run us on board, her jib-boom coming through the bulwark abaft the cross-tree, and broke short off. Having fresh way on the ship by putting the helm a port, we carried away his bowsprit, when between the main and mizen rigging, the enemy threw a number of men on board of us, and fell astern. By the time he was clear, we had drove the boarders from the deck, over the side and into the chains, where a number of them were killed and wounded, falling and jumping overboard, the enemy lying across our stern, still keeping up a smart fire. At one, his mainsail came down, we keeping up a hot and destructive fire from the stern guns right into him. At half-past he hauled off on the larboard tack, her fire slackening. On her coming to the wind, the foremast went over the starboard bow, taking with it the main topmast, and the head of the mainmast. I never saw so complete a wreck. She then came to an anchor, many of her crew swimming towards her. It was my first intention to have renewed the action, but on hauling to the wind for that purpose, I found all our sails and running rigging shot away, and a strong lee current running, I determined to run for this port, which I trust will meet your approbation.

"From one of the boarders, wounded on our quarter-deck (who died next day on board us), I gained the following account:—The schooner privateer *Snap Dragon,* Captain Murphy, of 112 men, six four-pounders, and one long twelve-pounder traverse gun, belonging to South Carolina, out four weeks, had taken nothing, and had been chaced by a line of battle ship and a frigate.

"My chief officer, Mr. Williams, and every one of my crew, behaved in a most gallant and daring manner, so much so, it is impossible to say to whom the preference is due. Mr. Williams, just at the conclusion of the action, was wounded by a pistol ball in the left jaw; he, together with three seamen also wounded, are doing well.

"The boarders took so precipitate a leave of us, that they left two pistols, one bayonet, and a cartouch box on our quarter deck.

"Enclosed I transmit you duplicates of two letters* I have received from Captain Muddle, of his Majesty's ship *Columbine.*

"Gentlemen, Your most obedient servant, WILLIAM HILL.

The following description of a desperate conflict between the Liverpool Letter of Marque ship *Fanny* and the American

"His Majesty's sloop *Columbine*,
Demerara, 16th March, 1814.

"SIR,—I have to acknowledge the receipt of your letter of yesterday's date, detailing the gallant defence of the ship *Liverpool*, under your command, on the 3rd inst. against the American privateer *Snap Dragon*, and send you herewith a pendant which you will be pleased to carry on this coast, during the time I may have the command, as a mark of distinction for your meritorious conduct; as also the enclosed protection for your gallant crew from impress by any of his Majesty's vessels under my orders. I am, Sir, your most obedient servant,

[Signed,] "R. H. MUDDLE, Captain.

"To Mr. Wm. Hill, Master of the ship *Liverpool*, of Liverpool."

"By R. Henry Muddle, Esq., Commander of his Majesty's sloop *Columbine.*

"In consequence of the very gallant defence of the ship *Liverpool*, Mr. W. Hill, master, against the American privateer *Snap Dragon*, on the 3rd of March, off Surinam River, I have thought proper to permit her to wear a pendant, during the time of my command on this coast, and to grant her a protection for her crew, during the said time.

"The commanders of his Majesty's vessels under my command are hereby required and directed to respect the same.

"Given under my hand, on board his Majesty's sloop *Columbine*, in the River Demerara, 16th March, 1814. "R. HENRY MUDDLE, Commander."

privateer *General Armstrong*, of New York, was addressed by Captain Laughton, of the *Fanny*, to his owners, Messrs. Brotherston and Begg, on the 30th of April, 1814:—

"On Monday, the 18th inst, about meridian, we discovered a schooner standing towards us, supposing her to be an enemy, we immediately prepared for action, but it being wet and squally, he did not think it right to engage us on that day, but kept sight of us the ensuing night, and about half-past seven a.m. bore down to us with two American ensigns flying, and when he had got about the distance of a pistol shot he commenced a most severe and destructive fire, which the *Fanny* with alacrity returned, but the wind having fallen almost to a calm, the *Fanny* would scarcely steer, and the enemy having the superiority in sailing, kept upon our quarters, notwithstanding we shot away his main-fore jib, and flying-jib halyards, when he fell alongside with only his topsails set. At this time was a desperate conflict, but his fire from a long French forty-two-pounder proved so tremendous, and his numerous musketry so galling, that the great part of the men on the main deck could not be kept to their quarters, notwithstanding the exertions of Mr. Bridge, the chief mate. I thought it my duty, though a painful one, to save the lives of the brave few that remained true, to haul down the colours, after engaging one hour, never out of pistol shot, to the *General Armstrong*, Champlin, from New York, and I trust you will not think the *Fanny* given away. She had scarcely a shroud left standing, nor one brace, the sails completely reduced, several gun carriages disabled, not a breeching left whole, one shot between wind and water, several others through different parts of her hull, the maintopsail and topgallant yards shot through, not a running rope but what was cut to pieces, a complete wreck on the quarter deck, the second mate, my brother, killed by my side, and six others wounded, five severely, one slightly. Amongst the former I am truly sorry to say is your Mr. Begg, whilst gallantly doing his duty with the musketry, but I hope his wound is doing well.

"The *General Armstrong* is a schooner of the largest class, say from 250 to 300 tons, armed with 1 French forty-two long-pounder, 6 long nine-pounders (King's), with G. R. upon them, and between 90 and 100 men. She has been chased within the last month by three different men-of-war, but always escaped by superior sailing, although at times nearly within gunshot. I remain, Gentlemen, your most obedient servant, JOHN LAUGHTON."

The *Fanny* (which was bound from Maranham to Liverpool when captured), was retaken by the *Sceptre*, M.W.

In August, 1813, the American sloop of war *Argus*, 360 tons, mounting 18 twenty-four-pounder carronades, two long twelve-pounders, and 149 men, commanded by Captain W. H. Allen, committed great ravages off the coast of Ireland, capturing and burning many valuable vessels. One morning, in sight of Lundy, in a very thick fog, she found herself in the midst of the Leeward Island fleet, eleven in number, several of which she captured. Captain Allen, while on board one of the prizes, said he had destroyed eleven vessels off the Shannon, and had orders to destroy all vessels they fell in with, in retaliation for the damage done by the British navy on the coasts of America. He had taken a great many other vessels. The *Argus* had actually a Pill pilot on board. She was captured by boarding, on the 12th of August, 1813, after an action of 43 minutes, by H.M.'s sloop *Pelican*. The commander of the *Argus* was dreadfully wounded in his leg and thigh, by one of the raking fires of the *Pelican*, which at the same time, carried away the leg of a midshipman, wounded the first lieutenant in the head, and killed several of the crew. Captain Allen suffered amputation after the *Argus* arrived at Plymouth, and received the most "humane and polite attentions," but he appeared to be aware of his approaching dissolution, spoke little, and seemed perfectly resigned. He was taken out of the *Argus* and carried to the hospital at Mill Prison. On leaving his ship, of which he must have been proud, the dying hero looked up for a moment and exclaimed, "God bless you all, my lads—we shall never meet again." His auditors were so deeply affected that not a man of them could articulate, "Farewell." Soon after he reached the hospital he expired. His remains were interred at the Old Church, Plymouth, with the most distinguished honours. "The funeral procession, as it moved from the Mill Prison, afforded a scene singularly impressive to the prisoners, who beheld with admiration the respect paid by a gallant, conquering enemy to the fallen hero. Five hundred British Marines first marched, in slow time, with arms reversed; the band of the Plymouth Division of Marines followed, performing the most solemn tunes. An officer of Marines, in military mourning, came after these. Two interesting black boys, the servants of the deceased, then preceded the hearse. One of these bore his master's sword, the other carried his hat. Eight American officers followed the hearse, and the procession was closed by a number of British naval officers. On the arrival of the body at the Old Church, it was met by the officiating minister, and three volleys over the grave—the tribute to departed heroism—closed the scene. Captain Allen was First Lieutenant of the *United States*, in her action with the *Macedonian*, and was made captain for his bravery in the action. Captain Decatur was much attached to the deceased, and made him a present of two brass guns from the *Macedonian*, which are now on board the *Argus*. He was highly esteemed in his profession, and was an officer of the most determined courage."* Such was the

*The *Waterford Mirror* brought a charge of "barbarity" against Captain Allen, on the testimony of a cattle dealer, passenger in the *Diana and Betsy*, one of the captured vessels who stated that there were 30 head of cattle on board, of which the enemy killed three for the use of his crew, *and burned the rest with the vessel.* Other papers bear testimony to the excellent conduct of Capt. Allen, and his courtesy and humanity towards the passengers and crews who fell into his hands.

tribute paid by the valour of old England to the gallantry of young America.

The celebrated American privateer brig *Yankee*, owned by Mr. James De Wolf, of Bristol, Rhode Island, was a most fortunate cruiser, and made a great many captures. She took the *Royal Bounty*, a British Letter of Marque ship, after a severe engagement, and ran all the war without being captured. In several of her first cruises, she was commanded by Captain Wilson, and subsequently by Captain Smith. The *Yankee* arrived at Newport, R.I. after a cruise of about 150 days, during which she had scoured the whole western coast of Africa, taken eight prizes, 62 guns, 196 men, 496 muskets, and property worth 296,000 dollars. She had on board 32 bales of fine goods, 6 tons of ivory, and 40,000 dollars' worth of gold dust. She looked in at every port, river, town, factory, harbour, etc., on the coast. Among her prizes were the following:—

Brig *Thames*, Captain Toole, of Liverpool, 8 guns and 14 men; with ivory, dry goods and camwood; worth 40,000 dollars.

Brig *Shannon*, Captain Kendall, from Maranham for Liverpool, 10 guns, and 15 men ; worth 50,000 dollars.

Portuguese ship *St Jose*, from Liverpool for Rio Janeiro, with dry goods, hardware, etc., valued at about 600,000 dollars, said to be British property, and sent into Portland, U.S.

The *Eliza Ann*, from Liverpool to Baltimore, sent into Boston.

The schooner *Alder*, Captain Crowley, of Liverpool, 6 guns (nine-pounders) and 21 men ; laden with 400 casks muskets, flints, bar lead, iron, dry goods, etc. ; vessel and cargo worth 24,000 dollars. In the engagement an explosion occurred, which blew up her quarter deck, and killed her captain and five of her men.

The *Yankee* also captured the *Despatch*, from Liverpool for Quebec, with a cargo invoiced at £80,000 sterling, and six other vessels, including the *Ann*, of Liverpool, which was afterwards retaken. The *Yankee* would have taken several other prizes but for the injury she received in capturing the barque *Paris*, of Liverpool, Captain Harrison, who with only 6 guns and 11 men fought the *Yankee*, which carried 18 twelve and nine-pounders, one bow gun, and 75 men, for 45 minutes, receiving great injury herself, and inflicting damage on the *Yankee*, which compelled her to put into port to refit.

The *Alexander*, Captain Newby, from St. Thomas's to Liverpool, was captured on the 2nd of August, 1814, in lat. 47. 13 N., long. 32. W., by the *Mammoth* American schooner privateer, of Baltimore, Captain Franklin, of 14 guns and 140 men. The *Mammoth* had been out seven weeks, and had made 16 captures! Captain Newby was seventeen days on board the privateer, cruising between the latitude of his capture and Cape Clear, and during the whole of that period did not see a single British ship of war. A meeting of the Committee of the Underwriters' Association of Liverpool was held on the 22nd of August, for the purpose of making some communication to the Admiralty respecting the numerous captures made by the Americans, when it was resolved, "as the most delicate and proper mode of proceeding," that a list of the captures made by the *Mammoth* be transmitted to J. W. Croker, Esq., for the information of the Lords Commissioners of the Admiralty. In a four months' cruise, the *Mammoth* took 21 prizes, 18 of which she destroyed, or gave up as cartels. Her cruising ground was principally on the coasts of Great Britain, and in the Bay of Biscay.

On the 29th of August, 1814, a meeting of merchants, shipowners, underwriters, etc., was held in the Liverpool Town Hall, to take into consideration a memorial to the Government on the subject of the numerous captures made by American cruisers. Mr. John Gladstone proposed an address to the Lords of the Admiralty, but this was opposed on the ground that representations had been made to that department without redress. Mr. Clare proposed an address to the Prince Regent, which, after warm opposition on the part of Mr. Gladstone, was carried, and the petition despatched on the 30th. The address conveyed a censure upon the Admiralty. At another meeting, held on the 30th, a counter address to the Admiralty was voted, very numerously signed, and sent off on September 1st. In this memorial, complaining of a want of sufficient naval protection against American captures, the memorialists spoke of privateers destroying vessels as a novel and extraordinary practice, which they were informed was promoted by pecuniary rewards from the American Government, and they wished measures adopted to prevent as much as possible the ruinous effects of this new system of warfare. Mr. Croker replied on behalf of the Admiralty, that an ample force had been under the orders of the Admirals commanding the western stations; and that during the time when the enemy's depredations were stated to have taken place, not fewer than three frigates and fourteen sloops were actually at sea for the immediate protection of St. George's Channel and the western and northern parts of the United Kingdom. In that case the vessels must have been totally unfit for the service required of them, for we cannot believe that the officers and men of the British navy shirked their duty.

Perhaps nothing will better illustrate the state of public feeling on this question, than the following resolutions of the merchants, manufacturers, shipowners, and underwriters of Glasgow, passed in public meeting, the Lord Provost in the chair, on the 7th of September, 1814:—

"That the number of American privateers with which our Channels have been infested, the audacity with which they have approached our coasts, and the success with which their enterprise has been attended, have proved injurious to our commerce, humbling to our pride, and discreditable to the directors of the naval power of the British nation, whose flag, till of late, waved over every sea, and triumphed over every rival. That there is reason to believe, that in the short space of less than twenty-four months, above eight hundred vessels have been captured by that power, whose maritime strength we have hitherto impolitically held in contempt. That at a time when we were at peace with all the rest of the world, when the maintenance of our marine costs so large a sum to the country, when the mercantile and shipping interests pay a tax for protection, under the form of convoy duty, and when, in the plenitude of our power we have declared the whole American coast under blockade, it is equally distressing and mortifying that our ships cannot, with safety, traverse our own Channels ; that insurance cannot be effected but at an excessive premium ; and that a horde of American cruisers should be allowed, unheeded, unresisted, and unmolested, to take, burn, or sink, our own vessels, in our own inlets, and almost in sight of our own harbours.

"That the ports of the Clyde have sustained severe loss from the depredations already committed, and there is reason to apprehend still more serious suffering, not only from the extent of the coasting trade and the number of vessels yet to arrive from abroad, but as the time is fast approaching when the outward bound ships must proceed to Cork for convoys, and when, during the winter season, the opportunities of the enemy will be increased both to capture with ease and escape with impunity.

"That the system of burning and destroying every article which there is fear of losing—a system pursued by all the cruisers, and encouraged by their own government—diminishes the chances of recapture, and renders the necessity of prevention more urgent.

"That from the coldness and neglect with which previous remonstrances from other quarters have been received by the Admiralty, this meeting reluctantly feels it an imperious duty at once to address the Throne, and that therefore a petition be forwarded to his Royal Highness, the Prince Regent, acting in the name and on behalf of his Majesty, representing the above grievances, and humbly praying that his Royal

Highness will be graciously pleased to direct such measures to be adopted, as shall promptly and effectually protect the trade, on the coast of this Kingdom, from the numerous insulting and destructive depredations of the enemy."

One of the lieutenants of H.M.S. *Leander*, writing from Fayal, on the 14th of January, 1815, gives the following curious account of an adventure which befell an American prize-master, in charge of a captured Liverpool brig:—

"In search of the American squadron, we saw a large brig the other day, which the captain ordered us to draw to, but under a moderate sail, so as not to show any particular anxiety, suspecting from circumstances she was a British vessel captured, and being desirous, if she should prove so, of getting hold of the American prize-master, and by imposing this ship upon him as an American frigate, obtaining information which otherwise we might not get. Nothing could have happened better. This brig proved to be the *John*, of Liverpool, lately captured by the *Perry* privateer; and the American prize-master, a high-blooded Yankee, hoisted out his boat, and without any hesitation came on board the *Leander*. The moment he got upon deck, he congratulated the officers on the squadron being at sea, and in a situation where they would, as he expressed it, do a *tarnation* share of mischief to the damned English *sarpents*, and play the devil's game with their rag of a flag. He then observed that he knew this ship the moment he saw her, by her black-painted masts and sides, and the cut of her sails, to be the *President*, as he was in New York just before she sailed. After these observations, in which the Yankee professed to be very well informed, he walked up to Sir George Collier;* and to the extreme amusement of us all, making his bow, addressed him as the American Com. Decatur, reminding him at the same time of having once seen him at New York. Sir George agreed to all this; when the Yankee presented the *John's* papers, to shew what she was, and complained of his

* Those who are familiar with the portraits of Sir George Collier, will readily believe that the scene must have been a fine comedy, highly enjoyed by the British commander.

crew, which he said were a set of such vile, mutinous sarpents, that his life was in his hand every night, and requested, therefore, some of them might be changed for so many of the supposed *President's* crew, and that one, in particular, might have a second flogging. All this Sir George promised, with great gravity, should be done, and ordered the First Lieutenant to have as many men ready in exchange for those complained of. The captain then asked Jonathan into his cabin, and retiring for a moment for a chart, returned with one in which the *Leander's* track was marked, over which was written "*President*, from New York, on a cruise," and placing his finger upon these words, as if by accident, they immediately caught the eye of the Yankee, who exclaimed, that he knew the *President* the moment he saw her, and Nick himself could not deceive him. He was then asked by Sir George, pointing to the *Acasta*, if he knew her. His reply was, that she was the *Macedonian;* and when asked what the *Newcastle* was, he said he did not know her; on which Sir George told him she was the *Constitution*. He replied, he recollected she was, though not painted as she used to be. He then asked the *Perry's* cruizing ground, and he said he had spoken the *Whig* privateer, who told him he would probably soon fall in with Commodore Decatur's squadron, which rejoiced his heart, as he knew he should then get rid of some of his mutinous crew. After he had no more to tell, Sir George recommended his returning to the *John*, and in great form returned him the ship's papers, wishing him a good voyage, and desiring he would not forget to let it be known he left Commodore Decatur and his squadron well. Jonathan took his leave with great apparent satisfaction, but when about to quit the *Leander*, our First Lieutenant M'Dougall stopped him, and apprized him of his real situation. For a long time he considered this a joke, but casting an eye upon the English Captain's uniform, in which Sir George Collier then appeared, he became almost frantic with disappointment."

Early in the spring of 1814, the peace, for which the people were wearying, at length arrived, and was thus announced in Liverpool: "Downfall of the tyrant! Peace! heavenly peace! the desire of all nations dawns on the world!! The Almighty's name be praised!" In May, the long-suspended commerce with France was renewed, by the importation of two cargoes of grain into Liverpool from Havre. Negotiations for the arrangement of the differences with the United States were opened at Ghent, in the following June, and, after a long delay, which cost many thousands of lives, were brought to a close on December 24th. The arrival of the first American ship in Liverpool after the peace, was thus announced in the paper of the 3rd of April, 1815 :—

> "Several hundred vessels left this port on Friday and the day before, which had been detained many weeks by adverse winds. The river afforded a most brilliant and interesting spectacle. A still more pleasing and interesting sight was witnessed on Thursday, about one o'clock, in the arrival of the ship *Milo*, the first belonging to the United States which has arrived since the restoration of peace. The day was remarkably bright, and she came up the river in very fine style, with the British flag flying at the mainmast head, the American colours at the mizenmast, which were lowered on passing H.M.S. *Argo*, lying in the river, and a beautiful signal-flag at her foremast. This first effect of the restoration of amity between two countries, designed by nature, habits, and mutual interests, to maintain uninterruptedly the relations of peace, was hailed with delight by a great number of spectators, who covered the piers and the shore. The *Milo* left Boston on the 12th ult., in company with the Liverpool packet, daily expected. The *Milo* arrived in ballast."

The arrival of the first British vessel at New York, on the 5th of May, 1815, was thus announced in one of the papers of that city:—"The regular British packet, after an absence of nearly three years, at length re-appears in our harbour, in token of returning amity. We hail with sensations of gladness the joyful omen, and may no inauspicious event ever occur again to banish her from our waters!"

This was the last war in which Letters of Marque and Reprisals were granted by the British Government. Without discussing the wisdom or otherwise of the Declaration of Paris, or attempting to prophecy its effect on British commerce in the event of a great naval war, we shall close this section, with a plain statement of the present attitude of the British and American governments, towards privateering.

On the 30th of March, 1856, on the conclusion of the war with Russia, there was signed the so-called Treaty of Paris. Subsequently the plenipotentiaries, who signed that treaty sat in conference, and on the preamble that "maritime law in time of war, had long been the subject of deplorable disputes," they adopted a solemn Declaration, which has since been known as the "Declaration of Paris," and which was appended to the treaty, on April 16th.

The Declaration ran as follows:—

> "1. Privateering is, and remains abolished.
>
> "2. The neutral flag covers enemy's goods, with the exception of contraband of war.
>
> "3. Neutral goods, except contraband of war, are not liable to capture under the enemy's flag.
>
> "4. Blockades, in order to be binding, must be effective, that is to say, maintained by a force sufficient in reality to prevent access to the coasts of the enemy.
>
> "The Declaration not to be binding except between the Powers acceding to it."

By this Declaration, Great Britain and the other states who signed it were, of course, bound; and all civilised nations have since acceded to it, except the United States, Mexico, and Spain. Accordingly the United States, in the deplorable event of a war with Great Britain, would be justified in using privateers, and Great Britain, though a signatory of the above Declaration, would also be justified in using them in a war against a state, which is not bound by it.

At first sight it seems extraordinary that so enlightened a country as the United States should be found associated with Spain and Mexico in upholding privateering. This has not always been the attitude of American opinion and practice on this question, for, we find that more than a century ago, in 1785, it was stipulated by a treaty, negotiated by Franklin, between the United States and Prussia, that in case of war, neither Power should commission privateers to depredate upon the commerce of the other. And here it is worth while quoting Franklin's opinion of privateering, as expressed in his printed works :—*

> "It is for the interest of humanity in general," says the venerable statesman and philosopher, "that the occasions of war and the inducements to it should be diminished. The practice of robbing merchants on the high seas, a remnant of the ancient piracy, though it may be accidentally beneficial to particular persons, is far from being profitable to all engaged in it, or to the nation that authorizes it. *Piraterie*, as the French call it, or privateering, is the universal bent of the English nation, at home and abroad, wherever settled. No less than seven hundred were, it is said, commissioned in the last (the American) war. These were fitted out by merchants, to prey upon other merchants who had never done them any injury. Methinks it well behoves merchants to consider well of the justice of a war, before they voluntarily engage a gang of ruffians to attack their fellow merchants of a neighbouring nation, to plunder them of their property, and perhaps ruin them and their families if they yield to it; or to wound, maim, and murder them, if they endeavour to defend it. Yet these things are done by *Christian* merchants, whether a war be just or unjust; and it can hardly be just on both sides. They are done by English and American merchants who, nevertheless,

* Franklin's Works, 12mo, 1793. II., 152-178.

complain of private theft, and hang by dozens the thieves they have taught by their own example. It is high time, for the sake of humanity, to put a stop to this enormity. The United States of America, though better situated than any European nation to make profit by privateering (most of the trade of Europe with the West Indies passing before their doors) are, as far as in them lies, endeavouring to abolish the practice, by offering in all their treaties with other powers, an article, engaging solemnly, that in case of future war, no privateer shall be commissioned on either side; and that unarmed merchant ships, on both sides, shall pursue their voyages unmolested. This will be a happy improvement of the law of nations. The humane and just cannot but wish general success to the proposition."

The United States may possibly say that the reason why they have not repeated and endorsed that stipulation in subsequent treaties, is all the fault of Great Britain. Great Britain has always maintained her right to destroy an enemy's private property at sea (not now by privateers, but by public war vessels). Constant representations have been made to us by other Powers, including the United States, asking us to abandon this right. But, as the greatest maritime Power, we have stood out for the right of destroying hostile private property at sea, as a method of warfare most effective and substantial, without inflicting a disproportionate amount of suffering upon individuals. Whether we are really the gainers in the end, by clinging to this principle, is another question. No doubt it gives a great advantage to the Power with the strongest fleet; but when that Power has also far the largest amount of private property at sea, it is clear that the compensating disadvantage is considerable. The protection of our own merchandise and the complete destruction of the enemy's will prove an exceedingly heavy task for our navy. But this is beside the point, for Great Britain rightly upholds the principle and must fight accordingly. What then was the actual position of the United States as to the Declaration of Paris? They said: "If England will not abandon the right to capture private property at sea, then we will not abandon the right to use privateers. We do not choose to be at the cost to maintain a large fleet, and must rely on privateers in case of war. We realise the abuses of privateering, but we are bound to perpetuate them because Great Britain will not give way on the first point." But it is obvious that there is no necessary connection between privateering and the right of capture by national vessels. It would appear that in the matter of privateering, the United States to-day are guided, not by the example and opinions of Benjamin Franklin, but by the vigorous sentiments of Jefferson, promulgated on the 4th of July, 1812:—

> "What is war? It is simply a contest between nations, of trying which can do the other the most harm. Who carries on the war? Armies are formed and navies are manned by individuals. How is a battle gained? By the death of individuals. What produces peace? The distress of individuals. What difference to the sufferer is it that his property is taken by a national or private armed vessel? Did our merchants who have lost nine hundred and seventeen vessels by British captures, feel any gratification that the most of them were taken by his Majesty's men-of-war? Were the spoils less rigidly exacted by a seventy-four gun ship, than by a privateer of four guns; and were not all equally condemned? War, whether on land or sea, is constituted of acts of violence on the persons and property of individuals; and excess of violence is the grand cause that brings about a peace. One man fights for wages paid him by the government, or a patriotic zeal for the defence of his country; another, duly authorised, and giving the proper pledges for his good conduct, undertakes to pay himself at the expense of the foe, and serve his country as effectually as the former, and government drawing all its supplies from the people, is, in reality, as much affected by the losses of one as the other, the

efficacy of its measures depending upon the energies and resources of the whole. In the United States, every possible encouragement should be given to privateering in time of war with a commercial nation. We have tens of thousands of seamen that without it would be destitute of the means of support, and useless to their country. Our national ships are too few in number to give employment to a twentieth part of them, or retaliate the acts of the enemy. But by licensing private armed vessels, the whole naval force of the nation is truly brought to bear on the foe, and while the contest lasts, that it may have the speedier termination, let every individual contribute his mite, in the best way he can, to distress and harass the enemy, and compel him to peace."

In the American Civil War, Congress authorised the President to issue Letters of Marque, but he did not avail himself of this power. The Confederates went so far as to offer their Letters of Marque to foreigners, but the acceptance of them would, of course, have been a gross infringement of the restriction of neutrality, and the Northern States threatened to treat foreign privateers as pirates.

It must be admitted that the contention of the United States Government that in the event of their engaging in hostilities with a country having a powerful navy, their adhesion to the Declaration of Paris would place them at a great disadvantage, is as true to-day as it was in 1856. In comparison with the royal navy of England—the most tremendous maritime force in the world—the American public navy is practically non-existent. It is difficult to see how the United States, in the improbable event of a conflict with this country, could keep the sea for any length of time, even if they converted every vessel in their mercantile marine into a privateer. Neither mercantile nor public navies can be created in a few weeks or months, and the present preponderance of Great Britain in both classes of vessels, could not be seriously reduced on the outbreak of hostilities, by even the clever and energetic American people. This statement is borne out by a comparison of British and American shipping, made in a *Times* article of December, 1896, from which we learn that the total tonnage of the United States mercantile marine is rather under 4¾ millions, more than 70 per cent. of which is of timber.

"It needs no elaborate parade of argument," says the *Times*, "to prove that, with a merchant marine in which timber still constitutes 70 per cent. of the total tonnage, the United States could not possibly hope to compete with a marine like our own where timber is almost entirely discarded, and where sailing vessels have ceased to be, as they once were, a dominating factor, more especially when we add that almost one-half of the American marine is still sail propelled. Little more need be said in explanation of the fact that during the first ten months of the current year—to take the latest figures available—of the tonnage that entered and cleared at British ports only some 625,000 were of American nationality, whereas 35½ million tons were of British origin. This preponderance of British tonnage is all the more striking when we remember that during the same ten months the total value of our direct trade with the United States was not less than 88 millions sterling, or about 20 per cent. of the value of our trade with all countries in that period. As against the American marine of nearly 4¾ million gross tons already alluded to, we have in the United Kingdom a mercantile marine of about twelve million tons, almost entirely built in iron or steel, and so largely modernised from year to year that it is necessarily at the highest point of efficiency. This is more than can be said of even the steel-built tonnage in the American marine, although much of that tonnage is superior to what could be shown a few years ago."

While it would be a rash thing to assert that the American merchant navy will never seriously compete with the British marine, it is safe enough to assume that the Union Jack is not likely to have anything to fear from the Stars and Stripes for a long time to come. The true patriots on both sides of the Atlantic will ever pray that nothing more bitter than friendly rivalry in the arts of peace may stimulate these two great nations, whose mission is, hand in hand, to scale the heights of civilization, and shower blessings upon mankind.

For more than eighty years, the merchant navy of the British Empire has sailed in every sea unmolested and unmolesting, until our merchants have become almost oblivious of those contingencies which require to be specially provided against during a naval conflict. Yet do the signs of the times indicate, that in the future, as in the past, the merchant vessels of Liverpool may play a distinguished part in the terrible game of war. On the 26th of June, 1897, there assembled in the historic waters of the Solent, in honour of Queen Victoria's record reign, and of days more truly "spacious" than those of "great Elizabeth," a magnificent naval pageant comprising nearly 200 war vessels, or twenty-five miles of fighting force. Never in the world's history has so stupendous an exhibition of naval supremacy been seen. So large an array, such strength, such powers of destruction, such speed were never before assembled together, to fill the mind with awe and admiration, and to teach the nations of the earth to cultivate peaceable habits. Conspicuous even among this unparelleled demonstration of sea power, were the magnificent representatives of Liverpool's armed merchant cruisers, and especially the *Campania*, "before which, with its towering bulwarks, its graceful lines, and its huge red funnels, even the largest of the war-ships seemed to dwarf." Perhaps it is not too much to say that if a single *Campania*, or *Teutonic*, were opposed to all the privateers that ever sailed out of Liverpool, she could destroy them all, and still be none the worse for the encounter, provided that her ammunition did not prematurely give out, or her guns wear out. Such is our appalling progress in deadliness, and such, too, is the capacity of Liverpool for keeping abreast of the times, and, in peace or war, holding its own on the "silver sea."

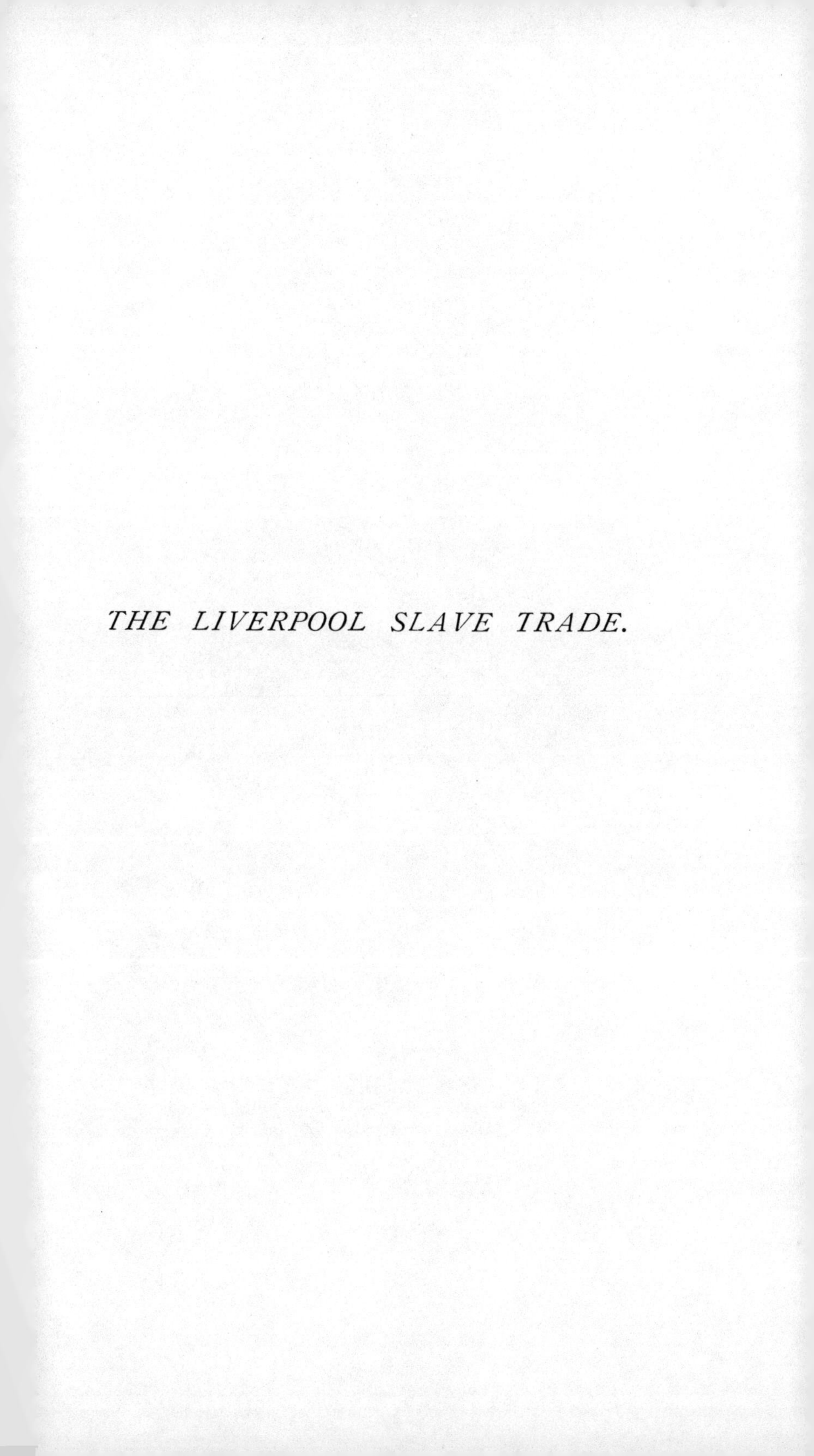

THE LIVERPOOL SLAVE TRADE.

THE LIVERPOOL SLAVE TRADE.

CHAPTER I.

How It Originated and Thrived.

> "Man finds his fellow guilty of a skin
> Not coloured like his own, and having power
> To enforce the wrong, for such a worthy cause
> Dooms and devotes him as his lawful prey."

THE British first began to trade with Africa in 1553. In August of that year, two vessels under the command of Thomas Windham, sailed from Portsmouth on a voyage to Guinea and Benin. The second voyage was made in 1554, by John Lok, who reported that he carried "five blacamoors" to England. To Sir John Hawkins, one of the great sea captains of the reign of Queen Elizabeth, belongs the infamous distinction of being the first Englishman who engaged in the importation of slaves from Africa. Elizabeth, at first, seems to have revolted at the very thought of the new British traffic, and to have foreseen the evils to which its continuance might lead. We find her sending for Captain Hawkins on his return from his first voyage to Africa and Hispaniola, whither he had carried slaves, and expressing her anxiety lest any of the negroes should be carried off without their free consent, declaring that "it would be detestable, and call down the vengeance of Heaven upon the undertakers." Captain Hawkins promised to respect the scruples of his royal mistress, but when he reached the coast of Africa, on his second voyage, the sight of so much "black ivory" proved too strong a temptation for him. He seized many of the inhabitants, carried them off as slaves, and sold them to the Spaniards to work in the mines and plantations. "Here," says Hill, the historian, "began the horrid practice of forcing the Africans into slavery, an injustice and barbarity, which, so sure as there is vengeance in heaven for the worst of crimes, will sometime be the destruction of all who allow or encourage it."

In the year of the Armada (1588), Queen Elizabeth by letters patent, limited the trade with Africa to a company, which was also encouraged by James I. and Charles I. In 1618, an African company was established, but a traffic in slaves formed no part of its objects. It was not until after the colonizing of Barbadoes and Antigua, in the years 1623-25, that the slave trade, though very profitable, commenced to be carried on in English ships. From that time the English merchants and shipowners plunged into the trade as eagerly as the Portuguese and Spaniards had done before, and as the French and Dutch did about the same time. In consequence of the great depredations committed upon the English traders by the Dutch, Charles II., in 1662, granted an incorporation to "The Company of Royal Adventurers of England to Africa," who, being unsuccessful, resigned their Charter in favour of "The Royal African Assiento Company." In 1689, this company entered into a contract to supply the Spanish West Indies with slaves. A large house, called the South Sea House, was built at Kingston, Jamaica, for the accommodation of factors, who were stationed there to conduct the business, and for the reception of the human cargoes which survived the horrors of the "middle passage." Although the company's Charter was abrogated by the Bill of Rights—the third great Charter of British freedom—the company carried things with a high hand, and seized the ships of private traders. Bristol, however, carried on the traffic under great difficulties from the time the monopoly was rendered illegal by the Bill of Rights, until the trade was thrown open in 1698 by the breaking up of the Assiento Company.

The great wealth of the merchants of London and Bristol, enabled them to enjoy a practical monopoly of the African slave trade for a long period prior to Liverpool having any share in it. Liverpool adventurers with a small capital were unable to equip vessels and purchase goods specially adapted to the African market and of no use outside of that market, nor could they afford to await the uncertain results of round voyages, sometimes prolonged to more than a year, and subject to terrible dangers unknown to any other description of trading adventures. Early in the eighteenth century, however, a successful rivalship with Bristol, in exporting provisions, and coarse checks and silk handkerchiefs of Manchester make, to the West Indies and the continent of America, eventually enabled the merchants of Liverpool to participate in the more lucrative slave traffic. While Liverpool obtained from this competition a sudden accession to her commerce, which stimulated the industrious and enriched the enterprising, multiplied the ships in her docks, and filled her warehouses with sugar, rum, and other West India produce, the trade of Bristol to the West Indies declined. The checks of Manchester, carried in Liverpool ships, ousted from that market the German, French, and Scotch osnaburgs exported from Bristol. Finding themselves out-distanced in the competition, the merchants of Bristol embarked with energy in the slave trade, and that so successfully, that the number of ships despatched by them to the slave coast from the year 1701 to 1709 was no less than 57 per annum. The effect of this upon the London slave traffic was enormous. The 104 vessels employed by that port in the Guinea trade in 1701, fell to 72 in 1702, and 50 in 1704, while in 1707 there were only 30 ships so employed. As yet, the hands of Liverpool were clean from negro blood. There was one easy gradient to descend before she entered upon the horrid traffic, and this was the contraband trade with the Spanish main. Spain shipped to her colonists in America, French and German checks, stripes and osnaburgs, under a duty of 300 per cent. To evade this exorbitant impost, the Spanish West India traders ran down in schooners and large canoes from the Havannah, Portobello, and Carthagena to Jamaica, to purchase Manchester goods, which they found not only cheaper, but superior in quality to those made in France and Germany. A growing demand resulted in ample returns in specie to Liverpool and Manchester, and in spite of the vigilance of the Spanish Guarda Costa, which were continually cruising between the south-end of Jamaica and the Spanish main, this trade flourished for about twenty years, and gave the Guarda Costa some excuse for the cruelties they practised in boarding and plundering British vessels, under the pretence of searching for contraband goods. The traffic received a check in 1740, in consequence of a remonstrance from the King of Spain, and finally received its death blow from the Grenville treaty of 1747. The cutting off of a branch of commerce, which, while it lasted, helped to establish the manufactures of Manchester, and lay the foundation of the fortunes of several mercantile houses* both in that town and Liverpool, threw a strong temptation in the way of the Liverpool merchants to employ their ships in the Guinea trade.

When the slave trade was thrown open in 1698, Parliament enacted that private traders should pay to the Assiento Company 10 per cent. for the repairs of the forts and the expenses of the factory. Disputes and dissensions arising

* In Edwards's History of the West Indies, it is stated that the value of the goods annually disposed of amounted to £1,500,000 sterling, while 4,000 tons of shipping were employed in this one single branch of commerce. When the contraband trade was at its height, the annual return to the town of Manchester for the first cost of the goods was estimated at £560,000, and the amount of the profits to the merchants of Liverpool at £273,467.

from this arrangement, the legislature, in 1730, granted a certain sum for such purposes, and enacted that persons trading to Africa should pay to the Chamberlain of London, the Clerk of the Merchants' Hall, Bristol, or to the Town Clerk of Liverpool, 40s. for the freedom of the new company, which should consist of all his Majesty's subjects trading between Cape Blanco and the Cape of Good Hope. London, Bristol, and Liverpool were each to send three committee men to manage the business, and to take charge of the forts and factories. Thus encouraged, the merchants of Liverpool, trebly qualified by the capital, spirit of adventure, and knowledge of the requirements of the West India Islands gained in the contraband trade, entered heartily into the new speculation. The merchants of London having almost relinquished the slave trade in 1720, the memorable year of the South Sea disaster, the only rival Liverpool had to fear in its fresh sphere of enterprise was Bristol. The predominance gained by that port over London in the Guinea trade, and the success obtained over Bristol by Liverpool in the West India market were not more remarkable than the success of Liverpool adventurers in a traffic which, however repugnant to the feelings of humanity, was productive of vast profits. For a period of 77 years, they carried on the trade with a characteristic vigour and ability that outdistanced every competitor, and won for Liverpool the unenviable distinction of being the chief slaving town of the Old World.

It must not be supposed from the fact that, between the year 1709 and 1730, only a single barque of 30 tons burthen sailed from the Mersey for Africa, and laid the foundation of a great but terrible commerce by conveying 15 slaves across the Atlantic, that Liverpool felt any repugnance to embark in a lucrative trade in which every seaport of Europe was engaged, which London and other English ports had carried on for nearly a hundred years, and which was winked at, if not sanctioned by, the British legislature.* No more scruple was then felt as to the "licitness" of the trade in slaves than as to the lawfulness of the trade in black cattle. "So totally different was the feeling which then prevailed on this subject," says Baines, "that whilst the article of the treaty of Vienna, denouncing the African slave trade, was regarded as the noblest article of the great pacification of 1815, the article of the treaty of Utrecht, giving England the privilege of importing negroes into the Spanish possessions in America as well as into her own, was regarded as one of the greatest triumphs of the pacification of 1713."

Immediately on the adoption of the new regulations in 1730, 15 vessels, of the average burthen of 75 tons each, were despatched from Liverpool to the coast of Africa. The number of slaves annually imported in the infancy of the trade cannot now be ascertained, but the encouragement must have been very great to increase the vessels more than double in seven years, 33 Guineamen having cleared for the coast in the year 1737.

The Liverpool merchants at length found it advantageous to have their own factors settled at Jamaica, as had long been the practice of London and Bristol. Liverpool hitherto had been compelled to dispose of her adventures by supercargoes, who were often obliged to sell their goods at a low advance on the invoice, to enable them to make their returns in the vessel, which was frequently an impediment to the sale of their goods. The planters soon discovered that they could purchase slaves from the Liverpool stores at four or five pounds per head less than from the London and Bristol factors, and yet get the same length of credit as had been given by the latter. The causes which enabled Liverpool merchants thus to dispose of Guinea cargoes of "prime negroes" at about 12 per cent. less than the rest of the

* The property in slaves was specifically acknowledged by statute of 5th, Geo. II., Cap. 7, and again by 13th, Geo. III., Cap. 14.

kingdom, and at the same time return an equal profit, are thus set forth by a well-informed local author of the eighteenth century:—*

> "The reason the port of Liverpool could undersell the merchants of London and Bristol, was the restriction in their outfits and method of factorage. The London and Bristol merchants not only allowed ample monthly pay to their captains, but cabin privileges, primage and daily port charges; they also allowed their factors five per cent. on the sales, and five per cent. on the returns, and their vessels were always full manned by seamen at a monthly rate. The Liverpool merchants proceeded on a more economical but less liberal plan, the generality of their captains were at annual salaries, or if at monthly pay, four pounds were thought great wages at that time, no cabin privileges were permitted, primage was unknown amongst them, and as to port allowances, not a single shilling was given, while five shillings a day was the usual pay from Bristol, and seven and six from London. The captains from these ports could, therefore, occasionally eat on shore, and drink their bottle of Madeira; whereas, the poor Liverpool skipper was obliged to repair on board to his piece of salt beef and biscuit, and bowl of new rum punch, sweetened with brown sugar. The factors, instead of a rate per centum, had an annual salary, and were allowed the rent of their store, negro hire, and other incidental charges; therefore, if the consignments were great or small, the advantages to the factor suffered no variation. Their portage was still more economical, their method was to take poor boys apprentice for long terms, who were annually increased, became good seamen, were then second mates, and then first mates, then captains, and afterwards factors on the islands. This was the usual gradation at the time, whereby few men at monthly pay were required to navigate a Liverpool vessel."

* A general and descriptive history of Liverpool, published anonymously in 1795.

In 1746, the *Fortune*, Captain Green, of Liverpool, on her voyage from Africa to Jamaica, was captured and carried to Porto Cavalla, with 354 slaves on board.

In 1747, the slave-ship *Ogden*, Captain Tristram, of Liverpool, bound from Africa to Jamaica, was taken by a Spanish privateer. The gallant resistance made by the crew so irritated the Spaniards, that, on boarding the *Ogden*, they killed all, both whites and blacks, during which the ship sunk, and all on board, except one man, five boys, and nine negroes, perished.

In 1751, the African trade, under legislative enactments, had swelled to a great volume, and in that year no fewer than 53 vessels, with an aggregate burthen of 5,334 tons, sailed from the Mersey for the slave-coast. Owing to the length of the round voyage, which sometimes occupied over a year, the returns of Guineamen that cleared annually for the coast from Liverpool do not represent all the vessels belonging to the port then actively engaged in the trade. We find, for instance, in the returns,* that 58 vessels cleared for Africa in the year 1752,† but from "Williamson's Liverpool Memorandum Book," published in 1753, we know that in 1752, Liverpool possessed no less than 88 vessels employed in the African trade,‡ all of which, with one exception, carried slaves. That exception was the *Eaton*, owned by Messrs. John Okill & Co., which traded in wood and teeth. The 87 slavers had a capacity for conveying about 25,000 negroes across that terrible belt of ocean in which

*See appendix.

† In the History of the County Palatine and Duchy of Lancaster, by Edward Baines, edited by Jas. Croston, F.S.A., it is stated that only 51 or 52 Liverpool vessels out of upwards of 300 were engaged in the slave-trade in the year 1752, an assertion contradicted, both by the clearance list and the list of vessels actually employed.

‡ They traded with the following places:—5 with Benin, 11 with Angola, 3 with New Calabar, 11 with Old Calabar, 38 with the Windward and Gold Coast, &c., 12 with Bonny, and 8 with Gambia. All these vessels were also engaged in the trade with America: for the living cargoes which they took in on the coast of Africa they conveyed either to the West Indies or the North American plantations, from Maryland and Virginia southwards, after which they returned to Liverpool, with cargoes of sugar, rum, and other tropical or colonial produce.

so many heart-broken captives found rest. The number shipped, if not actually delivered "in good order and condition," was probably much higher, as it was then customary to overload, with the most frightful results.

From the same "Memorandum Book" we learn that there were in Liverpool, in the year 1752, 101 merchants who were members of the Company trading to Africa* established by Act of Parliament in 1750 (the 23rd of George II.), entitled "An Act for extending and improving the trade to Africa belonging to Liverpool." In the same year there were in London, 135 African merchants, and in Bristol 157, though the African trade of the latter was less extensive than that of Liverpool.

From this time the man traffic set in with such a steady current, that it soon became one of the most lucrative branches of the commerce of the port. Fast sailing vessels, specially adapted for the trade, were built in the shipbuilding yards on the banks of the Mersey, where many a noble frigate for the king's navy was turned out in those days, and soon the odour of the human shambles began to mix with that of tar and rum in the docks of Liverpool. Here, as elsewhere, it was impossible to keep the pollution at a distance —the smoke of the evil genie followed the homeward bound Guineamen across the seas, and tainted the town, in spite of every effort to bottle it. The insignia of the men-stealers were boldly exhibited for sale in the shops and warehouses, and advertised in the papers.† Busts of blackamoors and

* In the appendix will be found a list of the Liverpool African merchants, and also of their 88 Guineamen, their commanders' names, and the slave-carrying capacity of each vessel.

† In 1756, the following articles suitable for a Guinea voyage were advertised to be sold by auction at the Merchants' Coffee-house :—One iron furnace and copper, 27 cafes(?) with bottles, 83 pairs of shackles, 11 neck collars, 22 handcuffs for the travelling chain, 4 long chains for the slaves, 54 rings, 2 travelling chains, 1 corn mill, 7 four-pound basons, 6 two-pound basons, 3 brass pans, 28 kegs of gunpowder, 12 cartouches boxes, 1 iron ladle, 1 small basket of flints. In the paper of May 27th, 1757, another lot was advertised: one large negro hearth with 2 iron furnaces, 1 copper ditto for 450 slaves, 1 decoction copper kettle, ditto pan, a parcel of shackles, chains, neck collars, and handcuffs, 1 iron furnace, 245 gallons, with a lead top, sufficient to boil 10 barrels of liquor.

elephants, emblematical of the African trade, adorned the Exchange or Town Hall.* One street in the town was nicknamed Negro Row, and negro slaves were occasionally sold by auction in the shops, warehouses, and coffee-houses, and also on the steps of the Custom House.† The young bloods of the town, when not engaged in more disgraceful pursuits, deemed it fine amusement to circulate handbills in which young ladies were offered for sale.

In an auctioneer's bill of the period, we find, "twelve pipes of raisin wine, two boxes of bottled cyder, six sacks of flour, three negro men, two negro women, two negro boys, and one negro girl."

Amongst the many curious advertisements which appeared in the Liverpool papers, while the slave trade was in full swing, were two side by side in *Williamson's Advertiser*, of August 20th, 1756. The first announced the hull of the snow *Molly*, to be sold by the candle at 1 o'clock noon at R. Williamson's shop, adding: "N.B.—Three young men slaves to be sold at the same time." Facing it in the next column we read, "Wanted immediately a negro boy. He must be of a deep black complexion, and a lively, humane disposition, with good features, and not above 15, nor under 12 years of age. Apply to the printer." The irony of contrast presented by these two advertisements was, of course, lost upon the most "lively and humane" reader of

* "Between the capitals runs an entablature or fillet, on which are placed in base relief the busts of blackamoors and elephants, with the teeth of the latter, with such-like emblematical figures representing the African trade and commerce." —"History of Liverpool."

† The Custom House, on the east side of the Old Dock, now Canning Place, was built about 1700, by Mr. Silvester Moorecroft, who was mayor in 1706. It was a meagre red brick building with two slightly projecting wings; the angles and windows being ornamented with stone. It had the royal arms carved in stone in front, and was entered by a wide flight of steps in the centre, through arches, into an arcade or piazza, out of which several doors opened, and a staircase led to the long room, which was above the piazza, and to several other offices. Ships loaded and discharged at the quay in front of the building, and at the back were the Custom House yard and warehouse (the latter fronting Paradise St.), access to which was obtained by a passage on the south side of the Custom House. The slave-auctions were held on the flight of steps leading to the main entrance.

that day. In the same paper, for June 24th, 1757, we read the following:—

"For Sale immediately, ONE stout NEGRO young fellow, about 20 years of age, that has been employed for 12 months on board a ship, and is a very serviceable hand. And a NEGRO BOY, about 12 years old, that has been used since Sept. last to wait at a table, and is of a very good disposition, both warranted sound. Apply to Robert Williamson, Broker. N.B. A vessel from 150 to 250 tons burthen is wanted to be purchased."

Among the wants advertised in December, 1757, by Robert Williamson aforesaid, who kept the Universal Register Office, near the Exchange, and, amongst other matters, registered "Persons of Ingenuity and Learning," were the following: "A French Horn for a Letter of Marque. A Black Boy that can beat a drum, for an officer in the Army. A person that can play on the Bagpipes, for a Guinea ship." We are not told whether the piper was required to discourse sweet strains to the crew, or to tame the mutinous negroes.

In the short-lived *Liverpool Chronicle*, James Parker, auctioneer, advertised for sale by the candle, at the Merchants' Coffee-house, a fine negro boy, 11 years of age, imported from Bonny, by Mr. Thomas Yates, a Guinea merchant, who lived in Cleveland Square.

The following is from *Williamson's Advertiser* of Feb. 17th, 1758:—

"For Sale a Healthful Negro Boy, about 5 feet high, well proportioned, of a mild, sober, honest disposition; has been with his present master 3 years, and used to wait on a table, and to assist in a stable."

On the 8th of September, 1758, the following appeared in the same paper:—

"Run away from Dent, in Yorkshire, on Monday, the 28th August last, Thomas Anson, a negro man, about

5 ft. 6 ins. high, aged 20 years and upwards, and broad set. Whoever will bring the said man back to Dent, or give any information that he may be had again, shall receive a handsome reward from Mr. Edmund Sill, of Dent; or Mr. David Kenyon, merchant, in Liverpool."

In 1765, we have another specimen from the same source:—

"To be sold by Auction at George's Coffee-house, betwixt the hours of six and eight o'clock, a very fine negro girl about eight years of age, very healthy, and hath been some time from the coast. Any person willing to purchase the same may apply to Capt. Robert Syers, at Mr. Bartley Hodgett's, Mercer and Draper near the Exchange, where she may be seen till the time of Sale."

In the paper of September 12th, 1766, was announced "to be sold at the Exchange Coffee-house in Water Street, this day the 12th inst. September, at one o'clock precisely, eleven negroes, imported per the *Angola*, * * * * Broker."

On December 1st, 1767, one negro man and two boys were advertised for sale at Mr. Robinson's office.

Thus the hateful traffic was not kept altogether at a distance, nor confined to those referred to by the poet—

> "But ah! what wish can prosper or what prayer,
> For merchants rich in cargoes of despair;
> Who drive a loathsome traffic, gauge and span,
> And buy the bones and muscles of the man."

It is, indeed, too often forgotten that while British ships were employed in transporting millions of "African labourers" to their doom in the mines, and on the sugar and cotton plantations of the New World, the traffic in human flesh and blood was polluting freedom-loving England itself: and in justice to Liverpool, a few facts must be stated under this head, lest the reader should imagine that her people were worse than their neighbours.

When Henry Esmond Warrington, Esq., of Virginia, landed at Bristol, in 1756, with his black slave Gumbo, he was only the type of thousands of others who landed upon our shores, and Gumbo, boasting in the servants' hall at Castlewood, and singing in church as loud as the organ, was but the idealised representative of thousands of black slaves held in bondage in England at that period. In 1764, the "Gentleman's Magazine" estimated that there were upwards of 20,000 black slaves then domiciled in London alone, and these slaves were openly bought and sold on 'Change. These unfortunate creatures were burnt with some distinguishing mark, and collars and padlocks were deemed a necessary part of their livery. That a collar was considered as essential for a black slave as for a dog, is clear from the *London Advertiser* for 1756, in which Matthew Dyer, working goldsmith, at the Crown, in Duck Lane, Orchard Street, Westminster, intimates to the public that he makes "silver padlocks for Blacks or Dogs; collars, &c." In the *London Gazette* of March, 1685, a reward was advertised for bringing back John White, a black boy of about 15 years of age, who had run away from Colonel Kirke's. He had a silver collar about his neck, upon which was the Colonel's coat-of-arms and cipher; he had also upon his throat a great scar, &c. King William III., "of glorious memory," had a favourite slave, a bust of whom may be seen at Hampton Court; the head is of black marble, and the drapery round the shoulders and chest of veined yellow marble, while the throat is encircled by a carved white marble collar, with a padlock, in every respect like a dog's metal collar. In the *Daily Journal*, of September 28th, 1728, is an advertisement for a runaway black boy, who had the legend, "My Lady Bromfield's black, in Lincoln's Inn Fields," engraved on a collar round his neck. A specimen of these slave collars is preserved in the Museum of the Antiquarian Society, in Edinburgh. The collar, which in this instance was worn by a white man, bears the following inscription: "Alexander Stewart, found guilty of death for theft, at Perth, December 5, 1701.—Gifted by the Justiciaries, as a perpetual servant, to Sir John Erskine, of Alva." The following advertisements show how common was the custom of buying and selling black slaves in England in the eighteenth century:—

In the *Tatler,* for 1709, a black boy, twelve years of age, "fit to wait on a gentleman," is offered for sale at Dennis's Coffee-house, in Finch Lane, near the Royal Exchange.

The *Daily Post,* of August 4th, 1720, contains the following: "Went away the 22nd July last, from the house of William Webb, in Limehouse Hole, a negro man, about 20 years old, called Dick, yellow complexion, wool hair, about five foot six inches high, having on his right breast the word 'Hare' burnt. Whoever brings him to the said Mr. Webb's, shall have half-a-guinea reward and reasonable charges."

In the *Daily Journal,* of September 28th, 1728, a negro boy, eleven years of age, was advertised for sale at the Virginia Coffee-house, in Threadneedle Street, behind the Royal Exchange.

The following appeared in the *London Advertiser,* of 1756: "To be sold, a Negro Boy, about fourteen years old, warranted free from any distemper, and has had those fatal to that colour; has been used two years to all kinds of household work, and to wait at table; his price is £25, and would not be sold but the person he belongs to is leaving off business. Apply at the bar of the George Coffee-house, in Chancery Lane, over against the Gate."

In the *Public Ledger,* of December 31st, 1761, "a healthy Negro Girl, age about fifteen years," is offered for sale; "speaks good English, works at her needle, washes well, does household work, and has had the small-pox."

In 1763, one John Rice, was hanged for forgery at Tyburn, and among his effects, sold by auction after his execution, was a negro boy, who fetched £32. The "Gentleman's Magazine," commenting on the sale of the boy, says that this was "perhaps the first custom of the kind in a free country."

At Lichfield, in 1771, there was offered for sale, by public auction, "A Negro Boy, from Africa, supposed to be ten or eleven years of age. He is remarkably stout, well proportioned, speaks tolerably good English, of a mild disposition, friendly, officious, sound, healthy, fond of labour, and for colour, an excellent fine black."

The *Stamford Mercury,* for 1771, states that "at a sale of a gentleman's effects at Richmond, a Negro Boy was put up and sold for £32;" adding, "a shocking instance in a free country!"

In March, 1752, the *Clayton* snow,* Captain Patrick, of Liverpool, 200 tons burthen, armed with 4 two-pounders and 10 swivel guns, was taken off Fernando Po, on the coast of Africa, by pirates, also from Liverpool. These proved to be nine men and a boy belonging to the *Three Sisters*, Captain Jenkins, who had run away with the ship's longboat. The pirate took the opportunity of luffing up under the lee quarter of the *Clayton* when all her hands were forward, except the captain and gunner, and then boarded with sword and pistol in hand, wounded the captain in several places, captured the ship, kept the crew in irons one night, and the next morning put them on board their own longboat and turned them adrift. The pirates had brought with them in their boat a bale of scarlet cloth and another of handkerchiefs, and told the *Clayton's* crew that if they "would go a-roving they should be clothed with scarlet." Four, unable to resist this dazzling proposal, voluntarily entered as rovers, and the chief mate and two boys were impressed into the pirate service. The rest of the crew were 12 days in getting into the river Bonny, where the king seized their longboat, and the men had to

* A common type of slaver at this time was a snow, of about 14c tons, square sterned, 57 feet keel, 21 feet beam, 5 feet between decks, 9 feet in the hold—a miniature Malbolge when crammed with slaves like sardines in a box.

enter on board different slavers trading there. The pirates carried the *Clayton* to Pernambuco, where a Portuguese man-of-war retook her and carried her to Lisbon. The *Three Sisters* was wrecked on the coast of Wexford, and Captain Jenkins, with most of the crew, perished.

In the summer of 1756, while a sloop commanded by Alexander Hope was making the middle passage with a cargo of slaves, being then 100 leagues from the coast, six or seven slaves, who were upon deck, watched an opportunity when the first mate (Mr. Ashfield) and some more of the crew were in the hold, rushed into the cabin, knocked Captain Hope's brains out, wounded the second mate (Mr. Charles Duncan) in several places, secured all the arms, and kept possession of the cabin for four hours. Duncan, with difficulty got out of the cabin, and, with the assistance of the first mate and the cooper, got the door shut upon the negroes. The blacks then fired all the muskets and blunderbusses at the crew through the door, but hurt none of them. The first mate and the cooper then rushed into the cabin, disarmed the slaves, and recovered the vessel. The ringleader of the slaves jumped overboard and was drowned. The first mate and the cooper received several wounds.

Captain Jenkinson, of the *Fanny*, writing to his owners in Liverpool, from Jamaica, on November 27th, 1756, says: "On the 19th we arrived here, with 110 slaves, 22 ct. of ivory, and —— ounces of gold, after a tedious passage of 13 weeks and 4 days. My slaves are sold from £50 to £48 per head."

In January, 1757, the *Nancy*, Captain Gill, with 72 fine slaves on board, was captured at Junk, while trading, by two French frigates. "Our usage, whilst on board them," says Captain Gill, "was cruel, no better lodgings than the decks, only short and bad allowance, and to be marooned without provisions was treatment beneath an European enemy, let alone the polite nation of France." In his letter Captain Gill mentions that the *Priscilla*, of Liverpool, had arrived at Barbadoes on the 30th of March, from the coast of Africa, having buried 94 slaves on the middle passage! Her complement was 350 slaves, which, allowing for overloading, shows a shocking rate of mortality.

Captain Baille, commander of the slave-ship *Carter*, writing to his owners in Liverpool from the River Bonny, Africa, on January 31st, 1757, reveals the method sometimes resorted to by slave-captains to compel the native chiefs to trade with them. He says:—

"We arrived here the 6th of December, and found the *Hector*, with about 100 slaves on board, also the *Marquis of Lothian*, of Bristol, Capt. Jones (by whom I now write), who was half slaved, and then paying 50 Barrs, notwithstanding he had been there 3 months before our arrival. I have only yet purchased 15 slaves at 30 and 35 Barrs; but as soon as the bearer sails, I propose giving more; for at present there is a dozen of our people sick, besides the two mates, some of whom are very bad, and I have been for these last 8 days in a strong fever, and frequently insensible. Yesterday morning I buried Thomas Hodge, and on the 13th James Barton. Capt. Nobler of the *Phœnix* arrived here the 3d, and on the 19th our trade was stopt (as it had often been before); upon which we all marched on shore to know the reason and applied to the King thrice, though he constantly ordered himself to be denied, and wou'd not admit us. However, we heard his voice in doors, and as he used us so ill, we went on board, and determined (after having held a Council), to fire upon the town next morning, which we accordingly did, in order to bring them to reason, but found that our shot had little effect from the river, upon which we agreed that the *Phœnix* and the *Hector* shou'd go into the Creek, it being nigher the town, whilst Captain Jones and I fired from the river. The *Phœnix* being the headmost vessel went in, and the *Hector* followed about a cable's length astern. The *Phœnix* had scarce entred the Creek before they received a volley of small arms from the bushes, which were about 20 yards distant from the ship, and at the same time several shot from the town went through him, upon which they came to anchor, and plied their carriage guns for some time; but finding there was no possibility of standing the decks, or saving the ship, he struck his colours, but that did not avail, for they kept a continued fire upon him, both of great and small arms. His people were thrown into the utmost confusion, some went down below, whilst others jumpt into the yaul which lay under the ship's quarter, who (on seeing a number of canoes coming down to board them) desired Capt. Nobler to come down to them, which he at last did, as he found the vessel in such a shattered condition, and that it was impossible for him to get her out of the Creek before the next ebb tide, in case he cou'd keep the canoes from boarding him. With much difficulty they got on board the *Hector,* but not without receiving a number of shot into the boat. The natives soon after boarded the *Phœnix*, cut her cables, and let her drive opposite the town, when they began to cut her up, and get out her loading, which they accomplished in a very short time. But at night in drawing off some brandy, they set her on fire, by which accident a great many of them perished in the flames. The *Phœnix's* hands are distributed amongst the other three ships, and all things are made up, and trade open, but very slow, and provisions scarce and dear." The *Marquis of Lothian* was afterwards taken and carried into Martinico.

The dangers to which slave-captains were exposed in war time is set forth in the following letter from Captain Jackson, of the ship *King George,* dated Surinam, June 6th, 1757:—

"On the 28th of March last, being at an anchor in Melimba Road, on the coast of Africa, in company with the *Ogden*, Captain Lawson, *Penelope*, Captain Wyatt, and the *Black Prince*, of Chester, Captain Creevey, two French men of-war (the *St. Michael* of 64 guns, and the *Leviathan* of 36 guns), stood directly in for us. As soon as we found it impossible to escape, we slipt, and run our ships on shore, choosing rather to lose all, than fall into the enemies' hands. I had then on board 390 slaves, who ran away and were for the most part taken by the natives. We have lost everything, except a few things I had in the factory on shore, and about 20 slaves, with whom I got on board the ship *Wolpenburg*, of Flushing, and took passage for this place. As she does not sell here, but sails for St. Eustatia in the morning, I propose going with her. Captain Creevey got his passage by way of Rotterdam: Captain Lawson was carried away by the French men-of-war, and sailed for Martinico ten days after they forced us on shore, and Captain Wyatt we left at Melimba, with some slaves that he had saved. My surgeon and Will Dawson are with me; part of my people propose staying on the coast of Africa, whilst others design going in the long boat to Island Princess, or St. Thomas's. I left my second and third mate at Melimba, who are well, but poor Tom Cross is dangerously ill, and I had the misfortune to bury Mr. Moncaster on the 29th of April."

Further details of the unfortunate affair were supplied by Captain William Creevey, on his arrival in Liverpool. They were to the following effect:—

"The *St. Michael* had 600 men on board, and the *Leviathan* 300. They first appeared in sight about 7 o'clock in the morning, under English colours, upon which all the boats then on shore, distrusting them, immediately repaired on board their respective ships, and made what preparations they could for an engagement. The frigate, being the headmost ship, stretched first in with them, upon which a smart engagement ensued between her and the Englishmen, whose metal were only 3 and 4 pounders, and hers 18 pounders. The engagement lasted till the 64-gun ship came within reach of them with her 24 pounders, which obliged them to slip their cables, intending to run their ships on shore. The Frenchmen dispatched two launches full of men after them, intending to cut them off the shore, and Captain Creevey's vessel being the sternmost, they attempted to board him, but received such a warm reception from his stern chase guns, loaded with musket balls, that they sheered off, and afterwards steered for and boarded the *Ogden*, Captain Lawson, whom with most of his people they carried on board the Commandant and used extremely ill. Next morning they burnt the *Black Prince* and *Ogden*, and after waiting two days, destroying all before them, they landed all Captain Lawson's people but himself and the Doctor, went to Cape Binda to wood and water, and sailed for Martinico. By their behaviour on the coast they seemed as if their only object was to destroy the trade; for they allowed 70 of the natives to plunder the *Ogden,* but fixed a fuzee to the powder magazine, which blew up the ship and all the black men on board. This wanton cruelty so exasperated the natives that they threatened to take revenge on the first French ship that fell into their hands. The blacks behaved extremely kind to all the Englishmen, and assisted them with what they wanted."

On the 6th of January, 1758, we read that the *Knight*, Captain William Boates, from Annamaboo, with 398 slaves, had touched at St. Kitts, all well, and had gone down for Jamaica. On the 17th of February, we read of his arrival at Jamaica with 360 Coromantee, Ashantee, Akin, and Whydah negroes, from which we gather that the sharks had banqueted on 38 prime negroes. Off the Leeward Islands, Captain Boates had a smart engagement with a French privateer sloop of 12 carriage guns, and full of men, which attempted to board him several times. Captain Boates armed several of his negroes, who behaved very gallantly with the small arms, and eventually the privateer sheered off, much disabled, and it was afterwards reported that she had sunk. The story of Captain Boate (or Boats) is a strange one. His real name will never b known, as he was a waif, found in a boat, hence th peculiar surname. He was brought up by the person wh found him, placed in the Blue Coat School, which ha turned out so many capable and worthy men in every walk of life, and afterwards apprenticed to the sea. He rose to be commander of a slave-ship, and prospered amazingly, becoming one of the leading merchants and shipowners of Liverpool. In the paper of June 6th, 1760, the marriage is announced of "Capt. Wm. Boates, formerly of the African trade, merchant, to Miss Brideson, daughter of Mr. Paul Brideson, of Douglas, Isleman." It is related that one of his vessels captured a Spanish ship with a large quantity of gold and silver bullion and specie on board. When the news was communicated to Mr. Boates, he ran along the Pierhead exclaiming "Billy Boates—born a beggar, die a lord!" Part of the structure known as Drury Buildings, Drury Lane, was formerly his residence. It was built in a superior style for that age, a large portion of the woodwork being mahogany. The Liverpool newspaper of November 3rd, 1794, records the death, at the age of 78, of "William Boates, Esq., whose extensive transactions in the commercial world rendered him a most useful member of society, and whose memory will be long revered by all who had connections with him." He was interred in the Old Churchyard. His daughter, the wife of Richard Puleston, Esq., died at Brighton, in September of the same year. His son, Henry Ellis Boates, of Rosehill, Denbighshire, died in January, 1805. "It is a remarkable fact," says Brooke, "that of the large number of Liverpool persons who made fortunes in the African slave trade, and some of them acquired by that odious traffic considerable wealth, it only remained in very few instances in their families until the third generation, and in many cases it was dispersed or disappeared in the first generation after the death of the persons acquiring it."

Captain Boates had retired from the sea, and settled down as a merchant, when he appended his signature to a very interesting document,* of which the following is a copy :—

* In the possession of Mr. C. K. Lace.

"LIVERPOOLE, 14 April, 1762.

"CAP[N]. AMBROSE LACE,

"SIR,—You being Master of the ship *Marquis of Granby*, and now cleard out of the Custom house, and ready to sail for Africa, America, and back to Liverpoole, the Cargoe we have shipd on Board is agreeable to the Annexed Invoice, which we consign you for sale, For which you are to have the usual Commission of 4 in 104 on the Gross Sales, and your Doctor, Mr. Lawson, 12d. ℔ Head on all the slaves sold, and we give you these our orders to be observed in the course of your intended voyage. With the First Favourable wind you must sail and proceed in company with the *Douglas*, Cap[n]. Finch, who has some Business at the Isle of Man, when you must accompany him not waiting longer for him than six days. When finished at the Isle of Man, you are to make the Best of your way in Company thro the So. Channell and as you are Both Ships of Force, and we hope Tolerably well mann'd you will be better able to Defend yourselves against the Enemy, we therefore Recommend your keeping a good Look out that you may be Prepaird against an attack, and shoud you be Fortunate enough to take any vessell or vessells From the Enemy, we recommend your sending them Home or to Cork whichever will be most convenient so as not to Distress your own ship, and on your arrival at Old Callebar if one or more ships be there you will observe to make an agreement with the Master or Masters so as not to advance the Price on each other and we doubt not you will use your utmost endeavours to keep down the Comeys which in Generall are to extravagant there and For which you have no Return at least not worth any thing to the Ownery and as your Cargoe is larger than we expected we hope will be able to Purchase 550 slaves, and may have to spare £400 to lay out in Ivory which we Recommend your Purchasing From the Beginning of your Trade and pray mind to be very Choice in your Slaves. Buy no Distemperd or old Ones, But such as will answer at the Place of Sale and stand the Passage and as Callebar is Remarkable for great Mortality in Slaves we Desire you may take every Prudent

Method to Prevent it, viz.—not to keep your Ship to Close in the Day time and at Night to keep the Ports shut as the night Air is very Pernicious. The Privilege we allow you is as Follows: yourself ten Slaves, your first mate Two, and your Doctor Two, which is all we allow except two or three Hundred wt. of screveloes amongst your Officers, but no Teeth, which you will take care to Observe, as we will not allow any thing more. When Finished at Callebar you are to make the Best of your way For Barbadoes, where you will Find Letters Lodged For you at the House of Messrs. Wood & Nicholas, how you are to Proceed which will be to Guadaloupe or Martinico or any other of the Leeward Islands, whichever is the best Markett which you may advise with the House of Messrs. Wood & Nicholas unto which place to Proceed, or any other Person you Can Confide in. We expect your Cargoe of Slaves will be taken up at £ *stg ⅌ Head, and what more they sell For to be For the Benefitt of the Owners and to have the Ship Loaden in the Following Manner viz: about One Hundred Casks good Mus$^{o.}$ Sugar for the Ground Tier, the Remainder with First and Second white Sugars, and Betwixt Decks with good Cotton and Coffee, and the Remainder of the neat Proceeds in Good Bills of Exchange at as short Dates as you can. If the aforementioned Prices cannot be obtaind For your Slaves at either Guadaloupe or Martinico, or the Leeward Islands as aforesaid we then desire as little time may be Lost as Possible, but proceed for Jamaica and on your arrivall there apply to Messrs. Cuthbert & Beans, Messrs. Hibberts, Messrs. Gwyn and Case, or any other House you think will do best for the Concern, unto whom Deliver your Cargoe of Slaves which you think will make the Most of them, if Possible, by a Country Sale and to have your agreement in writing and the Ship Loaden in the Following Manner; as much Broad Sound Mahogany as will serve for Dunnage, the Hold filld with the very Best Mus^{o} Sugar and Ginger and Betwixt Decks with good Cotton and Pimento and about Ten Puncheons Rum, the Remainder of the

* Obliterated.

neat Proceeds of your Cargoe in Bills of Exchange at as short Dates as you can get them. The House you are to sit down with must Fournish you with what money you may want for Payment of wages and other necessary Disbursements of your ship which we recommend the utmost Frugality. In annexd you have invoice of Slops for the use of the seamen and apprentices. What the seamen have you must lay an advance on to pay Interest of Money, &c. The Apprentices only Prime cost. We recommend your keeping Good Rules and good Harmony amongst your Crew and a good watch, Particularly whilst you have any Slaves on Board, and Guard against accidents of Fire, Particularly in Time of Action. Suffer no Cartridges to be Handed out of the Magazine without Boxes, which will Prevent any Powder being sprinkled on the Deck and in Case of your Mortality (which God Forbid) your First Mate, Mr. Chapman must succeed you in command. Pray mind to embrace every opportunity that Offers advising us of your Proceedings, For our Government as to Insurance &c. We wish you a Prosperous Voyage and Safe Return and are your assured friends.

CROSBIES & TRAFFORD — CHAS. GOORE
W[M] ROWE — WILL[M] BOATS
ROBERT GREEN — CHA[S] LOWNDES
THOS. KELLY

P.S. You and your officers' slaves
are to be equal Qy Male and Female."

The following letter, dated Barbadoes, February 28th, 1758, was written by Captain Joseph Harrison, commander of the slave-ship *Rainbow*, to his owners, Messrs. Thomas Rumbold & Co., of Liverpool :—

"We arrived here on the 25th inst. in company with Capt. Perkins from Bonny, and Capt. Forde from Angola, whom we fell in with at St. Thomas's. The packet arrived here from England the day after us. I expect to sail from hence for South Carolina in five days, having on board 225 slaves, all in good health except eight. On the 23rd of June last, I had the misfortune to fall in with a French brig privateer, of fourteen 6-pounders, to leeward of Popo. We engaged him four hours, and were so near for above four glasses, that I expected every moment we should run on board him, as he had shot away all my running rigging and the fluke of my small bow anchor. My standing rigging and sails were mostly cut to pieces, and the privateer was in a little better condition. Fifteen of his shot went through and through my sides, we being scarce the length of the ship from one another. I lost in the engagement, my boatswain — William Jackson — Robert Williams—and Henry Williams. My first and second mates, three landsmen, and one servant were wounded. The privateer being well satisfied sheered off. We were three days in repairing our rigging, &c., and on the 28th got over the Bar of Benin and found only one vessel there, viz. a Portuguese sloop at Warree. I purchased eight slaves on the windward coast, and 261 at Benin, besides 5400 weight of ivory. Leaving the river, Nov. 9th, we arrived at St. Thomas's Dec. 17th, from whence our three vessels sailed, Jan. 4th. I have buried all my officers, except my first and third mates and gunner. Having lost since left Liverpool, 25 white people and 44 negroes. The negroes rose on us after we left St. Thomas's; they killed my linguister whom I got at Benin, and we then secured them without farther loss. We have an account of five privateers being to windward of Barbadoes, by a retaken vessel brought in here this day, so that we shall run a great risk when we leave Barbadoes."

Writing on board the French 64-gun ship *Fortune*, at Isle Grand, on the coast of Brazil, on the 27th of June, 1758, Captain William Creevey gives the owners of the slave ship *Betty* a pathetic account of the capture of that vessel on her voyage to Gambia, and of his own misfortunes. Captain Creevey was the father of T. Creevey, Esq., M.P., and resided in School Lane:—

"Pursuant to your several orders, I sailed from Portsmouth under convoy of his Majesty's ships *Warwick*, *Nassau*,

Ray, and *Swan*, bound to Africa; about five leagues S.S.E. from Plymouth, the wind flew out at N. and began to blow, which made the King's Ships to out carry the merchantmen, and they got in shore, whilst I with several others were left about five leagues to the southward, after beating to windward there all day, my people greatly fatigued, night coming on, and no possibility of getting into any harbour on the coast of England, and knowing I was more exposed to the enemy's cruizers that might happen to be beating in the channel, than I should be in running to the Southward; therefore, with the approbation of my officers, I took the opportunity of the night and made what sail I could, leaving Ushant about twelve leagues to the eastward. The gale continued, and I got clear of the Bay of Biscay without being spoke to by any ship except the *Antelope* privateer of London, and one of our comrades that had parted with the fleet. After we had got as far as Lat. 39 and Long. 17 and thought that we were entirely out of danger, to my inexpressible mortification we fell in with a fleet of French Indiamen outward bound, escorted by the *Fortune* of 64 guns, and 630 men. We were taken by one of their best sailing frigates, who sunk your Snow *Betty*, with the greatest part of her cargo. The prisoners were distributed into different ships. It fell to my lot to go on board the Commodore, where I have been treated with great humanity and politeness, but must leave you to judge of the shocking prospect that's before me, in being carried to the Indies, where I have neither money nor credit, and where there has been such acts of cruelty committed to prisoners. All this I must submit to, it being the unhappy event of war, in which I have been very unfortunate, to be twice taken in less than twelve months. When, how, or whether I shall return to thank you for the confidence and trust you reposed in me, is only known to that all wise and merciful God, who hears the distressed cries of the unfortunate prisoners, in the remotest parts of the earth; and I hope in his good time, will return me to the British shore, there to partake of those inestimable blessings of liberty and religion, which I am at present entirely deprived of."

The *Molly*, Captain Timothy Wheelwright, from the Windward and Gold Coast of Africa, was taken by the French in the middle passage. The Frenchman took the slaves out of the *Molly* and carried them to St. Domingo. The vessel was afterwards retaken and carried into Jamaica.

The *Hazard*, Captain W. Parkinson, with 411 slaves from Africa, had an engagement on the 28th of November, 1758, with a privateer sloop of 12 four-pounders, and full of men, who soon sheered off. The next day he was attacked furiously by another privateer, of 8 six-pounders and 4 four-pounders, for above five hours, who made several attempts to board him, but having only his topsails set and being prepared, he gave her such a reception as made them alter her course. He had only one man wounded, and himself a little scratched. His people behaved well, and the next day they arrived at St. Kitts.

The *Achilles*, Captain Chr. Carus, with 325 slaves on board, was taken and carried into Guadaloupe, by a sloop and a schooner privateer. Captain Carus bravely defended himself for some time, but during the engagement one of his four-pounders (which had been bored into a six-pounder before he left Liverpool) burst, killed his third mate, and wounded seven or eight of his crew. The Frenchmen wrecked the *Achilles* on some rocks going into the harbour, but landed the slaves and prisoners safe. Captain Carus died on board the *Hazard* on his passage home.

Early in 1759, the *Hector*, Captain Lievsey, of 14 six-pounders, and 37 men, arrived at St. Kitts with 365 slaves. When off Deseada she fell in with a French privateer brig, of 4 nine-pounders and 12 six-pounders, with 270 men on board, with whom she had a very close and smart engagement for four hours and-a-half. Captain Lievsey's men quitted their quarters three different times, but he and his officers bravely rallied them, and obliged the privateer to sheer off. During the engagement he had four men wounded, and three negroes killed; the French suffered much, and the action was deemed one of the bravest fought in the West Indies during the war.

Captain Linnecar brought to Liverpool the melancholy news of the loss of the *Perfect*, Captain Potter, at Mana. He had purchased over 100 slaves, who rose upon him, killed all the crew on board, and run the ship on shore. Luckily, the mate and three or four of the seamen were on shore with the boat, and saved themselves. Further details came to hand in a letter from Mr. Henry Harrison, to his relations in Liverpool, dated Plantains, on the coast of Africa, April 23rd, 1759:—

"On the 12th of January, we had the misfortune to be cut off by the negroes; they killed Captain Potter, our surgeon, carpenter, cooper, and James Steward, a boy. Luckily, the captain had sent me on shore that morning to go to the King's town, about ten miles up the river, to fetch slaves down; but before I reached the town, met two of his servants bringing a slave down; returned with them; made a smoke on the shore as a signal for our boat, but before had well made it, saw her put off from the vessel with six of our people in her, being all left alive on board. I swam off to her and we rowed for the *Spencer*, Captain Daniel Cooke, then lying at Cape Mount. At one o'clock that night, Captain Cooke got under way, and made sail in order to attempt to recover our vessel; at day-light, finding her at anchor, he fired his guns into her for about an hour, but I could not persuade him to board her. That evening the slaves run the snow on shore. We had purchased 103 slaves, and had a pledge for two more on board. The slaves and natives would not give us the least article of wearing apparel. When this fatal accident happened, our chief mate was gone with the yawl to windward, and the boatswain with the long boat to leeward to purchase slaves. Mr. Eaton and the boatswain got on board Captain Nichols, and I heard that they saved 15 or 16 slaves that were due to us on shore, and left Mana, Mar. 30, designed for the West Indies. I am now got moved to the brig *Industry,* Captain Banks, and we intend to sail for Antigua to-morrow, having 122 slaves, all in good health, on board. They have buried Richard Worthington and three more of their people."

On the 11th of August, 1759, about 70 leagues to the eastward of Antigua, the snow *Mac*, Captain Edward Cropper, on her passage from Benin, with 232 slaves, 3 tons of ivory, a parcel of screveloes, and some gold, was attacked by the *Mars* privateer brig, of 14 carriage guns, 20 swivels, and 90 men. After an engagement of an hour and-a-half, Captain Cropper was obliged to strike, he having only six white men able to stand the decks, the rest being all sick or dead. The *Mac* was carried into Martinico, and the prisoners sent in a cartel to Antigua. Captain Cropper came home passenger in the *Prussian Hero*, Captain Kevish.

On the 25th of July, 1759, the brig *Betsey*, Captain Jones, sailed from Liverpool for Africa, and on the 14th of August, was taken, after a chase of three days, by the *Marquis de Jarvis*, French schooner privateer, Monsieur de Schoye, commander, mounting 10 carriage guns, 16 swivels, and 100 men, and carried to Bayonne. On the 1st of September, in the night, William Peel, the cook, and six others of the crew, broke out of prison, seized a large fishing boat, bound the five occupants, laid them at the bottom of the boat, and put to sea. On the 3rd, seeing a sail, they put up a handkerchief at the masthead, as a signal of distress. The vessel, which happened to be British, bore down, and took them on board, the Frenchmen and their boat being liberated. They were landed at Londonderry, and travelled to Carrickfergus, where they met Captain Hutchinson, of the cod smack—our old privateering friend—who gave them their passage to Liverpool.

The doctor of the ship *Glory*, Captain Thomas Chalmers, writing from Papau, on the coast of Africa, on August 18th,

1759, says :—"I left the ship at Cape Coast, and came down here to purchase 26 slaves, but am afraid we shall be at Whydaw 5 or 6 months before we can purchase 540, owing to the high prices, which our captain is determined to beat down, otherwise we might get off the coast in 2 months."

The *Knight*, Captain Jenkinson, left the coast of Africa for Jamaica, with 390 slaves, a ton of ivory, and 90 ounces of gold dust on board.

The black prince, Accra, was a passenger on board the *Spy*, Captain Creevey, on her voyage from Liverpool to the coast of Africa to purchase slaves, and was safely landed.

On the 21st of March, 1760, the slave-ship, *Francis*, of 18 guns, Captain Thomas Onslow, was wrecked on the island Fuerteventura, one of the Canary Islands, on her passage from Liverpool to Africa, 23, out of a crew of 58, being drowned. The misfortune happened for want of a look-out, the second mate being asleep on the deck during his watch, when the helmsman called out, "land ahead."

On the 30th of July, 1763, the *Charlotte*, Captain Lowndes, for Africa, was blown up at the Magazines, and only one man saved.

The peace of 1763 gave a great impulse to the trade of Liverpool, and two years later, we find the number of slave-ships clearing for the coast had increased to 83, with a tonnage of 9,382, carrying 24,200 slaves.

So great was the success of Liverpool in the hideous traffic, that the ports of London and Bristol began to feel an abatement of their accustomed exports in proportion as those of Liverpool increased. Bristol, in particular, found her demand of slaves for the plantations rapidly decrease, insomuch that, in the year 1764, she cleared out only 32 vessels for Africa, while Liverpool cleared 74.* To such a

* In the year 1764, the number of ships cleared to Africa from Liverpool directly was 74; from Bristol 32; the number cleared to America from Liverpool was 141, against 105 from Bristol. There entered the Port of Liverpool in the same year, 7

height had the African trade of Liverpool advanced at this period, that more than one-fourth of the entire shipping belonging to the port consisted of Guineamen, and more than one-half of the African trade of the whole kingdom was in the hands of Liverpool merchants. Thus we behold the remarkable commercial phenomenon of Bristol, a wealthy city, which had apparently established a lucrative branch of trade and enjoyed a long experience of foreign commerce, being gradually ousted from its position by the energetic policy and enterprise of a port that, at the beginning of the eighteenth century, was not only situated on the utmost verge of commercial activity, but without sufficient capital to support a vessel of 30 tons in the same trade. While the vending of human beings was thus in the full tide of prosperity, to the great delectation and enrichment of Liverpool, Manchester, Birmingham, Sheffield, and other centres of industry that supplied the requisite wares for a Guinea voyage, no warning voice on earth, except the feeble wail of a few obscure Quakers, had as yet been raised against the great iniquity, though many accusing spirits had doubtless flown up to heaven's chancery with an ever-swelling indictment.

During the period which we have just passed over, there was connected with the Liverpool slave trade one of the most extraordinary characters of the eighteenth century. We mean John Newton, afterwards the celebrated rector of St. Mary Woolnoth, who, in 1752-54, commanded the slave-ships *Duke of Argyle* and the *African*, belonging to Mr. Joseph Manesty. To the story of his strange and adventurous career we devote the next chapter.

vessels from Africa and 188 from America, against 137 entering Bristol from America, and none from Africa direct. Liverpool had inwards 766 ships; Bristol, 332. Liverpool had outwards 832 ships; Bristol, 343.

CHAPTER II.

Captain John Newton.

John Newton was born in London on the 24th of July, 1725 (O.S.). His mother died when he was seven years old, and with her died the pious teaching which was intended to prepare him for the ministry. After two years spent at a boarding school in Essex, he made several voyages with his father, a stern sea captain in the Mediterranean trade, who, having been educated himself at a Jesuit college in Spain, found a situation for his son at Alicant. The youth's unsettled behaviour and impatience of restraint, necessitated his removal after a few months trial. Before he was sixteen years of age, he had taken up a religious profession three or four times, his condition alternating between asceticism and the most horrid profanity, as the mood took him. After two years of strict Pharisaism, he met with Lord Shaftesbury's "Characteristics," and the fine words of "The Rhapsody" beguiled his heart and operated like slow poison. In 1742, Mr. Manesty, a merchant in Liverpool and a friend of his father's, offered to send young Newton to Jamaica, and take care of his future welfare. John was well pleased, and everything was prepared for his voyage. He was upon the point of setting out the following week, when his father sent him to visit his relations in Kent for a few days. Here he met Mary Catlett, a young lady not quite fourteen, who had been designed from her birth, by his mother and her mother, as his future wife. Being profoundly ignorant of this little matrimonial arrangement, Newton fell madly in love with the girl. He preferred a treasure in Kent to a fortune in Jamaica, stayed three weeks instead of three days, missed his passage, and encountered his father's wrath. Soon after this he made a voyage to Venice as a common sailor, and fell a prey to evil companionship. A remarkable dream startled his conscience about this time, but the impression soon faded away. In December, 1743, he visited his friends in Kent, and again, for love of Mary Catlett, frustrated his father's plans on his behalf. His thoughtless conduct at length led him into the meshes of the press-gang, and he found himself on board the *Harwich* man-of-war, at the Nore. War being daily expected, there was no hope of release. After a month's hardship he was, by his father's influence, taken upon the quarter-deck as a midshipman. Here, with his foot on the lowest rung of the naval ladder, he might have aspired to high command, for he had in him the stuff which makes British admirals. Providence, however, had not raised John Newton to be either a martyred Byng, or an idolised Nelson. He fell under the influence of a zealous atheist, who completed the ruin of his principles—no difficult task,—for the youth, while talking of virtue, delighted in all manner of wickedness.

In December, 1744, the *Harwich* was in the Downs, bound to the East Indies. Newton got a day's leave, took horse and rode off to see his Mary, with the usual result. He overstayed his leave, and forfeited the favour of the captain, who had overlooked such pranks more than once. The vessel having put into Plymouth through stress of weather, Newton heard that his father was at Torbay, and determined to desert in order to get into the African employ through his influence, a short Guinea voyage being preferable to five years in the East Indies, from a lover's point of view. "I was sent one day in the boat," he says, "to take care that none of the people deserted, but I betrayed my trust, and went off myself." In a day or two he was caught on the road to Dartmouth by a party of soldiers, marched through the streets of Plymouth guarded like a felon, his heart full of rage, shame and fear. After two days' confinement in the guard-house, he was sent on board the *Harwich*, kept awhile in irons, then publicly stripped, whipped and degraded. Thus was the proud, head-strong and profane Mr. Midshipman Newton brought down to a level with the lowest, and exposed to the insults of all. Who, outside of the counsels of eternity, would have been bold enough to prophecy that this disgraced deserter was destined to be the famous rector of St. Mary Woolnoth, the friend of the gentle and pious Cowper, the joint author of the "Olney Hymns," the spiritual father and teacher of Thomas Scott, the commentator, and one of the most remarkable Englishmen that ever lived. The mere suggestion of such a destiny would have been received with derision by the whole ship's company, and most of all by Newton himself. His condition was now pitiable. The ship sailed, the captain was implacable, and the culprit, friendless, tabooed, exposed every hour to some new insult and hardship, his breast torn by conflicting passions, eager desire, rage and despair, watched the receding coast of England with intense wistfulness and regret. When the last dim line had faded from his view, he was tempted to throw himself into the sea. "But," he says, "the secret hand of God restrained me." On the passage to Madeira, he was a prey to the most gloomy thoughts, and, by brooding over his imaginary wrongs, he came at last to form designs against the captain's life. Fortunately, his love for Mary Catlett acted as a strong cable to hold him from plunging into a terrible abyss. She had not yet accepted him as her affianced lover, but he could not bear that she should think meanly of him when he was dead. When the *Harwich* arrived at Madeira, Newton, through the intercession of the lieutenants, was allowed to exchange into a Guineaman, bound to Sierra Leone. The captain, who happened to be acquainted with the elder Newton, received the son kindly, but the youth soon gave free rein to his evil passions, lampooned his benefactor, and made it his study to corrupt others. "Let it be buried in eternal silence," he says, of this part of his career. Six months later the captain died, and Newton, fearing his successor would put him on board a man-of-war, entered the service of a trader on board, who was returning from a visit to England. Hoping to rise, as this man had done, from poverty to riches, by purchasing slaves in the rivers and selling them to the ships, Newton omitted to make a proper agreement for wages, trusting to his master's generosity. The consequence was, that when he landed upon one of the Banana Islands, with little more than the clothes upon his back, there began for him a period of virtual slavery—of unspeakable degradation and misery. It looked as if Providence, in its mercy, had almost banished him from human society at a time when he seemed like one infected with a pestilence, capable of spreading a taint wherever he went. He soon fell too low to have any influence even for evil. "I was rather shunned and despised than imitated," he says, "there being few even of the negroes themselves, during the first year of my residence among them, but thought themselves too good to speak to me. I was as yet an 'outcast lying in my blood,' and to all appearances exposed to perish."

His new master, who had formerly resided near Cape Mount, now settled at the Plantains, upon a low, sandy island, about two miles in circumference, and almost covered with palm trees. They built a house and entered on trade, and as Newton was inclined to be diligent, he might have done well with his employer. Unfortunately, the trader was under the influence of a black woman, who lived with him as his wife. She was a person of some consequence in her own country, and he owed his first rise to her interest. This woman took a dislike to Newton from the first. He had a severe fit of illness, and his master, before sailing in a shallop to Rio Nuna, left him in her hands. As he did not recover soon enough, she grew weary, neglected him, and led him the life of a dog. He could scarcely procure a draught of cold water when burning with fever, his bed was a mat spread on a board, his pillow a log of wood. When the fever subsided, he was left almost to starve, though the black woman kept a good table, much in the European style. Occasionally, when in high good humour, she would send him victuals in her own plate after she had dined, and these, so greatly was his pride humbled, he greedily devoured. Once, when called to receive this bounty from her own hands, he, from extreme feebleness, dropped the plate and lost his dinner. The table was covered with dishes, but the black woman cruelly laughed at his disappointment, and would give him no more. So great was his distress at times, that, at the risk of being punished as a thief, he went by night into the plantation to pull up roots, which he ate raw upon the spot, in fear and trembling, but which seldom missed to act like tartar emetic. He was sometimes relieved by strangers, and even the slaves in the chain secretly brought victuals from their own slender pittance, to keep the future slave-captain from starvation! But to pressing want were added scorn and contempt, almost harder to bear. When he was slowly recovering, the black woman came with her attendants to mock, revile, and torment him. She called him worthless and indolent, compelled him to walk, set her servants to mimic his motions, to clap their hands, laugh, and pelt him with limes, or occasionally with stones. When she was out of sight, however, he was "rather pitied than scorned by the meanest of her slaves." When his master returned, he complained of ill-usage, but was not believed. He accompanied the factor in his next voyage, and did pretty well for a while, till a brother trader persuaded his master that he stole his goods in the night, or when he was on shore. This, as he tells us, was about the only vice of which he was not guilty, but the charge was believed, and Newton condemned without evidence. Hard usage followed. He was locked upon deck, with a pint of rice for his day's allowance. He kept himself from starvation, by catching an occasional fish at slack water, his master allowing him the entrails of fowls to bait his hook with. His joy at seeing a fish on his hook was pathetic; such a fish hastily broiled, or half burnt, without sauce, salt, or bread, afforded him a delicious meal. If he caught none, he had to sleep away his hunger till the next return of slack water, and then try again. He suffered much from the inclemency of the weather and the want of clothes. Clad only in a shirt, a pair of trousers, a cotton handkerchief instead of a cap, and a cotton cloth, about two yards long, to supply the want of upper garments, he was exposed sometimes for thirty or forty hours to incessant rains and strong gales of wind, without the least shelter; and some of the effects of such exposure after a long sickness remained with him for the rest of his life, "as a needful memento of the service and the wages of sin."

In about two months they returned to the Plantains, Newton's haughty heart brought down, but not to wholesome repentance. His spirits were sunk, he lost all resolution, and almost all reflection. But the tiger was only sleeping. It is remarkable that during this period of semi-starvation and extreme wretchedness, he often beguiled his sorrows by devoting his mind to mathematical studies. He used to take Barrow's "Euclid"—the only book he had brought on shore—to remote corners of the island, by the seaside, drawing his diagrams with a long stick upon the sand. In this manner he fairly mastered the first six books of "Euclid."

His master and mistress one day stopped awhile to watch him planting some lime trees, no higher than a young gooseberry bush. "Who knows," said the trader, mockingly, "but by the time these trees grow up and bear, you may go home to England, obtain the command of a ship, and return to reap the fruits of your labours. We see strange things sometimes happen." What was intended as a cutting sarcasm, turned out a true prediction. The black woman lived to see it fulfilled.

He continued in this abject state for about a year, sending an occasional letter to Mary Catlett, for in his deepest misery he clung to the hope of seeing her again. When he made shipwreck of faith, hope, and conscience, his intense love for this girl was his only religion. He wrote to his father, at whose request Mr. Manesty ordered one of his captains to search out the prodigal and bring him home. Meanwhile, the prodigal entered the employ of another trader, who fed and clothed him decently, and made him joint manager of a factory at Kittam. Business flourished; Newton thought himself happy, and was in some danger of "growing black," not in colour but in disposition. While the infatuation was growing upon him, and his engagements with the natives becoming closer, he was saved from himself in a remarkable manner. One day in February, 1747, his fellow-servant walking on the beach saw a vessel sailing past and made a smoke in token of trade. She proved to be the very ship which had orders to look for Newton, and the first words uttered by her commander had reference to that young man. When the captain went on shore to deliver his message, he found Newton rather indifferent to his proposals.

The invitation which would have been received as life from the dead by the sick and starving wretch at the Plantains a few months before, fell flat on the ear of the comparatively prosperous trader, whose despair of ever seeing England had caused him to form other plans. The captain, unwilling to lose him, lied on a grand scale, telling Newton that a relative, lately deceased, had left him £400 per annum; that he (the captain) had express orders to redeem him, though it should cost one-half of his cargo. Other embellishments were added by this resourceful mariner, but the plausible story was only partially believed by Newton. Something more powerful than the captain's blarney turned the scale. The sweet face of Mary Catlett passed before him, fresh as a breeze from the Kentish hills, and in less than an hour, John Newton, freed from a captivity of about fifteen months, embarked as a passenger on board the Liverpool ship. During a tedious trading voyage on the coast, lasting about a year, he amused himself with mathematics, and in the invention of new oaths and imprecations. The captain, who was no saint, at last believed that he had a Jonah on board—that a curse attended his passenger wherever he went, and that all the disasters they met with were owing to Newton being in the ship.

They sailed for England in January, 1748, and on the 9th of March, eight days after leaving the Banks of Newfoundland, a great sea struck the vessel, and in a few minutes reduced her to a mere wreck. Newton awoke to find his cabin filled with water. In making for the deck he was turned back by the captain, who wanted a knife, and this trifle saved his life, for the man who went up in his room was instantly washed overboard. With almost superhuman exertions the ship was kept afloat, Newton assisting at the pumps and encouraging his companions. The 21st of March was an ever memorable day for him. "On that day," he says, "the Lord sent from on high, and delivered me out of the deep waters." Exhausted with pumping from three in the morning till noon, he lay down, almost indifferent whether he ever rose again. An hour later he was called, took the helm and steered the ship till midnight.

While thus employed, the whole course of his past life rose up before him in review. He reflected on his former religious professions, the extraordinary turns in his life, the calls, warnings, and deliverances, his licentious conversation, and his profane ridicule of the gospel history, of the truth of which he was not yet convinced. If true, there was no forgiveness for him, and he waited with fear and impatience to receive his inevitable doom. When he heard, about six in the evening, that the ship was freed from water, there arose a gleam of hope. He saw the hand of God displayed in their favour, and began to pray. "I could not utter the prayer of faith," he says; "I could not draw near to a reconciled God, and call Him father. My prayer was like the cry of the ravens, which yet the Lord does not disdain to hear. I now began to think of that Jesus whom I had so often derided." But he was not yet a believer. He wanted evidence, and especially an assurance of the Divine inspiration of the Scriptures. He soon found in the New Testament certain sayings which made him resolve for the present to take the gospel for granted, as embodying hope, while on every other side he saw nothing but unfathomable despair. Weeks of suffering passed, the ship was driven from her course, the pumps were kept going incessantly, provisions were running very short, starvation, or the prospect of feeding upon one another, loomed before them. The captain, whose temper was soured by distress, hourly reproached Newton as the sole cause of the calamity, and believed that if he were thrown overboard they should be preserved from death. The captain did not mean to carry his theory into practice, but Newton was very uneasy, as his own conscience confirmed the master's words. "I thought it very probable," he says, "that all that had befallen us was on my account. I was, at last, found out by the powerful hand of God, and condemned in my own breast." Their last victuals were boiling in the pot when they anchored in Lough Swilly, and a great storm immediately arose, which, had they been at sea, would have sent them to the bottom. "About this time," says Newton, "I began to know that there is a God that hears and answers prayer." He was no longer an infidel. He renounced profanity, and became a changed man, though ignorant of the spirituality of the hidden life of a Christian. While the ship was refitting, he visited Londonderry, went twice a day to church, received the Sacrament, and "with the greatest solemnity engaged himself to be the Lord's for ever, and only his."

He arrived in Liverpool in May, 1748, and after paying a visit to Mary Catlett, returned to Liverpool, and sailed again for Africa, in August, as mate of a new slave-ship belonging to Mr. Manesty, who had in fact offered him the command. Soon after his departure he relapsed into religious indifference, and by the time they arrived in Guinea he was almost as bad as before. His business on the coast was to sail from place to place in the longboat, to purchase slaves. The ship was at Sierra Leone, while he was at the Plantains, the scene of his former captivity. He was now in easy circumstances, courted by those who formerly despised him. The lime trees he had planted were growing tall, and promised fruit the following year, if he should return with a ship of his own. Here he was seized with a violent fever. Weak and delirious, he crept to a corner of the island to pray. The burden was removed from his conscience, peace and health came quickly, and in two days he stepped on board his ship perfectly restored. This was his last great declension. He employed his leisure hours in learning Latin, and under great disadvantages acquired a spice of classical enthusiasm. Writing to a friend, in March, 1749, from the coast, he says :—

"Though we have been here six months I have not been ten days in the ship, being continually cruising about in the boats to purchase souls, for which we are obliged to take as much pains as the Jesuits are said to do in making proselytes, sometimes venturing in a little canoe through seas like mountains, sometimes travelling through the woods, often in danger from the wild beasts, and much oftener from the more wild inhabitants, scorched by the sun in the day, and chilled by the dews in the night."

Notwithstanding the perils he passed through, he was never so happy in his life as he was now. Referring to the eight months passed on the coast, and to the treachery of the natives, he observes:—

"Several boats in the same time were cut off; several white men poisoned, and in my own boat I buried six or seven people with fevers. When going on shore, or returning from it, in their little canoes, I have been more than once or twice overset by the violence of the surf, or break of the sea, and brought to land half dead, for I could not swim."

How often must these wild scenes of African adventure have rushed into the mind of the Rev. John Newton in after years, while walking with his mild friend Mr. Cowper, through the peaceful glades of Olney? With what grim pleasure must the devil have presented to the eye of the pious rector of St. Mary Woolnoth, some photographic picture of his African experiences, while the congregation sang one of his own beautiful hymns.

They sailed from Africa with their living cargo for Antigua, and from thence proceeded to Charleston, where, when time permitted, Newton prayed and sang hymns in the woods by day, and at night joined "vain and worthless company," venturing on the brink of temptation. He arrived in Liverpool on the 6th of December, 1749. In February, 1750, he was married to Mary Catlett, at St. Margaret's Church, Chatham, and in the following August, he sailed from Liverpool as commander of the slave-ship *Duke of Argyle*, 150 tons burthen, and 30 men, belonging to his constant friend, Mr. Manesty. He established public worship on board ship twice every Lord's day, officiating himself. Having now much more leisure, he prosecuted his classical studies. His letters to his wife show a very striking gradual development of his religious life, during this and subsequent voyages.

Writing to Mrs. Newton, from the Bananas, in November, 1750, he says:—

> "I have lately had a visit from my quondam black mistress, with whom I lived at the Plantains. I treated her with the greatest complaisance and kindness, and if she has any shame in her, I believe I have made her sorry for her former ill-treatment of me. I have had several such occasions of taking the noblest kind of revenge upon persons who once despised and used me ill. Indeed, I have no reason to be angry with them. They were what they little intended—instrumental to my good."

Further details of this circumstance, are given by the Rev. John Campbell:—

> "Upon being asked whether he ever met again with the black woman who had treated him so harshly when he was in Africa, Mr. Newton replied, 'Oh, yes; when I went there as a captain of a ship, I sent my longboat ashore for her. This soon brought her on board. I desired the men to fire guns over her head in honour of her, because she had formerly done me so much good, though she did not mean it. She seemed to feel it like heaping coals of fire on her head. I made her some presents, and sent her ashore. She was evidently most comfortable when she had her back to my ship. I just recollect a circumstance that happened to me when I first stepped ashore on the beach at that time. Two black females were passing; the first who noticed me observed to her companion, that 'there was Newton, and, what do you think, he has got shoes!' 'Ay,' said the other, 'and stockings too!' They had never seen me before with either."

Writing from Shebar, he speaks of the raillery he encountered amongst the sea captains he met with:—

"They *think* I have not a right notion of life; and I *am sure* they have not. They say I am melancholy; I tell them they are mad. They say I am a slave to one woman;* which I deny, but can prove that some of them are mere slaves to a hundred. They can form no idea of my happiness; I answer, I think the better of it on that account."

Having completed his purchases on the African coast, Captain Newton crossed the sea with his human cargo to Antigua, where he heard of his father's death. Writing to his wife he gives her the following account of his position and authority as captain:—

"My condition when abroad, and even in Guinea, might be envied by multitudes who stay at home. I am as absolute in my small dominions (life and death excepted) as any potentate in Europe. If I say to one, come, he comes; if to another, go, he flies. If I order one person to do something, perhaps three or four will be ambitious of a share in the service. Not a man in the ship will eat his dinner till I please to give him leave—nay, nobody dares to say it is twelve or eight o'clock, in my hearing, till I think proper to say so first. There is a mighty bustle of attendance when I leave the ship, and a strict watch kept while I am absent, lest I should return unawares and not be received in due form. And should I stay out till midnight (which for that reason I never do without necessity) nobody must presume to shut their eyes till they have had the honour of seeing me again. I would have you judge from my manner of relating these ceremonies, that I do not value them highly for their own sake; but they are old-fashioned customs, and necessary to be kept up, for without a strict discipline the common sailors would be unmanageable. But in the midst of

* Mr. Cecil tells us that he "heard Mr. Newton observe, that, as the commander of a slave-ship, he had a number of women under his absolute command; and knowing the danger of his situation on that account, he resolved to abstain from flesh in his food, and to drink nothing stronger than water, during the voyage; that, by abstemiousness, he might subdue every improper emotion; and that, upon his setting sail, the sight of a certain point of land was the signal for his beginning a rule which he was enabled to keep."

my parade I do not forget—I hope I never shall—what my situation was on board the *Harwich*, and at the Plantains."

After passing through various scenes of danger and difficulty he reached home in November, 1751, after a voyage of fourteen months. In July, 1752, he sailed again from Liverpool, commander of the new slave-ship *African*. He is no sooner at sea than down in his diary goes the expression of his earnest desire to live wholly to the Lord. He elaborates a scheme of rules for his own conduct, prays for his wife, whom he almost worshipped, arranges for as much Sabbath rest as was possible for his crew, and even sets apart a day of fasting and prayer on their behalf. These were the high and holy purposes of a good man, made in simple and child-like faith in God, and we must not let our present enlightened prejudice against the slave trade lead us to imagine that John Newton was a hypocrite or a fanatic. He was simply for thoroughness in all he did, whether on the side of black or white angels. Formerly energetic as an atheist, he was now energetic for Christ. Let us return to his own narrative :—

"A sea-faring life is necessarily excluded from the benefit of public ordinances and Christian communion ; but my loss upon these heads was at this time but small. In other respects I know not any calling that affords greater advantages to an awakened mind, for promoting the life of God in the soul, especially to a person who has the command of a ship, and thereby has it in his power to restrain gross irregularities in others, and to dispose of his own time; and still more so in African voyages, as these ships carry a double proportion of men and officers to most others, which made my department very easy; and excepting the hurry of trade, &c., upon the coast, which is rather occasional than constant, afforded me abundance of leisure. To be at sea in these circumstances, withdrawn out of the reach of innumerable temptations, with opportunity and a turn of mind disposed to observe the wonders of God in the great deep, with the two noblest objects of sight—the expanded heavens and the expanded ocean—continually in view, and where evident interpositions of Divine Providence, in answer to prayer, occur almost every day ; these are helps to quicken and confirm the life of faith, which, in a good measure, supply a religious sailor the want of those advantages which can only be enjoyed upon the shore. My knowledge of spiritual things was at this time very small, yet I have sometimes looked back with regret upon those scenes. I never knew sweeter or more frequent hours of divine communion than in my two last voyages to Guinea, when I was either almost secluded from society on shipboard, or when on shore with none but natives. I have wandered through the woods, reflecting on the singular goodness of the Lord to me, in a place where perhaps there was not a person who knew Him for some thousand miles round me. Many a time, upon these occasions, I have restored the beautiful lines of Propertius to the right owner ; lines full of blasphemy and madness when addressed to a creature, but full of comfort and propriety in the mouth of a believer—

> " Sic ego desertis possim benè vivere sylvis
> Quò nulla humano sit via trita pede ;
> Tu mihi curarum requies, in nocte velatra
> Lumen, et in solis tu mihi turba locis.

" PARAPHRASED.

> " In desert woods with thee, my God,
> Where human footsteps never trod,
>   How happy could I be !
> Thou my repose from care ; my light
> Amidst the darkness of the night ;
>   In solitude my company "

" In the course of this voyage I was wonderfully preserved in the midst of many obvious and many unforeseen dangers. At one time there was a conspiracy amongst my own people to turn pirates, and take the ship from me. When the plot was nearly ripe, and they only waited a convenient opportunity, two of those concerned in it were taken ill in one day ; one of them died, and he was the only person I buried while on board. This suspended the affair, and opened a way to its discovery, or the consequence might have been fatal. The slaves on board were likewise frequently plotting insurrections and were sometimes upon the very brink of mischief, but it was always disclosed in due time. When I have thought myself most secure, I have been suddenly alarmed with danger; when I have almost despaired of life, as sudden a deliverance has been vouchsafed to me. My stay upon the coast was long, the trade very precarious, and in the pursuit of my business, both on board and on shore, I was in deaths often."

On one occasion he was hindered from going on shore at Mana by some strange dream and premonition of danger, and found afterwards that it was no idle fear. A trader named Thomas Bryan, who owed him £100, sent him the money in a huff, charging him, at the same time, with an intrigue with one of his women, and refusing henceforth to have anything to do with him. The charge, which affected his honour and interest in Africa and England, and might have touched his life had he landed, was afterwards acknowledged to have been a malicious calumny, without the least shadow of a ground.*

In a letter to his wife, Captain Newton thus describes a "Sea-Sunday":—

"The Saturday evening is a time of devotion when I especially beg a blessing on *your* Sunday, as I know, where you are, you are unavoidably exposed to trifling company. I usually rise at four o'clock in the morning, and after seeking a blessing on the day, take a serious walk on deck. Then I read two or three select chapters. At breakfast, I eat and drink

* Among Newton's papers were found the following notes addressed by Bryan "For Capt. John Newton, these:"

"SIR,—I have sent you one boy-slave on board, and I am going up to my town. I shall be down again in three days. I would not have you go from here till you hear further from me, for I intend to do what I can for you. I have no further commands at present, but remain your friend and well-wisher, Thomas Bryan."

"SIR,—Mr. Corker gives his service to you, and has sent you one girl-slave on board, and says he will do what he can for you."

more than I talk, for I have no one here to join in such conversation as I then choose. At the hour of your going to Church I attend you in my mind with another prayer; and at eleven o'clock the ship's bell rings my own little congregation about me. To them I read the morning service according to the Liturgy. Then I walk the deck and attend my observation (i.e. take the latitude of the ship). After dinner a brief rest, or I write in my diary. I think again upon you at the time of afternoon service, and once more assemble the crew for worship. I take tea at four, then follows a Scripture lesson, and a walk and private devotion at six."

Captain Newton drew up a written instrument devoting himself once more the servant of God, "absolutely and for ever, without any reserve or competition," and the covenant was signed, sealed, and dated as in the presence of God, at "New Shebar, on the Windward coast of Africa, on Sunday, the 15th of October, A.S.M., 1752."

He resumed his Latin studies when business permitted, and was regular in his habits, allotting about eight hours to sleep and meals, eight hours to exercise and devotion, and eight hours to his books. He sailed from Africa for St. Kitts, and on the voyage wept two or three times over some passages in the life of Colonel Gardiner. One would have thought that there was more cause for weeping to be seen in the hold, but the good captain puts us right to our confusion.

"One circumstance," he says, "I cannot but set down here, and which I hope I shall always take pleasure in ascribing to the blessing of the God of peace, I mean the remarkable disposition of the men-slaves I have on board, who seem for some time past to have entirely changed their tempers. I was at first continually alarmed by their almost desperate attempts to make insurrections. One of these affairs has been mentioned, but we had more afterwards; and when most quiet they were always watching for opportunity. However, from the end of February, they have behaved more like children in one family than slaves in chains and irons, and are really upon all occasions more observing, obliging and considerate than our white people. Yet, in this space, they would often in all likelihood have been able to do much more mischief than in former parts of the voyage."

Captain Newton arrived at St. Kitts on June 24th, 1753. On the 11th of July, he sailed for England, arriving in Liverpool in August. He remained only six weeks in England, sailing in the middle of October on his third and last voyage. He sets Wednesday, November 21st, apart for the special purpose of seeking a blessing upon his voyage, and for protection through its various difficulties and dangers. He also resolves to devote a certain proportion of his earnings to charity. A strange and a sad thing happened in connection with this voyage. Before he sailed, Newton met with a young man, formerly a midshipman, and his own intimate companion on board the *Harwich*. A sober youth at first, he had fallen a victim to Newton's libertine principles. They resumed their intimacy at Liverpool, Newton told his story, and earnestly set about undoing the evil he had done, but was reminded by his friend that he was the very first person who had preached the scepticism against which he was now arguing. This was a terrible stab to Newton's conscience. His friend's owner having failed just as he was about to sail as master of a Guinea ship, and left him without employment, Newton, more in the hope of undoing his own evil work than to serve the man in business, took him on board his own ship as a companion, the owners promising to provide him a ship on his return. He soon had reason to repent the step. The man was exceedingly profane, and grew worse and worse. "I saw in him," says Newton, "a most lively picture of what I had once been, but it was very inconvenient to have it always before my eyes." After being a sharp thorn in the captain's side upon the voyage, requiring all his prudence and authority to keep him under any degree of restraint, the man was sent away in a small vessel to trade on the ship's account. He was greatly affected at parting with the captain, who gave him good advice, but no sooner was he free of the controlling eye, than he "gave a hasty loose to every appetite, and his violent irregularities, joined to the heat of the climate, soon threw him into a malignant fever, which carried him off in a few days. He died convinced, but not changed."

> "The account I had from those who were with him," says the Captain, "was dreadful. His rage and despair struck them all with horror, and he pronounced his own fatal doom before he expired, without any appearance that he either hoped or asked for mercy. I thought this awful contrast might not be improper to adduce, as a stronger view of the distinguishing goodness of God to me, the chief of sinners."

On the passage from the Coast of Africa to St. Christopher's, Captain Newton was attacked with fever, which, for a while, gave him "a very near prospect of eternity." In this illness, a somewhat strange fancy disturbed him: "I seemed not so much afraid of wrath and punishment, as of being lost and overlooked amidst the myriads that are continually entering the unseen world. What is my soul, thought I, among such an innumerable multitude of beings! And this troubled me greatly: 'Perhaps the Lord will take no notice of me.'" It does not seem to have occurred to him, that some of his cargo might be ready to identify him in the spirit world, with a view to a settlement of accounts. However, he remembered that "the Lord knoweth them that are his," and his perplexity vanished, as also did the fever before his arrival in the West Indies. At St. Kitts, Newton met Captain Clunie, commander of a London ship, and a member of the church of the Rev. Samuel Brewer, of Stepney. "He was a man," says Newton, "of experience in the things of God, and of a lively, communicative turn.

For near a month, we spent every evening together on board each other's ship alternately, and often prolonged our visits till towards day-break. I was all ears; and what was better, he not only informed my understanding, but his discourse inflamed my heart." His intercourse with Captain Clunie broadened his religious views, and proved of lasting value to him. He arrived in Liverpool on the 9th of August, 1754, "having had," he says, "a favourable passage, and, in general, a comfortable sense of the presence of God through the whole, and towards the end, some remarkable deliverances and answers to prayer." Then he makes a remarkable statement:—

"I had the pleasure to return thanks in the churches (at Liverpool), for an African voyage, performed without any accident, or the loss of a single man; and it was much noticed and acknowledged in the town. I question if it is not the only instance of the kind. When I made my first appearance upon 'Change, a stranger would have thought me a person of great importance, by the various congratulations I received from almost every gentleman present."

"My stay at home was intended to be but short, and by the beginning of November I was ready again for the sea; but the Lord saw fit to overrule my design. During the time I was engaged in the slave trade, I never had the least scruple as to its lawfulness. I was, upon the whole, satisfied with it, as the appointment Providence had marked out for me; yet it was in many respects far from eligible. It is, indeed, accounted a genteel employment, and is usually very profitable, though to me it did not prove so, the Lord seeing that a large increase of wealth could not be good for me. However, I considered myself as a sort of gaoler or turnkey; and I was sometimes shocked with an employment that was perpetually conversant with chains, bolts, and shackles. In this view, I had often petitioned, in my prayers, that the Lord (in his own time) would be pleased to fix me in a more humane calling, and (if it might be) place me where I might have more frequent converse with his people and ordinances, and be freed from those long separations from home, which very often were hard to bear."

His prayers were answered. Two days before sailing, while sitting at tea with his wife, he was seized with a fit. By the advice of his physicians he resigned the command of the *Bee*, which Mr. Manesty had bought purposely on his account, and thus escaped a calamitous voyage, and terminated his connection with the slave trade. Through the influence of Mr. Manesty, he was appointed tide surveyor of Liverpool.*

In October, 1755, his wife, recovering from a great illness, joined him in Liverpool, and they lived "in health, love, peace and plenty." "I jog on very comfortably," he writes, "in my new pro-consulship, and have struck some bold strokes in my way, one of which will perhaps put from £100 to £150 in my pocket." In January, 1756, he took a house in Edmund Street, off Oldhall Street, and set apart a little room for study and prayer. In April, he printed his "Thoughts on Religious Associations," and sent copies to every minister in Liverpool.

In October, 1757, we find him entertaining his first definite thoughts of entering the ministry.

"This," he says, "was my dear mother's hope concerning me, but her death and the scenes of life in which I afterwards engaged seemed to cut off the probability. I thought I was above most living, a fit person to proclaim that faithful saying 'that Jesus Christ came into the world to save the chief of

* In a letter to his wife, written before she joined him in Liverpool, Newton thus describes his new situation, which was by no means an uncomfortable one.

"I entered upon business yesterday. I find my duty is to attend the tides one week, and visit the ships that arrive, and such as are in the river; and the other week to inspect the vessels in the docks; and thus alternately the year round. The latter is little more than a sinecure, but the former requires pretty constant attendance, both by day and night. I have a good office, with fire and candle, fifty or sixty people under my direction, with a handsome six-oared boat and a coxswain to row me about in form. Mr. W. went with me on my first cruise down to the Rock. We saw a vessel, and wandered upon the hills till she came in. I then went on board, and performed my office with all due gravity; and had it not been my business, the whole might have passed for a party of pleasure."

sinners ;' and as my life had been full of remarkable turns, and I seemed selected to shew what the Lord could do, I was in some hopes that perhaps sooner or later he might call me into this service."

In the year 1758, he made his first effort to enter the Church, but the Archbishop of York, through his secretary, gave him "the softest refusal imaginable."

His first attempts at preaching, made in dissenting meeting houses, were ignominious failures, with MS. as well as extempore, and his shame was so great that he could not see two or three persons conversing in the street, without imagining that he was the subject of ridicule.

In March, 1764, John Newton was offered the curacy of Olney, which he accepted. He was then in his 39th year, and on the 29th of April, he was admitted to deacon's orders at Buckden. On his return to Liverpool, he was asked to preach for two of the clergymen who had signed his testimonials. Then was seen the strange sight of an ex-slave-captain ascending the pulpit of St. George's Church, and preaching to "a crowded and various auditory" composed of the cream of slave-trading Liverpool.*

On the 21st of May, 1764, the Rev. John Newton and his wife left Liverpool, where they had resided for eight years, and took up their residence at Olney. On the 17th of June,

* Of this wondrous turn in his affairs he speaks thus:—"I hope I was enabled to speak the truth. Some were pleased, but many disgusted. I was thought too long, too loud, too much extempore. I conformed to their judgment, so far as I lawfully might, on the Sunday when I preached at the other church in the morning, and at the Infirmary in the afternoon. The next and last Sunday, I preached at Childwell, and was followed by many from town, both of my own friends and others."

In a letter to his old friend, Captain Clunie, he adds:—"The Lord was very gracious to me at Liverpool. He enabled me to preach His truth before many thousands, I hope with some measure of faithfulness, I trust with some success, and in general with much greater acceptance than I could have expected. When we came away I think the bulk of the people, of all ranks and parties, were very sorry to part with us. How much do I owe to the restraining and preserving grace of God, that when I appeared in a public character and delivered offensive truths in a place where I had lived so long, and there appeared a readiness and disposition in some to disparage my character, nothing could be found or brought to light on which they could frame an accusation!"

Mr. Newton was ordained a priest by the Bishop of Lincoln, and so, after weary waiting, the desire of his heart and his mother's wish were granted. In November, 1766, Mr. Newton heard of the bankruptcy of his staunch friend, Mr. Manesty, with whom he had deposited all his savings. He bore the blow with his usual trust in God, thankfully remembering all the good he had formerly received, at Mr. Manesty's hands.

At Olney, where he laboured as curate and vicar for close upon sixteen years, he lived in closest intimacy with the poet Cowper. To be the friend of Cowper was to be the friend of what was best, purest and most spiritual in the England of that time, and the wonder is not that John Newton, the ex-slave-captain, obtained admittance into the Church of England, as an ordained minister, but that he found his way into the heart of the gentle poet, who sang the wrongs of the negroes whom Newton had been buying and selling, and conveying across the sea in floating prisons, the horrors of which he probably locked in his own breast.

It is to the endeared friendship that existed between these two extraordinary men—so dissimilar in all but sincere devotion—that we owe the "Olney Hymns," the joint production of the divine and the poet—of the emancipated slave and the tuneful champion of the negro. These sacred lays form a lasting monument to the kinship of soul subsisting between the strong, clear-visioned pastor, and his gentle, melancholy parishioner, and, in the light of Newton's story, afford another instance that—

"God moves in a mysterious way
His wonders to perform."

From Olney, Mr. Newton was removed to the united parishes of St. Mary Woolnoth and St. Mary Woolchurch Haw, Lombard Street, London. He often spoke with great feeling of his new position, seldom passing a single day anywhere without referring to the strange event in one way or other.*

In one of the letters of the Cardiphonia Series, dated October 27th, 1778, is the following remarkable passage :—

> "Last Sunday a young man died here of extreme old age, at twenty-five. He laboured hard to ruin a good constitution, and unhappily succeeded, yet amused himself with the hopes of recovery almost to the last. We have had a sad knot of such poor creatures in this place, who labour to stifle each other's convictions, and to ruin themselves and associates, soul and body. How industriously is Satan served! I was formerly one of his most active under-tempters. Not content with running the broadway myself, I was indefatigable in enticing others; and, had my influence been equal to my wishes, I would have carried all the human race with me. And doubtless some have perished to whose destruction I was greatly instrumental, by tempting them to sin, and by poisoning and hardening them with principles of infidelity; and yet I was spared. * * Had my abilities been equal to my heart, I should have been a Voltaire and a Tiberius in one character, a monster of profaneness and licentiousness. 'O to grace how great a debtor!' A common drunkard or profligate is a petty sinner to what I was. I had the ambition of a Cæsar or an Alexander, and wanted to rank in wickedness among the foremost of the human race."

With regard to the peculiar traffic in which Newton was so long engaged, and which, to modern views, seems so strangely at variance with the first principles of Christianity, he speaks for himself, in 1763, as follows :—

> "The reader may perhaps wonder, as I now do myself, that, knowing the state of this vile traffic to be as I have

* "That one," said he, "of the most ignorant, the most miserable, and the most abandoned of slaves, should be plucked from his forlorn state of exile on the coast of Africa, and at length be appointed minister of the parish of the first magistrate of the first city in the world—that he should there not only testify of such grace, but stand up as a singular instance and monument of it—that he should be enabled to record it in his history, preaching, and writings to the world at large—is a fact I can contemplate with admiration, but never sufficiently estimate."

described"—the reference is to a letter in which he has been speaking of the condition of the slaves—"and abounding with enormities which I have not mentioned, I did not at the time start with horror at my own employment as an agent in promoting it. Custom, example, and interest, had blinded my eyes. I did it ignorantly, for I am sure had I thought of the slave-trade then as I have thought of it since, no considerations would have induced me to continue in it. Though my religious views were not very clear, my conscience was very tender, and I durst not have displeased God by acting against the light of my mind. Indeed, a slave-ship, while on the coast, is exposed to such innumerable and continual dangers, that I was often then, and still am, astonished that anyone much more that so many, should leave the coast in safety. I was then favoured with an uncommon degree of dependence upon the Providence of God, which supported me; but this confidence must have failed in a moment, and I should have been overwhelmed with distress and terror, if I had known, or even suspected, that I was acting wrongly. I felt greatly the disagreeableness of the business. The office of a gaoler, and the restraints under which I was obliged to keep my prisoners, were not suitable to my feelings; but I considered it as the line of life which God in His Providence had allotted me, and as a cross which I ought to bear with patience and thankfulness till he should be pleased to deliver me from it. Till then I only thought myself bound to treat the slaves under my care with gentleness, and to consult their ease and convenience so far as was consistent with the safety of the whole family of whites and blacks on board my ship."

In 1787, Mr. Newton's remarks on the African slave trade were given to the world. Mrs. Hannah More writes to tell him how much she is pleased with his sensible, judicious, well-timed, and well-tempered pamphlet on the slave trade, and that in a letter from Bristol, she had been informed that Mr. John Wesley named it with great commendation in a sermon he preached on the subject.

He also received a communication from Scotland, speaking most highly of his writings, and stating that they had made his name "savoury" in the most remote and distant parts of the country, and concluding with a request, that he would allow their reprint in that part of the kingdom—a sure earnest of fame. From America, India, and other quarters nearer home, came strong proofs that his works were now read with comfort and delight by thousands of people who revered the name of Newton. He refused to accept the degree of Doctor of Divinity from the University of New Jersey, and threatened that if any more letters came from Scotland, addressed to Dr. Newton, he would return them unopened. He supplied Mr. Clarkson with valuable information connected with the slave-traffic, and was one of the most important witnesses called before the Parliamentary Committee appointed to enquire into that subject.

In 1792, when Wilberforce had renewed his motion in the House of Commons for the abolition of the slave trade, Newton preached upon the subject, as he had done on a like occasion in 1791.

> "I regarded it," he says, in a letter to the Rev. W. Bull, "not in a political, but in a moral view. I consider myself bound in conscience to bear my testimony at least, and to wash my hands from the guilt which, if persisted in now that things have been so thoroughly investigated and brought to light, will, I think, constitute a national sin of a scarlet and crimson dye."

About the same time, he published his "Thoughts upon the African Slave Trade," in which he pointed out its fearful political and moral evils—its injury alike to the slaves and those who trafficked in them:—

> "If my testimony," he says, "should not be necessary or serviceable, yet, perhaps, I am bound in conscience to take shame to myself by a public confession, which, however sincere, comes too late to prevent or repair the misery and mischief to

which I have formerly been accessory. I hope it will always be a subject of humiliating reflection to me that I was once an active instrument in a business at which my heart now shudders. Perhaps what I have said of myself may be applicable to the nation at large. The slave trade was always unjustifiable; but inattention and interest prevented for a time the evil from being perceived. It is otherwise at present. The mischiefs and evils connected with it have been of late years represented with such undeniable evidence, and are now so generally known, that hardly an objection can be made to the almost universal wish for the suppression of this trade, save on the ground of political expedience."

Three thousand copies of this pamphlet were printed and distributed by the Abolition Society.

In the course of a sermon, preached on Friday, Feb. 28th, 1794, the day appointed for a general fast, Mr. Newton referred to the slave trade in these words :—

"I should be inexcusable, considering the share I have formerly had in that unhappy business, if upon this occasion, I should omit to mention the African Slave Trade. I do not rank this amongst our national sins, because I hope and believe a very great majority of the nation earnestly long for its suppression. But, hitherto, petty and partial interests prevail against the voice of justice, humanity, and truth. This enormity, however, is not sufficiently laid to heart. If you are justly shocked by what you hear of the cruelties practised in France, you would perhaps be shocked much more if you could fully conceive of the evils and miseries inseparable from this traffic, which I apprehend, not from hearsay, but from my own observation, are equal in atrocity, and perhaps superior in number, in the course of a single year, to any or all the worst actions which have been known in France since the commencement of their revolution. There is a cry of blood against us ; a cry accumulated by the accession of fresh victims, of thousands, of scores of thousands, I had almost said of hundreds of thousands, from year to year."

Preaching in the parish church of St. Mary Woolnoth, on December 19th, 1797, the day of general thanksgiving for the late naval victories, he said :—

> "Oppression is a national sin if the grievance be publicly known, and no constitutional measures adopted for prevention or relief. Charges of this nature have been brought against the exercise of our power, both in the east and in the west. I pretend not to say how far they were founded in truth, or exaggerated. I confine myself to a single instance, of which my own knowledge warrants me to speak. I have more than once confessed with shame in this pulpit the concern I had too long in the African slave trade. This trade, marked as it is with the epithet INFAMOUS by a vote of the House of Commons, is still carried on, and under the sanction of the legislature. Though the repeated attempts to procure the abolition of this trade have not succeeded, they have doubtless contributed to meliorate the condition of the blacks who are in a state of slavery in our West India Islands. The mode of their transportation thither from the African coast seems to be less tormenting and fatal than formerly. How far this trade may have been affected by the present war I know not. When I was engaged in it, we generally supposed, for an accurate calculation was not practicable, that there were not less than a hundred thousand persons, men, women, and children brought off the coast by the European vessels of all nations ; and that an equal number lost their lives annually by the wars and other calamities occasioned by the traffic, either on shore, without reaching the ship, or on shipboard before they reached the places of sale. It was also supposed that more than one-half, perhaps three-fifths of the trade was in the hands of the English. If the trade is at present carried on to the same extent, and nearly in the same manner, while we are delaying from year to year to put a stop to our part of it, the blood of many thousands of our helpless, much injured fellow-creatures is crying against us. The pitiable state of the survivors who are torn from their nearest relatives, connections, and their

native land, must be taken into the account. Enough of this horrid scene. I fear the African trade is a national sin, for the enormities which accompany it are now generally known; and though, perhaps, the greater part of the nation would be pleased if it were suppressed, yet, as it does not immediately affect their own interest, they are passive. The shop-tax, a few years since, touched them in a more sensible and tender part, and therefore petitions and remonstrances were presented and repeated, till the tax was repealed. Can we wonder that the calamities of the present war begin to be felt at home, when we ourselves wilfully and deliberately inflict much greater calamities upon the native Africans, who never offended us? That is an awful word—'Woe unto thee that spoilest, and thou wast not spoiled; when thou shalt cease to spoil, thou shalt be spoiled.'"

Not a word did the ex-slave-captain say of the negroes who behaved "like children of one family" forty-five years before!

A friend who called on him when he was seventy-nine, relates how "when the servant was employed putting on his shoes, he looked up, saying, 'I had not this trouble in Africa, for I had no shoes. Sir'—looking at his friend—'when I rose in the morning, and shook myself like a dog, I was dressed. For forty years past I have thought every waking hour on my former misery.'"

In January, 1806, his friend, Mr. Cecil said to him: "In the article of public preaching, might it not be best to consider your work as done, and stop before you evidently discover you can speak no longer?" "I cannot stop," said he, raising his voice. "What! shall the old African blasphemer stop while he can speak?"

An interesting reference to his African life occurs in a letter to a Mr. Campbell a few years before his death. Twenty African boys and girls had been brought over for instruction, and after five years were sent back to the coast of Guinea.

"Last week," he says, "I was at Clapham, and saw the twenty African blackbirds. The girls were at Battersea, out of my reach. When I went into the school, I said *Lemmi*, which is, being interpreted, How do you do? Two or three answered *Bah*, that is, I thank you; by which I knew that they had some knowledge of the language of Sherboro', the scene of my bondage. I am told the boys come forward apace, behave well, and seem very happy, and especially when they see Mr. Macaulay."

What strange thoughts must have passed through the old man's mind as he stood there in the presence of types of Africa's future civilization! And the children—the African blackbirds—had they any inkling that they were in the presence of a once typical enemy of their race, soon to be doomed to extinction?

As a minister of the gospel, there was such a zeal for the truth, such gentleness, candour, and forbearance in Mr. Newton, that conciliated enemies and made him beloved by his friends. On this head the testimony of Cowper is very clear. In a letter to Newton, the poet says, "A people will love a minister, if a minister seems to love his people. The old maxim, *simile agit in simile*, is in no case more exactly verified; therefore you were beloved at Olney; and if you preached to the Chickesaws and Chactaws, would be equally beloved by them."

For some months before his death, Mr. Newton was confined to his room. "I am," said he, "like a person going on a journey in a stage coach, who expects its arrival every hour, and is frequently looking out at the window for it;" and another time: "I am packed and sealed, and waiting for the post." Mr. Jay visited him near the closing scene. He was hardly able to speak, but said: "My memory is nearly gone; but I remember two things: that I am a great sinner, and that Christ is a great Saviour."

About a month before his death, he said to a lady who was sitting by him :—

"It is a great thing to die, and when flesh and heart fail, to have God for the strength of our heart, and our portion for ever. I know whom I have believed, and He is able to keep that which I have committed to him against that great day. Henceforth there is laid up for me a crown of righteousness, which the Lord, the righteous Judge, shall give me at that day."

At another time he said :—

"More light, more love, more liberty. Hereafter, I hope when I shut my eyes on the things of time, I shall open them in a better world. What a thing it is to live under the shadow of the wings of the Almighty! I am going the way of all flesh."

And so he "gradually sank as the setting sun, shedding to the last those declining rays which gilded and gladdened the dark valley." On the evening of Monday, December 21st, 1807—the year of the abolition of the Liverpool Slave Trade—he passed away in his eighty-third year. He was buried in his church of St. Mary Woolnoth. He composed the following epitaph for himself, which was inscribed on a plain marble tablet in the church:—

"John Newton, clerk, Once an Infidel and Libertine, A Servant of Slaves in Africa, was By the rich mercy of our Lord and Saviour—Jesus Christ, Preserved, Restored, Pardoned, And appointed to preach the Faith He had long laboured to destroy. He ministered Near XVI. Years as curate and Vicar of Olney in Bucks, And XXVIII. as Rector of these United Parishes. On Febry. the First MDCCL. he married Mary, Daughter of the late George Catlett, of Chatham, Kent, Whom he Resigned To the Lord Who Gave Her, On Decr. the XVth. MDCCXC."

Mr. Newton's conversational powers were remarkable.

He had wit, humour, ready thought, and expression, all tempered by cheerfulness, kindliness, and real piety. Sometimes he would have droll fancies, as when by a strong sneeze he shook off a fly which had perched on his gnomon, and immediately said: "Now, if this fly keeps a diary, he'll write, 'To-day, a terrible earthquake.'" Asked how he slept, he instantly replied: "I'm like a beef-steak—once turned, and I'm done." Once a little sailor boy, with his father, called on him. He took the boy between his knees and told him that he had been much at sea himself, and then sung him part of a naval song. His heart was as large as it was loving. There must have been something marvellously winning and uncommon about a man who had lifelong friendships with such people as Lord Dartmouth, and Wilberforce, the Thorntons, Charles Grant, Ambrose Serle, and Mrs. Hannah More, with the flower of the clergy of all denominations. He was no great preacher. He was not a Whitfield or a Wesley in the pulpit, trumpet-tongued to arouse the masses, but though he had neither music in his voice nor grace in his manner, great congregations hung upon his lips. The secret of his power was sincerity and earnestness. His whole soul was in sympathy with the truth, and with his hearers. He was a strong-minded man, a practical Christian, with a gift for turning his talents to the best account. His character was beautiful in its entireness. He never forgot what he had been, and if pride ever arose in his mind, he would say on such occasions he had only to mix a little Plantain sauce with his more savoury diet and the evil was at once suppressed. Prayer was his vital breath. As a pastor and house-preacher he was probably unequalled. For nearly half a century he was one of the best known and most valued ministers of the English church, and his character, which was peculiarly English, refined by grace, rendered him a man revered and loved better and more widely than most of his fellows. The story of his life and Christian experiences, his Letters, and some of his Hymns, will probably be remembered when the seaport from which he sailed as a slave-captain, is but a name like Tyre and Sidon, and when Africa, the scene of his degradation, is basking in the light of the great day of Restitution and Refreshment.

CHAPTER III.

The Massacre at Old Calabar.

In 1766, the *Vine*, Captain Simmons, returned from a voyage to Bonny, on the coast of Africa, and Dominica, in the West Indies, with 400 slaves, having accomplished the round voyage in seven months and ten days, and apparently broken the record. The market value of the cargo could not have been less than £13,000, as will be seen from the following table, showing the average price of negroes sold at Charleston, Jamaica, Grenada, Tortola, and Dominica, during seven years (1759, 1767-1772), and from the account sales of negroes imported in the ship *African*.

Average Price of Negroes.

						£	s.	d.
1759	Whydah	Negroes	averaged	at	Charleston	35	11	0
1767	Calabar	do.	do.	do.	Grenada	27	0	0
1767	Bonny	do.	do.	do.	do.	30	10	0
1768	Calabar	do.	do.	do.	Tortola	23	10	0
1769	Do.	do.	do.	do.	Jamaica	34	14	0
1770	Windward Coast	do.	do.	do.	Dominica	33	8	0
1770	Old Calabar	do.	do.	do.	Dominica	27	12	0
1771	Eboe	do.	do.	do.	Charleston	40	0	0
1771	Calabar	do.	do.	do.	Dominica	30	13	0
1772	Eboe	do.	do.	do.	Charleston	39	15	0
1772	Averages at Charleston from £50 to £54							
1773	Number of Negroes arrived at Charleston that year 4,500							

COPY OF ACCOUNT SALES OF NEGROES.

"Sales of 268 Negro slaves imported in the ship *African*, Captain Thomas Trader, from Malemba, on the acct. and risque of Messrs. John Cole & Co., owners of the said ship, merchants in Liverpool.

To whom Sold.	Men.	Women.	Boys.	Girls.	Total.	Price.	£	s.	d.
By James Fisher............	...	...	1	...	1	....	35	0	0
,, John Miller	...	...	1	...	1	....	35	0	0
,, Augustus Valtette......	...	...	1	...	1	....	40	0	0
,, George Richards	...	...	1	...	1	....	35	0	0
,, Ditto.	...	...	1	...	1	....	35	0	0
,, Papley & Wade	103	26	67	34	230	....	7820	0	0
,, Chambers & Mead ...	5	...	2	1	8	....	296	0	0
,, Sloop *Two Brothers*...	...	...	6	...	6	....	204	0	0
,, Monsr. Fontanelle ...	...	...	...	2	2	@ £36......	72	0	0
,, John Darey	...	...	2	...	2	@ £30......	60	0	0
,, Ditto.	4	3	2	3	12	@ £35......	420	0	0
,, Alexan. Forceston ...	...	1	...	1	2	Sickly	30	0	0
,, Sold at Vendue.........	...	...	1	...	1	C'pt to a/c for			
	112	30	85	41	268		£9082	0	0

CHARGES, viz. :—	£	s.	d.	£	s.	d.
To Cash paid Import Duty on 268 Slaves at 10/ and Bond 5/	134	5	0			
To ditto paid the Dr. his head money on ditto at 12/ ...	13	8	0			
To ditto paid Captain Trader, his Coast Commission, at £4 per 104 on £9082 gross sales	349	6	2			
To my Commission, at 5 per cent. on the gross sales ...	454	2	0			
				951	1	2
To Messrs. John Cole & Co., owners of the *African*, in account current for				£8130	18	10

Errors excepted.

KINGSTON, JAMAICA, 19*th September*, 1764. Per WM. BOYD.

Messrs. JOHN COLE & CO., Owners of the Ship "African," in Acct. Current with WM. BOYD & CO.

Dr.	£	s.	d.	*Cr.*	£	s.	d
To Amount of Sundries shipped in the *African*, per Invoice	6384	16	5¼	By Nt. Proceeds of the *African's* sales	8130	18	10
To Balance of the *African's* a/c of disbursements, per Capt. receipt	269	7	3½				
To my draft on Snell & Co., of London, for £1054 16s. 6d. sterling, Exc. at 40 per cent., payable at 60 days' sight	1476	15	1¼				
	£8130	18	10		£8130	18	10

Errors excepted.

KINGSTON IN JAMAICA, 20*th September*, 1764. Per WM. BOYD."

As few persons in this country ever saw a bill of lading for human beings, shipped on board a British vessel engaged in this odious traffic, we append a copy of an original bill of lading for slaves, shipped for Georgia*:—

"Shipped by the grace of God, in good order and well condition'd by James [surname illegible], in and upon the good Ship call'd the MARY BOROUGH, whereof is Master, under God, for this present voyage, Captain David Morton, and now riding at Anchor at the Barr of Senegal, and by God's grace bound for Georgey, in South Carolina, to say, twenty-four prime Slaves, six prime women Slaves, being mark'd and number'd as in the margin, and are to be deliver'd, in the like good order and well condition'd, at the aforesaid Port of Georgia, South Carolina (the danger of the Seas and Mortality only excepted), unto Messrs. Broughton and Smith, or to their Assigns; he or they paying Freight for the said Slaves at the rate of Five pounds sterling per head at delivery, with Primage and Avrage accustom'd. In WITNESS whereof, the Master or Purser of the said Ship hath affirm'd to three Bills of Lading, all of this tenor and date; the one of which three bills being accomplish'd, the other two to stand void; and so God send the good ship to her desir'd port in safety, Amen.

Marked on the Right Buttock
O
O

"Dated in Senegal, 1st February, 1766,

"DAVID MORTON."

It will be observed from the bill of lading, that those slaves were marked or branded with particular marks. The operation of marking slaves was performed on them by means of

* The original bill of lading was in the possession of the late Richard Brooke, Esq., F.S.A., who printed it in his "Liverpool as it was during the last quarter of the Eighteenth Century."

a heated iron, with as much indifference as if they had been merely cattle. Branding irons, with letters or marks for branding slaves, were exhibited for sale in the shops of Liverpool, and no doubt they were sold in the same manner in other seaport towns of the kingdom. Mr. Clarkson gives the following description of certain instruments which he bought during his sojourn in Liverpool:—

"There were specimens of articles in Liverpool, which I entirely overlooked at Bristol, and which, I believe, I should have overlooked here also, had it not been for seeing them at a window in a shop. I mean those of different iron instruments used in this cruel traffic. I bought a pair of the iron handcuffs with which the men slaves are confined. The right-hand wrist of one, and the left of another, are almost brought into contact by these, and fastened together by a little bolt with a small padlock at the end of it. I bought also a pair of shackles for the legs. The right ancle of one man is fastened to the left of another, by similar means. I bought these, not because it was difficult to conceive how the unhappy victims of this execrable trade were confined, but to show the fact that they were so. For what was the inference from it, but that they did not leave their own country willingly; that when they were in the holds of the slave vessels, they were not in the Elysium which had been represented; and that there was a fear, either that they would make their escape, or punish their oppressors. I bought also a thumb-screw at this shop. The thumbs are put into this instrument through the two circular holes at the top of it. By turning a key, a bar rises up by means of a screw, and the pressure upon them becomes painful. By turning it further, you may make the blood start from the ends of them. By taking the key away, you leave the tortured person in agony, without any means of extricating himself, or of being extricated by others. This screw, as I was then informed, was applied by way of punishment in case of obstinacy in the slaves, or for any other reputed offence, at the discretion of the captain. At the same place I bought another instrument which I saw. It was called a speculum oris. This instrument is known among surgeons, having been invented to assist them in wrenching open the mouth, as in the case of a locked jaw. But it had got into use in this trade. On asking the seller of the instruments on what occasion it was used there, he replied, that the slaves were frequently so sulky, as to shut their mouths against all sustenance, and this with a determination to die; and that it was necessary their mouths should be forced open to throw in nutriment, that they who had purchased them might incur no loss by their death."

The slave-captains sometimes got into awkward scrapes with the natives. Captain James Berry, of Liverpool, gives the following remarkable account of his being taken prisoner:—

"On board Brig *Dalrimple* Old Callabar April 3, 1763

This is to acquaint all gentlemen that it may fall into the hands of that on the 30 of Jany I arrived hear in a small vessell came too at 7 Fathom Point wrote up to Abashey finding no vessell their I imagined the might Lett me stay paying a small acknowledgement to the King the Duke and some of the Heads* Abashey came down and prevaild on me to go up the River I accordingly went up that night next morning according to custom went ashore to shake the Kings and the Rest of the getlemen Hands made my proposalls which was af first Refuse'd but after standing out about fifteen Day aggreed to pay 1000 Coppers among's them all the King and Duke each 65 Crs the rest of the Gentlemen in proportion I gott pledges out of the Kings Town Dukes and Tom Henshaws Likewise there caim as tokens of their Honour Robin John Town Refused me a son for pledge but thought I had sufficient security on the Second of March was much out of order and Had been for four Days before that was unfortunate

* The leading people of Old Town, Calabar, were the King, the Duke, Ephraim Robin John, Robin John Tom Robin, John Amboe, Orrock Robin John, Archibong Robin John, John Robin John, Otto Robin John, Abashey, Egbyoung, Tom Henshaw, Solomon Henshaw, Amboe Robin John, Ancona Robin Robin John.

enough to go down the River to gett a little air thinking their was no Danger of being molested by any Body haveing the Kings Sons Duks and Tom Henshaw's Egbyoung Antera in the Boat with me but no sooner gott the Lenght of Old Town but that Rouge Eph^{m}. Robin John Joined by R^{n}. John Tom R^{n}. Captn. John Ambo and the Rest of that Town sent ten canoos full of people and took me out of my Boat by force hauld me over nine into the Tenth the first vilain that I was rec^{d} by was Tom R^{n}. who told me it was very well I was come it had saved them the Trouble of fetching me out of my ship Eph^{m}. came on board my vesell the night before that with that design only I was at that time very bad but had intended to have come again in a day or two they haveing counted all my people and pitched upon their Boys for the seaing my people seeing so few and three or four of them at that time sick while the took me bo force and putt me in the canooe he kept me on shore 29 Day and obliged me to pay him and the Rest of the Scoundrells just what he pleas'd the amount of his imposition is 4251 Copper besid's him takeing in spight of all I cood do one of my great guns which I have given the Duke an order for if he can possably get it he Likewise has gott three of my musquetts two Blunderbusses 2 pistolls 2 cutlasses and two of my Jacketts the Black Boys had on that was in the Boat with me he oblig'd me to give severall Books and one to clear him of all palaver with me which for sake of getting on board my vesell wood have given him any Books he wanted but the air all of no signification I immagine any Gentleman wodd do the same was it their case. On the 22nd of Mar the King the Duke Solomon Henshaw and the Rest of the Gentlemen of the other party come on board with 98 slaves the seeing how I was Imposed on by those Rascalls made my mate count all the good in my ship Abashey made Trade and Bought me 47 slave all of which was good only one woman and I believe did me justice in every thing the Duke carried ashore with him 605 Copper to buy yammes which he sent me as fast as he cood gett them I doant Blame any of them for what the did seeing the vilanious intentions of the

Old Town Scoundrells but never will for give the injury Eph[m] and the rest of them did me till I have satisfaction.

I am the Gentleman

Hum[le] Serv[t]

JAMES BERRY."

In 1767, there was a strong competition between the ports of Bristol and Liverpool in the trade to the coast of Africa. The inconveniences and dangers attendant on that branch of traffic are described in the following extract of a letter from Old Calabar, dated August 12th, 1767:—

"We had a tolerable good passage of three weeks and five days. There are now seven large vessels in the river, each of which expects to purchase 500 slaves, and I imagine there was seldom ever known a greater scarcity of slaves than at present, and these few chiefly from the low country. The natives are at variance with each other, and, in my opinion, it will never be ended before the destruction of all the people at Old Town, who have taken the lives of many a fine fellow. Captain Hutton's chief mate had the misfortune to suffer under their vile hands; but I now flatter myself, I shall be an assistant in revenging the just cause of every poor Englishman that have innocently suffered by them.

"The river of late has been very fatal both to whites and blacks. There have three captains belonging to Bristol died within these few months, besides a number of officers and sailors. I assure you, I never saw a worse prospect in my life for making a voyage than at present. The major part of the vessels here have very dangerous disorders amongst the slaves, which makes me rejoice that I have very few on board. I do not expect that our stay here will exceed eight months. The adjoining coasts of trade seem all to be very much thronged with shipping, except the Gold Coast, the bad effects of which, I am afraid, the Liverpool gentlemen must feel this season."*

In the year 1767, a terrible affair, which seems to be hinted at in the preceding letter, known as the massacre

* "Troughton's History of Liverpool," p. 143.

at Calabar, took place. The details are drawn from copies of the original depositions, in the case of the King against Lippincott and others, supplied to Mr. Clarkson by Mr. Henry Sulgar, a Moravian minister at Bristol. The originals were sworn before Jacob Kirby and Thomas Symons, commissioners at Bristol for taking affidavits, by Captain Floyd, of the city of Bristol, who had been a witness to the tragedy, and of Ephraim Robin John and Ancona Robin Robin John, two African chiefs, who had been sufferers by it. It appears from these documents, that in the year 1767, the ships, *Indian Queen*, *Duke of York*, *Nancy*, and *Concord*, of Bristol, the *Edgar*, of Liverpool, and the *Canterbury*, of London, lay in Old Calabar river. A quarrel, originating in a jealousy respecting slaves, existed at this time between the principal inhabitants of Old Town, and those of New Town, Old Calabar. The captains of the vessels before mentioned joined in sending several letters to the inhabitants of Old Town, but particularly to Ephraim Robin John, who was at that time a grandee, or principal man of the place. The tenor of these letters was, that they were sorry that any jealousy or quarrel should subsist between the two parties; that if the inhabitants of Old Town would come on board, they would afford them security and protection; adding, at the same time, that their intention in inviting them was that they might become mediators and thus heal their disputes.

The inhabitants of Old Town joyfully accepted the invitation. The three brothers of the chief, Ephraim Robin John, the eldest of whom was Amboe Robin John, first entered their canoe, attended by twenty-seven persons, and being followed by nine canoes, directed their course to the *Indian Queen*. They were dispatched from thence the next morning to the *Edgar*, and afterwards to the *Duke of York*, on board of which they went, leaving their canoe and attendants alongside of the same vessel. In the meantime, the people on board the other canoes were either distributed on board, or lying close to the other ships.

This being the situation of the three brothers, and of the leading people of the place, the treachery now began to appear. The crew of the *Duke of York*, aided by the captain and mates, and armed with pistols and cutlasses, rushed into the cabin with an intent to seize the persons of their three unsuspicious guests. The unhappy men, alarmed at this violation of the rights of hospitality, and struck with astonishment at the behaviour of their supposed friends and peacemakers, attempted to escape through the cabin windows, but being wounded, were obliged to desist and to submit to be put in irons. While this atrocious act was in progress, an order was given to fire upon the canoe, which was then lying alongside of the *Duke of York*. The canoe soon filled and sank, and the wretched attendants were either seized, killed, or drowned. Most of the other ships followed the example. Great numbers were thus added to the killed and drowned on the occasion, while others attempted to escape by swimming to the shore. But at this juncture, the inhabitants of New Town, who had concealed themselves in the bushes by the waterside, and between whom and the commanders of the vessels the plan had been previously arranged, came out of their hiding places, and, embarking in their canoes, made for such as were swimming from the fire of the ships. The ships' boats also were manned, and joined in the pursuit. They butchered the greatest part of those whom they caught. Many dead bodies were soon seen upon the sands, and others floating upon the water. Including those who were seized and carried off, and those who were drowned and killed, either by the firing of the ships, or by the people of New Town, the number lost to the inhabitants of Old Town on that day was three hundred souls. The carnage was scarcely over when a canoe, full of the principal people of New Town who had promoted the massacre, dropped alongside of the *Duke of York.* They demanded the person of Amboe Robin John, the brother of the chief of Old Town, and the eldest of the three on board. The unfortunate man put the palms of his hands together, and beseeched the commander of the vessel that he would not violate the rights of hospitality by giving up an unoffending stranger to his enemies. But no entreaties could prevail. The commander received from the New Town people a slave, of the name of Econg, in his stead, and then forced Amboe Robin John into the canoe, where his head was immediately struck off in the sight of the crew, and of his afflicted brothers. As for them, they escaped his fate, but they were carried off, with their attendants, to the West Indies, and sold into slavery.

The action of the captains has never been defended; but we must not forget that they were dealing with a shifty, greedy, and treacherous lot of rascals, who made a practice of selling their own countrymen into slavery. The delays and subterfuges resorted to by the native chiefs to enhance the price of slaves, and to extract more "coomey," must have been extremely exasperating to the slave commanders, whose lives and cargoes were imperilled by a prolonged bargaining, owing to the climate, and the possible outbreak of disease among the slaves cooped up in the hold, before they left the coast and entered upon the horrors of the sea passage. The following copies of papers belonging to the commander of the *Edgar,* show that the chiefs were in his debt, and that they exonerated him from the charge of kidnapping a boy named Assogua. Moreover, certain letters from the chiefs of Old Town, Calabar, addressed to the captain, prove that they held him and his family in the highest esteem, notwithstanding the fact that the *Edgar* was present in the unfortunate tragedy of 1767. Whether this is to be attributed to the innocence of the captain, who was at all events a worthy citizen of Liverpool, or to the abnormal development of Christian charity and forgiveness in the African chiefs and man-stealers of Old Calabar, let the reader determine for himself:—

"An Acct. of Goods and Slaves Owing to the ship *Edgar* from the Traders of Old Town as under:

"Archibong Robin John five slaves Goods		Dr
	Co*	
20 Iron 5 Nicconees 5 Brawels	155	Recd Nothing
4 Romales 3 Cushtaes 2 Photes	106	
8 B.Pipes 5 Flagons 50 Rods	102	
3 Basons 4 Guinea stuffs	25	
3 Blunderbus's 8 Kegs	112	
	500	

24th July 1767 Goods for 5 slaves.
Received a further trust 10 rods 1 Nicconee 20

"Orrock Robin John			Dr
		Co	
23rd July 1767	To 1 Keg of Powder	8	
	By a boy left on board name Asuqu not stoped by me as Orrok says nor was Orrock's son		

"Ambo Robin John		Dr
August 7. 1767 To Goods for two men slaves		
	Co	
2 Blunderbss 3 Kegs 8 Iron 1 Nicconee	98	Recd Nothing
2 Brawels 1 Cushtae 2 Romales	44	
1 Photac 2 Flagons 2 basons 3 Pipe bds	44	
10 Rods 8 Chints	18	
	204	

* "Co." means cowries, small shells brought from the East Indies, and used by the natives as substitutes for coin. Meneles, maniloes, or manillas were ornaments for the wrists and ancles. Romalls (or romales), niccannees, cushtaes, photes, photacs, or photaes, chellos, and Guinea stuffs, were Manchester and

"Ephraim Robin John		Dr
Co	
July 23rd 1767 To 20 Rods 1 Romale 4 Basons 4 L. Meneles	48	Recd Nothing
1 Neganepaut 1 Blunderbus 20 Rods 1 Baft 12 Knives	74
122	
24th To Goods for 2 men slaves as under		
Co	
4 Kegs 8 Iron 2 Nicconees 2 Brawels 1 Cushtae	104	Recd Nothing
1 Romale 1 Photac 16 Chello 4 bg Pipe bds	60
2 bg Red 2 G. Stuffs 1 Flagon 14 rods	36
200	

"John Robin John		Dr
Co	
July 7th 1767 To 10 rods 1 Nicconee 6 Romale	26	Recd Nothing
Augt 2 To 8 Chello 1 hatt 1 Jug brandy	16
42	

	Co		
"Augt 1, 1767 Otto Rob. John Dr To 5 Rords	5	Recd	Nothing
do Tom Andrew Honesty, do	5	do	do
July 30th Robin John 6 L Meneles 1 Rom	18	do	do
Augt 1st Rob. Rob. Jno. 1 Keg 2 Caps 1 Shenda 1 Br	20	do	do

All Coppers makes. 240 and 9 slaves makes 11 slaves and 20 Copers Tom Robin had makes near 12 slaves"

Indian fabrics. Brass and copper kettles and pans, pewter basons, iron pots, bars of lead, bars and rods of iron, shallow brass pans, called "Neptunes," for preparing salt out of sea water; plates, dishes, mugs, basons, wine glasses, tumblers, decanters, knives, spoons, razors, soap, gunpowder, muskets, brandy, rum, beads, trinkets, worsted caps, laced hats, looking-glasses, cottons, calicoes, chintz, silks, slops, salt, fish-hooks, axes, hatchets, cutlasses, carpets, handkerchiefs, felt hats, scarlet jackets; all these formed part of a Guinea cargo, to be bartered for "prime negroes."

"OLD CALABAR, *August* 22, 1776

"This is to certify whom it doth or may concern that the within is a True List of Debts owing by the Natives of Old Town to Captain Lace of Liverpoole, and that the Boy named Assogua was not stoped by Captain Lace has as been Reported, but was put on board by Orrock Robin John unto whom he belonged, and that Captain Lace carried him of for the within debts, because we made no application for him nor did we even offer to Redeem him whilst the ship staid in the River, as Witness our hands

Witness
John Richards
James Hargraves

his
King X George
mark

his
Jno. X Robin John
mark

Otto Ephraim

his
Orrock Robin X John
mark"

Another signature is also appended which is undecipherable.

The following letter written by the former captain of the *Edgar* to Mr. Thomas Jones, a Bristol owner of slave-ships, seems to have some reference to the two brothers carried off after the massacre in 1767:—

"LIVERPOOL, 11*th November,* 1773

"MR. THOS. JONES,

"SIR,—Yours of 7th I received wherein you disire I will send an Affidavit concerning the two black men you mention, Little Ep^m. and Ancoy, in what manner the ware taken off the coast, and that I know them to be Brothers to Grandy Ep^m. Robin John; as to little Ep^m. I remember him very well, as to Ancoy Rob. Rob. John I cant recolect I ever saw him. I knew old Robin John the Father of Grandy Ep^m. and I think all the Family, but never found that little Ep^m. was one of Old

Robins sons, and as to Rob. Rob. John he was not Old Rob. Johns son. Old Robin took Rob. Rob. J$^{no.}$ mother for a wife when Robin Rob. J$^{no.}$ was a boy of 6 or eight years old, and as to Rob. Rob. J$^{no.}$ hen ever had a son that I heard of. You know very well the custom of that place whatever Man or Woman gos to live in any family the take the Name of the first man in the family and call him Father, how little Ep$^{m.}$ came into the family I cant tell, and as to what ship they came off the coast in I know no more than you, therefore, cant make Affidavit Eather to their being Brothers to Grandy Ep$^{m.}$ or the manner he was brought off the Coast, as to Grandy Ep$^{m.}$ you know very well has been Guilty of so many bad Actons, no man can say anything in his favour, a History of his life would exceed any of our Pirates, the whole sett at Old Town you know as well as me. I brought young Ep$^{m.}$ home, and had him at School near two years, then sent him out, he cost me above sixty pounds and when his Fathers gone I hope the son will be a good man. As to Mr. Floyd he says more then I ever knew or heard of hes in many Errors, even in the Name of the vessell I was in hes wrong, there was no such a ship as the Hector while I was at Callebarr, a man should be carefull when on Oath, how he knows the two men to be brothers to Ep^{m}. I cant tell, I have several times had the pedigree of all the familys from Abashey the foregoing acct. of Rob. Rob. was from him, but to prove the two men to be Ep$^{ms.}$ brothers I dont know how you will do it, I assure you I dont think they are, if you think to send a vessell to Old Town it might ans^{r} for you to purchas the two men I once bog^{t} (bought) one at Jamaica a man of no consiquance in family but it ansr^{d} the Expence.

I am Sir your hbl Serv$^{t.}$

—— ——.

"P.S.—I left the duke of York and Indian Queen at Callebarr."

Copy of a letter from "Grandy King George," King of the Old Town Tribe, addressed to "Mr. Ambrose Lace and Companey, Marchents in Liverpool":—

"OULD TOWN, OULD CALLABAR, *January* 13, 1773

"MARCHANT LACE, S[R],—I take this opertunety of Wrighting to you and to aquant you of the behaveor of Sum ships Lately in my water there was Capt Bishop of Bristol and Capt. Jackson of Liverpool laying in the river when Capt Sharp arived and wanted to purchese his cargo as I supose he ought to do but this Bishop and Jackson cunsoulted not to let him slave with out he payed the same Coomey* that thy did thy sent him out of the River so he went to the Camoroons and was away two munths then he arived in my water again and thy still isisted upon his paying the Coomey acordingly he did a Nuff to Blind them so I gave him slaves to his content and so did all my peeple, till he was full and is now ready to sail only weats for to have a fue afairs sattled and this sall be don before he sails to his sattisfection, and now he may very well Laffe at them that was so much his Enemeys before, for that same day thy sent him out of the River this Jackson and Bishop and a brig that was [tender?] to Jackson at night began to fire at my town without the least provecation and continued it for twenty-four hours for which I gave them two cows but it seemed as after words Jackson confirmed that Bishop and him was to cary away all our pawns as it was lickely true for Jackson did cary of his but more than that before he sailed he tould me that if I went on bord of Bishop I shuld be stoped by him and my hed cut of and sent to the Duke at Nuetown, but I put that out of his power for to cut of my hed or cary of the pawns by stoping his boats and sum of his peeple and so I would Jackson had I known his entent when he informed me of Bishop, but he took care not to divulge his own secrets which he was much to bleam if he did so my friend marchant Lace if you Send ship to my water again Send good man all same your Self or same marchant black.† No Send ould man or man want to be

* Coomey was the duty paid to the King for the privilege of trading.

† Patrick Black, one of the oldest sea captains of the port of Liverpool. Troughton, in his History, dedicated to him a view of Woodside Ferry. "An Old Stager" gives the following amusing account of "Marchant Black," when he lived in Duke Street: "Picture to yourselves a kind and venerable man, in a cloak enveloping his whole body from head to foot, a gold-headed cane in his hand, and

grandy man, if he want to be grandy man let he stand home for marchant one time, no let him com heare or all Same Capt Sharp he very good man, but I no tell before that time Capt. Sharp go to Camoroons he left his mate till he came back again, so they say I do bad for them but I will leave you to Jude that for if any ship fire at my town I will fire for ship again Marchant Lace S[r] there is Mr Canes Capt. Sharp and second mate a young man and a very good man he is very much Liked by me and all my peeple of Callabar, so if you plase to sand him he will make as quick a dispatch as any man you can send and I believe as much to your advantage for I want a good many ship to cum, for the more ships the more treade wee have for them for the New town peeple and has blowed abuncko for no ship to go from my water to them nor any to cum from them to me tho Bishop is now lying in Cross River but thy only lat him stay till this pelaver is satteled for I have ofered him 10 slaves to Readeem the Pawns and let him have his white people, but he will not for I dount want to do any bad thing to him or any ship that cums to my water but there is 4 of my sons gone allredy with Jackson and I dont want any more of them caried of by any other vausell the coomy in all for my water now is 24 thousand cop[rs] besidges hats case and ship gun, Marchant Lace I did as you bob me for Lett[rs] when this tend[r] com I no chop

a wig. Oh! such a wig, a regular wig of wigs, as white as the whitest of hair-powder could make it, of a transcendental cauliflower appearance, and in size far beyond the proportions of the largest Sunday wig assigned to Dr. Johnson in the pictures which have come down to us. We recollect once, when about some six years old, getting into an awful scrape about this said venerable gentleman and his megatherium wig. We were walking with a small friend of our own age and inches, when suddenly the apparition of Mr. Patrick Black, arrayed as we have described him, came in sight. Our admiration, as usual, burst forth in the far from respectful and almost profane exclamation, 'There goes old Black with his white wig.' Hardly were the words out of our mouth, when a gentle tap came upon our shoulders, and a soft whisper fell upon our ear. 'Master Aspinall, if it would be any particular pleasure to you, I will ask my father to wear a black wig in future.' We looked round, and O! horror of horrors! were we not thrown into real agonies and almost hysterics, when, in the person uttering this mild remonstrance, we recognised the daughter of the old gentleman, whose wig we had been blaspheming? We stammered and hammered at an excuse, and then ran for our life. And for many a long day we disappeared round the nearest corner as quickly as possible, if any of the Black family came in sight of us in our walks. The joke, however, got wind and it was long before our martyrdom and persecution ceased, even in our own circle, where 'Old Black with his white wig was thrown into our teeth whenever we were inclined to be obstreperous and naughty.'"

for all man for you bob me No Chop to times for bionbi I back to much Cop^r for Coomy so I do all same you bob me who make my father grandy no more white man so now marchant Lace send good ship and make me grandy again for war take two much cop^r from me who man trade like me that time it be peace or break book like me so Marchent Lace if you Send ship now and good cargo I will be bound shee no stand long before shee full for go away."

The following is another lucid passage from "Grandy King George's" correspondence :—

"And now war be don Wee have all the Trade true the Cuntry so that wee want nothing but ships to Incorige us and back us to cary it on so I hope you and marchant Black wount Lat ous want for that In Curigement O^r the other marchants of that Pleasce thut has a mind for to send their ships thy shall be used with Nothing but Sivellety and fare trade other Capt^ns may say what they Please about my doing them any bad thing for what I did was thier own faults for you may think S^r that it was vary vaxing to have my sons caried of by Captn Jackson and Robbin sons and the King of Qua son thier names is Otto Imbass Egshiom Enick Ogen Acandom Ebetham Ephiyoung Aset and to vex ous more the time that wee ware fireing at each other thy hisseted [hoisted?] on of our sons to the yard arm of Bishop and another to Jacksons yard arm and then would cary all of them away and cut of my hed if it had not been Prevented in time and yet thy say I do them bad only stoping Sum of thier peeple till I get my Pawns from them Marchant Lace when you Send a ship send drinking horns for Coomey and sum fine white mugs and sum glass tanckards with Leds to them Send Pleanty of ship guns the same as Sharp had I dount care if there was 2 or 3 on a Slave Send one Chints for me of a hundrerd yard ^o 1 Neckonees of one hundrd yards 1 photar of a hundrd y's 1 Reamall 1 hund. yards one Cushita of a hundred yds one well baft of the same Send sum Leaced hats for trade and Vicor bottles and cases to much [to match?] for all be gon for war Send sum Lucking glasses at 2 Cop^rs and 4 Cop^rs for trade and Coomey to and send Planty of hack and Bally for Trade and Comey and Small Bells Let them be good ones and send sum Lango Sum Large and sum small and sum Curl beads Send me one Lucking glass six foot long and six foot wide Let it have a strong woden freme Send two small Scrustones that their Leds may Lift up send Plenty of Cutlashs for Coomey of 2 Cop[rs] price Let your Indgey goods be Right good and your ship no stand long send me one table and six Chears for my house and one two arm Schere for my Salf to sat in and 12 Puter plates and 4 dishes 12 Nifes and 12 forcks and 2 Large table spoons and a trowen and one Pear of ballonses 2 brass Juggs with thier Cisers (?) to lift the same as a tanckard and two Cop[r] ones the same two brass falagons of two gallons each Pleanty for trade of puter ones Send Plenty of Puter Jugs for trade send me two Large brass beasons and puter ones for trade Send me one close stool and Send me one Large Red [illegible] Send me one gun for my own shuting 5 foot barill and two pueter p*** pots Send one good Case of Rezars for my Saveing Send me sum Vavey brade Iron bars of 16 foot long Send 100 of them Send Large caps of 2 Cop[rs] for Coomey &c Please to show this to Marchant black and shend sum Large Locks for trade Sum chanes for my Salf two brass tea kittles and two scacepang a fue brass Kittles 12 or fifteen Cop[rs] each Send Pleanty of canes for Coomey and one long cane for my self gould mounted and small Neals for Coomey you may pay your Coomy Very Reasonable Saws or aney tools No Send Small Iron moulds for to cast mustcats and sum small 3 pounders Send me sum banue* canvess to make sails for my canows and sum large Leg monelones† with hendges [hinges?] to thim to lock with a Screw and two large iron wans for two sarve in the Room of irons and Send me one whip shaw and one cross cut shaw Send red green and white hats for trade Send me one red and one blue coat with gould Lace for to fit a Large man Send butt[r] and Suger for to trade Send sum green sum red sum blue Velvet caps with small Leace and Send Sum files for trade, So no more at Preasent from your best friend

"GRANDY KING GEORGE

* Possibly, "brand new." † Maniloes or Manillas perhaps.

"give my Complements to the gentlemen owners of the brigg *Swift* Mr Devenport Marchant Black and Capt[n] Black and as allso Mr Erll.*

"Please to have my name put on Everything that you send for me."

Robin John Otto Ephraim writes to Captain Ambrose Lace, merchant, in Liverpool, as follows:—

"PARROT ISLAND *July* 19*th* 1773

"SIR,—I take this opportunity to write to you I send Joshua 1 Little Boy By Captain Cooper I been send you one Boy By Captain faireweather I ask Captain Cooper wether Captain faireweather give you that Boy or not he told me Captain fairewether sold the Boy in the West India and give you the money I desire you will Let me know wether faireweather give you money or not my mother Send your wife one Teeth By Captain Sharp I done very well for Captain Cooper and my father too I am going to give a Town of my own I dar say you knows that place I am going to Live Bashey Dukey there once send Gun Enough for Trad. I want 2 Gun for every Slave I sell Send me 2 or 3 fine chint for my self and handkerchief any thing you want from Callabar Send me Letter I think I come to see next voyage Send me some writing paper and Books my Coomey his 1600 Copper Send me 2 sheep a Life Sir I am your BEST friend Otto Ephraim†

"S.P. I will Sell Captain Doyle slave because he told me you have part for his ship I expect Captain Sharp here in 4 months time Remember me to your Wife and Mr. Chiffies."‡

"OLD TOWN CALLABAR *December* 24*th* 1775

"Captain Lace I take this opportunity to write to you by Captain Jolly that letter you Send me by Sharp you did not put your name as for Captain Sharp I will do anything hys in my power to obliged you when Captain Cooper comes Let him

* Mr. Earle.

† This letter was marked "The King's own handwriting."

‡ Captain Chaffers, probably.

Guns enough I want 2 Gun for every Slave I Sell and father we Dont want Iron only 2 for one slave so no more at present from your friend

"EPHRAIM ROBIN JOHN

"S.P. Remember me to your wife.

"To Captain Ambrose Lace merchant in Liverpool."

CHIEF'S LETTER.

From "Otto Ephraim, King of Old Town, Old Callebar," to Mr. Ambrose Lace, merchant, in Liverpool.

"OLD TOWN OLD CALLABAR *August 23the* 1776

"MR. LACE,

"SIR,—I take this opportunity to write to you I received by Captain Cooper one painted cloth one book in the box one gown one ink cake and some wafers I was in the country when Orrock send that letter to you now I put my hand and my that is enough what Orrock can do he can do anything without my father and I please I pay Egbo men yesterday I have done now for Egbo I received by Captain Sharp one lace hat I make monkey Captain Loan pay me for that cap I got one hundred Copper for it I put him in the iron 5 days in Quabacke sea he told me that Captain Barley give the Willy Honesty but I make him pay for all that I was on board Barley myself he never mention it to me that you Send me a cap by him I have sent you by Cooper one teeth 50 weight

"Your most obedent Humble Servant

"OTTO EPHRAIM"

"OLD TOWN OLD CALLABARR *March 20th* 1783

"MR LACE,

"SIR,

"I take this oportunity By Captain Faireweather we have no News here only Tom King John come Down to live with my father is here now with us Orrock Robin John is Dead May 24th 1783 (?) we give all his coppers to his both son George Orrock and Ephraim Orrock Send me some Writing papers and 1 Bureaus to Buy

"Your Humble Servant

"OTTO EPHRAIM

"P.S. Remember me to your Wife and your son Joshua*
Ambrose William and Polly
"Mr Ambrose Lace
"Merchant in Liverpool
"Sent by ship *Jenny*."

The Liverpool newspaper of June 16th, 1769, contained the following laconic announcement:—"The *John*, Captain Erskine, from Bonny, at Barbadoes, with 200 and odd slaves, buried 247, and gone to Dominica." There was no fuss made about this mortality of at least 50 per cent. on the number originally shipped, but so tender is the public conscience in 1897, that the death of 247 bullocks in crossing the Atlantic would immediately be the subject of questions in Parliament.

On the 11th of January, 1769, about eight o'clock in the morning, as the *Nancy*, Captain Williams, of Liverpool, was lying at anchor at New Calabar, with 132 slaves on board, the negroes rose upon the crew and wounded several, which obliged them to fire amongst the slaves, killing six and wounding others. "It was with great difficulty," says the paper, "though they attacked them sword in hand, to make them submit. As soon as the natives on shore heard the report of the guns, great numbers of them came off in canoes, and surrounded the vessel, and finding her weakly manned (having only five people but what were sick), immediately boarded her, took away all the slaves, with some ivory, and a large quantity of different kinds of goods; plundered the vessel of everything on board, stripped the captain and crew of books, instruments, and clothes, afterwards split the decks, cut the cables, and set the vessel adrift. Captain Labbar, who was lying in the river, sent his boat, and brought Captain Williams and his people from the vessel, which was then driving with the ebb, a perfect wreck."

* Joshua Lace was the founder and first President of the Liverpool Law Society.

The following is a fragment of instructions handed to a slave-captain who sailed from Liverpool on the 3rd of August, 1770:—

"to whom deliver your Cargoe of Slaves provided they will engage to turn them out @ £30 ⅌ head sterling round clear of the Island Duty and the advantage of the sale to us in bills not exceeding 6 9 and 12 months (or less if possible) in equal Sums —Could a freight to Porto Rica be procured on Advantageous terms we should be glad and perhaps it would be a good opportunity to dispose of the Brig which we limit you at £150 Sterling either there or at Dominica you taking out the Butts and Guinea Materials. We have liberty in our policys of Insurance to go to Porto Rica. You'll find Letters lodged for you at Lovell Morson & Co.'s for your Government to which we at present refer. We allow of no private adventures being carried out that all trade be on the owners Acc^{t} recomending humane treatment to your Crew, care of accidents by Fire and that a Diligent Watch be kept so that the unhappy Misfortunes of Insurrections may be prevented. We are wishing you health and a prosperous voyage.

"Y^{o} friends &ct

"JOHN & WM CROSBIE
"EDW^{D} CHAFFERS
"AMBROSE LACE."

The following curious particulars regarding the customs paid at Whydah when trading for slaves, appear to have been drawn up by Captain Ambrose Lace, for the guidance of one of his captains:—

"State of the Customs which the ships that make their whole trade at Whydah pay to ye King of Dahomey :

Eight Slaves for Permission of Trade gongon Beater and Broakers	Thes slaves paid to ye Caborkees after which he gives you two small children of 7 or 8 years old which the King sends as a return for the Customs.

1 Slave for Water and washerwoman	These slaves paid to whom supplies you
2 Do. for the Factory house	
7 Do. for the Conoe	These to the Fort

The above Slaves are Valued as under :—

6 Anchors Brandy is	1	Slave	And if any other good must be in proportion but you must observe to pass the Goods Least in Demand.
20 Cabess of Cowries is	1	Do	
40 Sililees	1	Do	
200 lb Gunpowder	1	Do	
25 Guns	1	Do	
10 Long Cloths	1	Do	
10 Blue Bafts	1	Do	
10 Patten Chints	1	Do	
40 Iron Barrs	1	Do	

"After the Customs are paid which should be done as soon as possable for the traders dare not trade till the Kings Customs are paid, the Vice Roy gives you the nine following Servants viz. one Conducter to take care of the goods that comes and go's to and from the waterside which you deliver him in count and he's obliged to answer for things delivred him he's paid 2 Gallinas of Cowries every time he conducts any thing whether coming or going and one flask of brandy every Sunday.

"Two Brokers which are obliged to go to the traders houses to look for slaves and stand Interpiter for the Purchas the are paid to each two Tokes of Coweres ℔ day and one flask of brandy every Sunday and at the end of your trade you give to each of them one Anchor of Brandy and one ps of Cloth.

"Two Boys to serve in the house the are paid each two tokees ℔ day at the end of your trade ℔ ps of cloth.

"One Boy to Serve at the tent water side 2 Tokees ℔ day.

"One Doorkeeper paid 2 Tokees ℔ day 1 ps Cloth for him and ye above.

"One Waterwoman for the factory 2 Tokees ℔ day at end of trade. One ps of Cloth.

"One Washer Woman 2 Tokees ℔ day and six Tokees everytime you give her any Linnen to Wash and one ps of Cloth at ye end of trade.

"N.B. the two last Servants are sometimes one if so you only pay one.

"To the Cannoe men for bringing the Captain on shore one Anchor Brandy and to each man a hatt and a fathom Cloth. To the Boatswain a hat ½ ps Cloth one Cabes Cowrees a flask of brandy every Sunday and a bottle every time the cross the Barr with goods or Slaves and every time the pass a white man and at the end of trade for carring the Captⁿ on board one anchor of Brandy and four Cabeses Cowrees.

"N.B. The above Bottles flasks &c was usely given to ye Conoemen but now the Captⁿ gives yᵐ one Anchor of Brandy and one Cabese of Cowrees every Sunday for the weeks work. To the Gong Gong Beater for anouncing trade 10 Gellinas of Cowrees and one flask of Brandy.

"To the Kings Messenger for Carring News of the ships Arrivell and Captⁿ'ˢ Compliments to the King ten Gallinas.

"To the Trunk keeper a bottle brandy every Sunday and a peice of Cloth when you go away if you are satisfied with his service.

"To the Captⁿ of the Waterside on your arrivell one anchor of brandy and at your Depᵗ· one ps Cloth and one anchor of brandy.

"To the six Waterrowlers two tokees ℔ day each and two Bottles Brandy besides which you pay them 2, 3 or 4 tokees of cowrees each Cask according to the size at the end of trade two ps Cloth and one anchor Brandy.

"To the Vice Roy who go's with his people to Compliment the Capᵗ· at his arrivell and Conduct him to the Fort one Anchor Brandy and two flasks but if Coke be their four flasks Brandy.

"To the Vice Roy for his owne Custom 1 ps Silk 15 yards 1 Cask of Flower one of Beef but if you are short of these you may give him some thing else in Lew of them.

"To making the Ten one Anchor Brandy 4 Cabess Cowrees.

"To the Cap$^{tn.}$ Gong Gong that looks after the house at night one bottle ⅌ day and one ps Cloth if your content.

"You pay 3 Tokees of Cowrees for every load such as one Anchor 40 Sililees 10 ps Cloth and so in proportion for small goods but when loads are very heavy you pay more as ten Gallinas for a Chest of pipes &c.

The Tokee is	40	Cowrees
The Gallina is	200	—
The Cabess is	4000	—

"N.B. their go's five tokees to one Gallina and twenty Gallinas makes one Cabess."

Letter from a Chief, addressed "see Capt. Brighouse":—

"FRIEND WILLIAM BRIGHOUSE,

"I have sent you one woman and girl by Shebol. I will come toomorrow to see you. Suppose you Some Coffee to spar. Please send me a Little.

"I am your Friend

"Decr 30th 1777 "EGOBOYOUNG COFFIONG

"Sunday"

In the year 1772, slavery in England received its death-blow. In 1729, Lord Talbot, Attorney-general, and Mr. Yorke, Solicitor-general, had given an opinion which raised the whole question of the legal existence of slaves in Great Britain and Ireland. They said that the mere fact of a slave coming into these islands from the West Indies did not make him a free man, and he could be compelled to return again to those plantations. On the strength of this decision, slavery continued to flourish in England for a period of forty-three years. Chief-Justice Holt, however, had expressed a contrary opinion to that of the law officers of the crown ; and, after a long struggle the matter was brought to a final issue in the case of the negro Somerset, so nobly fought by Granville Sharp. On May 22nd, 1772, Lord Mansfield in the name of the whole bench, delivered the memorable decision,* which, from that day to this, has been one of the glories of our land—that "as soon as a slave set foot on the soil of the British Islands, he became free," or in the words of Cowper:

> "Slaves cannot breathe in England; if their lungs
> Receive our air, that moment they are free;
> They touch our country, and their shackles fall."

Notwithstanding this ruling, we find in *Williamson's Advertiser,* of May 4th, 1780, the following curious advertisement:—

"RUN AWAY, on the 18th of April last, from PRESCOT, A BLACK MAN SLAVE, named GEORGE GERMAIN FONEY, aged twenty years, about five feet seven, rather handsome; had on a green coat, red waistcoat, and blue breeches, with a plain pair of silver shoe buckles; he speaks English pretty well. Any person who will bring the black to his master, Captain Thomas Ralph, at the Talbot Inn, in Liverpool, or inform the master where the black is, shall be handsomely rewarded. All persons are cautioned not to harbour the black, as he is not only the slave, but the apprentice of Captain Ralph."†

* "On May 22nd, 1772, the court of King's Bench gave judgment in the case of Somerset, the slave, viz. that Mr. Stuart, his master, had no power to compel him on board a ship, or to send him back to the plantations. Lord Mansfield stated the matter thus: The only question before us is, is the cause returned sufficient for remanding the slave? If not, he must be discharged. The cause returned is, the slave absented himself, and departed from his master's service, and refused to return and serve him during his stay in England; whereupon, by his master's orders, he was put on board the ship by force, and there detained in secure custody, to be carried out of the kingdom and sold. So high an act of dominion was never in use here; no master ever was allowed here to take a slave by force to be sold abroad, because he had deserted from his service, or for any other reason whatever. We cannot say the cause set forth by this return is allowed or approved of by the laws of this kingdom, therefore the man must be discharged." *Annual Register*, vol. 15, p. 110.

† In contrast to the above, we take the following from the Liverpool newspaper: "On Saturday, February 26th, 1780, died in the 79th year of his age, Thomas Crowder, a merchant who had acquired a large fortune in Jamaica; and on Tuesday died his faithful black servant, who had served him upwards of twenty years."

"On Jany. 4th, 1797, died William Patrick, a black, upwards of 36 years a servant in the family of William Gregson, Esq., of Everton, in which capacity he was honest and faithful becoming his situation.'

The rapid decline of commerce consequent upon the revolt of the North American Colonies, and the activity of the American Privateers, seriously interfered with the Liverpool slave trade. In 1773, the number of ships cleared to Africa was 105, burthen 11,056 tons, which carried to the West Indies 28,200 negroes. In 1775, the number of ships fell to 81, burthen 9,200 tons, while during the war this branch of traffic, in common with others, had declined so much that in 1779, only 11 vessels, burthen 1205 tons, sailed from the Mersey to the coast of Africa. One great blow to the trade was an Order in Council prohibiting the exportation of gunpowder, an article which formed a large portion of every Guinea cargo.

In August, 1775, a sailors' riot broke out in Liverpool, and continued for several days, threatening to lay the town and shipping in ashes. Some sailors, who had been engaged on board the *Derby* Guineaman, Captain Yates, fitting out in one of the docks, having finished the rigging, demanded their wages at the rate of 30s. per month, for which they had contracted; but the owners refused to pay more than 20s., as there were then about 3000 sailors in the port unemployed, and no fewer than forty sail of Guinea ships laid up. The men returned on board the vessel, and in a short time cut and demolished the whole of the rigging, and left it on the deck. A party of constables seized nine of the ringleaders, whom the magistrates committed to the Tower in Water-street, whereupon upwards of 2000 sailors, armed with handspikes, clubs, and other weapons, attacked the gaol and rescued their comrades. The rioters then marched about the docks till near midnight, terrifying the inhabitants and unrigging all the vessels that were ready to sail. This was on a Friday; on Saturday all was quiet, and on Monday, the sailors, in a body, waited on the magistrates, praying redress and support. They came to no terms, but met the following day, and the merchants agreed to give the wages they demanded. On this they dispersed and spent the day in the greatest festivity, but hearing that 300 able-bodied men had been hired at 10s. per day to apprehend those who had been most forward in the riot, the sailors again met at nine o'clock the same evening, unarmed, and went to the Exchange, which they surrounded. Some straggler of their party unfortunately broke a pane of glass, whereupon the special constables within fired upon the mob, killing seven and wounding about forty. A general attack upon the windows of the Exchange was made with stones, amid the dismal cries and groans of the wounded. On Wednesday morning, upwards of 1000 sailors again assembled, all with red ribbons in their hats. They went to Parr's, the gun-smith, took about 300 muskets, plundered other shops of powder, balls, &c., and at one o'clock, being all armed, some with muskets and others with cutlasses, they surrounded the Exchange, against which they planted six cannon, which they had brought from the vessels in dock.* Having hoisted the bloody flag, they blazed away at the building with great guns and small arms. The cannon in Castle-street was so large, and the street so narrow, that the houses shook till scarce a pane of glass was left whole in the neighbourhood. In this attack four persons were killed. It is said that much more damage would have been done to the Exchange by cannon balls if some one had not cried, "Aim at the goose," alluding to the cormorant, or liver, the heraldic device of the town, which formed one of the figures in the pediment. The gunners took the hint,

* "When the sailors were attacking the houses of the African merchants in 1775," says Stonehouse, "a cannon was obtained from the Old Dock by a party of the rioters. One of these fellows took a horse out of Mr. Blackburne's stable at the Salt Works, and attempted to harness it to a truck on which the cannon had been placed. The leader of the gang, in stooping down to fasten a rope to the truck, offered so fair a mark for a bite, that the horse, evidently having notions of law and order, availed himself of the opportunity of making his mark upon Jack's beam end, which sent him off roaring, leaving the gun in the possession of the saline Bucephalus."

and the cannon, being pointed high, did less mischief than it might otherwise have done. From the Exchange, they marched to Whitechapel, to the house of Mr. Thomas Ratcliffe, a Guinea merchant, the attack upon which is thus described by an eye-witness :—

"This day I have been so frightened as hardly to be able to do anything. Such scenes of distress as I have been eye-witness to, with the clattering of swords and cannon, have so terrified me, that I hardly know what I say or do. To inform you of the particulars : you must know that in Whitechapel, lived a merchant [Mr. Thomas Ratcliffe], who was said to be the first that fired upon the sailors ; in consequence thereof, a large number of them came with a drum, a flag, and armed with guns, blunderbusses, cutlasses, clubs, &c. who fired on the said merchant's house, which stands in sight of us, where they threw out the feather beds, pillows, &c. ripped them open, and scattered the feathers in the air, broke open the drawers, full of clothes, laces, linen, tore in pieces the house and bed furniture, together with the stoves, parchments, china, &c. and all that was in the house. We were all in a dreadful confusion, but they behaved very well to every one, excepting those to whom they owed a grudge. They then marched to a very large house behind us, [in Rainford Garden] belonging to another merchant, whose name is [William] James, and one of the greatest traders here.* The family having been apprised of their coming, had left it, and taken some of their most valuable effects with them to a country house they have ; but such good furniture they destroyed here, would have grieved any one to see. They destroyed also the compting-house, with

* Mr. William James had, at one time, 29 vessels engaged in the slave trade, but they were not of large dimensions. He died at his house in Clayton Square, in January, 1798, aged 67. "Mr. James," says one who knew him, "sat for some years in the House of Commons, and gave evidence of talent far beyond mediocrity. There was also a spice of originality about him which commanded attention whenever he spoke, which, however, was but seldom. There was another Mr. James, in Liverpool, in those days, rather a rough-spun and unhewn kind of person, and very eccentric and amusing in his way—a character, in short, amongst his own circle. His name was Gabriel James, or 'the Angel Gabriel,' as some of his waggish friends called him. He had a ready tongue and plenty of mother wit, and seldom came off second best in a tilt and tournament with words.'

all the papers, goods, &c. The household furniture was very rich, with abundance of china and chintz bed-furniture, all of which were torn to shivers, and linen, plate, &c. tumbled into the street, and thrown about in fragments immediately, in the air. During the whole time the cellars were kept quite open, and what liquor they did not drink, they threw away. Our poor Debby would go to see them, and has got her eyelash cut with a candlestick.

"It is not possible to form any idea of the distress this place has been in, all this day. The merchants get to the corner of the streets, where, methinks, I yet see them standing, with fear painted in their faces. The 'Change has all its windows broke, and frames forced quite out. They have been firing also at the walls the greatest part of this day, and are now gone to Cleveland Square. I suppose there is not a merchant who has wanted to lower their wages but will be visited by them; and God knows how long these riots may continue. You will not wonder, after reading this, that I was terrified. I am a coward, it's true, but I think this would have alarmed any one. They read the Riot Act last night, and then began to fire on them, when they killed three, and wounded fifteen. This has made them so desperate. I could not help thinking we had Boston here, and I fear this is only the beginning of our sorrows."

In destroying Mr. James's furniture, a little negro boy was discovered by the sailors, concealed in the clock case, whither he had fled for safety. Having got drunk at Mr. James's house, the mob marched to Mr. Thomas Yates' in Cleveland Square, and from thence to Mr. John Simmons' in St. Paul's Square, sacking both houses, after which they met at their rendezvous, the North Ladies' Walk, where they gathered daily under a leader they called "General Gage." Besides other acts of turbulence and disorder, which were committed during several days, the rioters marched about the streets in gangs, presenting pistols at the breast of every person they met, and demanding money from them. They also visited the houses of the merchants, levying contributions of money, among the rest, the residence of Mr. William Leece, a merchant, in Water-street. It happened that no one was within, except the merchant's daughter and the female servants. Miss Leece, with a fearlessness and self-possession that was completely wanting in the local authorities of Liverpool during the riot, went to the door, and, addressing the mob leader, who was a sailor, enquired what they wanted. Jack, struck with admiration at her courage and coolness, took off his hat, and remained uncovered while, in respectful language, he solicited, instead of demanding, a contribution. Having received it, he thanked her, and drew off the rabble without doing any mischief. This wise and high-spirited lady afterwards married Mr. James Drinkwater, who was mayor of Liverpool in 1810. Her eldest son, Sir George Drinkwater, was mayor in 1829; her second son, Mr. William Leece Drinkwater, of the Isle of Man, was a member of the House of Keys; her third son, Mr. John Drinkwater, was the father of Deemster Drinkwater. Her daughter, Margaret, married Mr. Peter Bourne, who was mayor of Liverpool in 1825. The riot was eventually quelled by a troop of light horse from Manchester,* and in April, 1776, fourteen of the sailors

* A gentleman, who accompanied the party of Lord Pembroke's Royal Regiment of Horse, that was sent for from Manchester to Liverpool, to quell the riot, writing on September 6th, says: "Last Wednesday, at three o'clock in the afternoon, an express was received at Manchester from the Mayor of this place, demanding the assistance of the soldiery, to put a stop to the riotings of the sailors; and in the evening two of the principal gentlemen in the town arrived, praying their immediate march, otherwise, Liverpool would be laid in ashes, and every inhabitant murdered. Upon this, the men were collected together with all speed, to the number of 100 privates and six officers; and about three o'clock in the morning, they marched. It rained very hard, and did not cease until they came within six miles of Liverpool, where they were met by the Mayor, who told them the rioters were drawn up in a body to attack them. Before they proceeded any further, they examined their arms, which being very wet, required a short time to put them in order and when done they loaded, then marched in six divisions with their horses on each side, to keep the flanks clear, intending to give the sailors the street fairing. They arrived at Liverpool about four o'clock in the afternoon, in good spirits, though somewhat fatigued, amidst the acclamations of the whole town, who now

concerned in the affair, "were suffered to go on board one of his Majesty's ships destined for America." With the exception of the rebellion and the Gordon Riots, the annals of the eighteenth century probably cannot mention a more extraordinary and formidable popular outbreak in England than these riots, arising from the greed of slave-merchants and the ferocity of their hirelings, and in which cannon, muskets, pistols, cutlasses and other deadly weapons were freely used by the mob.

The *True Briton*, Captain Dawson, which sailed from Bonny, for the West Indies, on the 14th of June, 1776, with upwards of 500 slaves, in coming out had an insurrection on board, in which the sailmaker was killed, and cooper wounded.

One of the most inexplicable facts in connection with the trade is, that when the slave-ships were in danger from an enemy on the middle passage, the captains frequently armed some of the negroes, who fought most gallantly to preserve the vessels and the lives of the men who were carrying them into perpetual and pitiless bondage. We have an instance of this in the case of the notorious slaver, the *Brooks*. Captain Noble, her commander, writing to the owners from Montego Bay, Jamaica, on the 26th of April, 1777, says:—

> "I can with a good deal of pleasure inform you that your ship *Brooks* has been the destruction of one of the American privateers. The next morning after we left Barbadoes, we were chased by her, and made all the sail we could to get from her, but to no purpose, for she came up with us very fast, and a little afterwards we saw another privateer right ahead, so that we had then nothing to do but either fight or be taken.

came out of their houses, which they had not done, nor even shewn their faces, for some time before. Immediately upon their appearance, the rioters dispersed, with the utmost confusion, hiding themselves in garrets, cellars, &c. and in short, anywhere they could. The soldiers then surrounded several houses, and in the course of Thursday and Friday, made about sixty prisoners, who were sent to Lancaster Jail, and now all remains very quiet."

We therefore, to prevent being engaged by them both at once, took in all our small sails, and made ready for an engagement. She came up right astern, would shew no colours till we fired two shot at her, which did great execution ; upon which she hoisted American colours, and gave us a broadside, which we returned with our two stern chasers, which never missed raking them fore and aft. After engaging her about an hour, we were so lucky as to shoot away her mast, just above the deck, by which time the other was almost up with us, but seeing the sloop's mast gone, she hauled away from us as fast as possible. The sloop and us exchanged many shot after her mast was gone, but I thought it the most prudent way not to attempt taking her for fear of the other (which was a schooner) altering her mind, and coming back, upon which we bore away in a tattered condition, our sails and rigging being very much torn to pieces, and a great many shot in the hull, but miraculously nobody killed or wounded on board us, except the Doctor, who received a musket ball in his belly, but has got the better of it already, as it came through the stern before it hit him. We killed a great number on board the privateer, as they stood quite exposed to our shot. She was a sloop of ten or twelve guns, a great number of swivels, and as full of men as she could stow. I believe the greatest part Frenchmen by their appearance. I had fifty of our stoutest slaves armed, who fought with exceeding great spirit. After I left the sloop, the schooner came to her, and, I suppose, took the people out of her ; she sunk about an hour after I left her. The engagement was within two miles of St. Vincent, on the S.E. part of the island. I went into Kingston Bay, and went on board the *Favourite*, sloop-of-war, to beg some powder, which they supplied me with very readily, and that evening made sail for Jamaica, kept a great way to the southward, and then hauled right over for Jamaica, by which means (I dare say) we escaped a good many of the Americans. We saw several small sail on our way down, but what they were, I cannot tell."

Captain Noble, when writing, had not heard the sequel.

Soon after the removal of the sloop's crew on board the schooner, the latter blew up, and fifty-five persons were drowned and thirteen saved, amongst whom was the captain of the sloop privateer. The captain and three men were lodged in gaol at St. Vincent's. We shall presently hear more of the *Brooks*, and her accommodation for compulsory passengers.

At a public meeting of the African Freemen, merchants and others, held in the Exchange, in Liverpool, on the 14th of July, 1777, a committee of merchants was appointed to take into consideration the state of the African trade, and to draw up some plan to be laid before the ensuing Sessions of Parliament for the better regulation of the said trade. The following merchants were present :—

Alderman Gregson,	Mr. Higginson,	Mr. Sparling,
Mr. Slater,	Mr. T. Hodgson,	Mr. Blundell,
Mr. Caruthers,	Mr. Heywood,	Mr. Brown,
Mr. Bold,	Mr. Greenwood,	Mr. Birch,
	and Mr. Grimshaw.	

It was resolved that the Committee be an open one, "to which any merchant or other person, trading to Africa from Liverpool, or any Freeman of the African Company there, or other merchant of the same place, should be allowed to come, be heard, and vote." The Committee sat at ten o'clock every Monday morning in the Town Hall, and was formed of the following gentlemen :—

William Crosbie, Esq., Mayor,

William Gregson,	John Dobson,	Alexr. Nottingham,
Gill Slater,	Joseph Brooks, Jun.,	Thomas Hodgson,
Thomas Case,	Benjamin Heywood,	Thomas Staniforth,
George Case,	Thos. Rumbold,	Thomas Birch,
Richard Savage,	James Caruthers,	Wm. Crosbie, Jun.,
	Francis Ingram.	

The Secretary was Francis Gildart.

On the 4th of December, 1777, the *Jane*, Captain Syers, and the *Gregson*, Captain Boyd, two Liverpool slave-ships, arrived at Barbadoes, from Africa, after a passage of seven weeks. Two days before their arrival, they exchanged a broadside with a small sloop, but the day following, the *Jane*, then a good way ahead of the *Gregson*, was grappled and attacked by a large sloop of 14 guns and well manned, who managed to throw five boarders into the *Jane*, but these were soon repulsed, and a close action ensued for about two hours, when the privateer cut her grapnel and sheered off, having caught fire, which, however, was extinguished. The *Jane* had five men and a negro boy killed, and six seamen dangerously wounded. "Captain Boyd," says the Liverpool newspaper, "crowded all he could, but was not able to get up and assist, otherwise 'tis likely the people of Barbadoes would have had the pleasure of seeing those two brave African heroes bring the Rebel Taxgatherer into Carlisle Bay." An account from St. Vincent's, dated December 27th, mentions that a Liverpool Guineaman had given a rebel privateer a severe drubbing, near Barbadoes, and that 33 of the privateer's crew were killed, and upwards of 47 wounded, and this, no doubt, was the action with the *Jane*.

In March 1779, in a cause tried before Earl Mansfield, at Guildhall, Amissa, a free black of Anamaboe, on the coast of Africa, was awarded £500 damages against the captain of a Liverpool slave-ship, under the following circumstances. In 1774, the defendant, wanting hands while on the coast, hired the plaintiff as a sailor, advancing part of his wages. When the ship arrived at Jamaica, the plaintiff was sent, with three other sailors, to row some slaves on shore, and, to his intense astonishment and grief, instead of being allowed to return to the ship, he was detained by the purchaser of the slaves, to whom the captain had sold him, and sent up to the mountains to work as a slave. When the heartless captain returned with his ship to Anamaboe, he gave out to Amissa's friends that he had died on the passage. A year or two later, however, a black returned to Anamaboe, who reported that he had left Amissa in slavery at Jamaica, whereupon the King, and other great people of the country, desired Captain E——, who was then on the coast with his ship, on his arrival at Jamaica, to redeem Amissa and send him back to his friends, they paying all expenses. The better to identify his person, they directed the son of one Quaw, a gold taker at Anamaboe, to accompany Captain E—— on his voyage. Soon after their arrival at Jamaica, they found out the man, redeemed him after a slavery of near three years, and brought him to London, where the matter was laid before the African Committee, who ordered the defendant to be prosecuted as a warning to other captains, with the result, as aforesaid, of heavy damages.

Early in 1781, the *Sally*, Captain Taubman, had the good fortune of capturing, and escorting to Barbadoes, a Dutch Guineaman with 350 slaves, which, taken at the average market price ruling in Jamaica for eleven years—£50 a head—would amount to £17,500.

On the 28th of April, 1781, Captain Stevenson of the slave-ship *Rose*, wrote to his owners, in Liverpool, from Old Harbour, Jamaica, in the following terms:—

> "This is to inform you of my safe arrival here on the 16th inst, after a passage of 48 days from Cape Coast, but had the misfortune the day before we got in here, to fall in with a French privateer of 14 guns, and 85 men, called the *Mould*, belonging to Cape Nichola Mole, off the S.E. end of this island. At first coming up with us, we gave her two broadsides with our great guns and small arms, which she returned in the like manner, but her intention was for boarding us, he at last came up on our starboard quarter, with a stinkpot fast to the end of his gaff, thinking to swing it on board, but one of the Trantee slaves shot it away with his musquet. He then grappled our main chains, and we lay together yardarm and yardarm for

above one glass, when he thought proper to sheer off, having got his belly full. I had about fifty men, black and white, on deck at great guns and small arms, halfpikes, boathooks, boat oars, steering-sail-yards, firewood, and slack ballast, which they threw at the Frenchmen in such a manner that their heads rattled against one another like so many empty callibashes.

"My people all behaved very well, both white and black. We lost a white man named Peter Cane; myself wounded, and five other white people, as likewise seven blacks, one of which is since dead, the other six I am in hopes will recover. The Frenchmen hove such a large quantity of powder flasks on board us, that the ship abaft was all in a blaze of fire three different times; this hurt the blacks much, having no trowsers on them. I had my own shirt burnt off my back. After that I received a ball through my right shoulder, but, thank God, it was in the latter part of the action, so that I did not lose much blood. On the doctor's examining my wound, he found the ball was gone clean through my shoulder."

The *Rose* carried 12 guns, three and four-pounders, and 30 white people. On the 12th of June in the following year, she was taken on the Coast of Africa, by two French frigates and a cutter, and sent to England as a cartel with prisoners.

The *Othello* (Letter of Marque) Captain Johnson, a slave-ship belonging to Messrs. Heywood & Co., on her voyage to the coast of Africa, took the *St. Anne*, 300 tons burthen, from Buenos Ayres for Cadiz, with a cargo of 8,500 dry hides, 180 boxes of Peruvian bark, and four sacks of fine Spanish wool, the whole valued at £20,000.* The prize-master put into Killybegs, in September, 1781, to await orders from the Messrs. Heywood, before venturing to proceed to Liverpool, on account of the swarm of the enemies' privateers on the coast and in the Channel.

* The unfortunate owner of a considerable part of the cargo, a Spanish gentleman, who spoke very bad English, was on board the prize when taken. He told a horrible tale of a rebellion which had broken out in several provinces of South America, particularly Cuzco, where the native Indians had hanged the governor, and driven 500 Christians into a church, to which they set fire, and destroyed them.

In the spring of the year 1783, the *Othello* was taken, on the coast of Africa, by the crew, and retaken by the second mate and the doctor, but not until Captain Johnson had lost his life in attempting to quell the mutiny. In July of the same year, we read of her being cast away on her passage from Africa to Tortola, on the east of that island; the cargo, consisting of 213 slaves, was saved.

On the 7th of December, 1781, the *Nelly*, Captain Fairweather, on her passage from Africa to Jamaica, with 429 slaves on board, was wrecked in the night upon the Grand Canaries; 108 of the slaves, and one of the crew perished. The remainder of the blacks were shipped in a vessel the captain purchased, and sent to Jamaica.

The peace of 1783 infused new life into the trade, which had been languishing for nine years; the number of slave-ships which cleared from Liverpool for the coast of Africa in that year being 85, burthen 12,294 tons, carrying 39,170 slaves.* Hitherto, no public demonstration hostile to the traffic had been made, though private opinion in many quarters was gradually strengthening against it. The time, indeed, was fast approaching when a small but devoted band of men were to win undying renown by grappling with, and, after a fierce and prolonged struggle, slaying a monster more hideous than the Gorgon, cruel as Moloch, and hydra-headed in its ramifications. In 1787, the little cloud, no bigger than a man's hand, appeared in the political sky in the shape of a petition to the House of Commons, from some members of the Society of Friends, praying for the suppression of the trade in human flesh.

* The reader is warned against accepting the figures of Sir James Picton on this question, as he repeatedly gives the amount of tonnage as the number of slaves carried; for instance, in this case, he puts the number of slaves at 12,294, instead of 39,170. The tables in the Appendix show at a glance the number of vessels, tonnage, &c., cleared for Africa from 1709 to 1807.

CHAPTER IV.

THE ABOLITION MOVEMENT.

"When Clarkson his victorious course began,
Unyielding in the cause of God and man,
Wise, patient, persevering to the end,
No guile could thwart, no power his purpose bend.
He rose o'er Afric like the sun in smiles,
He rests in glory on the western isles."

THE causes which led to the agitation against the slave-trade will now be briefly stated. The cruelty, the injustice, and the impolicy of the traffic had been exposed in Dr. Beattie's "Essay on Truth" (1770), in Adam Smith's "Wealth of Nations" (1776), in Paley's "Moral Philosophy" (1785), and in John Wesley's "Thoughts on Slavery." The pulpit began to denounce the evil, and the spirit moved the Quakers of America and England to the most vigorous and chivalrous crusade against a traffic so peculiarly revolting to their humane and pacific tenets. In the year 1776, Mr. David Hartley, member for Hull, brought the question before Parliament. It was reserved, however, for Granville Sharp, the champion of the negro Somerset, to call public attention to a case which did more than any collection of essays to stamp the horrors of the trade upon the minds of disinterested persons, and produced an earnest desire for its abolition. It was a cause tried at Guildhall, in the year 1783, in which certain underwriters were heard against Gregson and others, of Liverpool, owners of the slave-ship *Zong*, Captain

Collingwood. It was alleged that the captain and officers of that vessel had thrown overboard into the sea 132 living slaves, in order to defraud the underwriters, by claiming the value of the same, as if they had been lost in a natural way. It came out in the evidence that the slaves on board the *Zong* were very sickly; that sixty had already died, and several were ill and likely to follow, when Captain Collingwood proposed to James Kelsall, the mate, and others, to throw several of the negroes overboard, stating that if they died a natural death, the loss would fall upon the owners, but that if they were thrown into the sea, it would fall upon the underwriters. He accordingly selected 132 of the most sickly of the slaves, 54 of whom were immediately thrown overboard, and 42 on the following day. A few days later, the remaining 26 were brought upon deck. The first batch of 16 submitted to be thrown into the sea, but the rest, with a noble resolution, would not permit the officers to touch them, and leaped overboard after their companions.

In May, 1787, the Society for the Abolition of the African Slave trade was instituted in London. The Committee consisted of Granville Sharp (chairman, and father of the cause in England), William Dillwyn (an American Quaker), Samuel Hoare, George Harrison, John Lloyd, Joseph Woods, Thomas Clarkson, Richard Phillips, John Barton, Joseph Hooper, James Phillips, and Philip Sansom. With the exception of Sharp, Clarkson, and Sansom, all the members were of the Society of Friends. At a meeting held on the 7th of June, Mr. Barton informed the members that Mr. William Roscoe, of Liverpool, author of a poem entitled, "The Wrongs of Africa," had offered the profits that might arise from the sale of that work, to the committee towards furthering their cause. This offer, coming from the head-quarters of the iniquity, was deemed very encouraging, especially as the preface to the poem was written by Dr. Currie, another dweller in the tents of unrighteousness.

This poem, the second part of which was published the next year, was well received by the public, and afterwards translated into German. In his juvenile poem of "Mount Pleasant," written in 1771, and published in 1777, Mr. Roscoe had previously voiced his abhorrence of the unhallowed traffic. In 1787, he published a temperate and masterly pamphlet entitled, "A General View of the African Slave Trade, demonstrating its injustice and impolicy; with hints towards a bill for its abolition." "I rejoice," writes good Quaker Barton to him, "to find that thy pamphlet has occasioned a ferment amongst the African merchants at Liverpool, and I trust it will occasion a ferment amongst our senators likewise, and produce the conviction we so much wish them to feel."

William Roscoe, the man who had the courage to deliver this straight blow from the shoulder at the favourite sin of his native town, was the son of an innkeeper and market-gardener, on the slope of what is now Mount Pleasant. In his youth he worked in his father's garden, and carried the potatoes upon his head to sell in the public market. Then we trace him, as a boy, in a bookseller's shop, and from there to an attorney's office. He conquered the dead languages; he made himself master of many living tongues; and then, emerging as an attorney, banker, poet, and historian of a high order, displayed an erudition rare in those who have not enjoyed university training, combined with an elegance and originality of style, which achieved a world-wide reputation, and caused him to be sought out by the illustrious and learned of every land. In Italy, the name of the historian of Lorenzo the Magnificent, and of the Pontificate of Leo the Tenth, was a passport in all cultivated society, and even in his native town, where his sentiments regarding the slave trade were hateful to the majority of the people, he was not without honour; and when the stroke of undeserved misfortune bowed the noble head, it was admitted that he had "worn the white flower of a blameless life," and kept unsullied his escutcheon as one of Nature's noblemen. During the last hundred years the names of two illustrious sons of Liverpool have stood out on the roll of history, conspicuous for their combination of moral worth with rare intellectual and literary powers. One of them—William Roscoe—has long since passed away; the other—William Ewart Gladstone—is still amongst us, with mental forces unimpaired at eighty-eight. These two great men and great scholars were born within bowshot of each other.

In January, 1788, the Society for the Abolition of the African Slave Trade made its first appearance before the public of Liverpool with a well-written address, designed to prove that the traffic, which was then said to bring about £300,000 a year into the Port of Liverpool, was immoral and unjust, and one which ought to be abolished, as unworthy of a Christian people. A list of the members of the Society* was published in the same year, from which it appears that there were eight righteous persons still left in Liverpool, who had not bowed the knee to Baal. Their names, and the amount of their subscriptions, were as follows :—

	£	s.	d.
Anonymous, Liverpool	2	2	0
Dr. Jonathan Binns, Liverpool	1	1	0
Mr. Daniel Daulby, Liverpool	1	1	0
Mr. William Rathbone, Liverpool	2	12	6
Mr. William Rathbone, Junr., Liverpool ...	2	2	0
Mr. William Roscoe, Liverpool	1	1	0
Mr. William Wallace, Liverpool...	2	2	0
Mr. John Yates, Liverpool	2	2	0

These worthy men, however, were not all the enemies of the trade in Liverpool, even at that early period. The blind

*Baines, Picton, and others state that only two Liverpool names—those of William Rathbone and Dr. Binns—figured in the list of original members. In the printed list at the Picton Reference Library, we find eight subscribers, as above. "Anonymous" was probably Dr. Currie.

poet, Edward Rushton, sang the wrongs of the negro in vigorous, manly verse. Mr. Clarkson has inscribed his name, along with that of Roscoe, and Dr. Currie, in his map of the pioneers in the great cause of abolition, because each of the three, acting independently, published his work on behalf of the poor African, before any public or combined agitation had been commenced. Rushton, indeed, though residing in Liverpool, had been bold enough to affix his name to his work, entitled, "West Indian Eclogues." His story is a remarkable one. Born at Liverpool in 1756, educated at the Free School, and, when about eleven years of age, apprenticed to the sea, in the employ of Messrs. Watt & Gregson, merchants, he had an early experience of the horrors of the slave trade. When he was sixteen years of age, the vessel in which he served was in danger of shipwreck; the captain and crew gave up all for lost, and abandoned themselves to despair. Young Rushton seized the helm, saved the ship, and was promoted to the rank of second mate. He sailed as mate on a voyage to Guinea, and while on the coast, contracted a friendship for a negro named Quamina, whom he taught to read. Going one day to the shore with a boat's crew, of which Quamina was one, the boat upset. Rushton swam towards a small water cask, which point of safety Quamina had previously attained, and when the negro saw that his friend was too much exhausted to reach the cask, he pushed it towards him, bade him good bye, and sank to rise no more. Rushton afterwards lost his sight in an attempt to relieve the sufferings of a cargo of slaves, afflicted with ophthalmia. An ardent love of freedom constituted a leading feature in Mr. Rushton's character through life. The idea of the Liverpool School for the Blind is said to have originated with him. His son became stipendiary magistrate of Liverpool in May, 1839.

Another of the gallant little band of Reformers, who dared to unfurl, in the very stronghold of the enemy, the standard of truth, liberty and justice, was Dr. James Currie, a native of Annandale, and father of William Wallace Currie, who became first Mayor of Liverpool, under the Municipal Corporations Bill. Dr. Currie, then rising into notice as a physician in the town, had the courage to risk popularity and practice by writing in defence of the down-trodden African, and in reprobation of the slave trade, and it may be that the fact of his doing so will be remembered, when his biography of Burns, which disgusted Charles Lamb, is forgotten.

In March, 1788, two months after the Abolition Society had broken ground in Liverpool, the slave-merchants put forth or found a champion worthy of their cause, in the person of the Rev. Raymond Harris,* a Spanish Jesuit, of English extraction, who had settled in Liverpool. This reverend defender of iniquity put forth a pamphlet entitled, "Scriptural Researches on the Licitness of the Slave Trade, showing its conformity with the principles of Natural and Revealed Religion, delineated in the sacred writings of the Word of God." We need not do more than indicate the singular specimens of sophistry and perversity that characterised this production. The author contended "that no one could doubt the licitness of the slave trade, who believed that the Bible was the Word of God. In proof of this assertion he first laid it down as an axiom, that whatever practices were mentioned in either the Old Testament or the New, with implied approbation, were sanctioned by God, and would continue to be lawful through all time. This he contended, was the case with slavery and the slave trade. His first example was that of Hagar, slave of Sarah, Abraham's wife, who having fled from her mistress, in consequence of having been hardly dealt with by her, was

* He was born at Bilbao, 4th September, 1744; admitted S.J., 1758; expatriated from Corsica, April 1st, 1767; was afterwards chaplain at Walton Hall; removed to Liverpool, where he officiated at the Catholic Church, in Edmund Street. Whilst there he was thrice suspended by his Bishop.

ordered by God to return, and humble herself under the hand of her mistress. A second example was that of the patriarch Joseph, who had bought the whole people of Egypt for King Pharaoh, during the seven years of famine. A third was that of the Gibeonites, who had been condemned to be hewers of wood and drawers of water for ever to the Israelites. Many other instances, equally apposite, followed, and the general inference which the author drew from them all was, that the slave trade was a 'licit' occupation, and that those who did not believe it to be so, did not believe their Bibles."*

This astounding vindication of man-stealing so delighted the Corporation that they presented the reverend sophist with a gratuity of £100. Lord Hawkesbury (afterwards Earl of Liverpool) actually condescended to distribute some of Harris's precious "Researches," recommending them at the same time as containing unanswerable arguments in favour of the slave trade.

While those who were interested in the traffic admired the reverend gentleman's "bold attempt to degrade the noblest of all the attributes of the Deity—his justice and his mercy"—others were filled with indignation and loathing at its appearance. Dr. Currie, a man of cool and dispassionate judgment, thus speaks of the trade and its advocate in a letter to a friend:—

> "The general discussion of the slavery of the negroes has produced much unhappiness in Liverpool. Men are awaking to their situation, and the struggle between interest and humanity has made great havoc in the happiness of many families. If I were to attempt to tell you the history of my own transactions in this business, I should consume more time than I can spare. Altogether I have felt myself more interested and less happy than is suited to my other avocations. The attempts

*Thomas Baines's "History of Liverpool," pp. 472-3.

that are continually made to justify this gross violation of the principles of justice one cannot help repelling; and at the same time it is dreadful to hold an argument where, if your opponent is convinced, he must be made miserable. A little scoundrel—a Spanish Jesuit—has advanced to the assistance of the slave-merchants, and has published a vindication of this traffic from the Old Testament. His work is extolled as a prodigy by these judges of composition, and is in truth no bad specimen of his talents, though egregiously false and sophistical, as all justifications of slavery must be. I have prompted a clergyman—a friend of mine—to answer him, by telling him that if such be religion, I would 'none on't.'"

Harris's pamphlet was promptly answered by several writers, who had no great difficulty in showing that if his argument proved anything, "it proved a great deal too much; for it proved that the marrying of three or four wives at one time was a commendable practice; made it a matter of duty to stone all blasphemers to death; and justify true believers, not only in making slaves of heathen nations, but in exterminating them with fire and sword." Among those who published rejoinders were the Rev. Mr. Hughes and the Rev. Henry Dannett, M.A., minister of St. John's Church, whose pamphlet was entitled, "A Particular Examination of Mr. Harris's Scriptural Researches on the Licitness of the Slave Trade."

The most eloquent answer, however, to Harris's work was that of William Roscoe, entitled, "A Scriptural Refutation of a Pamphlet lately published by the Rev. Raymond Harris, entitled 'Scriptural Researches on the Licitness of the Slave Trade,' in four letters from the Author to a Friend." This rejoinder immediately attracted the attention of the London Abolition Committee, who pronounced it the work of a master, thanked the author for the important service he had rendered to the cause, and arranged with him for the issue of a new edition on his own terms.

Nothing daunted, the Rev. Raymond Harris, mounted on his good steed, "Sophistry," again entered the lists as the champion of "licit" man-stealing; but his death, very soon afterwards, left the controversy to be carried on by writers who argued the question on behalf of the merchants as one of profit and political expediency, rather than of right and wrong. The line of defence adopted in the newspapers and pamphlets of the day was the great importance and magnitude of the trade; the ruin of the West India Islands if the supply of "labourers from Africa" were discontinued; that slavery was rather a blessing than otherwise for the negroes themselves; and that the blacks were an inferior race, incapable of living as free men.

While collecting evidence on behalf of Abolition, the Rev. Thomas Clarkson visited Liverpool, and soon found himself in dramatic situations. He called upon Mr. William Rathbone, Mr. Isaac Hadwen, Dr. Binns (all of the noble people called Quakers), Mr. Roscoe, Dr. Currie, and Mr. Edward Rushton. Mr. Rathbone, than whom the negro had no better friend, had supplied Mr. Clarkson (through their fellow-worker, Mr. James Phillips, of London) with copies of the muster-rolls of Guineamen, from the Custom-house at Liverpool. The revelations of mortality among the seamen in the slave trade made by these muster-rolls, had an important bearing upon the agitation, and helped to explode the theory that the employment was "a nursery for seamen." Mr. Rathbone, desirous that Mr. Clarkson should have reliable information, introduced him to Mr. Robert Norris, a merchant, who had formerly been a slave-captain, and who was writing a history of Dahomey,

* Jonathan Binns, M.D., was for many years senior physician to the Liverpool Dispensary. He published, at Edinburgh, in 1762, *Dissertatio Medica in Auguralis de Exercitatione.* He superintended, for some time, the school belonging to the Society of Friends, in Yorkshire, and whilst there, published an English Grammar, and also a Vocabulary. He removed to Lancaster, where he practised as a physician until the time of his death, in 1812, aged 65 years.—"Smithers' History," p. 433.

and a life of the King, who sold his own subjects to the slave-captains. Mr. Norris, at this time, made no secret of the facts with which he was priming Mr. Clarkson, and even "answered in a solid manner" the arguments of a slave-merchant named Coupland, who fired up and defended the humanity and policy of the trade. Finally, Mr. Norris drew up certain clauses, which, if made law, would effect the abolition of the traffic. Strange to say, Mr. Norris afterwards completely changed his front, and became one of the champions of the trade, receiving the thanks of the Corporation for his exertions, which were further acknowledged, after his decease, by the grant to his widow of a life annuity of £100 from the Corporate funds.

Mr. Clarkson found a sympathiser in Captain Chaffers, who had been in the West India employ, and who offered to introduce him to Captain L——, whose long experience in the slave trade would be invaluable to the abolitionist. The two accordingly called upon Captain L——, who, in speaking of the productions of Africa, happened to mention that mahogany trees, the height of a tall chimney, grew at Calabar. Then a curious scene ensued :—

"As soon as he mentioned Calabar," says Mr. Clarkson, "a kind of horror came over me. His name became directly associated in my mind with the place. It almost instantly occurred to me, that he commanded the *Edgar* out of Liverpool, when the dreadful massacre there took place. Indeed I seemed to be so confident of it, that attending more to my feelings than to my reason at this moment, I accused him with being concerned in it. This produced great confusion among us. For he looked incensed at Captain Chaffers, as if he had introduced me to him for this purpose. Captain Chaffers again seemed to be all astonishment that I should have known of this circumstance, and to be vexed that I should have mentioned it in such a manner. I was also in a state of trembling myself. Captain L——— could only say it was a bad business. But he never defended himself, nor those concerned in it. And we soon parted, to the great joy of us all."

On his first arrival in the town, Mr. Clarkson found the people ready enough to talk about the slave trade. Horrible facts were in everybody's mouth, and it seemed to him that the inhabitants were no longer capable of being surprised at the extent of their own wickedness. From this callousness he expected to extract the damnatory evidence for which he yearned. After the argument with Mr. Coupland, and the scene with Captain L——, however, he found that attention of an unpleasant character had been drawn to the purpose of his visit, and information was no longer obtainable. Slave-merchants, slave-captains, and others, dropped into the "King's Arms," where he was staying, and gazed at him as an animal exceedingly rare and somewhat to be feared. Dale, the master of the tavern, was delighted with the custom his new guest attracted. Many of the callers dined there, and entered into warm arguments with the enemy of "the trade;" some provoked and insulted him, others hinted that men were going about to abolish the slave trade who would have done much better if they had stayed at home. One said that he had heard of a person turned mad, who had conceived the thought of destroying Liverpool and all its glory; while another, laughing boisterously, raised his glass and gave as a toast, "Success to the trade," watching if Clarkson would drink it. Mr. Clarkson, fortunately, had with him Mr. Falconbridge, of Bristol, who had been in the slave trade. He was an athletic and resolute looking man who, when the slave-captains ridiculed Clarkson's statements and asked if he had ever been on the Coast, used to strike in with, "But I have; I know all your proceedings there, and that his statements are true." This went on day by day, until the situation became dangerous. The friends of Abolition saw him privately, said he was right, and exhorted him to persevere; but fear of having their houses pulled down held them back from giving evidence publicly against the man-traffic. These were not idle fears, for Dr. Binns had nearly fallen into a plot laid against him because he was a subscriber to the Abolition Society, and was suspected of aiding Mr. Clarkson. That brave-hearted man, without absolutely leaving the "King's Arms," took rooms in Williamson-square, where he was visited by seamen from the Guinea ships, whose stories of wrong, murder, and ill-treatment, made his life a misery. The hostility against him increased. He received anonymous letters entreating him to leave the town immediately, or he should never leave it alive. The only effect these threats had upon him was to make him more vigilant when he went out at night, which he never did without Falconbridge, who, unknown to him, went about well armed. That Mr. Clarkson's life was in danger at this time seems undeniable. When he was standing on the pier-head one day, a gang of men closed upon him and bore him back within a yard of the precipice, when perceiving his danger, he darted forward, knocking one of the ruffians down, which broke their ranks, and enabled him to escape, "not without blows, amidst their imprecations and abuse." Amongst them he recognised the murderer of the steward of a Guinea ship, around whom he was drawing the coils of justice, and two others who had insulted him at the "King's Arms."

Mr. Clarkson paid a visit to Lancaster, where he learned that the slave-merchants of the place made their outfits at Liverpool, as a more convenient port. Lancaster, too, was then under a cloud in the African trade. The captain of the last slave-ship which sailed thence to the Coast, had taken off so many of the natives treacherously, that any other vessel known to come from Lancaster would be cut off. There were then only one or two superannuated slave-captains living in the town. On looking over the muster-rolls at the Custom-house, Mr. Clarkson found that the loss of seamen was precisely in the same proportion as at other ports.

On his return to Liverpool, he learned from Mr. Falconbridge that during his absence visitors had continued to call at the "King's Arms," to deliver their abuse of him, and that one of them had said that "he deserved to be thrown over the pier-head."

Mr. Clarkson had now collected in London, Bristol, and Liverpool, the names of more than 20,000 seamen who had made different voyages, and he knew what had become of each man. As the Committee in London were pressing him to write an Essay on the impolicy of the Slave Trade, he bade farewell to his few friends in Liverpool. To one of them he refers as follows:—

> "The last of these was William Rathbone, and I have to regret that it was also the last time I ever saw him. Independently of the gratitude I owed him for assisting me in this great cause, I respected him highly as a man. He possessed a fine understanding, with a solid judgment. He was a person of extraordinary simplicity of manners. Though he lived in a state of pecuniary independence, he gave an example of great temperance, as well as of great humility of mind. But, however humble he appeared, he had always the courage to dare to do that which was right, however it might resist the customs or the prejudices of men. In his own line of trade, which was that of a timber merchant on an extensive scale, he would not allow any article to be sold for the use of a slave-ship, and he always refused those who applied to him for materials for such purposes. But it is evident that it was his intention, if he had lived, to bear his testimony still more publicly upon this subject; for an advertisement, stating the ground of his refusal to furnish anything for this traffic, upon Christian principles, with a memorandum for two advertisements in the Liverpool papers, was found among his papers at his decease."

Mr. Rathbone resided in Liver-street, and afterwards in

Cornhill. He was a fine, venerable-looking man, with dark eyebrows and flowing, silvery hair. He was very highly respected, and it is said that he was not surpassed by any contemporary individual in Liverpool in acts of benevolence and charity. In 1805, he published "Memoirs of the Proceedings of the Quakers in Ireland." Mr. Rathbone died in February, 1809. It had been his custom to inscribe in a book devoted to that purpose the names of those of his family whom he had lost by death. In this volume, his bosom friend, William Roscoe, has, in his own hand, thus recorded the death of his friend :—

"11th FEBRUARY, 1809.

"William Rathbone died at nine o'clock in the morning, aged 51 years and 8 months.

"This domestic record which contains the brief memorials of his beloved and respected relatives, registered by his own hand, and endeared by the warm expression of his affection, now receives the honoured name of

"William Rathbone,*

"Of Liverpool, Merchant ;

"a name which will ever be distinguished by independence, probity, and true benevolence, and will remain as an example to his descendants of genuine piety, patient resignation, and of all those virtues which give energy to a community, adorn society, and are the delight of private life.

"Through life beloved ! O let this votive line
Unite in death its author's name with thine.

"WILLIAM ROSCOE."

The agitation of this question by a man like Clarkson, who, with invincible patience, zeal, and faith, hunted up the most astounding facts and made them public in his

* Mr. Richard Brooke, who died in 1852, in his 92nd year, knew five out of seven William Rathbone's in successive generations, namely :—a great-great-grandfather, a great-grandfather, grandfather, father, and son. What is more remarkable perhaps is, that virtue and benevolence appear to be hereditary in this old Liverpool family.

pamphlets, and by Wilberforce, who, in 1787, had become the parliamentary champion of abolition, could have only one possible termination; but the effects produced in Liverpool by the twenty years' contest between right and wrong were of a very demoralising character. "The secret consciousness," says Sir James Picton, "that the trade would not bear the light either of reason, Scripture, or humanity, combined with the conviction that the prosperity of the town depended upon its retention, produced an uneasy feeling of suspicion and jealousy, and a dread of all change, which could not but impart a peculiar character to those at least connected with the occupation."

CHAPTER V.

HORRORS OF THE MIDDLE PASSAGE.

"Freighted with curses was the bark that bore
The spoilers of the West to Guinea's shore;
Heavy with groans of anguish blew the gales
That swell'd that fatal bark's returning sails;
Old Ocean shrunk, as o'er his surface flew
The human cargo and the demon crew."

WE have now arrived at a period when a Parliamentary enquiry into the whole conduct of the slave trade enables us to give an authoritative account of the method of procuring slaves on the coast, together with a reliable sketch of the perils and horrors of the middle passage.

The great demand for slaves made it the interest of the princes and chiefs of Africa to procure supplies by any means—by war, by rapine, or perfidy. In their efforts to keep pace with the merchants' cry of "more, more!" these cruel panderers did not scruple to turn a naturally fine and productive country into one continued scene of devastation and slaughter, for more than three or four thousand miles along the coast.

With a hellish ingenuity the very crimes of the country seemed to have been made on purpose to serve the interests of slave-sellers and slave-buyers. Theft, adultery, witchcraft, and the removal of fetiches were falsely imputed for the sake of selling the accused into slavery, and some of the chief men were said to employ the best looking women they could find, well dressed, in order to entice the unwary into criminal situations, which ensured their conviction, or offered a pretext for selling them to Europeans. It was, in effect, argued by the defenders of the trade that the slaves procured in consequence of native wars would have been put to death, had not the slave-merchants humanely and providentially stepped in and relieved the native belligerent powers of the necessity of committing wholesale massacres. The abolitionists, however, maintained that what the slave-traders called war was nothing else but pillage, robbery, and kidnapping, of the most wanton, cruel, and sordid character. When slave-ships arrived on the coast, the petty princes of the country sent out their myrmidons in parties of from 300 to 3000, often on horseback, to attack and burn towns and villages in the dead of night, so that the panic-stricken inhabitants were the more easily seized and bound, while attempting to save themselves and those most dear to them from the flames. Every man, woman, and child that could be secured by this armed banditti were carried off without mercy; the men stripped quite naked, and chained together; the women and children loose. In this manner, they were all driven by their own countrymen, assisted sometimes by Europeans, towards the place of sale, like sheep for the slaughter, the distance to be travelled before they reached the coast being often two or three hundred miles. Thus the dearest relatives were torn from each other's arms, in all probability never more to meet on earth. Even children were separated from their parents, except the sucking infants, who were permitted, for obvious reasons, to accompany the mother. "What a moving scene," exclaims Clarkson. "Parents and children, husbands and wives, brothers and sisters, not only forced from their native country, but denied, in their exile and captivity, the small consolation of mingling their sighs and tears in mutual condolence and commiseration! Such a scene must exceed the powers of language to express, or of the human mind to conceive, where not felt or seen." When, as sometimes happened, native princes objected to pillage their subjects and sell their countrymen into bondage, the traders kept them in a state of intoxication till their end was attained. Some noted traders were a terror to the country, for they openly and avowedly kidnapped their brethren, whom they carried off gagged, lest their cries should alarm the country as they passed. This method was called "panyaring," and no questions were asked by the slave-captains, whose business was to make up their cargoes speedily. The slave-ships occasionally expedited this desirable object by capturing canoes at sea, and along the coast. They also decoyed the natives on board on pretence of traffic, seized them, and put them in irons. Another dastardly method was to make some leading native drunk, get him to sell some of his own relations, whom he redeemed, when sober, at any price insisted upon by the slave-captains or their agents. A son sold his father, who was a slave-owner, and he had to give twenty slaves to redeem himself. A trader, returning from a ship with the proceeds of four slaves he had just sold at a high price, was seized by a native chief, taken to the vessel and sold, thus becoming the companion in misery of those over whom he had a short time before held the power of life and death. Traders were occasionally invited to dine with the captain on board ship, when they were filled with drink, and awoke to find themselves out at sea. They were then stripped, branded, and thrust down the hold to share the fate of the other slaves, some of them possibly their own victims. The instances of wicked artifice and base treachery employed in procuring slaves, mentioned in the evidence, are almost innumerable.

The slaves having been procured by some, or all, of the foregoing methods, we must now find a ship to transport them to the scene of their future labours, or—to death; their own choice being probably the latter. We choose for our purpose a typical vessel, with whose fighting qualities we are already acquainted—the *Brooks*, Captain Noble.

From a return presented to Parliament in 1786 by Captain Parrey, who was sent to Liverpool by the Government to take the dimensions of the ships employed in the African slave trade from that port, it appears that the dimensions of the "*Brooks*" were as follow:—The length of the lower deck, with the thickness of the gratings and the bulkheads, was 100 ft.; her breadth of beam on the lower deck from inside to inside, 25 ft. 4 in.; the depth of the hold from ceiling to lower deck, 10 feet; height between decks, 5 ft. 8 in.; length of the men's room on lower deck, 46 ft.; breadth of the men's room on lower deck, 25 ft. 4 in.; length of the platform in men's room on the lower deck, 46 ft.; breadth of the same platform on each side, 6 ft.; length of the boys' room, 13 ft. 9 in.; breadth of the boys' room, 25 ft.; breadth of platform in boys' room, 6 ft.; length of the women's room, 28 ft. 6 in.; breadth of the women's room, 23 ft 6 in.; length of platform in women's room, 28 ft. 6 in.; breadth of platform in women's room on each side, 6 ft. The number of air-ports going through the side of the deck was 14; the length of the quarter deck, 33 ft. 6 in., by a breadth of 19 ft. 6 in.; the length of the cabin was 14 ft., by 19 ft. in diameter, and 6 ft. 2 in. in height; length of the half-deck, 16 ft. 6 in., by 6 ft. 2 in. in height; length of the platforms on the half-deck, 16 ft. 6 in., by 6 ft. in breadth. The vessel is described as frigate-built, without forecastle, and pierced for 20 guns. Nominal tonnage, 297; supposed tonnage by measurement, 320; number of seamen, 45. It appears from the accounts given to Captain Parrey by the slave-merchants themselves, that, when leaving the coast of Africa, she carried, besides her crew, 351 men, 127 women, 90 boys, and 41 girls, a total of 609, though only legally allowed to carry about 450. She lost by death, on her passage, 10 men, 5 women, 3 boys, and 1 girl. In the year 1782, she arrived at Jamaica with 646 slaves, but how many she had when she left the coast on that voyage is not stated.

Her provisions for the negroes were:—20 tons of split beans, peas, rice, shelled barley, and Indian corn ; 2 tons of bread ; 12 cwt. of flour ; 2,070 yams, averaging 7 lbs. each ; 34,002 gallons of water ; 330 gallons of brandy, rum, &c. ; 70 gallons of wine ; 60 gallons of vinegar ; 60 gallons of molasses ; 200 gallons of palm oil ; 10 barrels of beef ; 20 cwt. of stock fish ; with 100 lbs. of pepper. She was 49 days on the passage from the Gold Coast to the West Indies, the shortest passage of nine vessels reported being 42 days, and the longest 50 days.

The mind cannot realise, language cannot paint the sufferings of one day, nay, of one hour, passed under such circumstances, by the tightly-wedged human cargo in the hold of the best managed slaver. Dreadful must have been the agony under the most favourable conditions, with a humane captain, like John Newton, an able surgeon, fine weather, and a short passage, but what a circumscribed hell were they tormented in when, after several months spent on the coast to complete the cargo, they experienced, during a long passage to the West Indies, lasting from six to eight weeks, rough weather, inhuman treatment, and scanty rations of bad quality! In one instance, the slaves on board a schooner which carried only 140, were kept below, and the gratings covered with tarpaulin, during a gale of wind, which lasted eighteen hours, when no less than 50 slaves perished in that brief space of time. "One real view, one minute, absolutely spent in the slave rooms on the middle passage," says an officer employed in the trade, "would do more for the cause of humanity than the pen of a Robertson, or the whole collective eloquence of the British Senate."

To indicate the sanitary condition of the ships on the

middle passage, it is sufficient to quote the testimony of a surgeon employed in the trade :—

> "Some wet and blowing weather having occasioned the port-holes to be shut and the grating to be covered, fluxes and fevers among the negroes ensued. While they were in this situation, my profession requiring it, I frequently went down among them, till at length their apartments became so extremely hot as to be only sufferable for a very short time. But the excessive heat was not the only thing that rendered their situation intolerable The deck, that is, the floor of their rooms, was so covered with the blood and mucus which had proceeded from them in consequence of the flux, that it resembled a slaughter-house. It is not in the power of the human imagination to picture to itself a situation more dreadful or disgusting. Numbers of the slaves had fainted, they were carried upon deck, where several of them died, and the rest were, with difficulty, restored. It had nearly proved fatal to me also."

The men — except in sickness — were kept constantly chained two and two ; the right leg of one to the left leg of the other, their hands being secured in the same manner. In this miserable state they had to sit, walk, and lie, sometimes for nine or ten months, without any mitigation or relief till they reached their destination. It was impossible for them to turn or shift posture with any degree of ease, or without hurting one another. The effects of this severe treatment were assigned as the reason why the men died on the passage in double the proportion of the women and children, who went unfettered. Some who went below in the evening in apparent good health were found dead in the morning. The Rev. John Newton, in his evidence, said he had often seen a dead and a living man chained together. Such as were out of irons were packed spoon-ways, one on another, so that each had less room than a man in a coffin ; those who did not get into their places quick enough being stimulated by the cat-o'-nine tails.

In favourable weather the slaves were brought up on the main deck daily, about eight o'clock in the morning, and as each pair ascended from the horrible dungeon, a strong chain, fastened by ring bolts to the deck was passed through their shackles, a precaution absolutely necessary to prevent insurrections, but often ineffectual, as we have repeatedly had to record. They were allowed to remain on deck about eight hours, during which time they were fed, and their apartment below was cleaned—a terrible task. If the weather was bad, however, they were only permitted to come up in small parties of about ten at a time, to be fed ; after remaining on deck a quarter of-an-hour, each mess was obliged to give place to the next in rotation.

When feeding time was over, the slaves were compelled to jump in their chains, to their own music and that of the cat-o'-nine-tails, and this, by those in the trade, was euphemistically called "dancing." Those with swollen or diseased limbs were not exempted from partaking in this joyous pastime, though the shackles often peeled the skin off their legs. The songs they sang on these occasions were songs of sorrow and sadness—simple ditties of their own wretched estate, and of the dear land, and home, and friends they were never more to see. During the night they were often heard to make a howling, melancholy noise, caused by their dreaming of their former happiness and liberty, only to find themselves, on waking, in the loathsome hold of a slave-ship. The women, on these occasions, were often found in hysteric fits.

From statistics kept by several vessels, it appears that out of 7904 slaves purchased on the Coast, 2053 perished on the middle passage. In one document, the average is put at 20 per cent., and in the case of the *John*, already referred to, the rate of mortality was actually 50 per cent. Yet

"Fresh myriads, crowded o'er the waves,
Heirs to their toils, their sufferings, and their graves!"

The slave-ships were peculiarly constructed, with a view to prevent the negroes from ending their misery by plunging into the sea; nevertheless, the utmost vigilance was not able to frustrate such manumission, and a score have been known to muster up all their strength, burst from their chains, and leap overboard, exulting, with apparent joy, as they sank in the waves, or fell a prey to the procession of sharks that followed in the wake of the Guineamen.

One witness stated that his feelings were much hurt by being so often obliged to use the cat, to force the slaves to take their food; and that in the very act of chastisement they have looked up at him with a smile, saying in their own language, "Presently we shall be no more." Some of them endured voluntary starvation, frequent floggings, the torture of the thumbscrews—which made the sweat run down their faces, and their bodies to tremble all over as in an ague fit—and other cruel usage, for the space of eight or nine days. Then came death, the kind manumitter, for whom they had longed with a great longing. The very children sometimes chose rather to die than live. We have a remarkable instance of a young child that refused all sustenance. The captain, enraged at his obstinancy, flogged him, and, with horrid imprecations, threatened to kill him if he would not eat. This discipline was repeated several times without effect. The child's feet being swelled, were, by the captain's order, put in water, though the ship's cook told him it was too hot. This brought off the skin and nails. After about four days of this usage, the child died, just after the captain had done whipping him. The mother refusing to cast her son overboard when ordered to do so, was severely beaten, compelled to take up the body and go to the ship's side, where, turning away her face to avoid the sight, she dropped the child into the sea, and continued for many hours to cry bitterly.

A certain Liverpool captain, in a large company at

Buxton, related how a female slave on her voyage fretted herself to a very great degree on account of an infant child, which she had brought with her. "Apprehensive for her health, I snatched the child," said this monster, glorying in his unparalleled brutality, "I snatched the child from her arms, knocked its head against the side of the ship, and threw it into the sea."

When any of the Guineamen were driven out of their course by stress of weather, and provisions ran short, the slaves were sometimes compelled to "walk the plank," or, in other words, to jump overboard. In cases of shipwreck, they were either left to perish, entangled in their irons, or the seamen put them to death, to prevent their escape and to ensure their own safety. One terrible instance of this is given in the evidence. A ship from Africa, with about 400 slaves on board, struck in the night upon some shoals about eleven leagues distant from the south end of Jamaica. The officers and seamen landed in their boats, carrying with them fire-arms and provisions. The slaves were left on board, shackled together in their chains. Having somehow got out of their irons, they were discovered at daybreak busily making rafts of broken parts of the wreck, upon which they placed the women and children, while the men and others that could swim, accompanied the rafts as they drifted before the wind towards the island, where the crew had landed the preceding night. The seamen, fearing that the slaves would consume the water and provisions, which they had saved from the ship, came to the horrid resolution of destroying them, and accordingly fired upon them with such good effect, as they were attempting to make the land, that between three and four hundred were massacred; so that out of the whole cargo only thirty-three or thirty-four were spared, carried to Kingston, Jamaica, and exposed to public sale.

Many other instances, substantiated by affidavit, could be

given of the most shocking cruelties and murders perpetrated on both the negroes and the seamen by captains, whom the slave trade had converted into fiends incarnate.

On the arrival of the ships at their destined ports, the slaves were prepared for sale, much pains being taken to clean and anoint their bodies, that they might appear to the best advantage. There were three methods of sale, all more or less attended with circumstances harrowing to the feelings and degrading to humanity. The first was by private treaty between the merchants' factor and the planters, the latter selecting their "goods" at the factory; the second, and the most curious method, was by scramble, which was described by Mr. Falconbridge of the *Emila.* The ship, on its arrival at Jamaica, was darkened with sails and covered round. The men slaves were placed on the main deck, and the women on the quarter-deck, while an eager crowd of buyers, who had been supplied with tallies or cards, waited on shore for the sale to begin. When all was ready, a signal gun was fired, and the gangways thrown open, when the buyers rushed through the barricado door with the ferocity of brutes, seized as many slaves as they thought fit for their purpose, encircling them with handkerchiefs tied together. At Grenada, where a scramble sale took place, the women were so terrified that several of them got out of the yard, and ran about St. George's town as if they were mad. At Kingston, Jamaica, during a sale on board the *Tryal*, Captain Macdonald, forty or fifty slaves leaped into the sea, but were recovered. From the evidence of several witnesses, this appears to have been a very common method of sale in America and the West Indies. The slaves on these occasions were all at one price.

The slaves sold by the third method—public auction or vendue—were generally the refuse and sickly, some of whom died before the fall of the auctioneer's hammer. The prices sometimes fell so low as a single dollar per head. In one case, the ship *Lottery*, Captain Whittle, belonging to Mr. Thomas Leyland, of Liverpool, we find a blind negro given away! An officer of high rank testified that he once saw a number of slaves, who had been landed from a ship, brought into the yard adjoining the place of sale. Such as were not very ill were put into little huts, while the most sickly were left in the yard to die, for nobody gave them food or drink, and some of them lived three days in that situation. Slaves were often carried from the ship to the vendue-master in the agonies of death, and expired in his piazza. At these auctions the slaves were exposed to public view, naked as they came into the world, regardless of age or sex, and the slave-merchants and planters viewed and handled them as a butcher handles the cattle he is about to purchase for slaughter.

A terrible mutiny of slaves on the middle passage occurred in the year 1797. The ship *Thomas,* of Liverpool, belonging to Mr. Thomas Clarke, and commanded by a very brave, respectable, and intelligent man, Captain Peter M'Quie* took on board 375 picked slaves at Loango, and sailed for Barbadoes. On the morning of the 2nd of September, 1797, while the crew were at breakfast, two or three of the women-slaves discovered that the armourer had incautiously left the armour-chest open. They got into the after-hatchway, and passed the arms through the bulkheads to the men-slaves, about two hundred of whom immediately ran up the fore-scuttles, and put to death all of the crew that came in their way. The captain and a few of his men fought desperately with the arms remaining in the cabin; but they were eventually overpowered, the slaves gaining complete possession of the ship. Captain M'Quie and many of his

* A native of Minnigaff, in the County of Galloway, Scotland, and father of the late Mr. Peter Robinson M'Quie, merchant, of Liverpool, who communicated the details of the tragedy to Mr. Brooke, author of "Liverpool in the Last Quarter of the Nineteenth Century."

crew perished, being either killed in the conflict, butchered afterwards, or driven overboard. Twelve of the hands, however, escaped in the stern-boat, and after enduring the most dreadful hardships, two only survived to land in Barbadoes. A few were kept alive to steer the vessel back to Africa. Four of these escaped in the long-boat, and after being six days and nights without food or water, reached Watling's Island, one of the Bahamas, in a wretched condition. Five of the crew still remained on board the *Thomas*, the negroes not being able to steer the vessel without their assistance. After forty-two days of misery and dread, an American brig, laden with rum, came alongside, of which the negroes made themselves masters, her crew escaping in their boats. Rum casks were opened, and a scene of drunkenness and confusion ensued, during which several of the blacks were drowned. The remaining crew of the *Thomas* took immediate advantage of this occurrence and recaptured the brig—the boatswain, with the captain's cutlass, having first killed the ringleader of the negroes,—set sail for the nearest land, and reached Long Island, Providence. The *Thomas*, with the surviving negroes, was afterwards recaptured by H.M. frigate *Thames*, carried into Cape Nicola Mole, and sold there.

CHAPTER VI.

Emoluments of the Traffic—A Millionaire's Ventures.

"I own I am shocked at the purchase of slaves,
And fear those who buy them and sell them, are knaves;
What I hear of their hardships, their tortures, and groans,
Is almost enough to draw pity from stones.
I pity them greatly, but I must be mum,
For how could we do without sugar and rum?"—*Cowper.*

A drunken actor,* on the stage of the Theatre Royal, Williamson-square, on being hissed by the audience for presenting himself before them—not for the first time—in that condition, is said to have steadied himself, and vociferated, with offended majesty, "I have not come here to be insulted by a set of wretches, every brick in whose infernal town is cemented with an African's blood." This was a home-thrust which might have made the daring offender the hero of an unrehearsed tragedy. The taunt, however, would have been almost as applicable hurled at London, Bristol, or certain southern port audiences, whose bricks were more or less cemented in the same sanguinary fashion for fully one hundred years before the people of Liverpool ever soiled their hands and souls in the African slave trade. The brilliant success which crowned the shrewd enterprise of Liverpool merchants in this, as in all other branches of

* George Frederick Cooke, tragedian, 1756-1812; the predecessor of Kean in his peculiar line of characters.

commerce, has made them the focus of scorching censure, while the older offenders, left far behind in the race for pelf, are comparatively forgotten, and their exceeding weight of guilt overlooked. In a word, Liverpool, while sowing wild oats in its commercial youth, or leading a sort of double life—wedded to freedom at home, and courting slavery abroad,—took a hand in this dark game, swept the board, and, rather unjustly, has had to bear the concentrated odium attached to the whole of the play. Roscoe, all his life the firm, but statesman-like opponent of the man-traffic, speaking at a public dinner held at the "Golden Lion," to celebrate his election as one of the representatives in Parliament of his native town, thus referred to the national character of the iniquity:—

> "It has been the fashion throughout the Kingdom to regard the town of Liverpool and its inhabitants in an unfavourable light on account of the share it has in this trade. But I will venture to say that this idea is founded on ignorance, and I will here assert, as I always shall, that men more independent, of greater public virtue and private worth, than the merchants of Liverpool do not exist in any part of these kingdoms. The African trade is the trade of the nation, not of any particular place; it is a trade, till lately, sanctioned by Parliament and long continued under the authority of the Government. I do not make this remark in vindication of the character of any gentlemen engaged in the trade, who stand in need of none, but in order to shew that if any loss should arise to any individuals who are concerned in it, it is incumbent upon Government to make them a full compensation for the losses they may so sustain."

However we may detest the trade, and shudder at the horrors which necessarily accompanied it, even when most rigorously supervised, and conducted by the most humane instruments; though we know that no casuistry can convert wrong into right, yet must we remember that custom has a wonderful effect in blinding the moral perceptions; that men's standard of morality is being raised, as the leaven of Christianity spreads with power, and that ages, like individuals, are prone to

> "Compound for sins they are inclined to,
> By damning those they have no mind to."

Let us now endeavour to arrive at the approximate amount of the emoluments of the traffic; certainty as to the aggregate profits is impossible, but we are in a position to compare estimates, made while the trade was in full swing, with the real profits made by certain ships, whose accounts are forthcoming. It appears from one calculation* that, during the eleven years from 1783 to 1793, 878 slave-ships belonging to Liverpool, imported to the West Indies, etc., 303,737 slaves, whose estimated sterling value amounted to the enormous total of £15,186,850.

In order to arrive at the net amount returned to the port of Liverpool out of this sum, we must deduct factors' commission of 5 per cent. on the sales, in addition to 10 per cent. for contingencies, making a deduction of 15 per cent., or £2,278,027 from the gross sales of £15,186,850, leaving the net proceeds £12,908,823. But this sum is subject to a further deduction of £614,707, being the factors' commission of 5 per cent. on the actual amounts remitted by them to the merchants of Liverpool. It appears, therefore, that the net proceeds remitted to Liverpool, for the eleven years' slave-trading, from 1783 to 1793, amounted to £12,294,116† sterling, or on an average £1,117,647 per annum; expressed perhaps more clearly in the following manner:—

*See Table in the Appendix.

†Baines, evidently quoting from the same source, gives the gross amount brought into the port as "£12,908,823 in eleven years, or £1,117,647 a year," which on the face of it, is an error, the cause of which is the omission of the factors' commission of £614,707, which of course, never reached Liverpool.

Gross amount of Sales of 303,737 slaves, averaged at £50 per head		£15,186,850
Deduct Factors' Commission 5% on ditto	£759,342	
Deduct for Contingencies 10%	1,518,685	2,278,027
		£12,908,823
Deduct Factors' Commission 5% on real amount remitted...		614,707
Net proceeds remitted to Liverpool ...		£12,294,116

Average net proceeds remitted during eleven years £1,117,647 a year

Taking the number of slaves imported in the year 1786 as his basis, and allowing twelve months for the length of the voyage, instead of the average nine months, the same writer arrives at the appended statement of probable net gains on a Guinea cargo:—

The net proceeds on 31,690 negroes		£1,282,690
Gross value of goods exported to Africa...	£864,895	
Freight of 31,690 slaves ...	103,488	
Maintenance of 31,690 slaves at 10/- each	15,845	984,228
Profit on the whole		£298,462

Having shown by the above statement that the profits were upwards of 30 per cent., our author proceeds to analyse the aggregate sums and discovers—

The net proceeds of one slave to be		£40 9 6¾
The prime cost of one slave on the Coast	£27 5 10	
The freight of one slave ...	3 5 3¾	
The maintenance of one slave	0 10 0	31 1 1¾
Profit on the sale of one slave		£9 8 5

It appears then, from the preceding calculations, that there was a profit of upwards of 30 per cent. on the sale of each slave; that in the year 1786, the town of Liverpool pocketed £298,462 sterling from the importation of 31,690 negroes; and that during the eleven years, from 1783 to 1793, both inclusive, the gains on 303,737 slaves sold amounted to £2,861,455 13s. 1d. or on an average £260,132 6s. 8d. per annum.*

"This great annual return of wealth," says our author,† "may be said to pervade the whole town, increasing the fortunes of the principal adventurers, and contributing to the support of the majority of the inhabitants. Almost every man in Liverpool is a merchant, and he who cannot send a bale, will send a bandbox, it will therefore create little astonishment, that the attractive African meteor has, from time to time, so dazzled their ideas, that almost every order of people is interested in a Guinea cargo." The small adventurers, however, whose ships were despatched irregularly, could not, under the most favourable circumstances, derive a large income from the trade, although the returns might sometimes arrive in time "to prop a tottering credit." In the case of a ship importing 100 slaves, which by the preceding estimate would yield a profit of £942 1s. 8d., we find, on subdividing the amount amongst the shareholders, that an eighth is £117 15s. 2½d., a sixteenth, £58 17s. 7d., and a thirty-second, £29 8s. 10d. Spasmodic adventurers, then, were not greatly enriched in one voyage, but it was far otherwise with firms limited to three or four persons, who traded regularly in human misery. For

* The writer we are following gives the two latter amounts as £2,361,455 6s. 1d. and £214,677 15s. 1d., but if we multiply 303,737 (the number of slaves sold) by £9 8s. 5d. (the profit on the sale of each slave) we find the aggregate profit of £2,861,455 13s. 1d., or an average of £260,132 6s. 8d. yearly, for the eleven years. Troughton, accepting his predecessor's figures (£214,677 15s. 1d.) as "a late calculation made with great accuracy," calls it "an influx of wealth which, perhaps, no consideration would induce a commercial community to relinquish."

† "A General and Descriptive History of Liverpool," (1795,) p. 230.

instance, we find one firm importing 2850 in five ships, which, calculated on the previous scale, produced a net profit of £26,849 7s. 6d., or if divided between four shareholders, the sum of £6712 6s. 10½d. for each adventurer. The great wealth accruing from the traffic, however, went into the coffers of ten leading houses, who maintained a regular routine of slavers. From a summary* of eleven carefully prepared tables, covering the eleven years in question, we find that although 359 firms sent out no less than 878 Guineamen, yet ten houses despatched 502 out of that number, which was not only more than one-half the shipping employed, but proves that the 502 vessels were of greater aggregate burthen, for the number of slaves imported by the ten firms was nearly four-sixths of the whole number imported in Liverpool Guineamen during the eleven years under notice. Although a Guinea voyage might exceed twelve months, the instances were comparatively few, and the ten leading houses aforesaid must be allowed to have had yearly regular returns, or uniform successive annual adventures, producing successive annual remittances of a highly satisfactory and soothing character to the shareholders. From a report presented to the Privy Council, while the Slave Bill was depending, it appears that the number of negroes transported yearly from Africa to the West Indies, from 1783 to 1793, amounted to 74,000. Of this number, Great Britain imported 38,000, Holland 4000, Portugal 10,000, Denmark 2000, and France 20,000. Of the immense multitude of 814,000 negroes conveyed from Africa to the West Indies in eleven years, Liverpool had the profit and the disgrace of conveying 303,737.†

We now leave estimates and pass on to incontestable facts, commencing with the ship *Lottery*, Captain John Whittle, belonging to Mr. Thomas Leyland, banker and

*See Appendix.

† Baines puts the number at 407,000, but our tables show this to be wrong.

millionaire, thrice mayor of Liverpool. She sailed from the Mersey on the 6th of July, 1798, arrived at Bonny on the 22nd August, passed Barbadoes on the 27th of November, after a passage of 50 days, with 460 negroes. The following details, showing the result of the voyage, as far as Mr. Leyland's pocket was concerned, are taken from the original account books :—

Net proceeds of 453 Negroes sold by Bogle & Jopp, as remitted by bills of various dates after payment of all charges		£22,726 1 0
Deduct :—		
Cost of ship's outfit ...	£2307 10 0	
Cost of the cargo sent out to Africa	8326 14 11	10,634 4 11
Profit on the voyage		£12,091 16 1

But this was not all ; there would be, probably, some of the Guinea cargo and stores left over for future use.

The *Lottery*, Captain Charles Kneal, also belonging to Messrs. Thos. Leyland & Co., sailed from Liverpool on the 21st of May, 1802, on her sixth voyage, and carried from Africa to Jamaica 305 negroes. The amount of the outfit and cargo was £7982 2s. 6½d. When a final balance was struck on October 31st, 1811, the profits stood thus :—

Thomas Leyland	½ bal^ce^.	£9510 16 0
R. Bullin	¼ ,, ...	4755 8 0¼
Thos. Molyneux	¼ ,, ...	4755 8 0¼
		£19,021 12 0½*

In the next case, we subjoin a copy of the instructions penned by Messrs. Leyland & Co., for the guidance of the captain. These are taken from the original account book,

* There were a few hundreds more made on returned goods, and on rum, sugar, &c.

which begins with the following memoranda of the voyage :—

"Ship *Enterprize*, 1st. Voyage.

"Sailed from Liverpool, 20 July 1803

"August 26th detained the Spanish Brig *St. Augustin*, Capt. Josef Ant°. de Ytuno, in Lat. 22, 47 North, Long. 26, 14 West; bound from Malaga to Vera Cruz, which vessel arrived at Hoylake on the 25th October.

"September 10th Recaptured the *John* of Liverpool in Lat. 4, 20 North, Long. 11, 10 West with 261 Slaves on board, and on the 2nd November she arrived at Dominica.

"September 23rd the *Enterprize* arrived at Bonny, and sailed from thence on the

"December 6th the *St. Augustin* sailed from Liverpool.

"9th January 1804 the *Enterprize* arrived at the Havanna and sold there 392 Negroes. On the 28 March she sailed from the Havannah and arrived at Liverpool 26 April 1804."

"LIVERPOOL, 18 *July* 1803

"CAP. CÆSAR LAWSON,

"SIR,—Our ship *Enterprize*, to the command of which you are appointed, being now ready for sea, you are immediately to proceed in her, and make the best of your way to Bonny on the Coast of Africa. You will receive herewith an invoice of the Cargo on board her which you are to barter at Bonny for prime Negroes, Ivory, and Palm Oil. By Law this vessel is allowed to carry 400 Negroes, and we request that they may all be males if possible to get them, at any rate buy as few females as in your power, because we look to a Spanish market for the disposal of your cargo, where Females are a very tedious sale. In the choice of the Negroes be very particular, select those that are well formed and strong; and do not buy any above 24 years of Age, as it may happen that you will have to go to Jamaica, where you know any exceeding that age would be liable to a Duty of £10 ⅌ head. While the slaves are on board the Ship allow them every indulgence Consistent with your own Safety, and do not suffer any of your officers or Crew to abuse or insult them in any respect. Perhaps you may be able to procure some Palm Oil on reasonable terms, which is likely to bear a great price here, we therefore wish you to purchase as much as you can with any spare cargo you may have. We have taken out Letters of Marque against the French and Batavian Republic, and if you are so fortunate as to fall in with and capture any of their vessels Send the Same direct to this Port, under the care of an active Prize Master, and a sufficient number of men out of your ship; and also put a Copy of the Commission on board her, but do not molest any neutral ship, as it woud involve us in expensive Lawsuit and subject us to heavy Damages. A considerable part of our property under your care will not be insured, and we earnestly desire you will keep a particular look out to avoid the Enemy's Cruisers, which are numerous and you may hourly expect to be attacked by some of them. We request you will Keep strict and regular discipline on board the ship; do not suffer Drunkenness among any of your Officers or Crew, for it is sure to be attended with some misfortune, such as Insurrection, Mutiny and Fire. Allow to the ship's Company their regular portion of Provisions &c and take every care of such as may get sick. You must keep the ship very clean and see that no part of her Stores and Materials are embezzled, neglected, or idly wasted. As soon as you have finished your trade and laid in a sufficient quantity of Yams, wood, water, and every other necessary for the Middle Passage, proceed with a press of sail for Barbadoes, and on your arrival there call on M[essrs.] Barton Higginson & C[o.] with whom you will find Letters from us by which you are to be govern'd in prosecuting the remainder of the voyage. Do not fail to write to us by every opportunity and always inclose a copy of your preceding Letter.

"You are to receive from the House in the West Indies, who may sell your cargo, your Coast Commission of £2 in £102 on the Gross Sales, and when this Sum with your Chief Mate's Privilege and your Surgeon's Privilege, Gratuity and head money are deducted, you are then to draw your Commission of £4 in £104 on the remaining amount. Your Chief

Mate, Mr. James Cowill, is to receive two Slaves on an average with the Cargo, less the Island and any other duty that may be due or payable thereon at the place where you may sell your Cargo; and your Surgeon, Mr. Gilbt. Sinclair, is to receive two Slaves on an average with the Cargo less the Duty beforementioned, and one Shilling Stg head money on each slave sold. And in consideration of the aforementioned Emoluments, neither you nor your Crew, nor any of them, are directly or indirectly to carry on any private Trade on your or their accounts under a forfeiture to us of the whole of your Commissions arising on this voyage. In case of your Death, your Chief Mate, Mr. Cowill, is to succeed to the Command of the ship, and diligently follow these and all our further orders. Any Prize that you may capture, direct the Prize Master to hoist a white flag at the fore and one at the main top Gallant Mast-heads, on his approach to this Port, which will be answered by a signal at the light House.

"We hope you will have a happy and prosperous voyage, and remain

"Sir, Your ob Servs

"P.S.—Shoud you capture any vessel from the Eastward of Cape of Good Hope, Send her to Falmouth and there wait for our orders. In case of your Capturing a Guineaman with Slaves on board, Send her to the address of Messrs. Bogle, Jopp & Co. of Kingston, Jamaica."

"I acknowledge to have received from Messrs. Thomas Leyland & Co. the Orders of which the aforegoing is a true Copy, and I engage to execute them as well as all their further orders, the Dangers of the Seas only excepted, as witness my hand this 18 July 1803

"CÆSAR LAWSON."

The outfit of the *Enterprize* cost £8148 18s. 8d.; her cargo of trading goods, £8896 3s. 9½d.; total, £17,045 2s. 5½d. In January, 1804, Captain Lawson delivered to Messrs. Joaquin Perez de Urria, at Havanna, 412 Eboe slaves (viz., 194 men, 32 men-boys, 66 boys, 42 women, 36 women-girls, and 42 girls) to be sold on account of Messrs. T. Leyland & Co. Nineteen of the slaves died, and one girl, being subject to fits, could not be disposed of. The net profit on the round voyage, after selling the 392 remaining slaves, paying damages for detaining the *St. Augustine*, and crediting salvage of the *John*, profit on teeth, logwood, sugar, etc., amounted to £24,430 8s. 11d., which was divided between the partners as follows :—

Thomas Leyland for his half of balance	£12,215	4	5½
Thomas Molyneux for his ¼ of Do	6,107	12	2½
Rd. Bullin for his ¼ of Do	6,107	12	3
	£24,430	8	11

Our next example is the first voyage of the slave-ship *Fortune*, Captain Charles Watt, which sailed from Liverpool on the 25th of April, 1805, arrived at Congo River 16th July, sailed thence 10th November, arrived at Nassau 21st of December, 1805, sailed thence 29th of March, 1806, and arrived at Liverpool on the 2nd of May, 1806. The result of the round voyage was as follows :—

Remittances &c on account of 343 slaves sold by Hy. and Jas. Wood, Nassau, New Providence, after payment of all expenses, ...				£13,271	0	1
This was apportioned as under :—						
Thomas Leyland for his ⅔ds. balance... ...	£8847	6	9			
Wm. Brown for his ⅓rd. balance	4423	13	4			
				£13,271	0	1

The cost of the outfit in this case was	£4124	18	9
,, ,, ,, cargo ,, ,, ,,	7267	18	7
	£11,392	17	4

The profit appears from the foregoing statement to be very trifling, £1878 2s. 9d., but we must add to it the sum of £7609 7s. 6d. for slaves sold on credit, making £9487 10s. 3d. Nevertheless, the expenses seem to have been remarkably heavy.

The slaves sold very slowly, there being 100 left on the factors' hands on July 31st, 1806, and the last batch of these was sold in September. The result was a big bill for rent of store, doctor's attendance, provisions, brandy, wine, tobacco, heads and offals, oil, etc., for the slaves. The muster-roll shows that Captain Watt, the third mate, and six seamen died on the voyage; two sailors were drowned, the fifth or trading mate, and one of the men ran away, while 34 seamen entered or were impressed on board his majesty's ships on the station. This refers to the original crew of 66 officers and men shipped at Liverpool.

The slave-ship *Louisa*, on her fourth voyage, having sold 326 negroes at Jamaica for the sum of £19,315 13s. 6d., the profit (after adding interest on account sales, £1051 19s. 7d., and deducting £1234 2s. 8d. for disbursements and commission, etc., due to factors) amounted to £19,133 10s 5d., which was apportioned among the owners as follows:—Thomas Leyland, £9566 15s. 2½; R. Bullin, £4783 7s. 7¼d.; Thomas Molyneux, £4783 7s. 7¼d.

In 1784, the *Bloom*, Robert Bostock, master, carried 307 slaves from the Windward Coast of Africa to the West Indies, on account of Messrs. Thomas Foxcroft & Co., merchants, Liverpool, who also owned the *Bud* and the *Pine*. The shares were held as follows:—Thomas Foxcroft, 5/16ths; Wm. Rice, 2/16ths; A. Wharton, 2/16ths; Felix Doran, 2/16ths; Jas. Welsh, 2/16ths; Robert Bostock, 2/16ths; and Geo. Welch, 1/16th.

The result of the voyage, as disclosed by the original account books, may be summarised thus:—

Sale of 307 slaves (103 men, 51 women, 99 boys, 54 girls), the lowest price being £21 for a woman and £40 each for man, a boy and 2 girls,				£9858	2	10
Charges :—						
To 6 slaves at the average £32 2s. 2½d. freight for Captain Bostock	£192	13	3			
To 1 slave for James Oddie the second mate, sold by desire of Captain B.	30	0	0			
To Captain Bostock's privilege on £9635 9s. 7d. at 2 per cent.	192	14	2			
Ditto on extra privilege one average	32	2	2½			
To Mr Wm Cockerill's one privilege	32	2	2½			
To Capt. Bostock's coast commission on £9378 11s. @ 4 per 104	360	14	3			
To Factors' (Taylor & Kerr), Commission on do 5% ...	465	1	8			
To Mr John McCulloch, surgeon, for head money on 300 slaves @ 1 currency is £15 @ 82½ per cent.	8	4	4			
To store rent, advertising, liquor, &c.	15	0	0			
To sugar, rum, &c., shipped and cash advanced	918	4	8			
,, Drafts at 15 mos' sight ...	2401	14	2			
,, ,, ,, 18 ,, ,, ...	2401	14	2			
,, ,, ,, 21 ,, ,, ...	2401	14	2			
,, Factors' Commission on remittances, &c., 5%	405	3	7			
				£9858	2	10

The appended table shows at a glance the highly satisfactory result of the six voyages—to all but the negroes. The Chief Accountant of the oppressed may possibly have pigeon-holed another table, compiled on the basis of our fifth chapter.

SUMMARY.

Ships.	Slaves sold.	Net profits.			Average profit per slave.			Owners.
		£	s.	d.	£	s.	d.	
Lottery	453	12,091	16	1	26	13	10	T. Leyland & Co.
Lottery	305	19,021	12	0½	62	7	4	Do.
Enterprize	392	24,430	8	11	62	6	6	Do.
Fortune	343	9,487	10	3	27	13	2	Do.
Louisa	326	19,133	10	5	58	13	10	Do.
Bloom	307	8,123	7	2	26	9	2	T. Foxcroft & Co.
	2126	£92,288	4	10½	£43	8	3	

The net profits shown in the third column are the amounts actually divided between the partners after payment of all expenses incurred on the round voyage, and the sums, of course, include the profit made on the rum, sugar, and other commodities sent home in payment for the slaves. The proceeds were sometimes brought home in dollars.

If we multiply the number of slaves imported in Liverpool ships in the eleven years (from 1783 to 1793) namely, 303,737, by £43, we have a total of £13,060,691 or an annual net profit of £1,187,335 11s. od. to the merchants of Liverpool. The difference between these amounts, which are based on facts, and those in the estimate (*viz.*, £12,294,116 aggregate net proceeds, and £1,117,647 yearly net profits) is accounted for by the fact that the price of slaves had risen considerably in the later years of the trade, when Mr. Leyland's ships were engaged in it. It does not appear that the compiler of the estimate allowed for the profits on the sugar, rum, etc. However, his figures, so far, appear reasonable when tested by realities. But when he attempts to estimate the net gains on a Guinea voyage, putting the profit at £9 8s. 5d. per slave, we part company with him and appeal to the foregoing table of facts, which shows that, even when the ship returned to Liverpool without a cargo of West Indian produce worth naming, as in the case of the *Bloom*, the profit on each slave imported was £26.

According to an extract from the books of the Liverpool Custom-house, supplied to Mr. Elliot Arthy, a master mariner and surgeon engaged in the slave trade, it appears that between the 5th of January, 1798, and the 5th of January, 1799, there sailed for Africa from the port of Liverpool, 150 vessels,* whose total tonnage was 31,533 tons, their complement of slaves, as allowed by Act of Parliament, 52,557, and the total of their complement of seamen, as required by law, 5,255. Mr. Arthy made a series of elaborate calculations, and arrived at the conclusion that the merchants made a clear profit of £3850 per vessel, or £577,535 in the year. He estimated the net remittances at £2,511,535, and the freight made by the ships from the West Indies (at £800 a ship) £120,000, total £2,631,535. Against this he computed the probable cost of repairing, outfitting, storing, victualling, goods for purchasing slaves, seamen's wages, insurance, etc., at £2,054,000, leaving a balance, as above, of £577,535. This only shows a profit of about £12 per head on the 47,500 slaves, whom he assumes to have survived the passage. In his desire to show the collateral benefits flowing from the trade, he appears to have exaggerated the cost of repairs, insurance, victualling, etc. The cost of a slave on the coast was from £20 to £25, and the average price in the West Indies in 1798-99, trade being very brisk, was £70. This clearly left a margin for a handsome profit, as shown in our summary of six voyages.

* See Appendix for names, owners, commanders, etc.

CHAPTER VII.

THE CORPORATION AND THE SLAVE TRADE.

"To abolish that trade would be to 'Shut the gates of mercy on mankind.'"—*James Boswell.* (*Life of Dr. Johnson.*)

THE presentation of a petition to the House of Commons, in 1787, signed principally by members of the Society of Friends, praying for the suppression of the Slave Trade, and the formation of the Anti-slavery Society, greatly alarmed the Liverpool Common Council. On the 14th of February, 1788, during the mayoralty of Mr. Thomas Earle, the Council met, and adopted a petition to Parliament, drawn up by Mr. Statham, against the abolition of the trade. The petition, which is cunningly framed to propitiate the Government, to implicate the Commons for having encouraged the corporation in its outlay on wet docks for the African ships, and to alarm the landed interest and the capitalists, is as follows:—

"To the honourable the House of Commons, &c. The humble petition of the Mayor, &c., sheweth : That your petitioners, as trustees of the corporate fund of the ancient and loyal town of Liverpool, have always been ready, not only to give every encouragement in their power to the commercial interests of that part of the community more immediately under their care, but as much as possible to strengthen the reins of government, and to promote the public welfare. That the trade of Liverpool, having met with the countenance of this honourable house in many Acts of Parliament, which have been granted at different times during the present century, for the constructing of proper and convenient wet docks for shipping, and more especially for the African ships, which, from their form, require to be constantly afloat, your petitioners have been emboldened to lay out considerable sums of money, and to pledge their corporate seal for other sums to a very large amount for effectuating these good and laudable purposes. That your petitioners have also been happy to see the great increase and different resources of trade which has flowed in upon their town by the numerous canals and other communications from the interior parts of this kingdom, in which many individuals, as well as public bodies of proprietors, are materially interested. And that from these causes, particularly the convenience of the docks, and some other local advantages, added to the enterprising spirit of the people, which has enabled them to carry on the African slave trade with vigour, the town of Liverpool has arrived at a pitch of mercantile consequence which cannot but affect and improve the wealth and prosperity of the kingdom at large.

"Your petitioners therefore contemplate with real concern the attempts now making by the petitions lately preferred to your honourable house to obtain a total abolition of the African slave trade, which has hitherto received the sanction of Parliament, and for a long series of years has constituted and still continues to form a very extensive branch of the commerce of Liverpool, and in effect gives strength and energy to the whole; but confiding in the wisdom and justice of the British senate, your petitioners humbly pray to be heard by their Counsel against the abolition of this source of wealth before the honourable house shall proceed to determine upon a point which so essentially concerns the welfare of the town and port of Liverpool in particular, and the landed interest of the kingdom in general, and which, in their judgment, must also tend to the prejudice of the British manufacturers, must ruin the property of the English merchants in the West Indies, diminish the public revenue, and impair the maritime strength of Great Britain. And your petitioners will ever pray, &c."

On the 4th of June, 1788, the Council ordered that the freedom of the borough be granted to Messrs. John Tarleton, Robert Norris, James Penny, John Matthews, and Archibald Dalzell, who had been deputed by the Committee of the Liverpool African merchants to attend in London on the business, "for the very essential advantages derived to the trade of Liverpool from their evidence in support of the African slave trade, and for the public spirit they have manifested on this occasion." On the 20th of the same month, the freedom of the borough was presented to Lord Hawkesbury, chiefly for his support of the slave trade, and in May, 1796, when he was created Earl of Liverpool, the Corporation invited him to quarter the arms of Liverpool with his own, which was done to the great honour and edification of all concerned. Lord Liverpool was Prime Minister from 1812 to 1827.

In the year 1788, Wilberforce began his agitation to obtain freedom for the slaves in our West Indian Colonies; and a bill was passed for the better regulation of slave-ships. This measure immediately roused the hostility of the Corporation of Liverpool, who, on the 20th of June, 1788, petitioned the House of Lords to throw out the bill, or to allow them or their counsel to be heard against it. They stated "that the trade had been legally and uninterruptedly carried on for centuries past by many of his Majesty's subjects, with advantages to the country, both important and extensive; but had lately been unjustly reprobated as impolitic and inhuman."

The spectacle of the corporation, the members of which must have been perfectly well acquainted with the horrors of the slave trade, appealing to the House of Lords to uphold the infamy of the town, is a melancholy, but striking example of the power of usage and self-interest in blunting the moral vision of men otherwise distinguished for many excellent and even noble qualities. In judging them we must not forget that in our own day there are commercial practices and walks of trade that may call for the indulgent criticisms of posterity.

In April, 1789, the Corporation presented to the House of Commons a petition verbatim with that of the previous year, and on July 1st, another, similar in most of its statements, and praying that the further inquiry and the examination of witnesses might be postponed for another year. On the 2nd of December, 1789, the thanks of the Council were presented to Messrs. Norris and Penny for their diligent attendance on the House, and otherwise respecting the business of the African Slave Trade Bill. In the same year, through the efforts of Wilberforce, Fox, and Burke, resolutions condemning the slave trade were introduced in the House of Commons.

From its immense importance to the town, the slave trade at election times naturally acted as the touchstone or Ithuriel's sword, by which the true member was discovered and elected—provided his purse was equal to the value which the freemen set upon their votes. In the election of 1790, great credit was given to the old members, Mr. Bamber Gascoyne and Lord Penrhyn, for having, "in the late violent attempt to abolish the supply of the West India Islands with labourers from Africa, given the most convincing proofs of superior abilities, unremitted attention, and invincible perseverance." "Was not the African Trade in danger?" asked an admirer, signing himself "Common Sense." "Was not Mr. Pitt, the minister, against it? Was not Mr. Fox, the leader of the Opposition against it? Was not the House of Commons against it? Was not the whole nation against it? Who was there to stand up for it but Lord Penrhyn and Mr. Gascoyne? How then can any man be so ungrateful as to give his vote against them?"

The idea of the two Liverpool members successfully

opposing the whole nation in defence of the favourite traffic of their constituents, reminds us of the *Skibbereen Eagle* keeping a restraining eye on the Emperor of Russia, or Dame Partington mopping out the Atlantic. The Silas Wegg of the Gascoyne party, dropping into poetry, rose to the height of his great argument in the following fashion:—

> "Be true to the man who stood true to his trust,
> Remember our sad situation we must;
> When our African business was near at an end,
> Remember, my lads, 'twas Gascoyne was our friend.
>
> If our slave trade had gone, there's an end to our lives,
> Beggars all we must be, our children and wives;
> No ships from our ports their proud sails e'er would spread,
> And our streets grown with grass, where the cows might be fed."

The Corporation did not drop into poetry, but they presented their thanks to Colonel Gascoyne, for his general attention to the interests of the port, and particularly for his unwearied exertions on behalf of the African slave trade.

In 1791, Wilberforce continued the Anti-Slavery agitation, which, in the following year, was very violently opposed by H.R.H. the Duke of Clarence (afterwards William IV.).

At a special council, on the 24th of May, 1792, Mr. Henry Blundell being mayor, another petition against interference with the pet trade, was approved, and John Barnes, Richard Miles, and Peter W. Brancker, Esquires, were desired to wait upon H.R.H. the Duke of Clarence, and request that he would present the same to the House of Lords. On the 5th of the following December, when Mr. Clayton Tarleton was mayor, the council recognised the services rendered to the slave trade of the town by the late Mr. Robert Norris, Mr. James Penny, and Mr. Samuel Green, by granting to the widow of the former, an annuity of £100 for life, a piece of plate of the value of £100 to Mr. Penny, and the sum of

£300 to the widow of the late Mr. Green, and £117 for his public services and expenses disbursed. Messrs. John Barnes and P. W. Brancker were sent to join the delegates from various ports, who sat in London directing the opposition to the Abolition Bill.

In 1792 (the year preceding the Revolutionary War), the number of Liverpool ships engaged in the African trade was 136, the tonnage 24,544, or about a twelfth part of the tonnage which entered the port.

Not only in public but also in private did Roscoe use his great influence on behalf of the despised and down-trodden negro, as will be seen from the following letter, addressed at this period, to a slave-captain:—

"CAPTAIN WM. LACE, of the ship *Joshua*, Angola.
"Per CAPTAIN EVANS, ship *Mary*.

"DEAR WILLIAM,

"As I mist the opportunity before you sailed, I take the first occasion of reminding you that I shall think myself much obliged by your bringing me a small quantity of such seeds of African or West Indian plants as may conveniently fall in your way—or if you can employ any person to collect them on the coast, I will pay the expense attending it. As to plants growing, I fear it wou'd not be possible to preserve them, and wou'd be attended with much trouble; but if any bulbous (or onion-like) roots cou'd be obtained, they would probably keep so as to grow in a hot-house here on their arrival. Both the seeds and roots should be preserved from wet, which is all that will be necessary.

"I cannot omit this opportunity of expressing my hearty wishes for your return in safety and health to your friends, and I am sure you will excuse me, if I remind you that the employment you are now intrusted with is very weighty and important. To have the unlimited direction and controul of several hundreds of people who are to rely upon your care and management for their protection and support, places you in a situation of great responsibility, not only to your owners, but to the poor creatures committed to your charge, and to your own conscience. That you will discharge this serious duty with fidelity, and with as much humanity as is consistent with the nature of this business, I make no doubt. I have observed, with pleasure, that your natural disposition is kind and liberal, and you can never have a fitter opportunity of exerting these qualities than your present situation affords. I need not, I am sure, remark that any warmth or hastiness of temper (which, if ever you had it, is, I think, now well corrected by experience) might be productive of consequences which you might ever have to repent. Coolness, vigilance, compassion, attention to the necessities of all under your charge are essential requisites. Let these never be forgotten, and let the poor imprisoned African find that in all his distresses he is not without a friend.

"May God bless you and all under your care, whatever may be their complexion, and believe me, my dear friend, ever affectionately yours,

"W. ROSCOE.

"LIVERPOOL, 12*th July*, 1792."

Captain William Lace was the son of Mr. Ambrose Lace, merchant and ship owner, of St. Paul's Square, and brother of Mr. Joshua Lace, the founder and first president of the Liverpool Law Society. He had a life full of adventure, for in the time of the war with France he fitted out privateers, and took the command of one himself. After taking many prizes, he was himself captured by the French fleet, and carried a prisoner to France, from which country he afterwards escaped, after enduring great hardships. On another voyage he lost his ship, and was 14 days in a small boat, part of this time without water, and, when picked up, was one of the few survivors. He was one of the early African explorers, and, we believe, the first to give us an account of the gorilla, long before Du Chaillu. He was an enthusiastic botanist, and largely contributed to the founding of the Liverpool Botanical Gardens, the freedom of which was presented to him in recognition of his gifts. Members of his family repeatedly refused the office of Mayor, and the last Bailiff of Liverpool was his cousin, Ambrose Lace. One of our illustrations is a fac-simile of an original sketch made by Captain Lace of the palace and stockade of an African king, of whom he purchased slaves; and the private signal code of a slave-ship is reproduced from the original in his handwriting.

In May, 1794, the *Bolton*, Captain Lee, arrived at Dominica from Africa, after a passage of thirty-four days, with a remarkably healthy cargo of negroes, having left the coast with her full complement, and buried only one.

The continuation of the slave trade till the first of January, 1796, was carried in the House of Commons by a majority of 19. "This decision signed," as Mr. Fox justly observed, "the death-warrant of perhaps a hundred thousand of our fellow-creatures, or more, and doomed an unknown number to perpetual slavery, with their seed, and extended misery to a still greater number of their relatives in Africa, who were left to mourn their parents or children, husbands or wives, torn from them by a merciless banditti, to satiate the unbounded cravings of British avarice." A similar motion in the Lords was postponed to the following year, in order to give time for the examination of witnesses.

Not content with importing from Africa a supply of slaves for our own plantations, the Liverpool merchants were induced by love of gain to perform the same work for some of the neighbouring kingdoms. Knowing that the time and opportunity of making such gain was now limited, they used redoubled exertions to procure as many of the natives as possible. Secretary Dundas presented a petition for a Mr. Dawson, of Liverpool, stating that he had eighteen vessels in the slave trade, for the service of Spain, and that the whole of the property embarked altogether in it was five hundred and nine thousand pounds and upwards.

At a special council, held on the 12th of March, 1796, during the mayoralty of Mr. Thomas Naylor, it was unanimously agreed that petitions be sent up on behalf of the Corporation against the Bills before Parliament for the abolition of the African slave trade, and praying to be heard by counsel. The petitions were merely an echo of those previously presented. Among the toasts drunk at a gathering of the friends of John Tarleton, Esq., met at the "King's Arms," in 1796, to celebrate the anniversary of the King's birthday, was "Prosperity to the African Trade, and may it always be conducted with Humanity."

The largest vessel at this time engaged in the African trade was the *Parr,* of 566 tons burthen, launched from Mr. W. N. Wright's yard, in November, 1797.

On the 3rd of October, 1798, during the mayoralty of Mr. Thomas Staniforth, the thanks of the Town Council, and a piece of plate to the value of 100 guineas, were presented to Mr. Peter Whitfield Brancker, a member of the Council, for having, in his character as delegate, attending in London every session of Parliament, been very instrumental in securing a continuance of the slave trade under proper restrictions and regulations. Mr. P. W. Brancker was bailiff in 1795, and mayor in 1801. In 1803, at a period of national danger, all the boatmen of the river Mersey were formed into a regiment of artillery under his command, John Brancker being one of the captains. Somewhat blunt and bluff of bearing, he was a true-hearted man of the old school, and far before most of the merchant princes of that day in reading and intellectual attainments.

On the 1st of April, 1799, during the mayoralty of Mr. Thomas Leyland, the corporation petitioned the Commons against "a bill to prohibit the trading for slaves to the coast of Africa within certain limits," characterising it as impracticable in parts, injurious, partial, and oppressive, and so forth. On May 1st, 1799, another bill introduced into the Lords for regulating the shipping and carrying of slaves in British vessels from the coast of Africa, also drew a petition from the Town Council, who held that the health and comfort of the slaves had been already effectually secured.

On the 14th of October, 1799, the Recorder, and a Committee of the Council, attended at St. James's Palace, and presented H.R.H. the Duke of Clarence (afterwards the "Sailor King") with the freedom of the borough, in a gold box (costing £226) with an address (illuminated for 25 guineas) "in grateful sense of his active and able exertions in Parliament" on behalf of the slave trade. After all, the expenses of the deputation, together with the presents, only amounted to the price of two or three "prime and healthy negroes"—a reasonable return for royal eloquence and support.

In the year 1802, the question of the slave trade appears to have been too stale for effective electioneering treatment. The only reference to it is in a stanza by one of General Gascoyne's admirers :

> "For if he your member be, my boys,
> Provisions still must lower;
> And open trade be carried on
> Along the Afric shore.
> And a plumping we will go."

In the year 1804, a bill for the abolition of the slave trade, was carried by Wilberforce in the House of Commons, but was thrown out by the Lords. In the next session a similar bill was rejected by the Commons. The capture by Great Britain of the French and Dutch Colonies in the West Indies, increased the demand for slaves, which had been diminishing. The number of slaves imported in Liverpool, London and Bristol ships, in the year 1802, was 41,086; in 1803, the number had fallen to 24,925; and in 1804, it had reached 36,899. Out of this number the proportion carried by Liverpool was as follows: In 1802, 122 vessels, of 30,796 tons burthen, carried 31,371 slaves; in 1803, 83 vessels, of 15,534 tons burthen, carried 29,954 slaves; in 1804, 126 vessels, of 27,322 tons burthen, carried 31,090 slaves.*

In 1805, an order in council prohibited the importation of negroes, to the newly conquered colonies of the British crown. After the death of Mr. Pitt, in 1806, the coalition ministry, under Lord Grenville and Mr. Fox, carried a bill prohibiting British subjects from supplying slaves either to foreign settlements or to our own colonies. One of the last acts of Mr. Fox before he followed his great rival to the grave, was to carry a resolution in the Commons, pledging the House to the abolition of the trade in the next session. Meanwhile, a bill was passed through both Houses, forbidding the employment of any new vessel in the trade.

On the 20th of January, 1807, during the mayoralty of Mr. Thomas Molyneux, petitions to the Lords were adopted from the Corporation and the Dock Trustees, praying that the Abolition Bill be not passed, "but if from considerations foreign to their interests it should be thought expedient that the Bill should pass," the petitioners prayed for compensation for the depreciated value of houses, warehouses, land, etc.

Parliament was dissolved in November, 1806, and in its short-lived successor, called by the Grenville Ministry, the great Roscoe, then in the zenith of his fame, sat as one of the members for Liverpool. His political opinions, and his enmity to the slave trade, were opposed to the views of the majority in a constituency where the most shameless bribery

*Sir James Picton, in his "Memorials of Liverpool," vol I. p 277, has fallen into the curious error of giving the tonnage of the ships as the number of slaves carried, the passage being as follows:—

"The trade had previously been diminishing, the number imported in Liverpool ships having dropped from 30,796 in 1802, to 15,534 in 1803. Stimulated by the new colonial markets, the number in 1804, had risen to 27,322, being five-sixths of the whole number imported."

prevailed. Nevertheless, his high standing in the commercial,* literary, and political world, and the unsullied excellence of his private character, induced the burgesses to return him at the head of the poll. There was something of "the everlasting fitness of things" in the presence, in this parliament, of the man, whose youthful genius had sung the wrongs of Africa; who, in early manhood had confuted the sophistries of the Jesuit Harris, and who had, from first to last, shared the hopes and fears of Clarkson and Wilberforce, in their long and arduous struggle with the monster evil, against which, single-handed, he had stood forth, like young David of old. Short as was the parliament summoned by the "ministry of all the talents," it covered itself and them with imperishable glory, by finally declaring the slave trade illegal, and Mr. Roscoe had the gratification of contributing to this result by a speech, delivered on the second reading of the Bill, which received the royal assent on March 25th, 1807. The Bill enacted that no vessel should clear out for slaves from any port within the British dominions after the 1st of May, 1807, and that no slave should be landed in the Colonies after the 1st of March, 1808. Thus was ended a conflict of twenty years between truth and falsehood, justice and selfishness, humanity and cruelty; and the foulest blot which ever darkened the name of England was removed. During the last fifteen months of the trade, from January 1st, 1806, to May 1st, 1807, the number of Liverpool vessels engaged in the traffic was 185, measuring 43,755 tons,† and allowed to carry 49,213 slaves. The immediate effect of the Bill upon the commerce of Liverpool was injurious. The tonnage fell from 662,309 in

* Mr. Roscoe was connected with one of the first banks in Liverpool, of which Mr. Thomas Leyland, afterwards the richest man and most skilful banker in the town, was the head. The firm of Leyland, Clarke and Roscoe was dissolved on December 31st, 1806, the partners being Thomas Leyland, John Clarke and William Roscoe.

† Here again Sir James Picton has substituted the tonnage of the ships for the number of slaves carried.

1807 to 516,836 in 1808, and the amount of the dues from £62,831 to £40,638. This was principally owing to the general anxiety to "make hay while the sun shines," which swelled the tonnage of 1807, and the depression was not lasting, for the tonnage in 1810 had risen to 734,391, and the dues to £65,782.

When it became known in Liverpool that Parliament had decreed that England should no longer play a guilty part in perpetuating the horrors of the middle passage, prophets of woe and evil sprung up in every street, and with the exception of the small band of abolitionists and two or three shrewd land speculators, who afterwards reaped a great reward, the whole community was terror-stricken.* The docks were to become fish-ponds, the warehouses to moulder into ruins, grass was to grow on the local Rialto, the streets were to be ploughed up, "Bootle organs"† were to sing in the deserted mansions and pleasure grounds of the merchant princes, and Liverpool's glorious merchant navy, whose keels penetrated to every land, and whose white sails wooed the breeze on every ocean, was to dwindle into a fishing vessel or two, while the brave tars, who had made themselves the terror of England's enemies on the seas, were to die of starvation or in the workhouse. "And what became of Liverpool? Were the melancholy predictions of her prophets fulfilled? Were her docks turned into fish-ponds? Did the mower cut down hay, or the reaper gather in his harvest in her deserted streets?" Without entering at length into the fascinating romance of Liverpool's progress it is a sufficient answer to quote the following figures, indicating the enormous growth of her shipping trade:—

In 1764 the total tonnage of vessels that entered the port

*"The effect of the abolition of the slave trade began to be felt in the cessation of the demand for common rum, for which the coast of Africa was the principal vent; and also of the demand for all kinds of goods suited for the African market, such as gunpowder, coarse cloth, muskets, and trinkets of all kinds."—Baines' "History of Liverpool," p. 732.

† Frogs.

was 56,499 tons, in 1780 it was 112,000 tons, in 1796 it was 224,000 tons, in 1811 it was 611,190 tons, in 1827 it was 1,225,313 tons, in 1841 it was 2,425,461 tons, in 1857 it had reached 4,645,362 tons, so that by the same rule that doubled the tonnage of the port between 1749 and 1764, the tonnage doubled itself between 1841 and 1857. It occupied 134 years to produce an increase equal to that which had taken place between 1841 and 1857. In the year ending July 1st, 1897, the tonnage had reached 11,473,421 tons. The value of exports in the whole kingdom in 1857 amounted to £110,000,000 sterling, out of which £55,000,000 passed through Liverpool alone. Of the total exports of the United Kingdom in 1896, amounting to £296,379,214, those of Liverpool were valued at £93,298,954; those of London at £83,227,874. The total value of the imports of foreign and colonial produce into the United Kingdom in 1896 was £441,808,904, of which £146,852,558 was the value of London imports, and £103,512,255 those of Liverpool. This enormous growth of commerce could never have taken place, but for the continuous vigilance and enterprise of the inhabitants of Liverpool in the construction and extension of the dock system. Up to 1715, floating docks were unknown in England; and in 1795, the Liverpool docks were only about 1½ miles in extent.*

*In a Diary of a Tour through Great Britain, in 1795, by the Rev. Wm. MacRitchie, of Clunie, we find some interesting references to Liverpool. Passing through Ormskirk, on the 6th of July, he saw "large fields of potatoes, very well dressed, and country girls, with their petticoats tucked up, bestriding the drills and taking out every weed with their hands." He approached Liverpool from the north-west, and says, "Vast number of ships under sail, making their way out of the river. Put up at the 'Cross Keys,' near the Exchange, where dine; after dinner call upon Mr. Keay, and take the grace-drink with him. In the evening, Mr. Keay accompanies me out, and shows me the docks and the shipping. This infinitely the most wonderful scene of the kind I have ever seen; and one who has not seen it cannot possibly conceive any idea of it. Sup at the 'Cross Keys' (Mrs. Walker) with a number of travelling gentlemen; some of them very entertaining; Welch, Irish, English, Scotch, American, West Indies—variety of characters." . . "Visit again the greatest thing to be seen here, or perhaps anywhere else—the Docks. Storehouses, the largest of any in Britain—particularly the Duke of Bridgewater's, etc. One gentleman here has storehouses eleven stories high. Bathinghouses, ladies' and gentlemen's; coffee-rooms; vast number of windmills for grinding corn, flint for the potteries,

In 1897, the Liverpool dock systems fringe the estuary on the Lancashire shore for nearly seven miles, and penetrate the Wirral Peninsula on the Cheshire side of the Mersey for two or three miles. The Liverpool docks are unrivalled, not only by reason of their extent, but for the solidity and magnificence of their construction, and the facilities which they offer for the quick handling of cargo. The Dock Estate on the Liverpool side of the estuary contains an area of 1105 acres, with 25 miles 1679 yards of quays. On the Birkenhead side, the area is 506 acres, the length of quay space 9 miles 925 yards, making the total area of the Dock Estate, 1611 acres with 35 miles 844 yards of lineal quay space.

The total cost of the docks is estimated at £42,000,000. Of this sum, £19,000,000 was expended during the thirty-five years from 1861 to 1896. A large portion of the money spent upon the estate has been defrayed out of the revenue, and the present (July, 1897) bonded debt is £18,166,583. The annual income of the estate is (July, 1897) £1,400,152. The whole of this magnificent property is a public trust, under the management of the Mersey Docks and Harbour Board, which was instituted in 1857.

In 1897, the largest ocean steamers are able to come alongside the Liverpool Landing Stage, which is moored in about the centre of the seven-mile frontage of the finest

flax-seed for oil, logwood, etc." . . "The docks extend more than one and a half miles, and exceed all description. This war, however, has considerably affected the trade of Liverpool. Harbour difficult of access, the tract in the river narrow, and many sandbanks on each side; pilots necessary." . . "Walk out again to the Docks. The Glasshouse here upon a small scale. The Copper work discontinued here; removed to Wales on account of the nearness of the ore there. Number of the best ships belonging to this place taken during the present war. Ships of upwards of a thousand tons built here. 'An endless grove of masts!' It gives one a very high idea indeed of the immense trade of Liverpool, supposed superior to that of Bristol, and inferior only to that of London." . . . "Thursday 9th July.—Breakfast at the 'Cross Keys.' After breakfast make my escape from this large, irregular, busy, opulent, corrupted town; where so many men and so many women use so many ways and means of gaining and spending so much money, and meat, and drink, etc." And so he passes on to dine at Boldheath, and quaff good ale, at threepence a pint, and to muse on the extravagance and wickedness of Liverpool.

docks in the world. Gigantic and mysterious looking dredgers—which would have struck the bravest of our privateersmen with terror, had they fallen in with them on a cruise—have scooped away the shallow bar, which long obstructed the entrance to the port, and now 25 feet is the least depth at low water, and the largest vessels have free access at all states of the tide. In this the Mersey Docks and Harbour Board have displayed the pluck and enterprise of the early makers of Liverpool. By sending out their fleet of powerful dredgers to cruise successfully against the enemy which, in former ages, destroyed the port of Chester, the Board have confounded many sceptical engineers, and deserved the thanks of present and future citizens of Greater Liverpool. "England, England," said the Marquis of Halifax, "thou art like Martha, busy about many things, but one is necessary for thy salvation—look to thy moat; of an Englishman's creed, the first article is the sea." This true saying is also good for Liverpool, and the Docks Board have lived up to it. They, too, have had their battles to fight against fearful odds, and, figuratively speaking, to face the great guns, swivels, small arms, and stinkpots of innumerable enemies and critics, and still the good ship *Liverpool* sweeps the seas, sound in every plank, her officers and crew in great spirits, and the red flag of "no surrender" flying at the main, as in the days gone by.

There are many other indications that, far from having reached the meridian height of her glory and prosperity, Liverpool, like that dark continent with the sad history of which her own darkest record is linked, is but on the threshold of a more splendid future. In this present year of Jubilee, about 30,000 natives went down from the hinterland into the town of Lagos, and were conducted by the governor, Major M'Callum, down to the beach. Scarcely one of them had ever before beheld the sea, and their countenances were a perfect picture as they gazed at the endless expanse. It was not the vastness of the area that impressed them most, but the unceasing roll of the waves, which they could not understand. That great sea and its ceaseless roll is to us typical of that mighty civilising power which is now advancing over Africa. In this honourable work, the merchants of Liverpool compete against the world as vigorously and successfully as their predecessors did in an iniquitous traffic.

When Governor M'Callum, on the occasion previously referred to, conducted the two leading black kings to the Durbar, one on each arm, the enthusiasm of the natives was tremendous. The chiefs said they never knew the white man as they knew him then, and the friendship expressed by them before they left for their homes, showed that the jubilee celebration, at Lagos, was an unparalleled event in the history of the colony. It would appear from this that it is only now, ninety years after the abolition of the Liverpool slave trade, that the white man is learning the true way of dealing with the African. Soon may he

> "View the accomplish'd plan,
> The negro towering to the height of man."

The Act of 1807 had not the effect of stopping the importation of slaves, which continued to be carried on by British subjects under the cover of foreign flags. Consequently, in 1811, another Act was passed, which made such importation felony, punishable with fourteen years' transportation. In 1824, the trade was declared to be piracy, subject to the penalty of death. In 1833, the slaves in the British Colonies, to the number of 770,000, were emancipated, subject to an apprenticeship, which expired in 1838, the sum of £20,000,000 being paid to their owners as compensation. The part played by Liverpool in the agitation which brought about this result, does not fall within the scope of the present work.

CHAPTER VIII.

CAPTAIN HUGH CROW.

CAPTAIN HUGH CROW commanded the last slave-ship that cleared out of the port of Liverpool. He was one of the bravest, shrewdest, quaintest, and most humorous old sea dogs that ever breathed. He lost his right eye when very young, but as one of his employers said, the other was "a piercer," and he was known far and wide as "mind your eye, Crow." He was generally on the most friendly footing with himself, and ready to uphold his own merits, the beauty of his native Isle of Man, and the Manx language, wherever he went. He was justly proud of the estimation in which he was held by the merchants and underwriters of Liverpool and London, and it was a great sight to see him nodding and bowing with much urbanity when he met them in the streets or on 'Change.

We have it on the high authority of Mr. Hall Caine, that the Manxman is a born sailor, and Hugh Crow heard the call of the sea very early. When a mere child, walking with his mother on the shore at Ramsey, where he was born, in 1765, he prophesied to her that he would command a big ship some day. When that prophecy was fulfilled he did not forget to send the old lady substantial tokens of his love and success. Soon after being bound apprentice to the sea, in the employ of a Whitehaven merchant, he had to fight for his life with a vindictive fellow-apprentice, who, on a dark night, attempted to throw him from the maintop-gallant yard into the sea. He saved himself with difficulty, forgave the attempted murder, and held his tongue.

His early life was full of adventure. On one occasion, he left his ship at night, taking with him his quadrant and chest, having procured a situation as second mate of a fine ship, bound to Honduras. His old captain, suspecting his intentions, and anxious to retain so valuable a man, discovered his retreat, and, attended by bailiff, constables, and soldiers, boarded his new ship. After a scuffle with the crew, who also desired to retain their new acquisition, the captain, the law, and the army discovered Crow in the pump-well, nearly suffocated with filth and heat. When he was dragged upon deck, the captain threatened to cleave him with the cook's axe if he made any resistance. He was tightly handcuffed, bundled into a boat, with only his shirt and trousers on, taken on shore, and thrown into a noxious prison, amongst a number of dirty, runaway negroes. "There I lay," he says, "without any food, and tormented by rats, for forty-eight hours. It is but a grateful acknowledgment on my part to state that many of the poor negroes shed tears on seeing my distressed situation."

In December, 1787, he sailed as a passenger on board a ship bound from Cork to Kingston, Jamaica, paying one penny, as was the custom, otherwise a sailor (though a passenger) might claim wages. They met with a succession of dreadful gales, which greatly disheartened the crew, and Hugh Crow endeavoured to rally their drooping spirits on many a stormy night by singing sea songs, and especially "Ye Gentlemen of England," which he always found to have an animating effect on his shipmates on dark and stormy nights.

He had several offers to go as second mate to the coast of Africa, but, like many other sailors, he was prejudiced against the Guinea trade, and had an abhorrence of the very name of "slaves," "never thinking," he observes, "that at the time I was as great a slave as well might be; and I agreed, though with fewer advantages, to embark as second mate of the *Elizabeth*, the first ship bound to Jamaica." *

He tells the following anecdote, which we quote, because it shows the soft side of a strong character that knew no fear. It relates to a black boy they had on board named "Fine Bone," about fifteen years of age:—

> "When we got further north," says Crow, "the cold began to pinch him severely, and, being very fond of me, he one morning came shivering to the side of my cot, and said: 'Massa Crow, something bite me too much, and me no can see 'im, and me want you for give me some was mouth, and two mouth tacken.' I knew that 'wash mouth' meant a dram, and he soon gave me to understand, by getting hold of my drawers, what he meant by 'two mouth tacken.' I furnished the poor fellow with the needful, and as he had shoes, stockings, and jacket before, he was quite made up."

His repugnance to the slave trade was at length overcome. In 1790, he made his first voyage to Africa in the *Prince*, belonging to Mr. J. Dawson, and afterwards he sailed in one of Mr. Harper's ships, and in the *Jane* belonging to Mr. Boats, as second mate, which was equal to chief mate in any other employ. In June, 1794, he sailed as chief mate of the *Gregson*, a fine ship of 18 six-pounders, Captain W. Gibson, bound to Guernsey for spirits, and thence to Cape Coast. Three days after leaving Guernsey, they were attacked by the *Robuste*, a large French ship, of 24 long twelve-pounders, and 150 men. After a vigorous action of about two hours,

* It was with money granted by the Underwriters for services rendered to the *Elizabeth* when she took the ground in coming into dock, that Crow bought the first respectable suit of clothes he ever possessed. He was expert as a carpenter as well as a sailor, having served two years to the trade of a boat-builder.

in which several of the 35 men who manned the *Gregson* were severely wounded, Captain Gibson, to avoid useless loss of life, reluctantly struck. They were carried to L'Orient, and fairly treated as prisoners of war for some weeks. Here they saw 150 fine looking women, who had been caught with a priest at prayers in a field on a Sunday, brutally driven into the town and handed to the public executioner without a trial. Crow, with others, was removed to Quimper, where he experienced terrible hardships in prison. Lady Fitzroy and her brother, the Hon. Henry Wellesley (sister and brother of the Duke of Wellington), were also prisoners at the same place, and not having been plundered of their money, sent nourishment to the sick. Crow, like a careful Manxman, had a little cash, and raised £20 on his note of hand, which saved his life, as those who had no money perished of want or disease. He complains that the feeling of the day, in England, was exclusively devoted to the melioration of the black slaves, while not a word was said of the white slaves who were daily dying by scores in the prisons of France.

"Often" he says "in our indignation at this partiality, and indifference to our fate, did we wish that our colour had been black, or anything else than white, so that we might have attracted the notice and commanded the sympathy of Fox, Wilberforce, and others of our patriotic statesmen."

By the middle of November, nearly 2000 prisoners had died, and Crow and the rest who could walk were marched to the north of France. After marching five or six hundred miles he was put in hospital at Pontoise, in February, 1795, and here an English mate taught him arithmetic and logarithms. He also picked up a few French words, and one day in May, 1795, having fixed a large tricoloured cockade in his hat, and the vocabulary in his mind, he made his escape. Next day, when he had proceeded fifty miles, he was stopped at a bridge by an officer and a file of soldiers.

His newly acquired French suddenly deserted him and he stood mute.

> "The officer," he says, "followed up one stern inquiry by another, but all to no purpose. At length, as a random expedient, I bolted out all the words of the different languages I could remember, and of which I had obtained a smattering in my different voyages, mingling the whole with my native language, the Manks, with a copiousness proportioned to my facility in speaking it. The Frenchman was astonished and enraged, and as he went on foaming and roaring, I continued to repeat (in broken Spanish), 'No entiendo!' until worn out of all patience, he swore I was a Breton, and giving me a sharp slap with his sword, he exclaimed 'Va-t en, coquin!' I thanked him over and over again, as loud as I could, in Manks, and I assure the reader never were thanks tendered with more sincerity. After this escape, I became more cautious, and resolved henceforward to travel only by night. With the dawn I looked out for a place of shelter and repose, and every morning, as I lay down to rest among the green bushes, my drooping spirits were not a little animated by the delightful notes of the thrush and the blackbird that emerged from their nests to enjoy the wide freedom of the air, while I, to preserve myself from a prison, sought covert from the beams of day."

Missing his way one day, he found himself close to a camp of soldiers, and in the greatest alarm took to his heels. After running and walking sixty miles, his "poor old hull was in so sad a condition from stem to stern," that he "put into the first port," turned into a house, and submitted himself to the mercy of the people. While they were giving him brandy, and putting his feet in warm water, he fell insensible across the tub. He was put to bed, slept soundly, and in the morning, after a good breakfast, he proceeded on his journey, blessing his benefactors. He reached Rouen, and in two days arrived at Havre, where a generous Danish captain gave him a passage to Deal, on arriving at which place, he kissed the soil, in gratitude for his deliverance. The Dane paid his fare to London, and he again raised £10 on his note of hand and started for Liverpool, where he arrived "in great spirits," after an imprisonment of about twelve months. The first person known to him, whom he saw in Liverpool, was his brother William, who had gone out as chief mate of the *Othello*, Captain Christian, to Bonny, where the ship one night caught fire and blew up, several whites, and about 120 blacks on board perishing, amongst them being a brother of King Pepple. William Crow had scarcely left the vessel when the explosion took place. Captain Christian, in a subsequent voyage, met with a similar and more fatal accident. Hugh Crow mentions a little romance in connection with his next vessel:—

> "After a short stay in Liverpool, I shipped as chief mate of the *Anne*, a fine ship, mounting eighteen guns, commanded by my old master, Captain Reuben Wright, and bound to Bonny. While we lay in the river Mersey, a number of men one day came on board, and amongst them was a prepossessing young sailor of apparently about eighteen years of age, named Jack Roberts. This youth drank grog, sang songs, chewed tobacco, enjoyed a yarn, and appeared in all respects, saving the slenderness of his build, like one of ourselves. In a few days, however, we discovered that Jack's true name and designation was *Jane* Roberts, and a very beautiful young woman she was. She was landed with all possible gentleness, and I was informed soon after married a respectable young man. It is remarkable that about this time several handsome young women committed themselves in the same way, and some succeeded in probably eluding all discovery of their sex, and made a voyage or two to sea."

The *Anne* arrived at Bonny almost at the same moment as the *Old Dick* and the *Eliza*, from which she had been separated at sea for fourteen weeks. The town being full of slaves, the *Anne* soon completed her cargo, and in three weeks they sailed to the westward in good health and spirits. When about three days' sail to windward of Barbadoes, they took under their protection a Lancaster brig parted from the convoy, and defended her gallantly against a French privateer which poured three or four broadsides into the *Anne*, but met with a warm reception in the shape of broken copper dross made up in bags, which did terrible execution. After continuing the action off and on for nearly five hours, the Frenchman made off, leaving the *Anne* much damaged, and with several whites and blacks wounded. On her arrival at Barbadoes every man and boy worth taking were impressed, "a galling reception," says Crow, "after the manner in which we had defended ourselves and the Lancaster brig from the enemy." They sailed for Santa Cruz, sold the slaves, loaded at St. Thomas's, and in due course arrived in Liverpool.

His next voyage was as mate in the *James*, Captain Gibson, with liberal wages, besides a gratuity of £100 and the promise of a ship on his return. They sailed in October, 1796, but the ship got ashore on the Cheshire side, and Crow earned the thanks of the owners and underwriters, for his conduct on this occasion. Proceeding on their voyage, they arrived at Bonny, and after taking in a cargo of negroes, weighed anchor on Janaury, 15th, 1797. They had scarcely proceeded five leagues, when the ship grounded at half-ebb on a bank, and then, with six feet water in her hold, was carried over the tail of the bank by the tide and came to anchor in deep water. The captain went off in a boat to Bonny to get assistance, leaving Crow to do the best he could with the ship and about 400 blacks, and only 40 whites to superintend them. While the pumps were going, and the spirits and strength of the crew sinking, Crow went down into the hold with the carpenter, found the leak, and crammed it with pieces of beef. The slaves had got themselves out of irons, and when Crow unlocked the hatch they all gathered round him, shook him by the hand, and asked him to permit fifteen of their best men to come up and assist at the pumps, which was readily agreed to. The ship had to be stranded in Bonny Creek, and the slaves put on board of other vessels. At night, the natives plundered her, but Crow, who had stowed all his own property on the booms, and furnished himself with scores of six-pound shot, defended himself stoutly against all attempts to dislodge him. At length Kings Pepple and Holiday came alongside and commanded their people to desist.

"I was rejoiced at the truce," he says, "for although my ammunition was not yet expended, so desperate and destructive was my defence, that had my assailants not been called off, they would, in revenge, certainly have killed me in the end. Many whites and blacks were wounded on both sides. The ship, in a few days, was literally torn to pieces, and a demand was even made for half the number of the blacks we had on board. The fatigues I had undergone brought on a severe illness, which continued for several days. On my recovery, I was invited by the Kings and the great men to spend some time with them on shore. When I reached the town, all classes were lavish of their presents to me (for I was always on good terms with the inhabitants), and even the children, amongst whom I was well known, sang after me in the streets. A grand ceremony afterwards took place, and I was sent for to attend the Palaver-house, where I found both the Kings and all their great men sitting, attended by crowds of priests and people. The priests proceeded to lead to the sacrifice hundreds of goats and other animals, and the Kings were very active in performing the part of butchers on the occasion. All the musicians in the town were in attendance, and a horrible discordant din they made. I was given to understand that during the ceremony I must neither laugh nor smile, and I believe I kept my instructions by maintaining a suitable gravity of visage. The day was afterwards devoted to feasting and revelry, and this grand *doment* was intended as a thanks-offering to their god, for his goodness in casting our ship upon their shore. From Bonny, I took my passage to Kingston, and thence to Liverpool; where, notwithstanding the unfortunate issue of our last voyage, I met with a most friendly reception from my employers. I afterwards agreed to go mate of a ship called the *Parr*; but had cause to change my mind, and it was well I did so: for that ship was blown up at Bonny on the same voyage. She had at the time her full complement of slaves on board, most of whom as well as the whites lost their lives, and of the number was Captain Christian, whose former ship, the *Othello*, it will be remembered, was also blown up, and he and my brother were amongst the few survivors. At length, as the old proverb goes, 'long looked for come at last,' I had the good fortune to be appointed to the command of a very fine ship called the *Will*, belonging to Mr. W. Aspinall, one of the most generous merchants in Liverpool. She was about 300 tons burthen, carried eighteen six-pounders, besides small arms, and was manned by fifty men. The instructions I received were most liberal, and as a young man on my first voyage as master of a ship, I could not but be highly gratified by the friendly and confidential language in which they were conveyed. We sailed for Bonny, in July, 1798, and arrived safely, after a fine passage. One of our first occupations was the construction of a regular thatched house on the deck, for the accommodation and comfort of the slaves. This building extended from stem to stern, and was so contrived that the whole ship was thoroughly aired, while at the same time the blacks were secured from getting overboard. These temporary buildings would cost from £30 to £40, according to the size of the ship. We soon procured a cargo, and after a pleasant run arrived at Kingston in good health and spirits. Our voyage proved to be most successful. I sold nearly £1200 worth of return goods, which I had saved from my outward cargo, and received the bounty allowed by government for the good condition of the slaves on their arrival.

We sailed for England with the fleet, from which we parted in a gale of wind; but ours was, nevertheless, the first ship that arrived at Liverpool. Mr. Aspinall, my owner, who was fond of a good joke, happening to meet one evening with old Mr. Hodson, merchant, commonly called 'Count' Hodson, their conversation turned upon the voyage we had just accomplished. Mr. Hodson observed, 'I give my captains very long instructions, yet they can hardly make any money for us;' adding to Mr. Aspinall, 'What kind of instructions, Will, did you give *your* captain?' 'Why,' replied Mr. Aspinall, 'I took him to Beat's hotel, where we had a pint of wine together, and I told him—CROW! MIND YOUR EYE! *for you will find many ships at Bonny*!' Mr. Hodson immediately said, 'Crow! mind your eye!—Will, I know the young man well, he has only one eye.' 'True,' said Mr. Aspinall, 'but that's a piercer!' The joke travelled to London, and I could hardly cross the 'Change there afterwards without hearing some wag or other exclaim, 'Crow, mind your eye!' It is very probable that Mr. Aspinall had, in joke, told some of them that these words were the only instructions I had ever received; and as such a fancy on his part was complimentary to me, I may here state that I should have been as proud of that laconic injunction, and acted as faithfully for his interests under it, as under the lengthened instructions which he penned, in his proper anxiety as a trader who had much at stake."

He sailed for Bonny in July, 1799. When off Cape Palmas, a fast-sailing schooner brushed up alongside of them, hoisted French colours, and began to fire; but they cooled his courage with a few broadsides, and he sheered off before the wind. They lay in the river Bonny about three months, slaving, and had a two hours' fight with some French vessels. On this voyage, while in the latitude of Tobago, on the 21st of February, 1800, the *Will* was attacked by a large French privateer, of 18 guns, who gave her two broadsides, and with a loud yell attempted to board, but received such a destructive fire from the *Will's* guns, loaded with round and broken copper dross, that he sheered off, and fired from a greater distance. After fighting for about two hours, he came up a second time, and ordered Captain Crow to strike, or he would sink him. Crow replied that sooner than strike to such as him, he would go down with the ship. This exasperated the French captain, who took up a musket and fired it at Crow several times. With another yell, the enemy attempted to board, but failed, and the *Will* poured into him three broadsides, which produced much havoc and confusion among his men. The privateer dropped astern to refit, and again came up fiercely, the action being stoutly maintained on both sides for about two hours longer. At last, after an engagement lasting in the whole about four hours and-a-half, the privateer sheered off, leaving the *Will* in a very shattered condition. One of the enemy's shot went into the men's room below, and wounded twelve blacks, two of whom died next day, and two others had their thigh bones broken. Three of the crew were wounded, and a gun dismounted by another shot.

"As soon as we had finally beaten him off," says the captain, "I went into the cabin to return thanks to that Providence which had always been so indulgent to me in all my dangers and troubles. When the black women (who had rooms separate from the men) heard that I was below, numbers of the poor creatures gathered round me, and saluting me in their rude but sincere manner, thanked their gods, with tears in their eyes, that we had overcome the enemy. My officers and the ship's company conducted themselves throughout the action with the greatest coolness and determination, and we found a young black man, whom we had trained to the guns on the passage, to be both courageous and expert. In a few days after this rencontre we arrived at St. Vincent's, where we refitted, and proceeded to Kingston. We had scarcely let go the anchor at Port Royal when no fewer than eight men-of-war boats came alongside, and took from us every man and boy they could find.

The impressment of seamen I have always considered to be, in many points of view, much more arbitrary and cruel than what was termed the slave trade. Our great statesmen, however, are regardless of such evils at home, and direct their exclusive attention to supposed evils abroad.

"Our voyage proved very successful, and the blacks were so healthy, and so few deaths had occurred amongst them, that I was, a second time, presented with the bounty of £100 awarded by government. We returned home under convoy; and on our arrival off the N.W. Buoy, my owner and his brothers paid me the compliment of coming out to meet me. To add to my satisfaction, Mr. Aspinall appointed a fine ship, the *Lord Stanley*, to sail with me on the next voyage. To Mr. Kirby, my mate, was given the command of that vessel, and she was placed in every respect entirely under my orders. Both ships, together with some others that were to join us, being fitted for sea, and with valuable cargoes, I received my instructions, which were of the most liberal nature, and we sailed in October, 1800, for the coast of Africa. We encountered some severe gales of wind, and did not reach Bonny till after a passage of ten weeks. There, the ship *Diana*, having been cast ashore and become a wreck, we received on board the captain and crew. After completing our cargo, we sailed in company, all in good health, and arrived at Jamaica without losing a man. Indeed my friends at Kingston used to say—'Crow has come again, and, as usual, his whites and blacks are as plump as cotton bags.' Having concluded our business, we sailed from Port Royal on the 21st of May, 1801, under the convoy of the *York,* sixty-four, Commodore John Ferrier. The *Will* was appointed a pennant ship, and at the same time, one of the whippers-in of the fleet."

The *Hector*, of Liverpool, having gone down, the crew jumping overboard, they were all saved by the *Will*, Captain Crow personally rescuing several of them. "One of them, a Swede," he says, "was only saved by being caught hold of with a boat-hook. He had on a pair of heavy half-boots, and was moreover loaded with a quantity of doubloons sewed up in a belt. This fellow, when he came to himself, without thanking God, or us, for his preservation, only made anxious inquiry if his money were safe." The fleet, consisting of 164 ships, when all under sail presented a most beautiful appearance. Soon after his arrival in Liverpool, Captain Crow was presented by the merchants and underwriters with a handsome silver tray, bearing the following inscription:—

> "This piece of plate is presented by the Merchants and Underwriters of Liverpool to Capt. Hugh Crow, of the ship *Will,* in testimony of the high estimation they have of his meritorious conduct in the River Bonny, on the coast of Africa, on the 16th of December, 1799, when menaced by three French Frigates."

The Underwriters of Lloyd's Coffee-house also presented him with a sum of money and an elegant silver cup, of the value of £200, for his gallantry in defending the ship *Will* against the French privateer, on the 21st of February, 1800.

After this, Captain Crow commanded the *Ceres*, a fine frigate-built ship of 400 tons, well armed and manned. One day, at Bonny, King Pepple came on board, flushed with palm wine, and began to boast of the services he had rendered to Crow, who lay in great pain upon a mattress, unable to satisfy the king's greed. At length, Pepple worked himself into such a rage, that, going up to the captain with insulting gestures, he began to utter all manner of abuse against the Isle of Man, which he denounced as little and despicable, and finally roared out that Manxmen were a miserable race of people, as poor as rats, and unable to support a king. At this, Crow sprang up, seized a stick, and shouting "You villain; how dare you abuse my country," followed his majesty on all fours, and fairly chased him out of the ship. As he left the side, the king reiterated "Poor boy! you cant havey king." They became good friends afterwards.

In December, 1806, Captain Crow was in command of the slave-ship *Mary*, about 500 tons burthen, carrying 24 long nine-pounders on the main deck, and 4 eighteen-pound carronades on the quarter deck. She was manned by between 60 and 70 men, 36 of whom were qualified to take the wheel. He gives an amusing description of his preparations for receiving an enemy on the middle passage:—

"It was my constant practice to keep the ship in a state of readiness to receive any enemy we might chance to meet, and particularly when we drew near to the coast of Cayenne, which I had learned, by dearly bought experience, was infested by French cruisers. To this end my crew were frequently trained to work the great guns and small arms, and on the present voyage I selected several of the finest of the black men to join them in these exercises, as well as in passing along the powder, and in other minor duties that might become requisite in the hour of action. The blacks, who were very proud of the preferment, were each provided with a pair of light trowsers, a shirt, and a cap; and many were the diverting scenes we witnessed, when they were in a morning eagerly employed in practising firing at empty bottles, slung from the ends of the yard arms. Being but indifferent marksmen few of their shots took effect; and the falling countenances of those who had just missed formed a ludicrous contrast to the animated features of the next sanguine competitors. The first who struck a bottle was presented with a dram and a new cap. This small reward excited a strong emulation, and the morning's sport furnished matter of exultation to the victors, and of general merriment to all throughout the day.

"Meantime we made a rapid run to the westward, and though my confidence in my crew was such that I thought very few French or other privateers, or even sloops of war could successfully cope with us singlehanded, yet as I one day paced the quarter deck ruminating on the chances of being attacked by probably an unequal force, a project came into my head for the greater annoyance and destruction of an enemy, of which I determined, if occasion required, to make experiment. Having got my plan to bear in my own mind, I sent for the gunner and the armourer, who were both clever men, and having expressed to them the great satisfaction I had all along derived from the good conduct of the officers and crew, I informed them that I had before had two actions off the coast of Cayenne, and that as there was a probability of our soon falling in with some powerful French privateers from that quarter, I had resolved, being so well manned, should any one attack us, not to give them much chance at long bowls, but to slap them right on board, if possible; and to run them down rather than expose the lives and limbs of my crew by a long action. I then desired them to take half-a-dozen of two-gallon jars, of which we had a number on board; and first to put about two quarts of powder into each, and the same quantity of small flints; over these an additional quantity of powder; next about two quarts of pepper, and then to fill up with powder and cork them up. They were finally to insert a tin tube with a good match through the middle of the cork, to cover the jars with canvas, and coat them thickly with a composition of powder, brandy and brimstone. Each jar was to be put into a loose sack that it might be hauled up into either top when wanted. 'And,' I concluded, addressing myself to the armourer, 'as you are the strongest man in the ship, your station will be in one of the tops, with a lighted match, so that you may, on the word being given, heave these destructive jars right on board of any enemy that may dare to come to close quarters.' The gunner, on hearing these injunctions, exclaimed—'Sir, I have seen a deal of service both in men-of-war and in privateers, but I never heard of, or saw, so deadly a contrivance before; and if any French or Spanish privateers venture to come alongside of us, they will never be able to get away again.' He did not probably exaggerate the effects of these infernal bombs if thrown upon the crowded deck of an enemy's vessel; for as the jars would burst into sharp and irregular fragments, they would cut and mangle with as much execution as the flints; and the burning pepper, which was to be kept in a blaze by the combustible covering, no one could abide in the heat of action. This contrivance was, I confess, destructive, if not wicked; but when I recurred to the horrors of a French prison, I should not have hesitated, rather than run the hazard of undergoing a repetition of my sufferings, and involving my crew in a similar misfortune, to resort even to more desperate means of disabling an enemy, if occasion required."

He had not long to wait for an opportunity to prove the mettle of his crew. On the 1st of December, while they were running down with studding sails set in the latitude of Tobago, he saw two sail which with the help of the glass he took to be powerful vessels of war, and as they were crossing the very ground where French cruisers often attempted to intercept British ships bound to the West Indies, he judged they came from Cayenne, and tried to avoid them. They both tacked, and gave chase under a heavy press of canvas. The *Mary* and her crew were a match for any single cruiser, but the captain did not want to fight two. Night coming on, Captain Crow called all hands to quarters, and addressing his men in a rousing speech, said he was determined that rather than be taken and sent to a French prison, he would defend the ship to the last, and go down with her sooner than strike. To a man they promised to stick by him. "Commend yourselves, my brave fellows, to the care of Providence," said the captain. "Let us have no cursing or swearing, but stand to your quarters, and such is my opinion of your abilities and courage that I have no doubt but that, should even both vessels attack us, we shall triumphantly beat them off; and woe be to them if they attempt to board us." They were not long left in that silent and intense anxiety that immediately precedes an engagement at sea, for the captain had scarcely done speaking when one of the vessels loomed large in the obscurity astern, and hailed him in English—an old French trick, and ordered him to bring to; and soon after her consort came up and made the same demand. To both Captain Crow coolly replied that he was the *Rambler* off a cruise, and that no strange vessel should bring him to in those seas in the night. The ships again hailed, but the gabble of the sea and the bustle on board made it impossible to detect whether the words were spoken with a foreign accent or not. Captain Crow had made up his mind that they were French, and that was enough. One of the vessels rounded to and poured a broadside into him, and he fought her at close quarters for some time. She then took her station at some distance, and they fought for another half-hour, when her consort came up on the *Mary's* larboard side; both vessels closed, and simultaneously attacked the *Mary*. Captain Crow animated his men, who blazed away with right good will in the pitchy darkness, which hid from them the fact that their captain was partially disabled by a violent blow from a splinter. They soon grew callous to the flashing of the guns on both sides of them, and to the storm of balls. For a moment the man at the wheel deserted his post, stunned by the wind of a large shot, but soon flew back when the captain's ringing voice cried, "What! is it possible we have a coward in the *Mary?*" Meanwhile the stout armourer was stationed in the maintop ready to fling the infernal combustible jars on the enemy if they attempted to board. Captain Crow was here, and there, and everywhere, cheering up his men, who boldly stood to their quarters, and fought like heroes. It was now past midnight, and the din like continued peals of thunder. A large shot entered a gun-port, and took off both the boatswain's thighs. Another entered the men's room below, and wounded a great number of blacks, five of whom died soon after. The cries of the dying and the wounded were pitiable, and aroused the spirit of vengeance in the seamen, who fought like demons. After an action of nearly six hours, one of the ships dropped astern, and Captain Crow sung out, "I think, my brave fellows, we have sickened them both, and your names will be honourably mentioned by our friends in Liverpool for your resolute conduct in this action." The men wanted to give him three cheers, but he sent them back to their quarters. The ship again came up and resumed the action as fiercely as ever. The *Mary* continued to engage both vessels, tooth and nail, until the grey of the morning, when Captain Crow was struck by a splinter and fell senseless on the deck. The man at the helm sung out that the captain was killed, the crew, worn out by fatigue, lost heart, and when the captain revived he found that the colours had been struck. Raising himself on the deck, with true Viking spirit, he entreated them to hoist the colours and give the enemy "three or four more broadsides to conclude with." His hope was that "as a chance shot will kill the devil," he might inflict an injury that would turn the scales of battle. But it was all in vain, the force of the enemy was now seen to be so great that further resistance would have been madness. The captain was carried to his cabin and laid on a mattress, while the crew prepared to go on board the enemy's vessels as prisoners of war. When the boats came alongside, the poor fellows were standing at the gangway ready to surrender, but what was their astonishment when they found that those who boarded them were their own countrymen, and that they had been all the while fighting two British men-of-war! One was the *Dart* sloop of war, of 30 guns, (thirty-two-pounders); the other the *Wolverine*, of 18 guns, of the same calibre. This was astounding intelligence for Captain Crow, who, in his anguish and vexation, struck his head several times against the cabin floor, until the blood started from his mouth and nostrils, and the effects of which never quite left him. Friendly hands restrained his phrenzy, and a flow of tears relieved his grief. The lieutenants of the war vessels consoled him, and told him that their captains were astonished to find that a Liverpool Guineaman could sustain an action of seven hours with so superior a force. They had mistaken the *Mary* for a French privateer. Captain Spear of the *Dart*, presented Captain Crow with the following certificate:—

"His Majesty's Sloop, *Dart*, at sea,
"Dec. 1st, 1806.

"I do hereby certify that Hugh Crow, commanding the ship *Mary*, of Liverpool, and last bound from the coast of Africa with slaves, defended his ship in a running action, under the fire of his Majesty's Sloop, under my command, and also his Majesty's Sloop *Wolverine*, both carrying thirty-two pounders, from about ten p.m. till near daylight the next morning, in a most gallant manner (supposing us French cruisers from Cayenne), and did not give up till his rigging and sails were nearly cut to pieces, and several of his people* wounded. Latitude 11° 27′ N.; longitude 53° W.

(Signed,) "JOSEPH SPEAR, Commander."

When the slaves came to know that Captain Crow was wounded, and that he had been fighting friends instead of foes, they rushed up in groups from below, and gathered round him in the cabin. Some of them took hold of his hands, others of his feet, and on their knees expressed in their own way, their sorrow for the unfortunate affair, offering him their rude, but sincere condolence.

The *Mary* made the land of Jamaica in a few days. On passing Port Royal, the negroes to the number of about 400, were nearly all on deck. When they saw the bodies of about a dozen men-of-war's men, who had been executed for mutiny, hanging on gibbets and in chains, and some in iron cages, on the low lands called the "Keys," they became dreadfully alarmed, lest they should be sacrificed in the same manner, and many of them were with much difficulty restrained from jumping overboard. Admiral Dacres sent

* Six died of their wounds. It is worthy of note that from the afternoon before the action, until it was over, the crew of the *Mary* had not a single glass of spirits, nor did a murmur arise on that account.

on board a protection from impressment for all the crew. The Captain's friends hastened on board to bid him welcome; the cargo, after all, was fine and healthy, and was disposed of to great advantage by Mr. Thomas Aspinall.

> "On the first Sunday after our arrival at Kingston," says Captain Crow, "a circumstance occurred on board the *Mary*, which was the more gratifying to me as it was entirely unexpected. While I was lying in my cot, about nine o'clock in the morning, Mr. Scott, my chief mate, hurried into the cabin and said 'Sir! a great number of black men and women have come on board, all dressed in their best, and they are very anxious to see you; will you allow them to come down?' 'By all means,' said I, springing up, and hastily putting on my clothes to receive them, and in a moment they all rushed into the cabin, and crowding round me with gestures of respect, and with tears in their eyes, exclaiming—'God bless massa! how poor massa do? Long live massa, for 'im da fight ebery voyage'—and similar expressions of good will and welcome. I soon recognised these kind creatures as having been with me in one or other of the actions in which I had been engaged on former voyages, and though my attention to them when on board was no more than I had always considered proper and humane, I was deeply affected by this mark of their grateful remembrance. Poor Scott shed tears when he saw them clinging round me, and observed, 'How proud, sir, you must be to receive this grateful tribute of regard! and how few captains can boast of a similar proof of their good treatment of the blacks under their charge.' Indeed, I could not refrain from shedding tears myself, when I reflected that the compliment came from poor creatures whom I had brought from their own homes on the coast of Africa. The women were neatly dressed in calicoes and muslins. Their hair was tastefully arranged, and they wore long, gold earrings. The men appeared in white shirts and trousers, and flashy neckcloths, with their hair neatly plaited. The whole were at once clean and

cheerful, and I was glad from my heart to see them. When they left the ship, which was not till they had repeatedly expressed their happiness to see me again, I distributed amongst them a sum of money, and they bade me good bye with hearts full of thankfulness and joy. In a few days afterwards the Governor of the Colony with his suite did me the honour to pay a visit on board the *Mary*, a compliment seldom known to be paid to the master of a merchant ship."

The blacks in Jamaica composed a song in honour of Captain Crow, of which the following verses are a specimen :—

"Captain Crow da come again,
But em alway fight and lose some mans,
But we glad for see em now and den,
Wit em hearty joful gay, wit em hearty joful gay.
   Wit em tink tink tink tink tink tink ara.
   Wit em tink tink tink tink tink tink ara.
But we glad for see em now and den
Wit em hearty joful gay, wit em hearty joful gay ara."

* * *

"But did you eber the governor see
When em went on board of he.
Den em say Sir Hugh you must be,
Wit you hearty joful gay, wit you hearty joful gay.
   Wit em tink tink &c.
Den em say Sir Hugh you must be,
Wit you hearty joful gay, wit you hearty joful gay."

Captain Crow did not forget to visit daily the poor wounded sailors in the hospital, several of whom died. "It was a consolation to me," he observes, "to be informed that, even in the height of their sufferings, they frequently mentioned my name in terms of attachment and respect. The captain was one day met by a fine young black, who in a very polite manner accosted him ; "Cappy Crow, how you do?" "I do not know you, boy," said the captain. "Cappy Crow," rejoined the negro, "me sabby you bery well!" "When and where did you know me?" demanded the skipper. "Me sabby you very much when you live for you ship for big water—when you look ebery day wit crooked tick for find da pass;" meaning, he knew the captain when he was on board the slave-ship at sea, when he took daily observations with the quadrant to find out the way. After a little conversation, Crow gave him some money, and away he went, delighted with the present and the condescension of the captain.

Captain Crow tells a humorous story of a monkey, who wanted to take command of the *Mary,* showing that the middle passage had its comedies as well as its tragedies:—

"During my last visit to Bonny, I had purchased a monkey of the largest size, which was a source of amusement, but more frequently of annoyance on board of the *Mary.* This fellow attached himself particularly to me, and as he constantly kept at my side, considering me no doubt his protector in his new mode of life, we became in a short time pretty well acquainted. He was uncommonly expert in imitating any thing he saw done, particularly if it were mischievous. Although I was sometimes obliged to check his propensity to evil doing, we for some weeks continued to maintain a mutual good understanding as shipmates. But the best of friends, alas! will sometimes quarrel, and so it was with us. One day while we were in the middle passage, we were overtaken by a squall, and while I was busy ordering sail to be taken in, my gentleman snatched the speaking trumpet from my mouth, with intent no doubt, to assist me by making his own sort of noise upon it. Jealous of my prerogative I insisted upon a restoration of my instrument of office—the trumpet; this he resisted, and a scuffle ensued, which ended in my being obliged to knock him down with the end of a rope. Before I had time to look about me, the fellow sprang at my neck, and after chattering and making faces of great consequence, he bit me several times. This was beyond endurance; he received a drubbing, which made him so outrageous that we were obliged to chain him. It appears he never forgave me for this infliction, for one morning very early, whilst I was lying asleep, he by some means got loose, and thirsting for revenge, ran down to the cabin, where mounting the table near my cot, he made no ceremony in pulling off the whole of the clothes that covered me, and that with such alacrity that I had no time to stand upon the defensive. The fellow then sprung to the beaufet and began, as fast as he could, to pitch the wine glasses and tumblers, and whatever else he could lay hold of, out through the cabin windows. The steward, at length, luckily came in, and we secured him. Owing to these and similar pranks, I determined to part with him, and a few days after we arrived at Kingston, I had him advertised in the newspapers, by the name of Fine Bone, from Bonny, on the coast of Africa, to be sold, for high crimes and misdemeanours. Having equipped him in a jacket and trowsers made for the occasion by a fashionable tailor, and a cap of the newest cock, he was on the day appointed sent on shore to a vendue store, where several hundreds of persons were waiting, brimful of curiosity, to see what kind of a being he was. He was put up in due form, and after a good deal of merriment among the bidders, and particularly among the Jew gentlemen present, whom he seemed to scrutinize with very knowing looks, he went off for £5 6s. 8d. I must not say, in auctioneering phrase, that he was 'knocked down' for that sum, for he would have been a bold man who would have knocked him down, unless indeed in such a manner as to give him his quietus. For myself, I did not venture to go on shore on the day of sale, for if he had seen me in the street he would certainly have run after me, and claimed the privilege of an old acquaintance in a manner more earnest than welcome. Next morning the wags in the town reported that they had seen him, during the night, at West-street, assisting the press-gang; and others gave out that he had run off with two half firkins of butter from a provision store, and would certainly be tried and banished the colony for so grave an offence."

Captain Crow sailed from Jamaica in March, 1807, and arrived in Liverpool on the 2nd of May, after a pleasant passage of five weeks. Here he favours us with his sentiments on the new bill.

"I was received by Mr. Aspinall with his usual kindness and hospitality. We however, got home 'the day after the fair,' for the African slave trade was abolished on the day preceding our arrival. The abolition was a severe blow for England, and particularly as it affected the interests of the *white* slaves who found employment in the trade. It has always been my decided opinion that the traffic in negroes is permitted by that Providence that rules over all, as a necessary evil, and that it ought not to have been done away with to humour the folly or the fancy of a set of people who knew little or nothing about the subject. One thing is clear; instead of saving any of the poor Africans from slavery, these pretended philanthropists have through the abolition, been the (I admit *indirect*) cause of the death of thousands; for they have caused the trade to be transferred to other nations, who in defiance of all that our cruisers can do to prevent them, carry it on with a cruelty to the slaves, and a disregard of their comfort and even of their lives, to which Englishmen could never bring themselves to resort."

Self-interest evidently blinded him. The slave trade, like the Rontgen rays, caused an obliquity of vision when closely followed.

As the *Mary* could not again clear out for an African voyage, Captain Crow took command of the *Kitty's Amelia*, of 300 tons, and 18 guns, belonging to Mr. Henry Clarke, which had been cleared out previous to the passing of the Abolition Bill. Messrs. Kerwen, Woodman & Co., insurance brokers, London, wrote to Captain Crow that they would insure his commissions from Liverpool to Africa at 15 guineas per cent., adding, "we have never heard greater praise bestowed on any commander than the underwriters in general have expressed in consequence of your very gallant behaviour, which will always procure their decided preference to whatever vessel you sail in." The Liverpool underwriters insured ship and cargo at the same rate, which was 5 per cent. lower than the usual premium. The *Kitty's Amelia* sailed on the 27th of July, 1807, with a crew of between fifty and sixty men, four of the ablest of whom were soon after impressed, in spite of their protections, by H. M. frigate *Princess Charlotte*, Captain Tobin. Captain Crow had three commissions, or Letters of Marque, but although he chased and boarded several vessels, he took no prizes. They arrived at Bonny after a passage of about seven weeks, and were immediately boarded by his Majesty, King Holiday, who anxiously enquired if it was true that Captain Crow was in command of the last ship that would come to Bonny for negroes. Captain Crow gives a curious account of what passed at a long palaver. The King's sentiments regarding the abolition were as follows:—

"Crow," he remarked, "you and me sabby each other long time, and me know you tell me true mouth (speak truth); for all captains come to river tell me you King and you big mans stop we trade, and 'spose dat true, what we do? For you sabby me have too much wife, it be we country fash, and have too much child, and some may turn big rogue man, all same time we see some bad white man for some you ship, and we hear too much white man grow big rogue for you country. But God make you sabby book and make big ship—den you sen you bad people much far for other country, and we hear you hang much people, and too much man go dead for you warm (war). But God make we black (here the poor fellow shed tears) and we no sabby book, and we no havy head for make ship for sen we bad mans for more country, and we law is, s'pose some of we child go bad and we no can sell 'em, we father must kill dem own child; and s'pose trade be done we force kill too much child same way. But we tink trade no stop, for all we

Ju-Ju-man (the priests) tell we so, for dem say you country no can niber pass God A'mighty."

The last words he repeated several times; and Captain Crow thought his remarks not altogether destitute of sense and shrewdness.

There were ten or twelve vessels waiting for slaves at Bonny, and Captain Crow had long to wait for his turn. When he did begin to trade, a misfortune befell him. In the hurry of fitting out the vessel at Liverpool, before the passing of the Abolition Bill, some returned goods from a former voyage (when the ship was sickly), were repacked in damp water casks, and when these were opened at Bonny, a malignant fever and dysentery broke out amongst the crew. The rotten goods were thrown overboard, but the sickness retarded the slaving. Terrible storms broke over the vessel, and, altogether, the voyage of the last slaver was attended with misfortunes. Lucky was it for the owners that they had Captain Crow at the helm. Having completed his purchase, he sailed from Bonny for the last time, with "as fine a cargo of blacks, as had ever been taken from Africa," but the disease baffled the skill of the two doctors, and he was deeply afflicted to see both whites and blacks dying around him daily at an alarming rate. They put into the Portuguese island of St. Thomas, to recruit, and here Captain Crow, with Captain Toole, also of Liverpool, visited the ruins of the Bishop's palace and saw the torture chambers of the Inquisitors. The sick having recovered, the ship resumed her voyage with additional passengers in the shape of several monkeys presented by the Governor to Captain Crow. They had not been long at sea before the sickness broke out afresh, both whites and blacks dying daily. The death of the chief mate added greatly to the captain's anxiety, as he feared that if anything happened to himself, there was no one left on board capable of navigating the ship to port. When they were in this trying situation, the horrors of the voyage were intensified by an accident, which we shall let the captain himself relate:—

"One afternoon, when we were ten or twelve hundred miles from any land, and were sailing at the rate of seven or eight knots, the alarm was given that the ship was on fire, in the afterhold. I was in the cabin at the time, and springing upon deck, the first persons I saw were two young men with their flannel shirts blazing on their backs; at the same time I perceived a dense cloud of smoke issuing from below, and looking round me I found the people in the act of cutting away the stern and quarter boats, that they might abandon the vessel. At this critical juncture, I had the presence of mind to exclaim, in an animating tone, 'Is it possible, my lads, that you can desert me at a moment when it is your bounden duty, as men, to assist me?' And observing them hesitate, I added, 'Follow me, my brave fellows, and we shall soon save the ship.' These few words had the desired effect, for they immediately rallied, and came forward to assist me. To show them a proper example I was the first man to venture below, for I thought of the poor blacks entrusted to my care, and who could not be saved in the boats, and I was determined, rather than desert them, to extinguish the fire, or to perish in the attempt. When we got below we found the fire blazing with great fury on the starboard side, and as it was known to the crew that there were forty-five barrels of gunpowder in the magazine, within about three feet only of the fire, it required every possible encouragement on my part to lead them on to endeavour to extinguish the rapidly increasing flames. When I first saw the extent of the conflagration, and thought of its proximity to the powder, a thrill of despair ran through my whole frame; but by a strong mental effort I suppressed my disheartening feelings, and only thought of active exertion, unconnected with the thought of imminent danger. We paused for a moment, struggling, as it were, to determine how to proceed. Very fortunately for us our spare sails were stowed close at hand. These were dragged out, and by extraordinary activity we succeeded in throwing them over the flames which they so far checked, that we gained time to obtain a good supply of water down the hatchway, and in the course of ten or fifteen minutes, by favour of the Almighty, we extinguished the flames. Had I hesitated only a few minutes on deck, or had I not spoken encouragingly to the people, no exertions whatever could have saved the ship from being blown up, and as the catastrophe would most probably have taken place before the hands could have left the side in the boats, perhaps not a soul would have survived to tell the tale. I hope, therefore, I shall be excused in assuming to myself more credit (if, indeed, credit be due) for the presence of mind by which I was actuated on this occasion, than for anything I ever did in the course of my life. The accident, I found, was occasioned by the ignorance and carelessness of the two young men, whose clothes I had seen burning on their backs; through the want of regular officers, they had been intrusted to draw off some rum from a store cask, and who, not knowing the danger to which they exposed themselves and the ship, had taken down a lighted candle, a spark from which had ignited the spirit."

What must have been the terror and sufferings of the slaves, while the gallant captain and his true men fought the flames? He goes on as follows:—

"I shall never forget the scene that followed the suppression of the flames. When I got on deck, the blacks, both men and women, clung round me in tears—some taking hold of my hands, others of my feet, and all, with much earnestness and feeling, thanking Providence for our narrow escape, an expression of gratitude in which, I assure the reader, I heartily joined them."

Truly a strange slave-captain is this, well-beloved by the very people he is carrying to perpetual and cruel captivity; a very different man from the monster depicted by Montgomery. But we must remember that Captain Hugh Crow and Captain John Newton were the exceptions that prove the rule. Captain Crow, after all his experiences of the horrors of the middle passage, and familiarity with many sanguinary engagements, could still sympathise with a sick monkey.*

The sickness abated as they neared the West Indies, but, on their arrival at Kingston, after a passage of eight weeks from St. Thomas's, the two doctors died, and the deaths on the voyage amounted to 80 (30 whites and 50 blacks). Captain Crow, who had always prided himself upon keeping a clean ship, and taking the bounty for healthy cargoes, was overwhelmed with grief, but he found that the mortality on board the *Kitty's Amelia* was only one-half that on other ships. It appeared that the Liverpool slave trade was doomed to come to an end amid death and ruin on a large scale. The hurried manner in which ships had been sent out without proper cleansing, and the glutting of the market by the arrival of so many vessels, proved almost the undoing of many merchants. At Kingston, Captain Crow found sixteen slave-ships which had been there five or six months, their cargoes unsold, and their crews and slaves daily diminishing through deaths. This was a dark outlook, but the good luck which

* "On this passage," he says, "I witnessed a remarkable instance of animal sagacity and affection. We had several monkeys on board. They were of different species and sizes, and amongst them was a beautiful little creature, the body of which was about ten inches or a foot in length, and about the circumference of a common drinking glass. It was of a glossy black, excepting its nose and the end of its tail, which were as white as snow. This interesting little animal, which, when I received it from the Governor of the island of St. Thomas, diverted me by its innocent gambols, became afflicted by the malady which yet, unfortunately, prevailed in the ship. It had always been a favourite with the other monkeys, who seemed to regard it as the last born, and the pet of the family; and they granted it many indulgences which they seldom conceded one to another. It was very tractable and gentle in its temper, and never, as spoiled children generally do, took undue advantage of this partiality towards it by becoming peevish and headstrong. From the moment it was taken ill, their attention and care of it were redoubled, and it was truly affecting and interesting to see with what anxiety and tenderness they tended and nursed the little creature. A struggle frequently ensued amongst them for priority in these offices of affection, and some would steal one thing and some another, which they would carry to it untasted, however tempting it might be to their own palates. Then they would take it gently up in their fore paws, hug it to their breasts, and cry over it as a fond mother would over her suffering child. The little creature seemed sensible of their assiduities, but it was wofully overpowered by sickness. It would sometimes come to me and look me pitifully in the face, and moan and cry like an infant, as if it besought me to give it relief; and we did everything we could think of to restore it to health, but in spite of the united attentions of its kindred tribe and ourselves, the interesting little creature did not long survive."

usually attended Captain Crow did not fail him now. His friends inserted a paragraph in the newspaper stating that Captain Crow had arrived with the finest cargo of negroes ever brought to Kingston. The puff did its work, and in five days the cargo of the *Kitty's Amelia* had been sold at higher prices than those obtained by any other ship. In spite of the disasters and sufferings undergone, the voyage turned out very profitable. It was Sunday morning when Captain Crow landed at Kingston. He found a number of his black friends, all neatly dressed, waiting on the wharf to receive him. They crowded round him, took hold of his hands, and said, "God bless massa! How massa do dis voyage? We hope massa no fight 'gen dis time." While they were thus congratulating him, another black, one of their party, and a wag, exclaimed, "Who be dis Captain Crow you all sabby so much?" Then the rest cried, "What dat you say, you black negro? Ebery dog in Kingston sabby Captain Crow, and you bad fellow for no sabby him." With that they fell to and beat him until Captain Crow interfered, and he was shrewd enough to guess that the scene had been contrived before hand. Yet he was pleased with their visit. He remained behind at Kingston, to transact some business, taking command of the *King George* schooner. The blacks came down to the wharf at night, and, hailing the schooner, asked for Captain Crow. When he made his appearance on deck they used to sing out, "Captain Crow, you have bery fine ship now—'one pole and half' ship," and after this humorous sally they took to their heels, the jolly captain doubtless pretending to be highly insulted.

From the captain's stand point, the consequences of the abolition of the slave trade were pernicious to England; it destroyed her nursery of seamen, and drove her young men, whose prospects at home were blighted, into the American service, where they afterwards fought against their own country. He is very severe on the abolitionists, or "pretenders to humanity," who should have begun their reforms at home. "Let them look to Ireland," he says, "which is in a most deplorable state of slavery and disaffection, for which no politician has yet discovered an adequate remedy."*

Captain Crow was only forty-three years of age when, in 1808, he retired from active service, having made a competent fortune by commanding several of the "crack ships" out of the port of Liverpool. About this time, his friend, Admiral Russel, wrote to the Rev. Dr. Kelly:— "Tell the warlike Crow to send me his son, that I may train him up to emulate his father." Young Crow, a very handsome, amiable, and brave boy of fifteen, had, however, been taken by Captain (afterwards Sir Robert) Mends on board the *Arethusa* frigate. "After fixing him, as I trusted, permanently," says Captain Crow, "I bore up for my native land, the Isle of Man, thinking to moor there in peace and security for life." He bought an estate near Ramsey, where he resided for some years, engaged in agricultural pursuits and improvements on the property.

In June, 1812, he was "proposed and appointed a member of the House of Keys," but declined the honour, wishing, after all his trials and hardships, to spend the remainder of his life in retirement. His heart, too, was well-nigh broken by the death of his gallant son, who had been taken by the French in one of the ship's boats, while in the act of cutting out some vessels from a French harbour, and who escaped from Verdun in a very daring

* When Captain Crow wrote those words, there was at Oxford a Liverpool youth of rare gifts, who was destined in after years to make an historic effort to pacify Ireland, not with the iron rod, but with the olive branch of brotherhood and friendship, yet he failed; with all his wisdom, learning, eloquence, ripe experience, statesmanship rarely if ever excelled, spotless character, and immense political influence, this great lawgiver of our time failed in his supreme endeavour, because the majority of people can only look at the question of Ireland as Captain Crow looked at the slave trade—from one side, and that the side of self.

manner, but contaminated with the wickedness and debauchery of the prison. On the way to join the *Arethusa* frigate, the lad fell into bad company, and enlisted in the 9th Light Dragoons. This nearly killed his father. The youth's discharge was procured, but he died i a few days after at Lisbon—of a broken heart, it was said,—thus blasting the captain's fondest hopes.

Finding life in the Isle of Man too monotonous, Captain Crow returned to Liverpool in the year 1817, to enjoy the society of kindred spirits, his favourite haunts being the Lyceum News Room, and the quays. After dinner, he foregathered with his cronies in the African trade, when each fought his battles over again, but such was the discipline among this knot of veterans that, at one of their rendezvous, the striking of a particular hour was the signal for a general separation, when they hurried out, helter skelter, often leaving the tale half told, and the glass unfinished. It was in scenes like this that the humour and originality of Captain Crow were seen at their best, rather than in his writings.

In 1827, the captain went to live in apartments at Preston, in a lovely spot, which, in one of his letters, he calls "Paradise Found." Here he wrote his memoirs, which he regarded as "one of the first things of the kind ever got up by a Manxman." He dearly loved the "oilan," and it was his constant custom when his ship lay at Bonny, to show his patriotism on holidays by hoisting the Manx flag at the mast head, to the amusement of King Pepple and the chief men there, who were greatly diverted by the strange device of the "Three Legs of Man."

Captain Crow died in 1829, in his 64th year, and his remains were interred in the burial ground of his ancestors in Maughold churchyard, where he lies entombed with his venerable parents, for whom, throughout his eventful life, he exhibited the strongest affection and the tenderest care.

In the long roll of Liverpool slave-captains, there were, doubtless, a few as humane, and scores as brave as Hugh Crow; but when we attempt to realise the total amount of misery and injustice which they and their employers systematically inflicted upon their black brethren, for the greater part of a century, it is some satisfaction to remember the lines of one,* whose father and mother died while preaching to the poor slaves in the West Indies the doctrine of another and a better world:—

"When the loud trumpet of eternal doom
Shall break the mortal bondage of the tomb;
When with the mother's pangs the expiring earth
Shall bring her children forth to second birth;
Then shall the sea's mysterious caverns, spread
With human relics, render up their dead:
Though warm with life the heaving surges glow,
Where'er the winds of heaven were wont to blow,
In sevenfold phalanx shall the rallying hosts
Of ocean-slumberers join their wandering ghosts,
Along the melancholy gulf, that roars
From Guinea to the Caribbean shores.
Myriads of slaves that perish'd on the way,
From age to age the sharks' appointed prey.
By livid plagues, by lingering tortures slain,
Or headlong plunged alive into the main,
Shall rise in judgment from their gloomy beds,
And call down vengeance on their murderers' heads!"

*James Montgomery.

APPENDIX No. I.

List of Vessels, trading to and from Liverpool, Captured by the Spaniards and French in the War of 1739-1748. The list is necessarily incomplete owing to the circumstances of the times:—

Ship's Name.	Master's Name.	Voyage.	Where carried, etc.
Snow Mary	Benson	Liverpool to Jamaica	Porto Rico
St. Michael	John Thompson	Jamaica to Liverpool	Plundered off Cape Antonio
Unity	Henan	Do.	St. Sebastian
Mar & Mary	Wilcox	Virginia to Liverpool	Do.
Thomas	Murray	Liverpool to Oporto	Do.
Endeavour	Whaley	,, ,, Lisbon	Paniche
Dove	Lee	,, ,, Africa	Do.
Priscilla	Cullen	,, ,, Antigua	Do.
Hannah	Holmes	Virginia to Liverpool	Do.
Three Sisters	Cardwell	Jamaica to Liverpool	Do.
Philippa	Dewhurst	Liverpool to Gibraltar	Vigo
Byrne	Walker	Figuera to Liverpool	St. Sebastian
Blackamore	Bradley	Liverpool to Gibraltar	Cadiz
Sarah	Idle	Liverpool to London	Helvoetsluys
Betty	Biddy	,, to Cape de Verde	St. Sebastian
Tryton	Thompson	,, to Leghorn	Ceuta and Algevire
Cape Coast	Green	From Liverpool	Taken on the Coast of Africa by Spanish Privateer
Swallow	Hughes	Do.	
Angola	———	Jamaica to Liverpool	Retaken
Success	Lewis	Do.	Leeward Islands
Jean	Bradley	Liverpool to Gibraltar	Cadiz
Ellen & Mary	John Simon	From Liverpool	Ransomed for 47 gs.
Mary & Anne	Rush	Do.	Taken by French privateer
Stafford	Perry	Gottenburg to L'pool	Ransomed for £225
Stafford	Perry	Liverpool to London	Havre de Grace
———	Barnes	Virginia to Liverpool	Morlaix
Mulberry	Barton	Jamaica to Liverpool	Brest
Martin	Wilmot	L'pool for Montserrat	Taken by French
Houghton	Postlethwaite	Liverpool for Antigua	Martinico
Thorn	Carter	L'pool for Harwich	Havre de Grace
Chester	Frith	———	Havre
———	———	From Liverpool	Do.
M'Grieque	———	Oporto to Liverpool	St. Malo
Jane	Hyth	Liverpool for Virginia	Ransomed
Vine	Walker	Maryland for L'pool	Do. by the French
Thomas	———	Liverpool for Africa	Rochelle

APPENDIX No. I.—Continued.

Ship's Name.	Master's Name.	Voyage.	Where carried, etc.
Content	Cooper	L'pool for Barbadoes	Martinico
Morecroft	Batty	Do. Leeward Islands	By French Privat'rs
Mary	Godsalve	Liverpool to Lisbon	Dieppe
Anne & Mary	Falkner	Liverpool to Antigua	Martinico
Lively	Dwyer	,, for Africa	Port Louis
———	Clark	,, to Colchester	Taken by the French
Robert	———	L'pool and Africa for Jamaica	Do.
Hare	———	Do.	Do.
Enterprise	———	Do.	Do.
Black Prince	Derbyshire	From Liverpool (50 guns, 400 men)	Do.
Two Ships	Names unknown	From Liverpool	Martinico
Recovery	Coates	London for Liverpool	Dunkirk
Two Ships	Names unknown	Virginia for Liverpool	St. Malo
Dolphin	Postlethwaite	L'pool to Philadelphia	Dunkirk
Vernon	Bannister	Jamaica for Liverpool	Bayonne
Benson	Rawlinson	L'pool for West Indies	Retaken
Bella	Foster	,, for Tortola	Do.
Rosendale	Hodson	Maryland for L'pool	St. Malo
Cleveland	Robinson	Virginia for L'pool	Bayonne
Earl of Derby	Penkett	Liverpool to Jamaica	Bilbao
Fortune	Gardiner	London for Liverpool	St. Malo
Pretty Peggy	Rankin	Liverpool for Oporto	———
Fortune	Green	L'pool and Africa for Jamaica	Porto Cavallo (354 slaves on board)
Leopard	Williams	L'pool for Rotterdam	Bergen
Graham	Naylor	L'pool for St. Kitts	Bilbao
Fanny	Thompson	Liverpool for Africa	By French Privateer
Brunswick	Sturke	New York for L'pool	St. Augustine
Elizabeth	Steward	Do.	Do.
Antigua Packet	Gardiner	Liverpool for Leeward Islands	Martinico
Goodwill	Darby	Virginia for Liverpool	Bayonne
John & Thomas	Brownhill	Liverpool for St. Kitts	By French
Diligence	Strong	From Liverpool	———
James	Matthews	L'pool to Barbadoes	Martinico
Blackburne	Robinson	From Liverpool	———
Elijah	Hornby	For Liverpool	St. Jean de Luz
Black Prince	Woodhouse	Liverpool for Gibraltar	Rochelle
Mary	St. Leger	From Liverpool	Old Gibraltar
Molly	Clegg	Liverpool to Antigua	Ransomed for £600
Blandenburg	Lookerman	Virginia for Liverpool	Bilbao
Defiance	Drape	Jamaica for Liverpool	Morlaix
Susanna	Pierce	For Liverpool	Dieppe
Bridget	Norton	Jamaica for Liverpool	Rans'm'd for £1100
Anne	Strong	Konigsberg for L'pool	Rans'm'd for £400
Union Galley	Frith	L'pool for Carolina	Ret'k'n with 12 ships
Liverpool Merchant	———	Virginia to Liverpool	———
Anne	———	From Liverpool	———
Occupation	Saunders	Do.	———
North Carolina	Everard	N. Carolina for L'pool	Havanna

APPENDIX No. I.—Continued.

Ship's Name.	Master's Name.	Voyage.	Where carried, etc.
St. George	Grayson	Liverpool to Africa	St. Malo
Charming Betty	Barnes	Liverpool to Jamaica	Guadaloupe
Benson	Brown	Liverpool for Antigua	Do.
Anne & Mary	Johnson	Liverpool to Tortola	Martinico
Martha	Wilson	From Africa to St. Kitts	Guadaloupe
A snow & a brig	———	Liverpool and Africa to America	Do.
Blackburn	Robinson	Africa for Jamaica	San Domingo
Sea Nymph	Whitesides	Liverpool for Africa	Ransomed
Mary	Haynes	L'pool to Barbadoes	Martinico
Benin	———	Liverpool for Africa	Do.
Trygarn	Kaye	Liverpool and Africa	St. Jago de Cuba
Mary Anne	Murthland	Jamaica for Liverpool	Do.
Charming Je'ny	Chivers	L'pool for Montserrat	Guadaloupe
Nancy	Pemberton	Liverpool and Africa	St. Kitts
Kings of Brentford		Liverpool for Carolina	By Spanish privat'r
	———	———	———
The Boss	White	L'pool for New York	Sunk by French [man-of-war
Qn. of Hungary	———	Of Liverpool	

APPENDIX No. II.

The *Enterprize* Privateer, Captain James Haslam, commander. Cost of Outfit, list of Officers, &c. September 1779.* The *Enterprize*, on her first cruize, was manned as follows:—

Names.	Stations.	Wages ℔ Month.		Cash advanced as a Privateer		Notes and Cash Paid.			Total Advanced.		
		£	s.	£	s.	£	s.		£	s.	d.
James Haslam	Captain	0	0	21	0	0	0		21	0	0
John Cotter	1st Lieutenant	0	0	9	0	0	0		9	0	0
George Pearson	2nd Do.	0	0	12	17	0	0		12	17	0
James Green	3rd Do.	0	0	8	10	0	0		8	10	0
Sam Robinson	Sailing Master	0	0	9	0	0	0		9	0	0
Henry Kermitt	Master's Mate	0	0	8	0	0	0		8	0	0
John Armstrong	Do.	0	0	9	0	2	2	W	11	2	0
Francis Lake	Prize Master	0	0	9	0	0	0		9	0	0
Henry Barr	Surgeon	0	0	10	10	0	0		10	10	0
Rob. Madgett	Capt'n Marines	0	0	10	10	0	0		10	10	0
James Gowdy	Do. Mate	0	0	7	7	0	0		7	7	0
John Cooper	Carpenter	0	0	10	0	0	0		10	0	0
	Carried forward	£0	0	124	14	2	2	W	126	16	0

*Summarised from the original accounts, in the possession of — Hampson, Esq., and not Mr. Dixon, as stated in a foot-note, page 31. Matter was kindly left at the publisher's by both these gentlemen, hence the error.

APPENDIX No. II.—CONTINUED.

Names.	Stations.	Wages ℔ Month. £	s.	Cash advanced as a Privateer £	s.	Notes and Cash Paid. £	s.	Total Advanced. £	s.	d.
	Brought forward	0	0	124	14	2	2	126	16	0
Francis Gill	C'rp'nt'r's Mate	3	10	0	0	7	0	7	0	0
Edward Hodge	Boatswain	0	0	11	5	0	0	11	5	0
Henry Cowet	Do. Mate	3	15	0	0	7	10 W	7	10	0
Richard Armstrong	Gunner	4	5	0	0	8	10 W	} 15	10	0
John Sharpe Run	———	0	0	0	0	7	0 W			
John Browne	Do. Mate	3	15	0	0	7	10 W	7	10	0
David Kenny	Cook	0	0	7	10	7	10	7	10	0
William Mack	Gunner's Mate	3	15	0	0	7	10 W	7	10	0
John McCloud	Do.	3	15	0	0	7	10	7	10	0
Thomas McDonald	Cooper	3	0	0	0	0	0	6	0	0
James Armstrong	Do. Mate	3	0	0	0	0	0	6	0	0
Lewis Hughes	Prize Master	4	10	0	0	0	0	9	0	0
John Maddock	Quarter Do.	4	0	0	0	0	0	8	0	0
John Hudson	Do.	3	15	0	0	0	0	7	10	0
Morris Jones	Do.	3	15	0	0	0	0	7	10	0
Rob. Wedgwood	Do.	3	15	0	0	7	10	7	10	0
William Walton	Armourer	3	10	0	0	0	0	7	0	0
James Morton	Captain's Clerk	3	0	0	0	0	0	6	0	0
John Bryan	Ship's Steward	3	0	0	0	0	0	6	0	0
Robert Yates	Cabin Do.	3	0	0	0	0	0	6	0	0
Timothy Lee	Do. Do.	2	15	0	0	0	0	5	10	0
Herbert Davis	Sailmaker	3	15	0	0	7	10 W	8	0	6
Lans. Devley or Boyle	Boats'in's Mate	3	10	0	0	0	0	7	0	0
20 Seamen (@ 60/- to 70/-) ..		58	10	22	10	28	0	143	2	6
6 ¾ Do. (@ 45/- to 65/-) ...		17	10	0	0	0	0	41	15	9
13 ½ Do. (@ 35/- to 50/-) ...		28	10	0	0	2	8	57	3	0
9 ¼ Do. (@ 20 to 40/-) ...		14	10	0	0	0	0	30	10	0
18 Landsmen (@ 20/- to 40/-) ...		29	10	3	10	4	0	62	7	10
3 Boys and 3 Apprentices ...		1	15	3	3	0	0	15	7	10½
	£	221	5	172	12	111	10	645	8	5½

Tradesmen's Notes for the *Enterprize* Outfits.	1st. Cruise. £	s.	d.	2nd. Cruise. £	s.	d.	3rd. Cruise. £	s.	d.
Henry Clarkson, Boards, Sawing, etc.	1	7	0	3	14	0	14	14	6
James Aspinall, Glazier	2	4	0	0	16	6	18	1	0
John Parr, Arms	15	5	4	1	9	0	11	11	0
James Leigh, Medicines	12	13	9	3	4	9	14	11	0
William Earle & Son, Iron Work ...	36	4	8	11	12	6	0	0	0
Robert Tyrer, Joiner	3	15	6	0	0	0	0	0	0
Thomas Staniforth, Cordage	64	3	0	6	16	0	221	10	6
William Neale, Blockmaker	10	18	0	6	18	6	18	11	9
George Worrall, Painter	12	11	8	0	7	5	13	3	0
Hulton & Foxcroft, Brandy	0	0	0	41	18	0	0	0	0
Carried forward £	159	2	11	76	16	8	312	2	9

APPENDIX No. II.—Continued.

Tradesmen's Notes for the *Enterprize* Outfits.	1st. Cruise.			2nd. Cruise.			3rd. Cruise.		
	£	s.	d.	£	s.	d.	£	s.	d.
Brought forward	159	2	11	76	16	8	312	2	9
Anthony Mollineux, Brazier, Copper Nails, &c....	6	6	0	1	5	10	15	4	6
Edgar Corrie & Compy., Bottled Beer	13	10	0	4	19	0	16	2	4
Edward Grayson, Carpenter	9	17	0	11	7	0	0	0	0
James Carruthers, Cooper	24	1	0	3	11	4	30	17	0
Joseph Matthews, Sailmaker	58	1	0	23	16	0	84	19	7
John Kaye, Slops	20	4	0	9	16	0	0	0	0
Joseph Yates, Grocer	24	2	6	3 0	11 11	2 8	}24	2	10
Peter Rigby & Sons, Iron Hoops ...	2	19	4	0	0	0	8	13	2
*Egerton Smith, "Stationary" ...	7	12	9	1	2	9	3	0	8
John Eaton, Cartage	4	5	11	0	19	9	7	8	10
Thomas Ryan, Wine	7	19	3	0	0	0	6	15	4
Baker & Dawson, Rum	95	4	5	0	0	0	0	0	0
Captain Haslam, Mr. Dillon, and Mr. Carruthers' disbursements at Whitehaven	126	6	5	0	0	0	16	3	0
Paid for Seamen going to and from Whitehaven and Chester ...	29	18	6	0	0	0	7	7	0
Crimpage, shipping seamen and board wages, pilotage and boatage ...	46	8	2	28	19	0	17	11	8
Paid Carpenters, Joiners, riggers, labourers, &c.	48	0	8	11	12	4	176	16	6
Alex. Anderson, French and Spanish Commissions, &c.	41	17	4	0	0	0	0	0	0
Mathew Ligoe and Wm. Corf, Fresh Beef	61	4	0	72	8	0	47	9	6
Jas. Johnson and John Coleman, Bread	135	3	3	21	19	4	68	19	4
Dillon & Leyland, Beef, Pork, &c. ...	290	19	0	0	0	0	237	5	9
Cazneau & Marlin and David Shannon, Pork	44	3	2	0	0	0	48	0	0
Cheese, flour, pease, barley, butter, potatoes, greens, fowls, fish, candles, water, salt, coals ...	87	10	8	15	12	6	69	6	6
Poles, white cooperage, glasses, hoops, priming, powder, &c. ...	14	18	4	7	11	10	33	14	4
Pitch, tar, oil, ballast, graving dock, boats, cables	28	2	6	58	19	10	40	3	6
Gill Slater and Francis Holt & Co., Anchors	0	0	0	0	0	0	27	13	9
Earle & Molyneux, Ironwork	0	0	0	0	0	0	68	14	8
Grayson & Ross, Carpenters	0	0	0	0	0	0	162	15	0
John Clowes & Co., Copper Sheathing	0	0	0	0	0	0	11	12	9
Warrington Copper and Brass Co., Do.	0	0	0	0	0	0	46	18	4
John Sparling, Rum	0	0	0	0	0	0	169	14	3
Miscellaneous expenses	0	7	2	1	19	3	49	13	9
Gratuity to Seamen's mothers impressed	0	0	0	2	2	0	0	0	0
Seamen's advance wages, first cruise ...	645	8	5½	0	0	0	0	0	0
Joseph Rathbone, Shot	0	0	0	13	2	4	0	0	0
Carried forward £	2033	13	8½	372	3	7	1809	6	7

*Father of the founder of the *Liverpool Mercury*.

APPENDIX No. II.—CONTINUED.

Tradesmen's Notes for the *Enterprize* Outfits.	1st. Cruise.			2nd. Cruise.			3rd. Cruise.		
	£	s.	d.	£	s.	d.	£	s.	d.
Brought forward	2033	13	8½	372	3	7	1809	6	7
John Stanton & Son, Gunpowder ...	0	0	0	20	11	3	14	12	7
Jos. Brookes, Junr., Gunpowder, Turpentine, and Oakum £14 10 10 Less paid Sailor for Spying the Prize ... £1 1 0	0	0	0	13	9	10	0	0	0
Paid Seamen's Advance Wages, Third Cruise	0	0	0	0	0	0	589	5	0
EXPENCES ON FRENCH PRISONERS.									
To Paid Seddan Chair for the French Captain	0	0	0	0	3	0	0	0	0
,, Madam Pennant's Board ...	0	0	0	2	12	6	0	0	0
,, John Carver, Cloths for the 2nd Captain...	0	0	0	4	3	5	0	0	0
,, Tho: Seyers, making Do. Do.	0	0	0	0	16	0	0	0	0
,, Present to 2nd Captain ...	0	0	0	10	10	0	0	0	0
,, Ditto to the Captain... ...	0	0	0	10	10	0	0	0	0
,, Jas. Leigh, Medicines, Captain	0	0	0	1	10	0	0	0	0
,, John Gladhill, Prisoners' Board	0	0	0	23	1	8	0	0	0
,, John Glover, ditto, and his trouble	0	0	0	6	6	0	0	0	0
,, A present to passenger Mr. Page, and his board ...	0	0	0	11	3	8	0	0	0
,, John Kaye, shirts and cloths for Sundrys	0	0	0	7	12	3	0	0	0
,, Present to the Dispensary ...	0	0	0	21	0	0	0	0	0
Seamen's advance Wages, second cruise	0	0	0	63	4	0	0	0	0
£	2033	13	8½	568	17	2	2413	4	2

TOTAL OUTFITS.

	£	s.	d.
First Cruise	2033	13	8½
Second Do.	568	17	2
Third Do.	2413	4	2
	£5015	15	0½
Value of the Ship ...	2050	0	0
	£7065	15	0½

A LIST OF THE OWNERS:—

Thomas Earle ...	3/16 the above
Edgar Corrie... ...	2/16 ,,
Francis Ingram ...	2/16 ,,
William Earle ...	2/16 ,,
Dillon & Leyland ...	2/16 ,,
Peter Freeland ...	1/16 ,,
Thomas Eagles ...	1/16 ,,
Edward Chaffers ...	1/16 ,,
James Carruthers ...	1/16 ,,
William Denison ...	1/16 ,,

APPENDIX No. III.

List of Vessels trading to and from Liverpool, Captured by the Enemy during the Seven Years' War, 1756-1763.

Ship's Name.	Master's Name.	Voyage.	By whom taken, and where carried.
York	Fowkes	Jamaica to Liverpool	Retaken by Guernsey Privateer
Betty	Logan	L'pool for Philadelphia	Cape Breton
Mary	Richmond	Do. Virginia	———
Landovery	Johnston	Do. Jamaica	France
Annabella	Settle	From Cape Fear	Morlaix
Fanny	Henderwell	Lyme for Liverpool	Havre
True Love	King	Do.	Do
Happy Return	———	L'pool for Carolina	Bayonne
———	Fisher	Maryland for L'pool	Do.
Anne	Ford	Rye for Liverpool	By French in Romney Bay
Orrell	Winter	Saloe for Liverpool	Marseilles
Hougwart	Martin	N. Yarmouth for L'pool	Boulogne
Schemer	Nichols	From Africa	Martinico
Austin	Holme	L'pool for Barbadoes	Do.
Margaretta	Hornby	Liverpool for London	Havre
Dolly & Nancy	Isaac Winn	Dartmouth to L'pool	Dieppe
Mary	Printon	Malaga for Liverpool	Malaga
York	———	L'pool for New York	By B'rdeau Priv't'er
Falmouth	Pole	Liverpool for Boston	Brest
Grampus	Corbett	Liverpool for Gambia	Bayonne
Lady Strange	Harrison	L'pool for Barbadoes	Do.
Snow Hesketh	Thos. Onslow	L'pool for Jamaica	Guadaloupe
Eliz'b'th & M'ry	Caruthers	Do. Africa, &c.	On Coast of Africa
King George	Jackson	Do. Do.	———
Ogden	Lawson	Do. Do.	———
Lloyd	Sweeting	Maryland for L'pool	Cape Breton
Lady Charlotte	Oakes	Barcelona for L'pool	By Pt. Mahon Pr'te'r
Anson Privateer	Cuthbert	From Liverpool	Rochefort
Crown Point	Lawrence	L'pool for New York	Norway
Patterson	Cole	Do.	Bayonne
Success	Catterwood	L'pool for Jamaica	Dieppe
Quester	Potter	L'pool and Africa for Jamaica with 87 slaves	St. Eustatia
Cavendish	———	Do. with 170 slaves	Guadaloupe
———	Jones	Do.	On Coast of Guinea
The Pickering	———	Do.	Guadaloupe
Two Snows	(Names unknown)	Do.	Do.
Hankinson	Dodgson	St. Pet'rsb'rg f'r L'pool	Norway
Ellis	Simpson	Jamaica for L'pool	Taken by the French
A Snow	———	Do.	St. Jean de Luz
Eliza	Parker	From Liverpool	Cadiz
Swan	Cowan	Liverpool for Africa	By the French
A Brig	Barnes	From Liverpool	Do. on the Coast of Africa
Adventure	Geo. Washington	L'pool for Jamaica	Port Louis

APPENDIX No. III.—Continued.

Ship's Name.	Master's Name.	Voyage.	By whom taken and where carried, etc.
Whidah	Hammill	Liverpool for Africa	By the Mauchault privateer of 24 guns, & 300 men, from Granville
Salisbury	Key	Do. Do.	
A Schooner	Hendrickson	From Liverpool	By the French
Adlington	Frierson	L'pool for Barbadoes	Retak'n by Guerns'y Privateer
Henry	Bond	Do. Do.	Martinico
Leghorn Trader	Hooper	Leghorn to Liverpool	Dieppe
Ch'rm'ng Rach'l	Scott	New York for L'pool	By Louisburg Pr't'rs
———	Marshall	Virginia for Liverpool	Do. Do.
Pemberton	W'lt'r Kirkpatrick	L'pool for Africa, &c.	Bayonne
Aurora	Josiah Wilson	———	Do.
A Ship	———	L'pool for Jamaica	By French Priv'te'r
Dragon	———	Do. St. Petersburg	Do. and Ransomed
Hopewell	Ford	Arundel for L'pool, with Corn	France
Success	Clare	Liverpool to Leeward Islands	Guadaloupe
Phœnix	Nobler	L'pool for Africa, &c.	By negroes on the Coast, & set on fire
Snow Betty	Wm. Creevey	Do. Gambia	Sunk by the French
Betty & Peggy	Hollingsworth	Stockholm for L'pool	Dunkirk
Henry	Thornton	Virginia for L'pool	By Belleisle Pr'te'r
Judith	Hayes	L'pool and Africa for America	Granada
Molly	Timothy Wheel-Wright	From Windward and Gold Coast	By the French
John	Peter Gibson	Liverpool to Virginia	Port Louis
Rose	Bashaw	Do. for Tortola	Martinico
Betsey	Watt	Do. for Virginia	Quebec
Biddy	Hamilton	Fr'm Windward Coast of Africa	Guadaloupe
Charming Betty	Colley	Do. Do.	Do.
Salisbury	John Sacheverell	From the Cameroons, with slaves	Do.
Lievsley	T. Onslow	From Calabar, with 323 slaves, 2 tons of teeth and other goods	Do.
Nelly	Hickhins	———	Taken by French
Providence	Clare	From Liverpool	Do. Do. in the West Indies
Defiance	Campbell	L'pool for Africa, &c.	———
America	Nicholson	L'pool for the Baltic	Christiansund
Catherine	Seth Houghton	L'pool for Montserrat	Vigo
Granville	Spears	From North Carolina	Ransomed for £500
Perfect	Potter	L'pool and Africa, &c.	Lost at Mana by mutiny of slaves, who killed nearly all the crew on board

APPENDIX No. IV.—Continued.

Ships.	Commanders.	Owners.	Tons.	Guns.	Men.
Brilliant	Wm. Priestman	John Sparling	600	20	—
Bellona	Fairweather	Bolden & Co.	250	24	140
Brooks	Noble	———	320	20	45
Basil	Robt. Leake	Peter Freeland	—	12	—
Bess	Perry	Slater & Co.	270	18	100
Benson	Ball	Rawlinson, Chorley & Grierson	360	20	79
Catcher	Fletcher	Salisbury & Co.	110	14	80
Commerce	Woods	———	—	14	42
Dragon	Briggs	Warren & Co.	112	14	80
Dreadnought	Taylor	Wagner & Co.*	200	20	120
Delight	Dawson	Rawlinson & Co.	120	12	39
Ellis	Washington (or Jolly)	Boats & Gregson	340	28	130
Eagle	Bond	Salisbury & Co.	110	14	80
Enterprise†	Pearce	Brooks & Co.	250	20	70
Ellen	Fell	France & Co.	200	20	70
Fly	Briggs	———	—	14	70
Griffin	Grimshaw	Hall & Co.	130	14	90
Greenwood	Reid	Crosbie & Greenwood	250	16	50
Gregson	Boyd (or Jolly)	Boats & Gregson	250	24	120
Hornet	Naylor	Liversley & Co.	120	16	90
Hawke	Bradley	Mason & Co.	120	16	70
Hercules	Wright	Whitaker & Co.	1200	30	100
Harlequin	Fayrer	Earle & Co.	180	20	—
Hope	Melling	Crosdale, Barrow & Co.	250	16	—
Industry	Jno. Moore	Meyer, Wilckens & Co.	200	10	—
Isabella	Wiseman	Gill Slater	300	16	80
James	Jno. Amery	James France	—	20	—
Jenny	Adams	Chorley & Co.	130	14	35
Jenny	Wade	Thos. Moss & Co.	250	14	70
Jenny	Ashton	Ashton & Co., of Tortola	80	12	30
Jenny (ship)	Walker	Tarleton & Co., or Daniel Backhouse	140	14	40
Jenny (brig)	Gill	Daniel Backhouse	—	16	—
Jamaica	Fletcher	Birch & Co.	350	18	110
Juno	Beaver	Hartley & Co.	90	14	40
Knight	Wilson	Hindley, Leigh & Co.	220	18	80
Liverpool	Wilcox	S. Shaw & Co.	210	16	45
Little Ben	Bostock	Radcliffe & Co.	110	14	50
Livingston	———	Rawlinson & Co.	400	20	—
Lady Granby	Powell	Ashton & Co.	45	10	60
Marchioness of Granby	Rogers	Marquis of Granby and Nicholas Ashton, Esq.	260	20	130
Molly	Kendall	Gregson & Co.	260	16	70
Mary Fearon	Caton	France & Co.	280	16	60
Mentor	Jno. Dawson	Baker & Co.	400	28	102
Mermaid	Smith	Sparling & Co.	250	16	50
Mary	Bonsall	Drinkwater & Co.	130	16	40
Molly	Woods	Rawlinson & Co.	240	14	40

* Mr. B. P. Wagner was the maternal grandfather of Felicia Dorothea Hemans.

† Afterwards owned by Francis Ingram & Co., and commanded by Captain Haslam.

APPENDIX No. IV.—Continued.

Ships.	Commanders.	Owners.	Tons.	Guns	Men.
Mersey	Gibbons	Whitaker & Co.	1400	28	100
Nancy	Hammond	Fowden & Berry	250	20	59
Nancy	Nelson	Pringle & Co.	150	16	50
Nanny	Harrison	Watts & Rawson	220	14	70
Nanny	Beynon	Hindley, Leigh & Co.	220	14	50
Patsey	La'ren'e Dooling	Rawlinson & Co.	—	18	—
Pole	Maddocks	Nelson & Co.	320	24	100
Pallas	Townsend	I. & R. Slinger	—	16	—
Queen	Gee	Richard Kent	800	20	100
Rose	Jackson	J. Zuill & Co.	120	14	40
Richard	Lee	Rawlinson & Co.	150	16	70
Resolution	Beard	Holme & Co.	400	22	100
Retaliation	Townsend	Syers & Co.	160	16	100
Revenge	Ramsey	Hughes & Co.	120	14	80
Rover	Bancroft	Kennion & Co.	120	14	80
Rumbold	Fayrer	Caruthers & Co.	250	20	57
St. George	Hanley	Warren & Co.	110	14	75
Sturdy Beggar	Cooper	Davenport & Co.	160	16	100
Sarah Goulburn	Lewtas	Brown & Jones	340	26	120
St. Peter	N. Holland	Holme, Bowyer & Kennion	320	22	147
Spy	Rigmaiden	J. Zuill & Co.	120	14	40
Spitfire	Bell	Do.	200	16	100
Success	Nevin	Crosbie & Greenwood	120	12	30
Sparling	Plato Denny	John Sparling	400	18	—
Do.	Ed. Forbes	Do.	300	14	—
Sally	Rimmer	Watts & Rawson	180	16	70
Sisters	Webster	Whitaker & Co.	800	20	100
Townside	Watmough	Mitton & Co.	130	16	90
Terrible	Ash	Nottingham & Co.	250	20	130
Tom	Davies	Mr. Clement	100	12	36
Tartar	Allanson	J. Backhouse & Co.	90	18 or 14	80
Tom	Lee	———	—	12	—
Two Brothers	Fisher	Roberts & Co.	150	16	39
Tyger	Qualtrough	———	200	14	70
Tyger	Amery	James France & Co.	300	16	60
Tryal	Eagle	———	—	14	80
Ulysses	Briggs	Baker & Dawson	250	16	—
Viper	Cowell	Birch & Co.	160	18	80
Viper	Philip Cowell	Gregson & Bridge	—	16	40
Wasp	Byrne	Kennion & Co.	220	14	95
Watt	Coulthard		—	32	164
Young Henry	Corrie	Hartley & Co.	270	18	60

APPENDIX No. V.

The *Swallow*, Letter of Marque Against the French, Dated 12th July 1796.

"GEORGE the Third, by the Grace of God of Great Britain, France, and Ireland, King Defender of the Faith; To all persons to whom these presents shall come Greeting. Whereas divers injurious proceedings have lately been had in France, in derogation of the honor of our Crown, and of the just rights of our subjects, and whereas several unjust seizures have been there made of the Ships and Goods of our subjects, contrary to the laws of nations, and to the faith of treaties. And whereas the said Acts of unprovoked hostility have been followed by an open declaration of war against us and our ally, the Republic of the United Provinces. We therefore, being determined to take such measures as are necessary for vindicating the honor of our Crown, and for procuring reparation and satisfaction for our injured subjects, did by, and with the advice of our Privy Council, order that general reprisals be granted against the Ships, Goods, and Subjects of France, so that as well our Fleets and Ships as also all other Ships and Vessels that shall be commissionated by Letters of Marque or general reprisals or otherwise, shall and may lawfully apprehend, seize, and take all Ships, Vessels, and Goods belonging to France, or to any persons being subjects of France, or inhabitating within the Territories of France, and bring the same to judgment in our High Court of Admiralty of England, or in any of our Courts of Admiralty within our Dominions, for proceedings and adjudication and condemnation to be thereupon had according to the course of Admiralty, and the laws of Nations. And, whereas by our commission under our Great Seal of Great Britain, bearing date the Fourteenth day of February, One Thousand Seven hundred and Ninety-three, we have willed, required, and authorized our commissioners for executing the office of Lord High Admiral of Great

Britain, or any person or persons by them empowered and appointed to issue forth and grant Letters of Marque and reprisals accordingly, and with such powers and clauses to be therein inserted, and in such manner as by our said commission more at large appeareth. And, whereas our said Commissioners for executing the office of our High Admiral aforesaid, have thought JOHN MACIVER fitly qualified, who hath equipped, furnished, and victualled a ship called the *Swallow*, of the burthen of about *two hundred and fifty-six* tons, *British built, square stern, scroll head*, and *two* masts, mounted with *eighteen* carriage guns carrying shot of *six* pounds weight, and no swivel guns, and navigated by *thirty-five* men, of whom one third are landsmen, and belonging to the *Port of Liverpool*, whereof he the said JOHN MACIVER is commander, and that THOMAS TWEMLOW, PETER MACIVER, SAMUEL McDOWALL, and IVER MACIVER, *of Liverpool, Merchants, and him the said* JOHN MACIVER are the owners. And, whereas he the said JOHN MACIVER hath given sufficient bail with sureties to us in our said High Court of Admiralty, according to the effect and form set down in our instructions made the Fourteenth day of February, One Thousand Seven Hundred and Ninety-three, in the Thirty-third year of our reign, a copy whereof is given to the said Captain, JOHN MACIVER. Know ye therefore, that we do by these presents issue forth and grant Letters of Marque and reprisals to, and do license and authorize the said JOHN MACIVER to set forth in a warlike manner the said ship called the "*Swallow*," under his own command, and therewith by force of arms to apprehend, seize, and take the Ships, Vessels, and Goods belonging to France, or to any persons being subjects of France, or inhabiting within any of the territories of France, excepting only within the harbours or roads of Princes and States in amity with us, and to bring the same to such port as shall be most convenient in order to have them legally adjudged in our said High Court of Admiralty of England or before the Judges of such other Admiralty Court as shall be lawfully authorized within our Dominions, while being finally condemned it shall and may be lawful for the said JOHN MACIVER to sell and dispose of such Ships, Vessels and Goods finally adjudged and condemned, in such sort and manner as by the Court of Admiralty hath been accustomed. Provided, always, that the said JOHN MACIVER keep an exact journal of his proceedings, and therein particularly take notice of all prizes which shall be taken by him, the nature of such prizes, the times and places of their being taken, and the values of them as near as he can judge, as also of the station, motion and strength of the French as well as he or his mariners can discover by the best intelligence he can get, and also of whatsoever else shall occur unto him or any of his officers or mariners, or be discovered or declared unto him or them, or found out by examination or conference with any mariners or passengers of or in any of the Ships or Vessels taken, or by any other person or persons, or by any other ways and means whatsoever, touching or concerning the designs of the French, or any of their Fleets, Vessels or Parties, and of their stations, ports and places, and of their intents therein, and of what Ships or Vessels of the French bound out or home, or to any other place, as he or his officers or mariners shall hear of, and of what else material in these cases may arrive to his or their knowledge, of all which he shall, from time to time, as he shall or may have opportunity, transmit an account to our said commissioners for executing the office of our High Admiral aforesaid, or their secretary, and keep a correspondence with them by all opportunities that shall present. And, further, providing that nothing be done by the said JOHN MACIVER, or any of his officers, mariners, or company, contrary to the true meaning of our aforesaid instructions, but that the said instructions shall by them, and each and every of them, as far as they or any of them are therein concerned, in all particulars be well and truly performed and observed. And we pray and desire all Kings, Princes, Potentates, States, and Republicks, being our friends and allies, and all others to whom it shall appertain, to give the said JOHN MACIVER all aid, assistance and succour in their ports with his said ship, company and prizes, without doing or suffering to be done to him any wrong, trouble or hindrance, we offering to do the like when we shall be by them thereunto desired. And we will, and require all our officers whatsoever, to give him succour and assistance as occasion shall require.

"In witness whereof we have caused the great seal of our High Court of our Admiralty of England to be hereunto affixed. Given at London the *Twelfth* day of *July* in the year of our Lord One Thousand Seven Hundred and Ninety-six, and in the Thirty-sixth year of our reign.

"ARDEN,

"Registrar."

APPENDIX TO SLAVE TRADE SECTION.

APPENDIX No. VI.

A List of the Company of Merchants trading to Africa (established by an Act of Parliament passed in the 23rd of George II., entituled an "Act for Extending and Improving the Trade to Africa"), belonging to Liverpool, June 24th, 1752.

- Armitage, John
- Atherton, John
- Ashton, John
- Bostock, John
- Bulkeley, William
- Blundell, Jonathan
- Backhouse, John
- Blundell, Bryan
- Blundell, Richard
- Blackburn, John
- Bradley, George
- Brooks, George
- Benson, Wm.
- Ball, Thomas
- Bridge, Edward
- Blundell, William
- Brooks, Joseph
- Brooks, Jonathan
- Bird, Joseph
- Crowder, Thomas
- Crosbie, James
- Cunliffe, Foster
- Cunliffe, Ellis
- Cunliffe, Robert
- Campbell, George
- Clay, Robert
- Craven, Charles
- Clayton, John
- Crompton, John
- Clews, George
- Chalmer, Thomas
- Davis, Joseph
- Dean, Edward
- Dobb, William
- Dunbar, Thomas
- Earl, Ralph
- Eddie, David
- Ellams, Elliott
- Forbes, Edward
- Farmer, Joseph
- Ford Richard
- Fletcher, Potter
- Gildart, Richard
- Goodwin, William
- Goore, Charles
- Gorrell, John
- Gildart, James
- Gordon, James
- Goodwin, John
- Hardman, John
- Heywood, Arthur
- Heywood, Benja.
- Hesketh, Robert
- Hughes, Richard
- Hardwar, Henry
- Higginson, William
- Hallhead, Robert
- Hughes, John Capt.
- Kendall, Thomas
- Knight, John
- Leatherbarrow, Th.
- Laidler, George
- Lee, Pierce
- Lowndes, Edward
- Lowndes, Charles
- Mears, Thomas
- Manesty, Joseph
- Nicholas, Richard
- Nicholson, John
- Ogden, Samuel
- Ogden, Edmund
- Oldham, Isaac
- Okill, John
- Pritchard, Owen
- Parr, John
- Parr, Edward
- Pardoe, James
- Penket, William
- Pole, William
- Parker, John
- Rowe, William
- Reed, Samuel
- Strong, Matthew'
- Shaw, Samuel
- Savage, Richard
- Seel, Thomas
- Strong, John
- Smith, Samuel
- Seel, Robert
- Smith, Rob., Broad-st., London
- Tarleton, John
- Townsend, Henry
- Townsend, Richard
- Trafford, Edward
- Tarleton, John
- Unsworth, Levinus
- Williamson, Wm.
- Whytell, Christo
- Whalley, William
- White, Hen., Lanc.
- Williamson, John

Total 101.

"N.B.—There are One Hundred and thirty five merchants free of the African company in London, and One Hundred and fifty seven in Bristol, whereas their Trade to Africa is not so extensive as the Merchants of Liverpool."*

* From "Williamson's Liverpool Memorandum Book, 1753," in the possession of Richard Cyril Lockett, Esq.

APPENDIX No. VII.

A LIST OF THE GUINEAMEN BELONGING TO LIVERPOOL IN THE YEAR 1752, with their Owners' and Commanders' names and the number of slaves carried by each :—*

Ships.	Commanders.	Where Bound.	Owners.	No. of Slaves.
Africa	Harrison	Benin	Jno. Welsh & Co.	250
African	John Newton	Win'd and Gold Coast	J. Manesty & Co.	250
Annabella	Wm. Harrison	Do.	W. Dobb & Co.	260
Antigua Merchant	Robt. Thomas	Angola	Jas. Gildart & Co.	200
Anglesey	James Caruthers	Win'd and Gold Coast	Tine, Farrar & Co.	180
Alice Galley	Rich. Jackson	Do.	R. Cheshyre & Co.	350
Ann Galley	Neh'm'h Holland	Calabar	Wm. Whalley & Co.	340
Adlington	Tho. Perkin	Win'd and Gold Coast	J. Manesty & Co.	320
Allen	Jas. Strangeways	Do.	J. Brooks & Co.	250
Achilles	Thomas Patrick	———	Hen. Hardwar & Co.	450
Betty	Sm. Sacheverelle	———	John Robinson	100
Blake	Alex. Torbet	Calabar	Jo. Bird & Co.	460
Barbadoes Merchant	John Wilson	Angola	G. Campbell & Co.	500
Boyne	Wm. Wilkinson	Bonny	E. Forbes & Co.	400
Beverley	William Lowe	Angola	E. Lowndes & Co.	200
Brooke	Thomas Kewley	Old Calabar	Roger Brooks & Compy.	400
Barclay	John Gadson	Do.	Jno. Welsh & Co.	450
Bulkeley	Chris. Baitson	Win'd and Gold Coast	Foster Cunliffe, Sons & Co.	350
Britannia	Jas. Pemberton	Do.	Thos. Leatherbarrow & Co.	300
Bridget	Anth'ny Grayson (or Hayston)	Do.	Foster Cunliffe, Sons & Co.	250
Clayton	Patrick	———	———	—
Cumberland	John Griffin	Gambia	E. Deane & Co.	260
Chesterfield	Patrick Black	Old Calabar	W. Whalley & Co.	440
Charm'g Nancy	Tho. Roberts	Win'd and Gold Coast	W. Davenport & Co.	170

* Compiled from a rare copy of *Williamson's Liverpool Memorandum Book*, published in 1753, in the possession of Richard Cyril Lockett, Esq., and exhibited on his behalf before the Historic Society of Lancashire and Cheshire, by J. Paul Rylands, Esq., F.S.A., March, 1897.

APPENDIX NO. VII.—CONTINUED.

Ships.	Commanders.	Where Bound.	Owners.	No. of Slaves.
Cavendish	Robert Jennings	Win'd and Gold Coast	Nicholas & Co.	170
Cecilia	Rd. Younge	Gambia	Fr. Green & Co.	120
Duke of Cumberland	John Crosbie	Bonny	J. Crosbie & Co.	450
Dolphin	Joseph Pederick	Win'd and Gold Coast	Ed. Forbes & Co.	200
Elizabeth	William Heys	Gambia	Sam. Shaw & Co.	200
Elijah	———	Win'd and Gold Coast	E. Lowndes & Co.	200
Enterprise	Sam. Greenhow	Gambia (miss'g)	John Yates & Co.	130
Ellis and Rob't.	Rich. Jackson	Win'd and Gold Coast	F. Cunliffe & Sons	32[illegible]
Eaton	John Hughes	Angola	John Okill & Co. (Wood & Teeth) 550	
Fanny	Wm. Jenkinson	Win'd and Gold Coast	J. Knight & Co.	120
Ferret	Joseph Welch	Do.	Jno. Welch & Co.	50
Florimel	Samuel Linaker	Calabar	Rich. Townsend & Co.	320
Frodsham	James Powell	Angola	Nich. Torr & Co.	450
Fortune	Hugh Williams	Bonny	Hy. Townsend & Co.	480
Foster	Edward Cropper	Benin	Foster Cunliffe, Sons & Co.	200
George	Charles Cooke	Angola	G. Campbell & Co.	250
Grace	———	Old Calabar	Ed. Forbes & Co.	400
Greyhound	Maurice Roach	Win'd and Gold Coast	Rd. Savage & Co.	120
Hesketh	James Thompson	New Calabar	R. Nicholas & Co.	260
Hector	Brook Kelsall	Do.	W. Gregson & Co.	480
Hardman	Joseph Yoward	Win'd and Gold Coast	J. Hardman & Co.	300
Jenny	Thos. Darbyshire	Do.	Jno. Knight & Co.	450
Judith	Nich. Southworth	Bonny	Jno. Welch & Co.	350
James	Jno. Sacherevelle	Win'd and Gold Coast	James Gildart	120
Knight	Wm. Boats	Do.	Jno. Knight & Co.	400
Lintott	Ralph Lowe	New Calabar	R. Nicholas & Co.	400
Lord Strange	Edward Smith	Benin	Wm. Halliday & Co.	230
Lovely Betty	George Jackson	Win'd and Gold Coast	Geo. Campbell & Co.	140
Little Billy	Thos. Dickenson	Do.	J. Knight & Co.	60
Mersey	John Gee	Benin	J. Kennion & Co.	300
Middleham	John Welch	Old Calabar	R. Gildart & Sons	320
Methwen	John Coppell	Win'd and Gold Coast	J. Crosbie & Co.	280
Minerva	Thomas Jordan	Gambia	Jas. Pardoe & Co.	400
Mercury	John Walker	Win'd and Gold Coast	Kennion & Holme	100

APPENDIX No. VII.—CONTINUED.

Ships.	Commanders.	Where Bound.	Owners.	No. of Slaves.
Molly	Richard Rigby	Win'd and Gold Coast	R. Golding & Co.	320
Neptune	Tho. Thompson	Old Calabar	Joseph and Jona. Brooks & Co.	450
Nelly	Joseph Drape (or Jno. Simmons)	Do.	Wm. Williamson & Co.	320
Nancy	John Honeyford	Bonny	T. Kendal & Co.	400
Nancy	Robert Hewin	Do.	Pet. Holme & Co.	400
Nancy	Thos Midgeley	Gambia	Knight, Mairs, & Co.	300
Orrel	James Griffin	Do.	W. Whalley & Co.	120
Ormond Success	———	Angola	Wm. Williamson & Co.	300
Pardoe	———	Win'd and Gold Coast	Jas. Pardoe & Co.	240
Priscilla	Wm. Parkinson	Angola	Jno. Welch & Co.	350
Phoebe	Wm. Lawson	Win'd and Gold Coast	Arthur and Ben. Heywood & Co.	280
Prince William	John Valentine	Angola	R. Gildart & Sons	200
Rider	Michael Rush	Do.	R. Gildart & Sons	300
Ranger	James Sanders	Win'd and Gold Coast	W. Farington & Co.	300
Sarah	Alex. Lawson	Bonny	T. Crowder & Co.	550
Salisbury	Thos. Marsden	Old Calabar	Robert Armitage	350
Sterling Castle	Charles Gardner	Bonny	John Backhouse & Co.	300
Samuel and Nancy	James Lowe	Win'd and Gold Coast	R. Savage & Co.	220
Swan	Peter Leay	Bonny	John Tarleton & Compy.	400
Sam'y & Biddy	Rob. Grayson	Win'd Coast	J. Blundell & Co.	120
Schemer	Robt. Grimshaw	Do.	T. Chalmers & Co.	120
Stronge	Thomas Cubbin	Bonny	Mat. and Jno. Stronge & Co.	300
Tarlton	Jas. Thompson	Do.	J. Tarlton & Co.	340
Triton	Chas Jenkinson	Do.	Levinus Unsworth & Co.	240
Thomas	Jas. Hutchinson	Gambia	G. Campbell & Co.	200
True Blue	Benjamin Wade	Benin	J. Cheshyre & Co.	300
Thomas and Martha	Jn. Gillman	Win'd and Gold Coast	G. Campbell & Co.	200
*Vigilant	Wm. Freeman	Do.	J. Bridge & Co.	160
Union	Tim. Anyon	Do.	J. Pardoe & Co.	350
William and Betty	Thos. Barclay	Angola	S. Shawe & Co.	400

* Missing.

Total 88 Ships carrying upwards of 24,730 Slaves, and 550 wood and teeth.

APPENDIX No. VIII.

The number of ships which cleared out from the port of Liverpool to the coast of Africa, from the earliest date to the time of the trade being abolished in May, 1807.* The majority of these vessels were employed in the slave trade, the rest carrying only wood and teeth. For instance, during the period covered by Appendix IX., (1783-93) 43 ships carried wood and teeth, while 878 carried slaves.

Year.	Ships	Tons.	Year.	Ships.	Tons.	Year.	Ships.	Tons.
1709	1	30	1768	81	8,302	1789	66	11,564
1730	15	1111	1769	90	9,852	1790	91	17,917
1737	33	2756	1770	96	9,818	1791	102	19,610
1744	34	2698	1771	105	10,929	1792	132	22,402
1751	53	5334	1772	100	10,159	1793	52	10,544
1752	58	5437	1773	105	11,056	1794		
1753	72	7547	1774	92	9,859	1795	59	
1754	71	5463	1775	81	9,200	1796	94	
1755	41	4052	1776	57	7,078	1797	90	20,415
1756	60	5147	1777	30	4,060	1798	149	34,937
1757	47	5050	1778	26	3,651	1799	134	34,966
1758	51	5229	1779	11	1,205	1800	120	33,774
1759	58	5892	1780	32	4,275	1801	122	28,429
1760	74	8178	1781	43	5,720	1802	122	30,796
1761	69	7309	1782	47	6,209	1803	83	15,534
1762	61	6752	1783	85	12,294	1804	126	27,322
1763	65	6650	1784	67	9,568	1805	117	26,536
1764	74	7978	1785	79	10,982	1806	111	25,949
1765	83	9382	1786	92	13,971	1807	74	17,806
1766	65	6650	1787	81	14,012			
1767	83	8345	1788	73	13,394			

N.B.—From the first day of January, 1806, to the first day of May, 1807, there had sailed from the port of Liverpool 185 African ships, measuring 43,755 tons, which were allowed to carry 49,213 slaves.

APPENDIX No. IX.

A list of the houses that annually imported upwards of 1000 slaves, the number of ships employed, and slaves by them imported from 1783 to 1793, both inclusive, whereby is seen the proportion which they held to all the slave-

* Troughton's "History of Liverpool,' p. 265.

vessels that annually sailed from the port of Liverpool during that period :—*

Years.		Houses.		Ships.		Houses.		Ships.		Slaves.
In 1783	there were	42	and	85	of which	13	employed	47	and imported	26,820
,, 1784	,,	33	,,	59	,,	10	,,	28	,,	13,590
,, 1785	,,	37	,,	73	,,	9	,,	36	,,	18,020
,, 1786	,,	37	,,	87	,,	13	,,	53	,,	21,520
,, 1787	,,	27	,,	72	,,	8	,,	39	,,	17,130
,, 1788	,,	28	,,	71	,,	8	,,	35	,,	13,606
,, 1789	,,	29	,,	62	,,	6	,,	32	,,	10,752
,, 1790	,,	30	,,	89	,,	10	,,	58	,,	19,089
,, 1791	,,	38	,,	101	,,	10	,,	56	,,	19,027
,, 1792	,,	33	,,	133	,,	14	,,	94	,,	29,905
,, 1793	,,	25	,,	46	,,	6	,,	24	,,	7,325
		359		878		107		502		196,784

APPENDIX No. X.

A list of the Company of Merchants trading to Africa (established by an Act of 23 of George II., Cap. 31, entitled: "An Act for the extending and improving the trade to Africa, 1750, for the port of Liverpool"), in 1807 :—

John Bridge Aspinall
James Aspinall
William Aspinall
Daniel Backhouse
John Backhouse, Wavertree
John Barnes, London
Ralph Benson
Robert Bent, London
Patrick Black
Jonas Bold
John Bolton
P. W. Brancker
Thomas Brancker
Joseph Brooks
John Brown
George Brown, Wales
James Carruthers
George Case
Henry Clarke, Belmont, Cheshire
Thomas Clarke
Samuel Clough
Edgar Corrie
William Crosbie
James Thompson Cukit
John Dawson
Edward Dickson
James Dickson
William Dickson
Thomas Earle
William Earle
William Forbes
James Gregson
James Gildart
Thomas Golightly
John Greenwood
William Harding
William Harper
B. A. Heywood
Thomas Hinde
Thomas Hodgson
John Hodgson
H. Blundell Hollinshead
Francis Ingram, Wakefield
John Chambres Jones, Wales
Peter Kennion, London
John Langton, Kirkham
Roger Leigh
George Lewis
William Neilson
Thomas Parke, Highfield
Thomas John Parke
Thomas Parr
Thomas Parr, Junr.
James Penny
Jonathan Ratcliffe
William Rigg
John Sanders
Christopher Shaw
John Shaw
Bryan Smith
George Spencer, London
Samuel Staniforth
Thomas Tarleton
John Tarleton
Thomas Moss Tate
William Thompson
James Watkinson
Richard Willis
William Watson
Richard Wilding
William Woodville, Havana
Richard Woodward

* "A General and Descriptive History of Liverpool" (1795).

APPENDIX No. XI.

Comparative Statement of Ships cleared out from the Ports of London, Liverpool, and Bristol, to the Coast of Africa, for ten years, from 1795 to 1804 inclusive.*

	London.		Bristol.		Liverpool.		Total.		Each Ship.
Year.	Ships.	Slaves Allowed.	Ships	Slaves Allowed.	Ships.	Slaves Allowed.	Ships.	Slaves.	M'dium Slaves.
1795	14	5,149	6	2,402	59	17,647	79	25,198	317
1796	8	2,593	1	393	94	29,425	103	32,411	315
1797	12	4,225	2	801	90	29,958	104	34,984	336
1798	8	2,650	3	1,433	149	53,051	160	57,104	356
1799	17	5,582	5	2,529	134	47,517	156	55,628	356
1800	10	2,231	3	717	120	31,844	133	34,722	261
1801	23	6,347	2	586	122	30,913	147	37,846	259
1802	30	9,011	3	704	122	31,371	155	41,086	266
1803	15	3,616	1	355	83	29,954	99	24,925	253
1804	18	5,001	3	798	126	31,090	147	36,899	244
10 years		46,405		10,718		323,770		380,893	

* Troughton's History of Liverpool, p. 266.

APPENDIX No. XII.

PAID FOR A NEGRO MAN AT BONNY, IN 1801 :—

One piece of Chintz, eighteen yards long.
One piece of Baft, eighteen yards long.
One piece of Chelloe, eighteen yards long.
One piece of Bandanoe, seven handkerchiefs.
One piece of Niccannee, fourteen yards long.
One piece of Cushtae, fourteen yards long.
One piece of Photae, fourteen yards long.
Three pieces of Romalls, forty-five handkerchiefs.
One large Brass Pan, two muskets.
Twenty-five kegs of powder, one hundred flints.
Two bags of shots, twenty knives.
Four iron pots, four hats, four caps.
Four cutlasses, six bunches of beads, fourteen gallons of brandy.
These articles cost about £25.

APPENDIX No. XIII.

LIST OF GUINEAMEN BELONGING TO THE PORT OF LIVERPOOL WHICH SAILED FOR AFRICA, from the 5th of January, 1798, to the 5th of January, 1799,* with owners' and commanders' names, and the complement of slaves allowed to each:—

Ships.	Captains.	Destinations.	Owners.	No. of Slaves.	Sailed.
Fair Penitent	John Gardner	Win'd Coast	S. McDowall & Co.	261	Jan. 8
Union	Robert Dowie	Gabon	J. Rackham & Co.	162	
Mercury	John Mill	Win'd Coast	Wm. Begg & Co.	370	9
Pilgrim	Robert Pince	Do.	R. Leigh & Co.	425	18
Mary	P. Henshall	Angola	J. Rackham & Co.	285	Feb. 2
Favourite	H. Bennet	Bonny	Neilson & Heathcote	666	
Kitty	George Walker	Old Calebar	J. & H. Clarke & Co.	505	
Lord Stanley	W. Murdock	Do.	Do.	394	
James	John Miller	New Calebar	W. Dickson & Sons	337	
Thomas	G. Farquhar	Angola	Neilson & Heathcote	442	
Penelope	Luke Mann	Bonny	W. Thompson & Co.	389	
Prince	John Kendall	Angola	J. Smith & Co.	435	
Parr	D. Christian	Bonny	Thos. Parr & Co.	700	5
Fame	Thomas Brade	New Calebar	A. Joseph & Mozely	250	18
Abigail	W. Williams	Angola	J. Tarleton, junr.	302	
Amelia Eleanor	Edward Duncan	Do.	W. Brettargh & Co.	440	
Triton	John Corran	Do.	W. Corran & Co.	448	28
Anne	John Muir	Cameroons	W. Begg & Co.	300	
John	N. Ireland	Do.	Tarleton & Backhouse	265	Mar. 1
Britannia	John Walker	Angola	Do.	238	
Unity	E. Lovelace	Do.	Jos. Greaves & Co.	100	11
Cecilia	James Blake	Do.	W. Thompson & Co.	285	12
Crescent	Thomas Huson	Do.	Do.	389	
King George	S. Hensley	Do.	J. Bolton & Co.	550	20
Betsy	Edward Mosson	Do.	John Bolton	317	
Lune	James Taylor	Do.	Geo. Case & Co.	317	25
Sally and Rebecca	Thomas Harold	Gold Coast	W. Begg & Co.	360	
George	Alex. Hackney	Lagos	F. Ingram & Co.	275	
Resource	Edward Clarke	Angola	Tarleton & Rigg	370	Apr. 2

* Extracted from the books of the Custom-house at Liverpool for Elliot Arthy, master mariner, and Surgeon in the African Slave Trade. The orthography of the original is followed.

APPENDIX No. XIII.—Continued.

Ships.	Captains.	Destinations.	Owners.	No. of Slaves.	Sailed.
Joshua	Edward Mentor	New Calebar	S. Lythgoe & Co.	243	
Telegraph	J. Maginnis	Angola	T. & Wm. Earle	140	Apr. 10
Enterprize	John Heron	Do.	J. Leyland & Co.	363	
Adriana	William Hewett	Do.	T. & Wm. Earle	424	
Louisa	William Brown	Do.	T. Leyland & Co.	465	
Fanny	John Adams	Gold Coast	Neilson, Heathcote & Co.	441	12
Margaret	E. Richardson	Angola	Jas. Penny & Co.	225	22
Indian	R. Pearson	Do.	John Shaw & Co.	344	
Betsy	D. Hayward	Win'd Coast	R. Leigh & Co.	317	
Elliot	J. Parkinson	Angola	J. & H. Clarke & Co.	505	May 2
Augustus	John Smith	Bonny	Titherington & Co.	363	
Friendly Cedar	Wm. Williams	Do.	Do.	245	
Swallow	Robert White	Angola	R. Abram & Co.	341	4
Edward	James Davies	New Calebar	Taylor, Clare & Co.	270	5
Rosamond	John Foulkes	Angola	Jas. Penny & Co.	323	14
Plumper	J. Corbett	Do.	C. Angus & Co.	143	
Maria	Robert Martin	Win'd Coast	M. Cullen & Co.	230	21
Africa	J. M. Smerdon	Angola	S. McDowall & Co.	146	
Don	H. Hamilton	Do.	Robert Johnson	118	
Mersey	Thos. Molyneux	Do.	Do.	346	
Henry	Matth. Cusack	Gabon	S. McDowall & Co.	165	
Molly	John Tobin	Angola	Geo. Case & Co.	436	27
Adventure	O. Pritchard	Do.	R. Buddicomb & Co.	280	
Sundett	W. Maxwell	Do.	W. Harper & Co.	312	
Diana	J. Ainsworth	Bonny	Moses Benson & Co.	409	
Britannia	Jos. Carshore	Old Calebar	T. & E. L. Hodgson & Co.	340	31
Mary	Thomas Flint	New Calebar	Hardman & Wright	280	
Diana	Robert Hume	Angola	J. & J. Parr & Co.	387	June 8
Annan	Andr. Davidson	Do.	Neilson, Heathcote & Co.	250	
Martha	Thomas Taylor	Do.	R. Fisher & Co.	384	
Iris	John Spencer	Do.	Do.	420	
Polly	J. Ainsworth	Do.	C. Fairclough	296	
Erl. of L'pool	Geo. Bernard	Bonny	T. Leyland & Co.	353	10
Amacree	R. Kendall	Do.	W. Harper & Co.	363	
Lottery	John Whittle	Do.	Thos. Leyland	462	
Fly	John Jones	Angola	John Shaw & Co.	615	25
Kingsmill	Thos. Mullion	Bonny	Mullion, Lenox & Co.	650	July 4
Friendship	Miles Booth	Angola	C. Angus & Co.	120	
Agreeable	James Seddon	Bonny	T. Hinde & Co.	300	
Nancy	Thomas Kerkley	Angola	Do.	500	
Arth'r Howe	Henry Booth	Do.	Geo. Case & Co.	314	

APPENDIX No. XIII.—Continued.

Ships.	Captains.	Destinations.	Owners.	No. of Slaves.	Sailed.
Bonwick	Benj. Cuite	Bonny	Flesher, Kelshaw & Co.	330	
Eagle	Michael Mills	Angola	Tarleton & Backhouse	300	July 5
Aurora	James Bowie	Do.	T. Parr & Co.	287	
James	W. Engledeu	New Calebar	W. Dickson & Sons	327	20
Maria	Thos. Phillips	Angola	P. Fairweather & Co.	120	24
Lightning	W. Quarrier	Do.	Tarleton & Rigg	390	
Jane & Sarah	B. Armstrong	Gold Coast	J. & E. L. Hodgson	287	
Annabelle	Thomas Cubben	Do.	Do.	350	
Tarleton	R. Shimmins	Bonny	Tarleton & Rigg	394	30
Christopher	J. Watson	Gold Coast	J. Bolton & Co.	390	
Nelly Anne	John Young	Gabon	W. Lenox & Co.	400	
Will	Hugh Crow	Angola	J. & J. Aspinall & Co.	420	
Ariel	John Guin	Do.	Jonathan Ratcliffe	153	Aug. 7
Windsor Castle	Thomas Jones	Old Calebar	Geo. Case & Co.	450	10
Enterprize	John Bovin	Angola	Do.	355	15
Active	D. Hayward	Old Calebar	Do.	434	
Catherine	W. Kewley	Bonny	France, Pool & Co.	339	
Bridget	Jos. Threllfall	Do.	Tarleton & Rigg	360	20
Fisher	Thos. Payne	New Calebar	J. Ward & Co.	339	22
Catherine	John Morrison	Bonny	J. Tarleton, junr.	277	
J'hn & S'rah	Thomas Brade	New Calebar	A. Joseph, Mozley & Co.	305	Sep. 7
London	Thomas Briscoe	Bonny	Do.	398	8
Charlotte	William Crow	Do.	Bailey, Taylor & Co.	344	
Dart	William Neale	Angola	John Bolton	384	8
Kitty	James Backhope	Bonny	John Smith & Co.	444	
Forbes	John Pince	Do.	Bailey, Taylor & Co.	368	
Sarah	Thomas Rives	Do.	W. Dickson & Sons	520	
Budd	Robert Tyror	Do.	J. Tarleton, junr.	281	
Hamilton	Rd. Durack	Win'd Coast	Robert Johnson	195	
Hannah	James Good	Angola	R. Abram & Co.	325	
Grace	D. Mc Elkeran	Gold Coast	W. Begg & Co.	215	9
Mary	Andr. Erskine	Angola	B. Thomas & Co.	293	14
Bess	Cæsar Lawson	Win'd Coast	Henderson, Sellar & Co.	208	
Eolus	John L. Neale	Bonny	Staniforth, Sons & Co.	355	20
Betsy & Susan	John Curry	Whydah	P. W. Brancker & Co.	296	
Mona	P. Mawdsley	Angola	J. Penny & Co.	293	Oct. 1
Fanny	James Irvin	Do.	John Shaw	380	4
Princess Augusta	Thomas Oliver	Do.	Jas. Penny & Co.	365	

APPENDIX No. XIII.—Continued.

Ships.	Captains.	Destinations.	Owners.	No. of Slaves.	Sailed.
King Pepple	James Phillips	Bonny	Neilson & Heathcote	476	Oct. 5
Ann	Thomas Lee	Do.	Timperon, Litt & Co.	358	6
Friendship	Robt. Catterall	Angola	Bell, Gibb & Blake	337	21
George	Richard Kellsall	Gold Coast	Jos. Ward & Co.	271	
Alexander	Wm. Cockrall	Angola	T. Sherington & Co.	517	23
Goodrich	H. Kennedy	Do.	J. & H. Clarke & Co.	210	29
H rriot	William Lace	Do.	Do.	313	
Trelawney	James Lake	Do.	T. Parr & Co.	467	
Neptune	James Williams	Old Calebar	Do.	350	
Otter	Alex. Grierson	Angola	W. Molyneux	417	
Beaver	William Murray	Do.	Do.	396	
Blanch	Rd. Andows	Benin	J. Gibbons & Co.	230	
Gascoyne	Jenkin Evans	Angola	Thomas Parr	444	
Perseverance	John Lawson	Benin	J. Gibbons & Co.	526	
Hannah	Andrew Arnold	Angola	Thomas Clare	523	
Bolton	J. Boardman	Bonny	John Bolton	432	Nov 12
Elizabeth	E. Neale	Do.	Do.	461	
Jack Park	John Little	Do.	J. & J. Aspinall & Co.	416	13
Mary	James Herd	Do.	W. Forbes & Co.	419	20
May	P. Callum	Do.	Do.	364	
Blanchard	Geo. Cormack	Do.	W. Thompson & Co.	419	
Sarah	R. Jones	Win'd Coast	T. & E. L. Hodgson	459	
Princess Amelia	J. Levingston	Bonny	J. Deare & Co.	464	21
Bird	J. Flint	Do.	Bailey, Taylor & Co.	368	
Tonyn	James Towers	Angola	Do.	326	
Harlequin	J. Topping	Do.	T. & W. Earle	275	22
Mary Ann	R. Taylor	New Calebar	Neilson, Heathcote & Co.	329	23
Expedition	W. Murdoch	Do.	J. & H. Clarke & Co.	354	Dec. 6
Hector	W. Stringer	Bonny	Thomas Clare	591	9
Fanny	Thos. Croaker	Gold Coast	Tarleton & Backhouse	300	11
Favourite	N. Evans	Angola	Jas. Penny & Co.	275	13
Penny	H. Kesack	Do.	Do.	360	
L'd. Duncan	John Hudson	Benin	S McDowall & Co.	242	
Cecilia	John Roach	Angola	Thompson & Co.	285	19
Mary	John Askeu	New Calebar	Pole & Gardner	200	21
Hind	Thomas Nuttal	Angola	Mullion, Lenox & Co.	355	
Adventure	Thomas Warren	Do.	Hardman, Wright & Co.	307	
Mercury	John Mills	Win'd Coast	Wm. Begg & Co.	376	24
Sarah	John Nerl	Angola	J. Ward & Co.	316	31
Ellis	James Soutar	Win'd Coast	T. & E. L. Hodgson	437	Jan. 2
L'pool Hero	Alex. Hackney	Gold Coast	F. Ingram & Co.	370	4
King George	Jas. Meckleghon	Do.	Do.	360	5

APPENDIX No. XIII.—CONTINUED.

On summing up the above account, the total of ships and of their complement of slaves, together with the number of ships destined for each place of trade in Africa, and the number of slaves supplied at those places respectively, stands thus :—

Places of Trade.	No. of Ships.	No. of Slaves.	Places of Trade.	No. of Ships.	No. of Slaves.
Angola	69	23,303	Amount brought up	140	49,696
Bonny	34	14,078	Benin	3	998
Gold Coast	11	3,587	Gabon	3	727
Windward Coast	10	3,278	Cameroons	2	565
New Calebar	10	2,977	Whydah	1	296
Old Calebar	6	2,473	Lagos	1	275
Amount carried up	140	49,696	Total	150	52,557

NOTE.—At that time the Guineamen were allowed by Act of Parliament to carry five slaves for every three tons of their burthen ; and required by law to take a proportion of ten people for each hundred of slaves, according to which the above number of slaves makes the total tonnage of the 150 ships 31,533 tons, and the total of their complement of seamen, 5,255.

APPENDIX No. XIV.

Summary of the aggregate number of Liverpool ships employed in the Guinea trade, together with the number and value of the slaves imported to the West Indies from 1783 to 1793 :—*

Years.	Gross No. of Ships.	Ships carrying Wood and Teeth.	Slave-Ships.	Number of Slaves.	Sterling Value.†
1783	90	5	85	39,170	1,958,500
1784	64	5	59	25,320	1,266,000
1785	77	4	73	29,490	1,474,500
1786	92	5	87	31,690	1,584,500
1787	80	8	72	25,520	1,276,000
1788	74	3	71	23,200	1,160,000
1789	66	4	62	17,631	881,550
1790	90	1	89	27,362	1,368,100
1791	105	4	101	31,111	1,555,550
1792	136	3	133	38,920	1,946,000
1793	47	1	46	14,323	716,150
	921	43	878	303,737	15,186,850

* "A General and Descriptive History of Liverpool," p. 222.

† The author of the calculations in arriving at the value of a slave, takes the average price in the West India market for eleven years—£50 sterling a head.

APPENDIX No. XV.

Extract from "A Log of the proceedings on board the Brigg *Mampookata*, on a voyage to Ambrize, on the coast of Angola," in the year 1787.

Year 1787. Week Days.	Day. Month.	Winds.	Remarks, &c., in Ambrize Road.
Saturday.	8th	Variable	Fresh breezes and clear sent the Long Boat with Mr. Smith and Brown to Marsoola to trade for slaves. Employed occasionally. Received on board one woman and one boy slave.
Sunday.	9th	Variable	Modrate breezes and clear weather. Employed starting Beens, &c. Received 3 woman slaves and one child. No. on board 5.
Monday.	10th	Variable	Light airs & clear. Employed clearing the fore Hould &c. Received on board 5 slaves, viz. 1 man 1 woman 1 boy & 2 girls. No. 10.
Tuesday.	11th	Variable	Light breezes and clear. Employed with sundries &c. Received on board 6 slaves, viz. 1 man, 2 woman with childer 2 boys & 1 girl. Total 16.
We'n'sday.	12th	Variable	Ditto weather. Employed occasionally. Received on board 5 slaves, viz. 1 boy & 4 girls. Total 21.
Thursday.	13th	Westward	Modrate and clear. Employd stowing casks in the Fore Hould. Returned from Marsoola the Long Boat without any slaves. Received from the Factory 3 slaves, viz. 1 man boy & 2 girls. Total on board 24.
Friday.	14th	Variable	Light breezes & cloudy. Employed laying the men's platform. Received one woman slave. No. on board 25.
Saturday.	15th	Variable	Ditto weather. Sent the boat for Marsoola with Mr. Smith and Brown. The carpenter employed making the Woman's room bulkhead. Received one woman slave. No. on board 26.
Thursday.	20th	Variable	Modrate breezes & cloudy. Arrived in the Road ye *Union*, Capt. Lawson from Burdux. Employed wooding & watering. Received 6 slaves, viz. man 2 men boys, 2 boys, and 1 girl. Total 39.

APPENDIX No. XV.—Continued.

Year 1787. Week Days.	Day. Month.	Winds.	Remarks, &c., in Ambrize Road.
Friday.	21st	Variable	Ditto weather. Employed as before. The carpenter building the Barricado on the main deck. Received two men boys slaves. No. on board 41.
Monday.	24th	Variable	Ditto weather. Received one man boy slave and one tooth weigh^g 112lb. 1 two pieces 25lb. Slaves on board 43.
Thursday.	Oct. 4th	Variable	Modrate and cloudy. Sailed a French ship, Captain Granier, for Cape Francois with 250 slaves. Received 1 man and man boy. Total on board 49, all well.
Monday.	Nov. 26th	Variable	Fresh breezes and cloudy with rain. Employed with sundries &c. Died one man boy slave of the fever after a sickness of 6 days. Slaves on board 153, all well.

H	K	F	Courses.	Wind.	Remarks, &c.
10	6	5	WSN	SSW	January 1st, 1788. Modrate breezes & cloudy. Employed occasionally. Buried 2 men slaves No. 3. Slaves on board 193.* Latt$_d$ pr Obs^n 5° 36″ south.
8	7	3	WBN ½N	SEBS	Sunday Jany 6th 1788. Buried a man slave of the flux and fever &c.
9	1	2	N by W	SSE	Light airs inclinable to calms. Washed the rooms and slaves, &c. Jany 11th 1788. Sunday February 3rd, 1788, anchored in Carlisle Bay in 8 fathom water. Slaves on board 192 all well. Received a quantity of vegetables on board for the slaves, &c. Tuesday Feby 5th 1788. Got under way and made sail.
6					Feby 13th. In Woodbridge Bay Dominica several gentlemen came on board to look at the slaves. Feb 14th. Sold 188 slaves to Mr. Forbes of St Christopher's. At 10 a.m. delivered 80 men & women on board a sloop. Remains on board 112. Saturday 16th. Delivered to Mr. Forbes 108 slaves, total, 188. Remains on board 4 viz. 2 men, 1 woman, & 1 boy.

* She left the coast on December, 21st, 1787, with 195 slaves.

APPENDIX No. XV.—CONTINUED.

Having taken in a cargo of sugar, cotton, coffee, and cocoa, the ship sailed for England, and arrived in Liverpool on the 10th of April, 1788. The logbook is adorned with water-colour drawings of the brig, and of the coast scenery, together with a pencil sketch of a gentleman in a cocked hat and pigtail, forming an exceptional specimen of maritime caligraphy. It is now in the possession of T. H. Dixon, Esq., The Clappers, Gresford.

APPENDIX No. XVI.

CHARACTER OF THE SEAMEN IN THE SLAVE TRADE.*

"With respect to the mortality amongst the crews of African ships, it must be taken into account that many of the individuals composing them were the very dregs of the community. Some of them had escaped from jails: others were undiscovered offenders, who sought to withdraw themselves from their country, lest they should fall into the hands of the officers of justice. These wretched beings used to flock to Liverpool when the ships were fitting out, and, after acquiring a few sea phrases from some crimp or other, they were shipped as ordinary seamen, though they had never been at sea in their lives. If, when at sea, they became saucy and insubordinate, which was generally the case, the officers were compelled to treat them with severity; and having never been in a warm climate before, if they took ill, they seldom recovered, though every attention was paid to them. Amongst these wretched beings I have known many gentlemen's sons of desperate character and abandoned habits, who had either fled for some offence, or had so involved themselves in pecuniary embarrassments, as to have become outcasts, unable to procure the necessaries of life. For my own part I was always very lucky in procuring good crews, and consequently the charge of great mortality could not apply to my ships. The deaths in the *Kitty's Amelia* were attributable to the culpable neglect of others, the consequences of which we could neither foresee nor control."

* Memoirs of Hugh Crow.

APPENDIX No. XVII.

Food of the Slaves.*

"It may not be uninteresting to the reader to learn with what kind of provisions the negroes were supplied. We frequently bought from the natives considerable quantities of dried shrimps to make broth; and a very excellent dish they made, when mixed with flour and palm oil, and seasoned with pepper and salt. Both whites and blacks were fond of this mess. In addition to yams we gave them, for a change, fine shelled beans and rice cooked together, and this was served up to each individual with a plentiful proportion of the soup. On other days their soup was mixed with peeled yams, cut up thin and boiled, with a proportion of pounded biscuit. For the sick we provided strong soups and middle messes, prepared from mutton, goats' flesh, fowls, &c., to which were added sago and lilipees, the whole mixed with port wine and sugar. I am thus particular in describing the ingredients which composed the food of the blacks, to show that no attention to their health was spared in this respect. Their personal comfort was also carefully studied. On their coming on deck, about 8 o'clock in the morning, water was provided to wash their hands and faces, a mixture of lime juice to cleanse their mouths, towels to wipe with, and chew sticks to clean their teeth. These are generally pieces of young branches of the common lime, or of the citron of sweet lime tree, the skin of which is smooth, green, and pleasantly aromatic. They are used about the thickness of a quill, and the end being chewed, the white, fine fibre of the wood soons forms a brush, with which the teeth may be effectually cleaned by rubbing them up and down. These sticks impart an agreeable flavour to the mouth, and are sold in the public markets of the West Indies, in little bundles for a mere trifle. A dram of brandy bitters was given to each of the men, and clean spoons being served out, they breakfasted about nine o'clock. About eleven, if the day were fine, they washed their bodies all over, and after wiping themselves dry, were allowed to use palm oil, their favourite cosmetic. Pipes and tobacco were then supplied to the men, and beads and other articles were distributed amongst the women to amuse them, after which they were permitted to dance, and run about on deck to keep them in good spirits. A middle mess of bread and cocoa nuts, was given them about mid-day. The third meal was served out about three o'clock, and after everything was cleaned out and arranged below, for their accommodation, they were generally sent down about four or five in the evening. Indeed I took great pains to promote the health and comfort of all on board, by proper diet, regularity, exercise, and cleanliness ; for I considered that on keeping the ship clean and orderly, which was always my hobby, the success of our voyage mainly depended. "

* From this gracious picture, drawn by Captain Crow, a model commander, under the more humane regulations of the closing years of the trade, the reader must not imagine that such paternal care was general.

INDEX TO NAMES OF PERSONS

MENTIONED IN THIS WORK.

A

B

H

I

J

K

L

M

N

O

P

Q

R

S

T

U

V

W

(SEE ALSO LISTS OF NAMES IN THE APPENDIX.)

INDEX TO SUBJECTS.

A

B

D

E

F

G

H

I

J

K

L

M

N

O

P

Q

R

S

T

U

V

W

Y

Z

LIVERPOOL EDWARD HOWELL CHURCH STREET